International Conference on Structural Nano Composites

(NANOSTRUC 2012)

IOP Conference Series: Materials Science and Engineering
Volume 40

Bedforshire, United Kingdom
2 - 4 July 2012

Editors:

Sophia Sachse **James Njuguna**

ISBN: 978-1-63439-851-0
ISSN: 1757-8981

Printed from e-media with permission by:

Curran Associates, Inc.
57 Morehouse Lane
Red Hook, NY 12571

Some format issues inherent in the e-media version may also appear in this print version.

Printed by Curran Associates, Inc. (2015)

For permission requests, please contact the Institute of Physics at the address below.

Institute of Physics
Dirac House, Temple Back
Bristol BS1 6BE UK

Phone: 44 1 17 929 7481
Fax: 44 1 17 920 0979

techtracking@iop.org

Additional copies of this publication are available from:

Curran Associates, Inc.
57 Morehouse Lane
Red Hook, NY 12571 USA
Phone: 845-758-0400
Fax: 845-758-2634
Email: curran@proceedings.com
Web: www.proceedings.com

Table of contents

Volume 40

International Conference on Structural Nano Composites (NANOSTRUC 2012)
2–4 July 2012, Cranfield University, Bedfordshire, UK

Accepted papers received: 9 August 2012
Published online: 13 September 2012

Preface

Papers

Relationship between the Morphology and Physico-Mechanical Properties of Polyethylene/Clay Nanocomposites R Rezanavaz and M K R Aghjeh pg. 352

IOP Conf. Series: Materials Science and Engineering **40** (2012) 011001 doi:10.1088/1757-899X/40/1/011001

NANOSTRUC 2012

International Conference on Structural Nano Composites
July 2-4, 2012 Cranfield University, Bedfordshire

In conjunction with:

nephh
SEVENTH FRAMEWORK PROGRAMME
Cranfield UNIVERSITY
ECOSHELL
Agilent Technologies
jm BIOCONNECT
EATON
gerhardt uk lab supplies
FRITSCH
Knowledge Transfer Network Nanotechnology
NIA Nanotechnology Industries Association
AMERICAN ELEMENTS
CRANFIELD NANO

IOP Conf. Series: Materials Science and Engineering **40** (2012) 011001 doi:10.1088/1757-899X/40/1/011001

Editors: Sophia Sachse & James Njuguna

Cranfield University, Cranfield, Bedfordshire MK43 0AL

T: +44 (0)1234 75 4186 or +44(0)123475 0111

E: s.sachse@cranfield.ac.uk, j.njuguna@cranfield.ac.uk

W: www.cranfield.ac.uk

Conference Website: www.nanostruc.info

Linked In: http://www.linkedin.com/groups/NanoStruc-4255826

IOP Conf. Series: Materials Science and Engineering **40** (2012) 011001 doi:10.1088/1757-899X/40/1/011001

Preface

Dear Colleagues,

It is a great pleasure to welcome you to Nanostruc2012 at Cranfield University. The purpose of the 2012 International Conference on Structural Nano Composites (NanoStruc 2012) is to promote activities in various areas of materials and structures by providing a forum for exchange of ideas, presentation of technical achievements and discussion of future directions.

NanoStruc brings together an international community of experts to discuss the state-of-the-art, new research results, perspectives of future developments, and innovative applications relevant to structural materials, engineering structures, nanocomposites, modelling and simulations, and their related application areas. The conference is split in 7 panel sessions, Metallic Nanocomposites and Coatings, Silica based Nanocomposites, safty of Nanomaterials, Carboin based Nanocomposites, Multscale Modelling, Bio materials and Application of Nanomaterials.

All accepted Papers will be published in the IOP Conference Series: Materials Science and Engineering (MSE), and included in the NanoStruc online digital library. The abstracts will be indexed in Scopus, Compedex, Inspec, INIS (International Nuclear Information System), Chemical Abstracts, NASA Astrophysics Data System and Polymer Library.

Before ending this message, I would like to acknowledge the hard work, professional skills and efficiency of the team which ensured the general organisation.

As a conclusion, I would like to Welcome you to the Nanostruc2012 and wish you a stimulating Conference and a wonderful time.

On behalf of the scientific committee,

James Njuguna, Conference Chair

IOP Conf. Series: Materials Science and Engineering **40** (2012) 011001 doi:10.1088/1757-899X/40/1/011001

Scientific committee

Dibakar Bandopadhya - Indian Institute of Technology - Guwahati, India
Aravid Dasari -Nanyang Technological University, Singapore
Rob Dorey - Cranfield University, UK
Andrzej Gałęski – Polish Academy of Science, Poland
Susheel Kalia - Shoolini University of Biotechnology & Management Sciences, India
Dirk Lehmhus - University of Bremen, Germany
James Njuguna - Cranfield University, UK
Sergio Pezzin - State University of Santa Catarina, Brazil
Krzysztof Pielichowski - Cracow University of Technology, Poland
Davide Roncato - Centro Ricerche FIAT (CRF), Italy
Yong Sheng - Leeds University, UK
Huijuan Su - Northampton University, UK
Nikos Tsourveloudis - Technical University of Crete, Greece
Raquel Verdejo - Consejo Superior de Investigaciones Científicas (CSIC), Spain
Paul Wambua - Moi University, Kenya
Jianqiao Ye - Lancaster University, UK
Shaik Zainuddin - Tuskegee University, USA
Huijun Zhu - Cranfield University, UK

Session Chairs

Panel session 1.1	**Metallic Nanocomposites and Coatings (Chaired by Krzysztof Pielichowski, Cracow University of Technology, Poland)**
Panel session 1.2	**Silica based Nanocomposites (Chaired by Rob Dorey, Cranfield University, UK)**
Panel session 1.3	**Safety of Nanomaterials (Chaired by James Njuguna, Cranfield University, UK)**
Panel session 2.1	**Cabon based Nanocomposites (Chaired by Aravind Dasari, Nanyang Technological University, Singapore)**
Panel session 2.2	**Multiscale Modelling (Chaired by Jianqiao Ye, University of Leeds, UK)**
Panel session 2.3	**Bio materials (Chaired by Kambiz Kayvantash, CADLM, France)**
Panel session 3.1	**Application of Nanocomposites (Chaired by Davide Roncato, Centro Ricerche FIAT (CRF), Italy and Dirk Lehmhus, University of Bremen, Germany)**
Panel Discussion	**Chaired by Dirk Lehmhus, University of Bremen, Germany**
Poster session	**Keith Westwood, EATON, UK**

Organising committee

James Njuguna (Organising Secretary) - Cranfield University, UK
Sophia Sachse - Cranfield University, UK
Jinchun Zhu - Cranfield University, UK
Francesco Silva - Cranfield University, UK
Elias Nassiopoulos - Cranfield University, UK
Huijun Zhu - Cranfield University, UK
Jenny Kay - Cranfield University, UK

IOP Conf. Series: Materials Science and Engineering **40** (2012) 012001 doi:10.1088/1757-899X/40/1/012001

Silver nanowires/polycarbonate composites for conductive films

I. Moreno, N. Navascues, S. Irusta, J. Santamaría

Instituto de Nanociencia de Aragón. Universidad de Zaragoza. Mariano Esquillor s/n. 50018. Zaragoza. Spain

Abstract. Silver nanowires (AgNW) with an aspect ratio of 85 were synthesized by a solvothermal process. The AgNW were characterized by SEM and XRD techniques. Nanocomposites of these silver nanowires in a polycarbonate matrix were prepared by simple solution mixing procedure in a concentration filler range 0-4.35 wt%. The obtained films were around 18 μm thick, optical microscopy and SEM characterization showed good dispersion of the nanowires in the polymeric matrix. The obtained composites presented low percolation threshold (0.04 wt%) and the maximum conductivity at 4.35 wt% filler loading was $2.3x10^{-2}$ S/cm.

1. Introduction

Electrically conductive polymer composites resulting from the association of an insulating matrix with conductive fillers are routinely employed in various commercial applications: batteries, electrodes, anti-static and corrosion resistant materials [1-3]. Besides, they have been studied by many groups for their sensing ability because of the resistivity variation with thermal, mechanical or chemical solicitations [4].

Common conductive fillers include carbon black, graphite, carbon fibers, metal nanoparticles and aluminum flakes. Metal nanowires/polymer nanocomposites are alternative materials to those of carbon nanotube-based polymer composites and the final applications of these materials would be determined by their properties, cost and ease of processing. Silver nanomaterials are of current importance due to its unique electrical, thermal, catalytic and optical properties, sensing characteristic and antimicrobial effects. The preparation of conductive composites using different silver shape materials: flake, nanowires, or spherical, was previously reported [5, 6]. There is great interest for one-dimensional conductive nanomaterials with high aspect ratio since it was demonstrated that increasing the aspect ratio of the filler will increase the conductivity of the composite [7]. Conducting polymer structures may be developed at low loading of 1D fillers due to the lower percolation thresholds needed for the high aspect ratio structures. Silver nanowire (AgNW)/polymer composites were repeatedly reported as effective materials for electrodes. AgNW electrodes on cross-linked polyacrylate subtract have shown low surface resistance [8] while a composite obtained by embedding Ag nanowires into the conducting polymers PEDOT:PSS were probed to be high-efficiency anodes [9].

Polycarbonate (PC) is an amorphous thermoplastic resin with very low electrical conductivity yet with a wide range of useful properties like excellent toughness, very good dimensional stability, transparency and thermal stability. The wide range of PC films applications should be further extended by the addition of electrical properties.

In this work silver nanowires were synthesized by a solvothermal method and then highly conductive films of these nanowires in an insulating polycarbonate matrix were prepared by a simple solution mixing process. The obtained films were characterized by optical microscopy, scanning electron microcopy and the electrical conductivity was measured.

2. Experimental

Silver nitrate (AgNO3, 99,8%, Panreac), polyvinylpyrrolidone (PVP, Mw-55,000, Aldrich), sodium chloride (NaCl, Aldrich), ethylene glycol (EG, anhydrous, 99,8%, Aldrich), acetone, poly(Bisphenol A carbonate) (PC, Aldrich) and tetrahydrofuran (THF, 99% Aldrich) reagents were used without further purification.

Silver nanowires were prepared by using a solvothermal process in a modification of the method proposed by Chen et al. [10]. Briefly, 10 ml ethylene glycol solution of 0,15M PVP is injected drop by drop into 10 ml of magnetically stirred EG solution of 0,1M AgNO3 / 0,1mM NaCl. The solution is magnetically stirred during 5 minutes and then it was put into a 45 ml Teflon-lined autoclave. It is sealed and heated up to 160°C for 2.5 hours. Afterward natural cooling at room temperature the solid was isolated by centrifugation (3000 rpm, 10 min) and thoroughly washed with acetone

Nanocomposite films were prepared using a solution mixing procedure, typically, 1g of polycarbonate was dissolved in 30 ml of THF in an ultrasonic bath during 15 minutes. The Ag nanowires were centrifuged, the supernatant is eliminated and THF was added to disperse the AgNW. The Ag nanowires dispersion was added drop by drop to the polycarbonate solution under magnetic stirring. The mixture was then heated at 70°C under magnetic stirring to evaporate the solvent. When the suspension is reduced to 10 ml, it is spread in a glass plate and drag with a rod. Vacuum treatment (50mbar / 80°C / 24 hours) was applied to the nanocomposite film in order to eliminate the residual solvent. AgNW/PC composites with theoretical weight fractions in the range 0.01-4.25wt% were prepared.

For Scanning electron microscopic (SEM; FEI Inspect) measurements, a few drops of the AgNW dispersion were place onto the carbon-coated copper grids and dried at room temperature. X-ray diffraction measurements of dried AgNW were obtained using D-Max Rigaku 2500 X-ray diffractometer.

Optical images of AgNW/PC composites were taken on Olympus BX-41 microscope. For Scanning electron microscopic measurements (backscattered electrons) of composites, samples were coated with a thin film of gold. Conductivity measurements were carried out placing the sample between two gold electrodes and measuring the complex impedance (Agilent 4294A Precision Impedance Analyzer).

3. Results and discussion

3.1. Silver nanowires

The synthesized silver nanowires were characterized by SEM microscopy (Figure 1). The morphology of the product is wire-like with a very broad diameter size distribution centred in 117 nm (based on the measurement of 200 nanowires). The length is also in a broad range (from 3 to 45 μm) with an average at 10 μm. The average aspect ratio is around 85. The wire-like shape is highly dominant, although a small amount of particles can also be seen.

The crystalline nature of the bulk nanowires was examined by XRD analysis; Figure 2 shows a typical XRD pattern of dried samples. The peaks can be indexed as reflections of face-centred cubic silver (JCPDS Card File N° 04-0783). The lattice constant *a* calculated from this pattern is 4,092 Å, which is consistent with the standard value of 4.086 Å [11]. The high intensity ratio between the (111) and (200) diffraction peaks would indicate that the faces of the wires are mainly composed of (111) planes

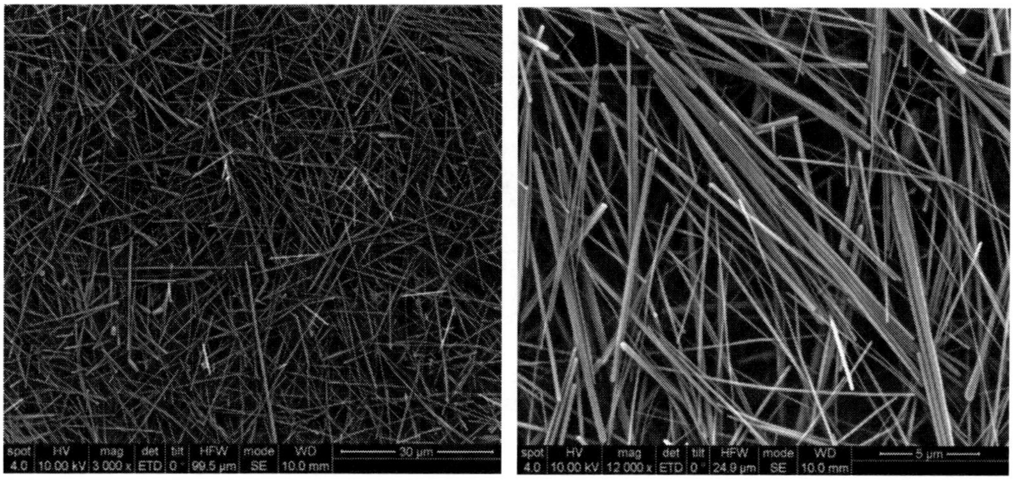

Figure 1. SEM images of synthesized Ag nanowires

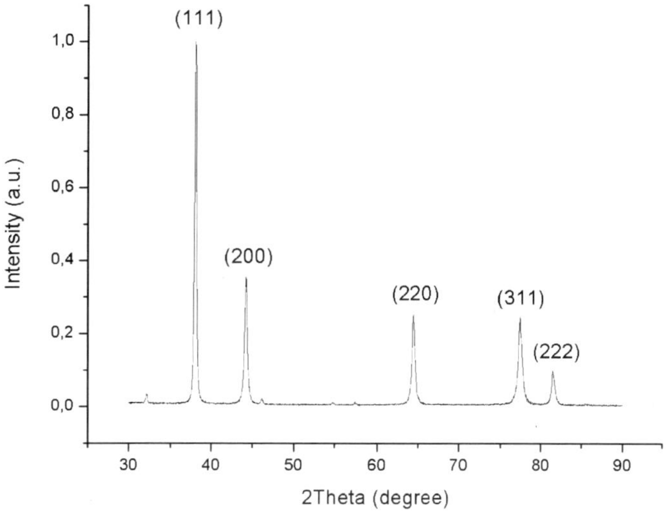

Figure 2. XRD diffractrogram of silver nanowires

3.2. Composites

All the prepared films have a thickness around 18 μm. In order to know the distribution of the silver nanowires in the polymeric matrix the prepared films were characterized by optical microscopy, typical images are displayed in Fig. 3. Good degree of wires dispersion was achieved even for the films with the higher nanowires concentration. Individual nanowires are randomly oriented and form a network for higher concentrations; the presence of some particles can also be observed. SEM analysis showed the presence of wires in the surface of films (Fig. 4).

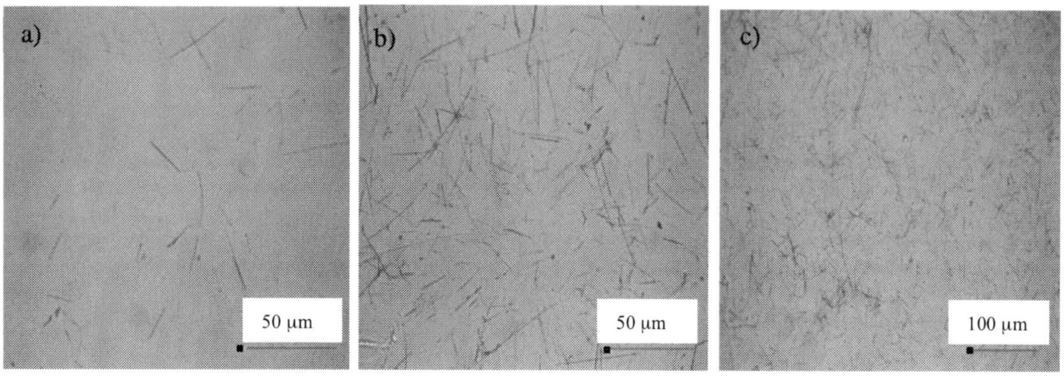

Figure 3. Typical optical images of the prepared composites a) 0.04 wt%, b) 0.16 wt%, c) 1.42 wt%

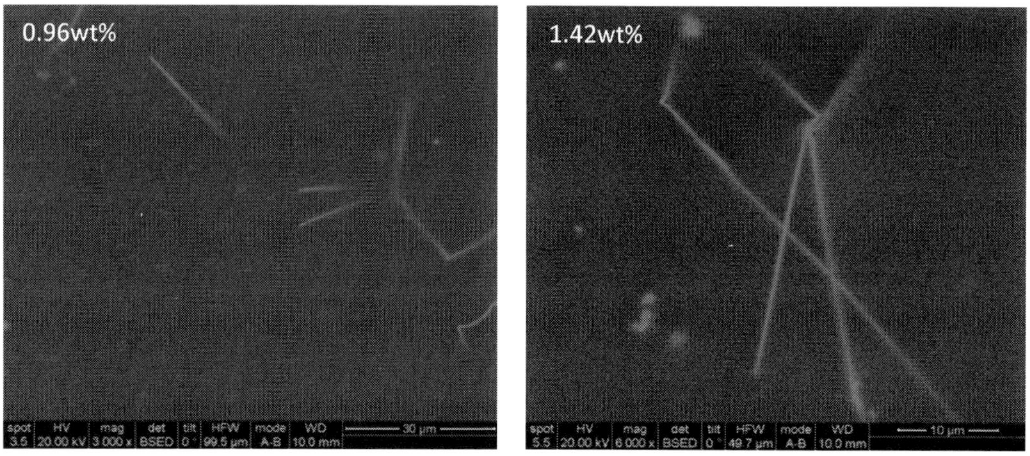

Figure 4. Typical SEM images of the surface of prepared composites

Due to the high aspect ratio, nanowires can form an interconnected network at very low volume fractions. The lowest nanotube concentration at which such networks are first formed is known as the percolation threshold. Figure 5 shows the electrical conductivity of the AgNW composites as a function of nanowires content. A typical percolation behavior can be observed, i.e. at a critical wire loading the composite electrical conductivity increase by several orders of magnitude indicating the formation of the interconnected network. From the experimental curve it can be seen that the introduction of 0.04 wt% of silver nanowires drastically increase the electrical conductivity of the composites films up to 6 orders, from 10^{-10} S/cm to 10^{-4} S/cm. At this AgNW load a very high percentage of electrons are permitted to flow through the sample at an applied electric field due to the formation of interconnecting conducting channels. Values of the percolation threshold reported in the literature for carbon nanotubes/polycarbonate composites are higher than 0.1 wt%; on the other hand the conductivity at a filler load of 4.35 wt% was $2.3\text{x}10^{-2}$ S/cm, higher than most of the reported values for carbon nanotubes/polycarbonate composites [13].

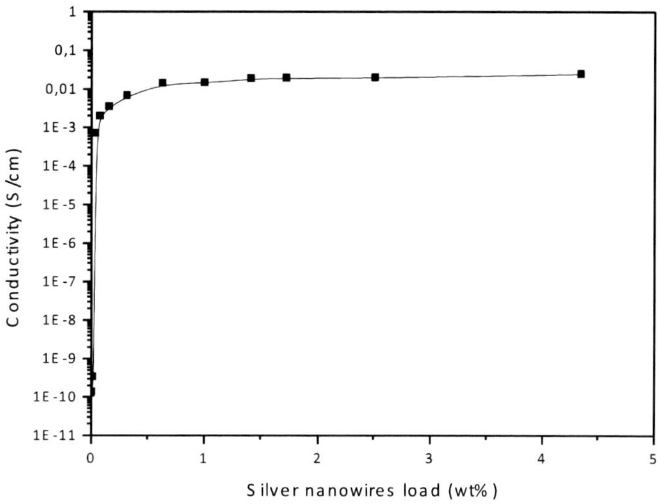

Figure 5. Conductivity of the prepared films vs. weight fraction of AgNW

4. Conclusions

The silver nanowires/polycarbonate composites films obtained by solution mixing showed a conductivity of 2.3 10^{-2} S/cm at filler loading of 4.35 wt%. The introduction of 0.04 wt% of silver nanowires in the polycarbonate matrix drastically increase the electrical conductivity of the composites films up to 6 orders, from 10^{-10} S/cm to 10^{-4} S/cm. These results would indicate that silver nanowires could be an interesting alternative to carbon based nanomaterials to obtain conductive polymer composites.

References

[1] Prosini, P.P. and S. Passerini, *The role of conductive carbon in PEO-based composite cathodes.* European Polymer Journal, 2001. **37**(1): p. 65-69.

[2] Liang, X., et al., *A composite of sulfur and polypyrrole–multi walled carbon combinatorial nanotube as cathode for Li/S battery.* Journal of Power Sources, 2012. **206**(0): p. 409-413.

[3] Yu, Z., et al., *Highly Flexible Silver Nanowire Electrodes for Shape-Memory Polymer Light-Emitting Diodes.* Advanced Materials, 2011. **23**: p. 664–668.

[4] Feller, J.F. and Y. Grohens, Evolution of electrical properties of some conductive polymer composite textiles with organic solvent vapours diffusion. Sensors and Actuators B: Chemical, 2004. **97**(2–3): p. 231-242.

[5] Yang, X., et al., Preparation and properties of a novel electrically conductive adhesive using a composite of silver nanorods, silver nanoparticles, and modified epoxy resin. Journal of Materials Science: Materials in Electronics, 2012. **23**(1): p. 108-114.

[6] Liu, J., et al., *High-performance electrically conductive silver paste prepared by silver-containing precursor.* Applied Physics A: Materials Science & Processing, 2010. **100**(4): p. 1157-1162.

[7] Silva, J., et al., *The influence of matrix mediated hopping conductivity, filler concentration, aspect ratio and orientation on the electrical response of carbon nanotube/polymer nanocomposites.* Composites Science and Technology, 2011. **71**(5): p. 643-646.

[8] Yu, Z., et al., *Highly Flexible Silver Nanowire Electrodes for Shape-Memory Polymer Light-*

Emitting Diodes. Advanced Materials, 2011. **23**: p. 664–668.

[9] Gaynor, W., et al., *Smooth Nanowire/Polymer Composite Transparent Electrodes.* Advanced Materials, 2011. **23**(26): p. 2905.

[10] Chen, D., et al., *Convenient synthesis of silver nanowires with adjustable diameters via a solvothermal method.* Journal of Colloid and Interface Science, 2010. **344**(2): p. 286-291.

[11] Li, Z., et al., *Large-scale synthesis of silver nanowires and platinum nanotubes.* Colloid Polymer Science, 2010. **288**: p. 1185-1191.

[12] Castillo, F.Y., et al., Electrical, mechanical, and glass transition behavior of polycarbonate-based nanocomposites with different multi-walled carbon nanotubes. Polymer, 2011. **52**(17): p. 3835-3845.

[13] Bauhofer, W. and J.Z. Kovacs, *A review and analysis of electrical percolation in carbon nanotube polymer composites.* Composites Science and Technology, 2009. **69**(10): p. 1486-1498.

IOP Conf. Series: Materials Science and Engineering **40** (2012) 012002 doi:10.1088/1757-899X/40/1/012002

3-dimensional free standing micro-structures by proton beam writing of Su 8-silver nanoParticle polymeric composite

H Igbenehi[1,*] and S Jiguet[2]

[1] Surrey Centre for Ion-beam Applications, Advanced Technology Institute,University of Surrey
[2] Gersteltech 1009 Pully Switzerland

Abstract. Proton beam lithography a maskless direct-write lithographic technique (well suited for producing 3-Dimensional microstructures in a range of resist and semiconductor materials) is demonstrated as an effective tool in the creation of electrically conductive freestanding micro-structures in an Su 8 + Nano Silver polymer composite. The structures produced show non-ohmic conductivity and fit the percolation theory conduction model of tunneling of separated nanoparticles. Measurements show threshold switching and a change in conductivity of at least 4 orders of magnitude. The predictable range of protons in materials at a given energy is exploited in the creation of high aspect ratio, free standing micro-structures, made from a commercially available SU8 Silver nano-composite (GMC3060 form Gersteltec Inc. a negative tone photo-epoxy with added metallic nano-particles(Silver)) to create films with enhanced electrical properties when exposed and cured. Nano-composite films are directly written on with a finely focused MeV accelerated Proton particle beam. The energy loss of the incident proton beams in the target polymer nano- composite film is concentrated at the end of its range, where damage occurs; changing the chemistry of the nano-composite film via an acid initiated polymerization - creating conduction paths. Changing the energy of the incident beams provide exposed regions with different penetration and damage depth – exploited in the demonstrated cantilever microstructure.

1. Introduction

Proton beam writing is a direct write lithographic technique demonstrated to produce 3 dimensional microstructures. Mechanical and micro-optical structures have been demonstrated in polymers such as SU 8 using this technique [1 - 3]. SU 8 is a chemically amplified epoxy based resin that polymerises on exposure to irradiation such as ultra violet light and high energy proton beams. Polymerization is photo-acid initiated. [4]

In order to create MEMS type micro-electromechanical structures in SU 8 using proton beam writing; modification of the photoresist's electrical properties or the electrical properties of any structures created is required. We have earlier described a novel method of making proton beam written structures on SU 8 conductive by a selective electroless plating process, whereby gold is plated to modified surface of the created structures [5].

The methods of modifying the electrical properties of polymers such as SU 8 as described in literature fall broadly under two main categories; mixing SU 8 with a conductive polymer and introduction of conductive metallic nano-particles. Such conductive polymers have found uses in optical and gas sensing applications [6].

The properties of polymers and that of SU 8 specifically have been modified to induced changes by the addition of nano-particles. T. Whitesides report the SU 8 exhibiting ferro-magnetic properties due to the addition of Nickel nanosphares [7] while S. Jiguet reports a reduction in mechanical stress by incorporation of nano-silica particles [8].

1.1 Modification of SU 8 to Induce Electrical Conductivity.

To create micro-electromechanical structures from SU 8 the conductivity of the polymer must be modified. Leeuw et al. describes a method of creating conductive SU 8 by a mixture with a semiconductor doped organic polymer PANI (Polyaniline) [6]. However M. Nordström reports that due to the acidic nature of the constituent Polyaniline mixture, cross-linking is initiated in SU 8 and a thick inhomogeneous mixture results. This mixture presents difficulty in spincoating and producing structures lithographically [9].

Conductive properties of SU 8 can be modified by the introduction of nano particles into SU 8 involving the physical dispersion of electrically conductive nano-particles into the polymer. L. Gammelgaard describes the addition of carbon black nano-particles to SU 8 to create an electrically conductive photo-plastic/carbon cantilever for sensor applications [10]. H. Chiamori et al. describe a method of physical incorporation of optically transparent and electrically conductive diamondoids by dispersion in SU 8 [11].

Multi-walled carbon nanotubes have also been chemically bonded to SU 8 and dispersed into the resist by ultrasonic irradiation. Zhang et al. describes a process where hydroxyl groups in functionalised carbon nanotubes react with open epoxide groups within the SU 8 polymer, to create a dark uniform mixture that is conductive and curable by exposure to ultraviolet radiation [12]. Chiamori et al. also describe incorporation of single walled carbon nanotubes and conductivity in the resultant composite when single walled nano particles comprise between 1% -5% by weight [11].

Jiguet et al. describe a method of incorporating silver nano particles into SU 8 and fabricating electrically conductive composites. However they also report difficulties in spincoating the resultant nano-composite such that spreading is required over the substrate. The resultant nano-composite resist is thick and not easily curable by UV lithography as the silver particles are not transparent to ultraviolet radiation, and as such transparent substrates with backside exposures were required. The structures produced had a maximum height of 5 micrometres [13, 14].

The difficulties associated with other lithographic techniques in patterning SU 8 containing metallic nano-particles due to dispersions and opacity are not present in proton beam lithography. We now report a process for creating electrically conductive microstructures in SU 8 /silver nano composite by proton beam lithography. However and understanding of the effects energetic proton beams have on polymers containing metallic nano-particles will be required to explain conduction mechanisms and features of the structures produced and characterized.

1.2 Percolation Theory and Conduction of Polymers

Percolation theory provides a model for explaining electrical conductivity of polymer films containing randomly dispersed nano particles of varying sizes. Conduction is achieved when an electron overcomes coulomb charging energy and "tunnelling of [these] single electrons between neighbouring metal nano particles" [15] these tunnelling is via the linker molecules of the resist.

When a voltage is applied to the film or structure containing metallic nano particles, the current flow induces a process of electro-migration which creates current channels or percolation paths. These are reported to be nano particles separated by nano metres distances [16, 17]. According to percolation theory, the percolation threshold is the per-volume ratio of nano particles to the polymer resist required for the creation of a continuous conduction path. This is reported to be 12% for uniformly spherical nano particles such as Silver and 25% for simple cubic lattices [9, 18] however

S. Jiguet et al. report electrical conduction in Silver filled SU 8 with volumes of 6% determined experimentally [14].

The discrepancies in determining percolation thresholds theoretically vs. experimentally probably arise since the percolation model does not factor any contribution to the conductivity from the polymer matrix. P. McCklusky et al. describes a matrix iterative model for the electrical conductivity of polymers containing dispersed silver nano-particles. The matrix iterative model provides an accurate experimental fit and is a modification to the percolation model which is reported does not adequately describe the conduction of dispersed nano-particles in polymers [19]. While ohmic conductivity has been reported in conducting polymers such as Polyaniline [20, 21], nonlinear electrical conductivity have been reported in silver nano-particle polymer composite where tunnelling occurs on approaching the percolation threshold.

A Kiesow et al. report a reversible but drastic change of up to six orders of magnitude in current to voltage behaviour in a silver nano-particle composite material. This behaviour is termed threshold switching [22].

1.3 Effect of MeV Ions on Nano Particles

As mentioned earlier exposure to MeV accelerated protons, create a photo-acid initiated cross-linking thus bond formation in the SU 8 resist. The cross linked regions have higher molecular mass and thus are insoluble to development solution, yielding 3 dimensional structures. A change in the electrical conductivity of Polyaniline based on fluence when exposed to MeV ions is reported [23].

However when polymers that contain nano-particles and are exposed to high energy ion beams, both the polymer and the nano-particles are modified. Shape transformation of spherical nano-particles into ellipsoids, varied deformations as well as melting along the ion tracks have been reported [24].

Aggregation or cluster formation is also reported when silver nano-particles in a polymer-nano composite is exposed to MeV ion irradiation irrespective of fluence, gold nano-particles however reportedly formed helical patterns [25].

2. Experimental

Spincoating is an essential step in depositing SU 8-Ag Nano-composite polymer (15% by volume Ag/Su8 nano-composite GCM3060 [26] sourced from Gersteltec Inc. Switzerland) unto a substrate. It involves using a spinning resist upon a substrate at speeds which dictate their film thickness. Due to SU 8 sensitivity to light, spincoating is performed under yellow light conditions in the class 100 cleanroom, and the following method is followed:

1. A dehydration bake is performed on "Piranha etched" cleaned cleaved silicon substrate on a level hotplate at 120° C for five minutes.
2. While the spinner is stationary, SU 8-Ag is aspirated via a syringe and spread evenly over the substrate. A two- staged spinning program should be used; a 30 second ramp, followed by 150 seconds at a steady speed. For a 75Micron resist layer, a 30 second ramp to 500 rpm is followed by a 150 second spin at 2300rpm.
3. In order to remove edge beads, and to prevent the sample sticking to the hotplate during baking, rotate the spinner chuck by hand and use a cleanroom wipe moistened in acetone on the edges of the sample.
4. A prebake stage is performed by bringing the sample to 65°C for 20 minutes. Before ramping up the hotplate to 95°C for 2 hours and 30 minutes. During this softbake process, check the sample is dry; by inserting a needlepoint into the film edges until the film is hard. As the sample to cool to room temperature; a reflow is visible. The prepared composite is now ready for proton beam lithography.
5. After proton beam lithography has been performed, SU 8-Ag is developed using a chemical developer (Micropropsite developer from MicroChem Inc.), Sonification in an ultrasonic bath for 15 seconds removes all un-irradiated regions and leaves behind a structure created.

A post exposure bake while not necessary is performed for 30 Minutes prior to development as this reportedly improves conductivity [27].

3. Results and Discussion

Patterned 3 Dimensional structures were created on a 75 micron thick Ag/SU8 nano-particle composite, with 2.5 MeV protons with a fluence of 30nC/mm^2. Irradiations were performed using the University of Surrey's tandem particle accelerator.
Scanning electron micrographs as seen in figure 1a, b, show a 3-dimensional test structures produced from the composite.

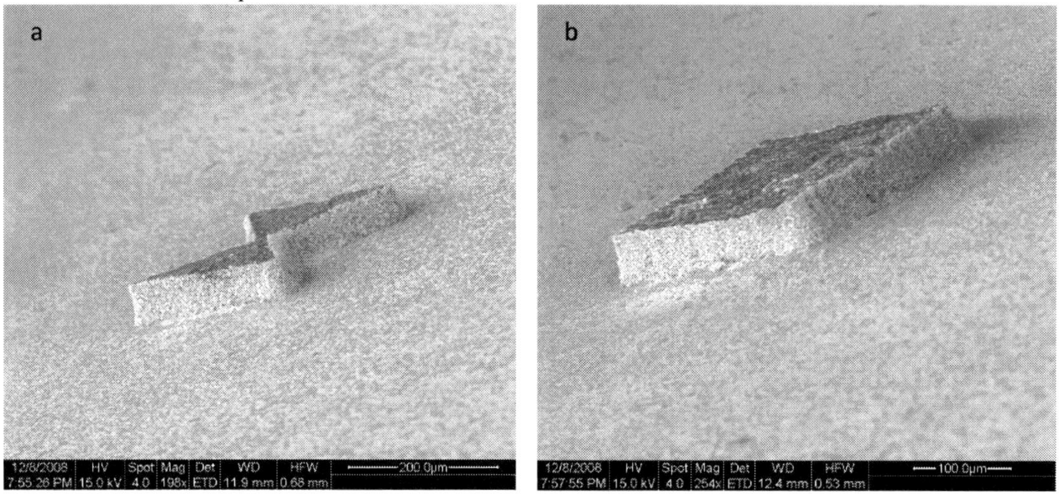

Figure 1a, b: 75 micron tall 3 Dimensional test structures created by 2.5 MeV Proton beam lithography of Ag/Su8 nano-composite.

The structures produced show what seem to be aggregations or clumps silver particles, leaving micro voids. This is probably a clustering effect of the silver nano particles on exposure to MeV ion beams [25].

When compared to structures produced in the same polymer composite using UV lithography, a marked improvement is seen in the structure height as well as distinctness of features such as edges. The micrographs in figure 2 shows contrasts the structural differences based on the lithographic method 2a UV lithography [27] 2b Proton beam lithography.

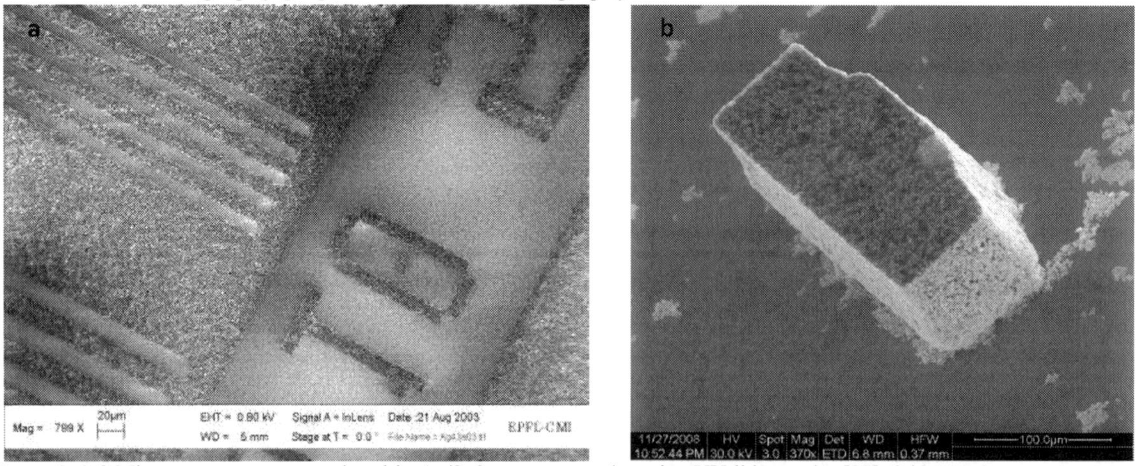

Figure 2.a) 8 Micron test structure produced in Ag/Su8 nano-composite using UV lithography [27]. 2 b) 75 micron test structure produced in Ag/Su 8 nano-composite using proton beam lithography.

3.1 Electrical Conductivity

The I-V characteristics of the resultant proton beam written composite structure was determined by 2 probe measurement using a Keithley probe. Unlike Polyaniline the structures exhibit non-ohmic conductivity [20, 21] and also exhibit current to voltage change / switching in 4 orders of magnitude [22]. This suggests that the conduction mechanism is based on nano tunnelling as explained by the percolation conduction model. The experimental measurement has been current limited to 0.01A to limit the amount of energy applied to the resist.

4. Conclusion and Future Work

We have demonstrated the creation conducting of 3-dimensional microstructures in Ag/Su 8 nano-composite using proton beam lithography. The structural features and feature heights are improved when compared to structures created using other lithographic techniques. The structures created are electrically conductive and conduct electricity via tunnelling, while exhibiting current switching where the resistance of structures drop to at least four orders of magnitude.

Proton beam lithography is unique due to the predictability of the penetration depth of protons in a target at specific energy. This feature could be exploited in creating freestanding conductive microstructures that can be actuated or utilized for radio frequency applications.

The figure 3a is a micrograph a test structure showing a conductive 3 Dimensional freestanding structure created by 1.0 and 2.5MeV proton beams on a 75micron thick layer of 15% Ag/Su8 nano-composite.

a b

Figure 3a) 3 Dimensional freestanding cantilever structure created by 1.0 and 2.5MeV proton beams on a 75micron thick layer of 15% Ag/Su8 nano-composite. 4b) Test pillars structure created by 2.5MeV proton beams on a 75micron thick layer of 15% Ag/Su8 nano-composite

The micrograph in Figure 3b highlights the changeability of structural patterns as proton beam lithography is a direct write process that does not require mask fabrication. Further work would also include modelling the interaction of ion beams within a nano-particle polymer composite target containing randomly distributed nano-particles with known volume ratios, and actuation of devices produced using this technique.

5. Acknowledgements

The authors acknowledge the contributions of PhD supervisors Professors K. Kirkby and R. Webb and Dr. M. Merchant, Dr. V.Palitsin and Dr. A. Smith of the Surrey Ion beam Centre, and the EPSRC funded Tandem microbeam line.

References

[1] F. Watt, A. A. Bettiol, J. A. Van Kan, E. J. Teo, M. Breese; Nuclear Instruments and Methods in Physics Research B vol. 210, (2003) pg. 14.

[2] J. A. Van Kan, F. Watt, A. A. Bettiol, E. J. Teo, International Journal of Nanotechnology 1(4) (2004).: p. 464 -479.

[3] F. Watt, A. A. Bettiol, J. A. Van Kan, E. J. Teo, M. Breese.; International journal of nanoscience. vol 4, No. 3 (2005) pg 269 - 286 .

[4] MDDS, SU 8 -100 Negative tone Epoxy Resin Photo resist. Published by MicroChem Ltd. (1999)

[5] H.Igbenehi, V. Palitsin M. Merchant. Electroless Plating of Proton beam Written Microstructures. (In press)

[6] D. Leeuw, G. Gelinck, M. Matters Method of producing vertical interconnects between thin film microelectronic devices and products comprising such vertical interconnects. United states patent number 6635406.

[7] G Whitesides, N. Damean, B. Parviz, J. Lee, T. Odom/ Composite ferromagnetic photoresist for the fabrication of microelectromechanical systems. J. Micromech. Microeng. 15 (2005) 29–34 doi:10.1088/0960-1317/15/1/005

[8] Jiguet, S., Bertsch, A., Judelewicz, M., Hofmann, H., Renaud, P., *Microelectron. Eng.* 2006, *83*, 1966–1970.

[9] M. Nordström , S. Keller 1, M. Lillemose , A.Johansson , S. Dohn, D. Haefliger, G. Blagoi, M. Havsteen-Jakobsen, A. Boisen: SU-8 Cantilevers for Bio/chemical Sensing; Fabrication, Characterisation and Development of Novel Read-out Methods. *Sensors* 2008, *8*, 1595-1612

[10] L. Gammelgaard, P. A. Rasmussen, M. Calleja, P. Vettiger, and A. Boisen Microfabricated photoplastic cantilever with integrated photoplastic/carbon based piezoresistive strain sensor. Appl. Phys. Lett. 88, 113508 (2006); DOI:10.1063/1.2186396.

[11] H. Chiamori, J. Brown, E. Adhiprakasha, E. Hantsoo, J. Straalsund, N. Melosh and B. Pruitt. Suspension of nanoparticles in SU-8: Processing and characterization of nanocomposite polymer. Microelectronics Journal Volume 39, Issue 2, February 2008, Pages 228-236.

[12] N. Zhang, J. Xie, M. Guers, V. Varadan. Chemical bonding of multiwalled carbon nanotubes to SU-8 via ultrasonic irradiation 2003 *Smart Mater. Struct.* 12 260-26.

[13] P. Renaud, S. Metz, S. Jiguet, A. Bertsch, Composite photopolymer microstructures: from planar to 3D devices TRANSDUCERS, Solid-State Sensors, Actuators and Microsystems, 12th International Conference on, 2003 Volume 2, Issue , 8-12 June 2003 Page(s): 991 - 994 vol.2.

[14] S. Jiguet, A. Bertsch, H. Hofmann, P. Renaud, Conductive SU8-silver composite photopolymer Micro Electro Mechanical Systems, 2004. 17th IEEE International Conference on. (MEMS) Volume , Issue , 2004 Page(s): 125 – 128.

[15] Müller, K.-H.; Herrmann, J.; Raguse, B.; Baxter, G.; Reda, T. Percolation model for electron conduction in films of metal nanoparticles linked by organic molecule s. Physical Review B, vol. 66, Issue 7, id. 075417. 2002.

[16] R. Fedorovich, A. Naumovets and P. Tomchuk . Electronic phenomena in nanodispersed thin films. J. Phys.: Condens. Matter 11 (1999) 9955–9967.

[17] B. Alekseenko R.Fedorovich *Thin Solid Films* 92 252 (1982).

[18] D. Stauffer, "Introduction to Percolation Theory," Chapter 5. Taylor and Francis, 2nd Edition. London, UK 2003. ISBN 0-203-21159-6.

[19] McCluskey, P.; Morris, J.; Verneker, V.R.P.; Kondracki, P.; Finello, D. Models of electrical conduction in nanoparticle filled polymers Proceedings of 3rd International Conference paper. 28-30 Sep 1998 Page(s):84 – 89 DOI: .10.1109/ADHES.1998.742007

[20] P. Jacosalem, M. Vequizo, D. Pamatong, H. Miyata. Conductivity measurements if chemically prepared Polyaniline film sample: Mindanao National Physics Conference-Workshop held on October 28-30, 2002.

[21] J. Díaz-de Leon Electrospinning Nanofibers of Polyaniline andPolyaniline / (Polystyrene and Polyethylene Oxide) Blends. Proceedings (NCUR)University of Kentucky(2001).

[22] A. Kiesow, J. E. Morris, C. Radehaus, and A. Heilmann, Switching behavior of plasma polymer films containing silver nanoparticles J. Appl. Phys. 94, 6988 (2003), DOI:10.1063/1.1622990.

[23] S. Klaumu¨nzer. Modification of nanostructures by high-energy ion beams. Nuclear Instruments and Methods in Physics Research B 244 (2006) p1–7.

[24] A Sharma, A Srivastava Ion beam induced modifications in nitroso substituted polyaniline: Spectral and electrical studies Current Applied Physics 7 (2007) 650–654

[25] A. Biswas , D.K. Avasthi, D. Fink , J. Kanzow ,U. Sch€urmann, S.J. Ding a, O.C. Aktas, U. Saeed, V. Zaporojtchenko , F. Faupel, R. Gupta, N. Kumar Nanoparticle architecture in carbonaceous matrix upon swift heavy ion irradiation of polymer–metal nanocomposites Nucl.Instr.and Meth.in Phys.Res.B 217 (2004) 39–50

[26] Functional Su8 PhotoEpoxy Datasheet GCM3060 Gersteltech Inc. Switzerland . Accessible on the web on http://www.gersteltec.ch/userfiles/1197841690.pdf first published May 2005.

[27] S. Jiguet, A. Bertsch, H. Hofmann, P. Renaud, SU8-silver photosensitive Nano-composite. Advanced Engineering Materials Vol 6. No. 9 (2004) Page 723 DoI: 10.1002/adem 200400068.

Effect of silver-doping on the crystal structure, morphology and photocatalytic activity of TiO_2 nanofibers

N A M Barakat[1,2], H Y Kim[1]

[1] Organic Materials and Fiber Engineering Department, Chonbuk National University, Jeonju 561-756, Republic of Korea
[2] Chemical Engineering Department, Faculty of Engineering, El-Minia University, El-Minia, Egypt

Tel: 0082 63 270 2363 E-mail: nasser@jbnu.ac.kr

Abstract. In this study, effect of sliver-doping on the crystal structure, the nanofibrous morphology and the photocatalytic activity of titanium oxide nanofibers have been investigated. Silver-doped TiO_2 nanofibers having different silver contents were prepared by calcination of electrospun nanofiber mats consisting of silver nitrate, titanium isopropoxide and poly(vinyl acetate) at 600 °C. The results affirmed formation of silver-doped TiO_2 nanofibers composed of anatase and rutile when the silver nitrate content in the original electrospun solution was more than 3 wt%. The rutile phase content was directly proportional with the $AgNO_3$ concentration in the electrospun solution. Negative impact of the silver-doping on the nanofibrous morphology was observed as increase the silver content caused to decrease the aspect ratio, i.e. producing nanorods rather nanofibers. However, silver-doping leads to modify the surface roughness. Study of the photocatalytic degradation of methylene blue dye clarified that increase the silver content strongly enhances the dye oxidation process.

1. Introduction

Enormous efforts have been devoted to the research of TiO_2 material, which has led to many promising applications in areas ranging from photovoltaics and photocatalysis to photo-/electrochromics and sensors [1-3]. New physical and chemical properties emerge when the size of the TiO_2 becomes smaller and smaller, and down to the nanometer scale. The main advantage of the nanostructural materials is the high surface to volume ratio, however much literature indicated that the one-dimensional (ID) nanostructures have special characteristics due to the axial ratio aspect [4-6]. Among the ID shapes, nanofibers have novel features because of the long axial ratio; consequently good reports were introduced about fabrication of TiO_2 nanofibers [7, 8]. Recently, metal-doping strategy has been invoked to improve the physical and chemical properties of TiO_2 nanoparticles/nanofibers [9-11]. Noble metal incorporation into the titania dielectric provides an absorption feature due to the surface Plasmon resonance (SPR) occurring over the visible range of the spectrum [12]. In particular, silver and gold metals are the most popular materials due to the strong SPR character [13]. Metal-doping may also have an influence on the titania crystal structure leading to distinct improving in the application fields [10, 14].

Here, we study effect of the silver-doping on the crystal structure, the morphology and the photocatalytic activity of TiO_2 nanofibers. Several techniques have been introduced to fabricate the nanofibers, however electrospinning is the most widely utilized process due to its simplicity, high yield and low cost [15, 16]. Accordingly, silver-doped TIO_2 nanofibers were prepared by calcination of electrospun mats consisting of silver nitrate, titanium isopropoxide and poly(vinyl acetate) at 600 ^{0}C. Control of silver content in the prepared nanofibers could be achieved by modifying the silver nitrate concentration in the original electrospun solution. Physiochemical characterizations indicated that the silver content does have distinct influences on the morphology, crystal structure and photocatalytic activity of the TiO_2 nanofibers.

2. Experimental section

2.1 Materials

Silver nitrate (99.8 assay), methylene blue dihydrate (95.0 assay), acetic acid (99.7 assay) and N,N-dimethylformamide (DMF, 99.5 assay) were obtained from Showa, Co. Japan. Titanium (1V) isopropoxide (Ti(Iso)), 98.0 assay) was purchased from Junsei Co. Ltd., Japan. Poly(vinyl acetate) (PVAc, MW = 500,000 g/mol) obtained from Aldrich USA. These materials were used without any further purification.

2.2 Electrospinning set-up

A sol-gel composed of titanium isopropoxide (Ti(Iso)) and poly(vinyl acetate) (14% wt, in N,N-dimethylformamide; DMF) was prepared by mixing these ingredients for 10 min, and then few drops of acetic acid were added till the solution became transparent. The final solution contained 60 wt% Ti(Iso). Ten gram of the prepared solution was mixed with silver nitrate/DMF solution ($AgNO_3$ was dissolved in a minimum amount of DMF) to prepare the silver-doped titanium oxide nanofibers [11], several solutions containing different contents of silver nitrate have been prepared. In the electrospinning process, a high voltage power supply (CPS-60 K02V1, Chungpa EMT Co., Republic of Korea) was used as a source of electric field. The sol-gel was supplied through a plastic syringe attached to a capillary tip. A graphite rod connected with the positive electrode (anode) was inserted into the sol-gel while the negative one (cathode) was attached to a metallic collector covered with polyethylene sheet. Briefly, the solutions were electrospun at 6 kV voltage and 15 cm working distance (the distance between the needle tip and the collector). The formed nanofiber mats were initially dried for 24 h at 80 °C under vacuum and then calcined in air atmosphere at 600 °C for 1 h with a heating rate of 5 °C/min.

The photocatalytic degradation of methylene blue dye was carried out in a simple photo reactor. The reactor was made of glass (1000 ml capacity, 23 cm height and 15 cm diameter), covered with alumina foil and equipped with 365 nm (wavelength radiation) ultra-violet lamp. The initial dye solution and the photocatalyst were placed in the reactor and continuously stirred for complete mixing during the photocatalytic reaction. Typically, 100 ml of the dye solution (10 ppm, concentration) and 50 mg of the catalyst were used. At specific time intervals, 2 ml samples were withdrawn from reactor and centrifuged to separate the nanofibers (catalyst), and then the absorbance intensity was measured at wavelength of 664 nm 1[11].

2.3. Characterization

The phase and crystallinity were characterized by using Rigaku X-ray diffractometer (Rigaku Co, Japan) with Cu Kα (λ = 1.54056 A°) radiation over a range of 2θ angles from 20 to 80^0. Surface morphology of the nanofibers was studied by JEOL JSM-5900 scanning electron microscope, JEOL Ltd, Japan and field-emission scanning electron microscope (FESEM, Hitachi S-7400, Japan). High resolution image was observed by JEOL JEM 2010 transmission electron microscope (TEM) operating at 200 kV, JEOL Ltd., Japan. The concentration of the dyes during the photodegradation study was investigated by spectroscopic analysis using HP 8453 UV-visible spectroscopy system, Germany. The spectra obtained were analyzed by HP ChemiStation software 5890 series.

3. Results and discussion

Titanium oxide has three popular crystal structures; antase, rutile and brookite. Anatase is a polymorph with the two other minerals. The minerals rutile and brookite as well as anatase all have the same chemical formula, TiO_2, but they have different structures. Rutile is the more common and the more well known mineral of the three, while anatase is the rarest. Anatase shares many of the same or nearly the same properties as rutile such as luster, hardness and density. However due to structural differences, anatase and rutile differ slightly in the crystal habit. The phase change from anatase polymorph of titania to rutile one has been the subject of considerable interest and the focus of many groups' activities over the years.

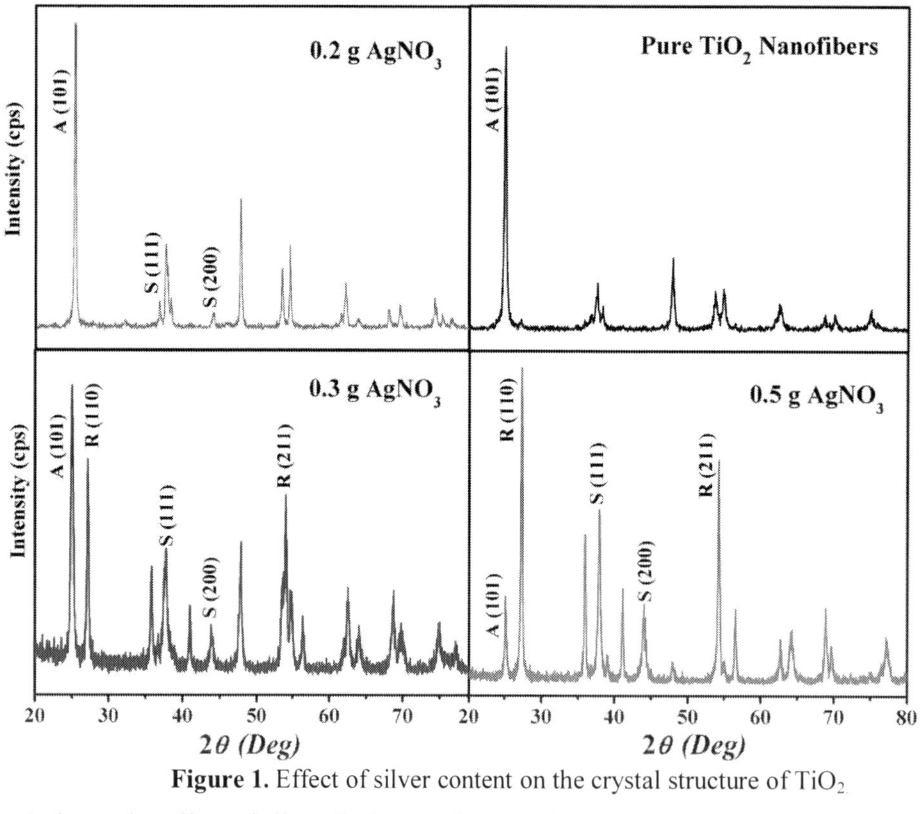

Figure 1. Effect of silver content on the crystal structure of TiO_2

Figure 1 shows the effect of silver-doping on the crystal structure of the obtained powder after the calcination process. In case of silver-free solution, the results affirm formation of pure anatase titanium dioxide, existence of strong diffraction peaks at 2θ values of 25.09°, 37.65°, 38.44°, 47.89°, 53.89°, 55.07°, 62.40°, 68.70°, 70.04° and 75.00° corresponding to the crystal planes of (101), (004), (112), (200), (105), (211), (204), (220), (220) and (215), respectively indicates formation of anatase titanium dioxide [JCPDS card no 21-1272]. Addition of small amount of silver nitrate (2 wt%) did not affect the crystal structure of the titania, however pure silver metal was formed due to decomposition of the silver nitrate upon heating. The extra peaks at 2θ values of 38.11°, 44.29°, 64.43° and 77.48°, corresponding to the crystal planes of (111), (200), (220) and (311), respectively affirm presence of silver metal [JCPDS card no 04-0783]. In Figure 1, the main peaks of anatase and silver were marked as A and S, respectively. Increase the silver nitrate content to 3 wt% in the original electrospun solution led to partial phase change in the titania crystal structure. The standard peaks of rutile can be observed at 2θ values of 27.45°, 36.09°, 41.23°, 54.32°, 56.64°, and 69.08° corresponding to the crystal planes of (110), (101), (111), (211), (220), and (301), respectively [JCPDS card no 21-1272]. Interestingly, the results indicate that increase the amount of silver enhances formation rutile phase as shown in Figure 1 (the main rutile peaks were marked by R letter). Addition of 5 wt% silver nitrate led to increase the rutile content to be higher than the anatase. According to the XRD data, the ratio of rutile to anatase is ~ 75 % and 440 % when the added silver nitrate was 3 wt% and 5 wt%, respectively.

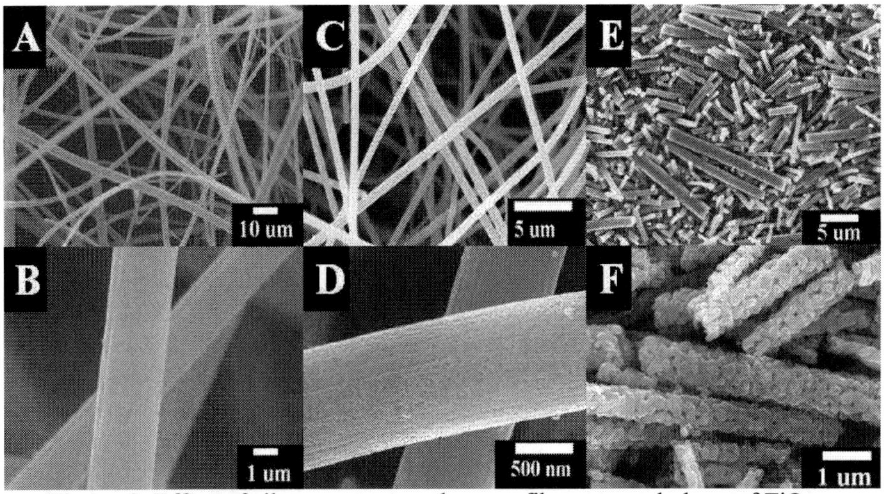

Figure 2. Effect of silver content on the nanofibrous morphology of TiO_2.

The electrospinning technique involves the use of a high voltage to charge the surface of a polymer solution droplet and thus to induce the ejection of a liquid jet through a spinneret. Due to bending instability, the jet is subsequently stretched by many times to form continuous, ultrathin fibers. It is widely used for production of many polymeric nanofibers. Moreover, the electrospinning process has been exploited to produce metal oxides nanofibers by calcination of electrospun mats obtained from completely miscible sol–gel solutions. Electrospinning of a sol–gel composed of Ti(IsO) and PVAc/DMF solution has been carried out in our lab in previous study [17], the resultant electrospun nanofibers have good morphology. Addition of silver nitrate to the sol-gel does not affect the morphology [11]. Accordingly, well morphology electrospun nanofibers mats were obtained from $AgNO_3$/Ti(Iso)/PVAc solutions at all the utilized $AgNO_3$ contents (data are not shown). Figureure 2 shows the morphology of the resultant powder after the calcination process. As can be observed in Figure 2A and B which demonstrate the obtained product from silver-free electrospun solution, well and smooth TiO_2 nanofibers were obtained, these nanofibers composed of pure anatase as could be concluded from the XRD results (Figure 1). Incorporation of small amount of silver does not have considerable impact on the nanofibrous morphology as shown in Figure 2C and 2D which represent the FE-SEM images of the powder obtained from calcination of an electrospun solution containing 2 wt% $AgNO_3$. However, the surface became little rough (Figure 2D).

Figure 3 demonstrates the TEM analysis results for TiO_2 nanofibers originated from an electrospun solution containing 2 wt% silver nitrate. Due to the large atomic radius of the silver atom, they cannot interpenetrate in TiO_2 crystals to form solid solution alloy. Therefore, silver metal distributes in the form of separated nanoparticles as shown in the normal TEM and HR TEM images. Of course, the average size of those nanoparticles is directly proportional with the amount of silver. Therefore, more sharp and strong peaks were observed in the XRD spectra representing the nanofibers containing high silver content. This might be an acceptable explanation of obtaining nanorods rather than nanofibers when the amount of the utilized silver nitrate was 3 wt% (Figure 2E and 2F), in other words, the big size silver nanoparticles led to break the nanofibers. However, the surface became much rougher (Figure 2F) compared with the pristine and low silver content TiO_2 nanofibers. More increase in the silver nitrate content in the electrospun solution led to complete annihilation of the 1D structure.

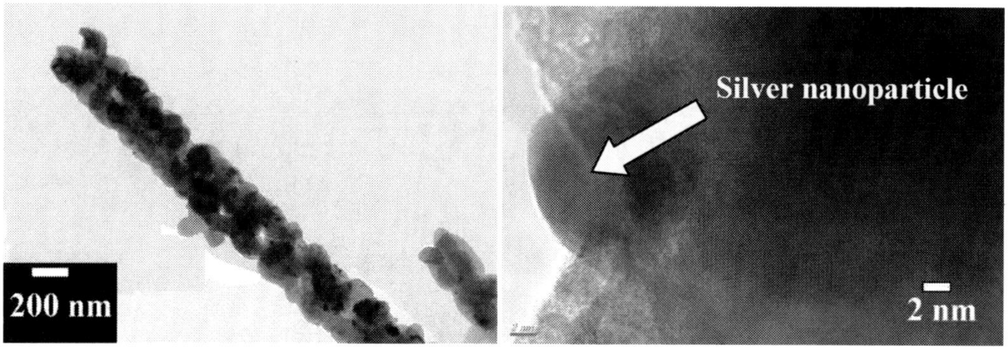

Figure 3. TEM image for the TiO2 nanofibers originated from electrospun solution containing 2 wt% silver nitrate.

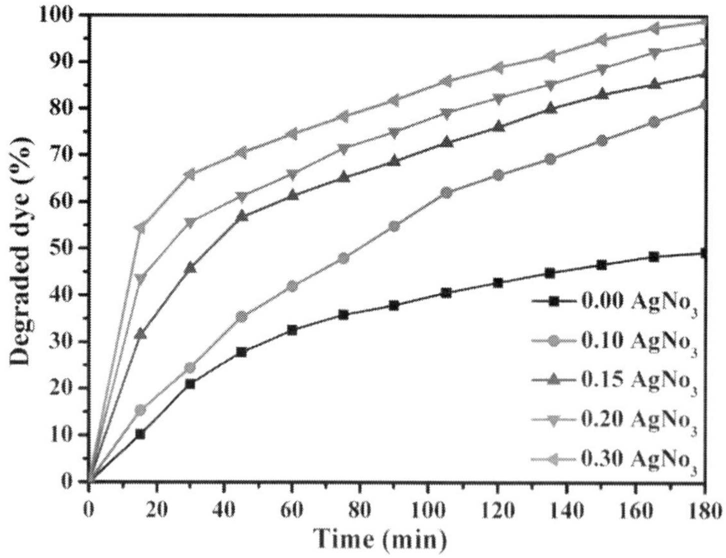

Figure 4. Effect of silver content on the photocatalytic activity of TiO_2 nanofibers.

Much literature has been reported explaining the positive effect for silver incorporation on the photocatalytic of TiO_2 nanostructures. Generally, these studies have drawn this conclusion. The Ag nanoparticles deposited on TiO_2 surface act as electron acceptors, enhancing the charge separation of electrons and holes and consequently the transfer of the trapped electron to the adsorbed O_2. The sufficient dye molecules are adsorbed on the surface of Ag-TiO_2 more than on the TiO_2 surface, increasing the photoexcited electron transfer from the sensitized dye molecule to the conduction band of TiO_2 and consequently increasing the electron transfer to the adsorbed O_2. Figure 4 affirms this conclusion. As shown in Figure 4 which represents the effect of silver content on photcatalytic degradation of methylene blue dye, the rate of degradation enhances with increasing the silver content in the TiO_2 nanofibers. Moreover, the surface roughness might have an influence due to increase the total surface area.

4. Conclusion

Calcination of electrospun nanofiber mats composed of silver nitrate, titanium isopropxide and poly(vinyl acetate) in air at 600 ^{o}C leads to produce silver-doped TiO_2 nanofibers having both of antase and rutile phases if the silver nitrate content is more than 3 wt %. Rutile phase content can be increased by increasing the silver content in the original electrospun solution. However, the nanofibrous morphology is strongly affected by the silver-content, excess silver causes to destroy the 1D structure. Silver-doping of the titanium oxide nanofibers greatly enhances the photocatalytic activity of this interesting material.

5. Acknowledgement

This paper was supported by research funds of chonbuk National University 2010. We thank Mr. T. S. Bae and J. C. Lim, KBSI, Jeonju branch, and Mr. Jong- Gyun Kang, Centre for University Research Facility, for taking high-quality FESEM and TEM images, respectively.

References

[1] Chen X, Mao S D, *Chem. Rev.* Titanium Dioxide Nanomaterials: Synthesis, Properties, Modifications, and Applications, 2007; **107:** 2891-2959.

[2] Grätzel, M, *Nature* "Photoelectrochemical Cells, 2001; **414**: 338-344.

[3] Millis A, Le Hunte S, *J. Photochem. Photobiol. A,* An Review of Semiconductor Photocatalysis, 1997; **108**: 1-35.

[4] Fraune M, Rüdiger U, Güntherodt G, *Appl. Phys. Lett.,* Size Dependence of the Exchange Bias Field in Ni/NiO Nanostructures 2000; **77**: 3815-3817.

[5] Barakat N A M, Kim B, Park S J, Jo Y, Jung M H, Kim H Y, *J. Mater. Chem.,* Cobalt Nanofibers Encapsulated in Graphite Shell with Thermally Independent Magnetic Properties 2009; **19**: 7371-73-78.

[6] Barakat N A M, Khalil K A, Mahmoud I H, Kanjwal M A, Sheikh F A, *J. Phys. Chem. C*, CoNi Bimetallic Nanofibers by Electrospinning: Nickel-Based Soft Magnetic Material with Improved Magnetic Properties, 2010; **114**: 15589-15593.

[7] Li D, Xia Y, *Nano Lett.*, Fabrication of tatania nanofibers by electrospinning, 2003; **3**: 555-560.

[8] Yuan Z Y, Su B L, *Colloids Surf. A*, Titanium oxide nanotubes, nanofibers and nanowires, 2004; **241**: 173-183.

[9] Di Paola A, Marci G, Palmisano L, Schiavello M, Uosaki K, Ikeda S, Ohtani B. *J. Phys. Chem. B*, Preparation of Polycrystalline TiO_2 Photocatalysts Impregnated with Various Transition Metal Ions: Characterization and Photocatalytic Activity for the Degradation of 4-Nitrophenol, 2002; **106**: 637-645.

[10] Arbiol J, Cerdá J, Dezanneau G, Cirera A, Peiró F, Cornet A, Morante J R, *J. Appl. Phys.*, Effects of Nb doping on the TiO_2 anatase-to-rutile phase transition 2002; **92:** 853-861.

[11] Kanjwal M K, Barakat N A M, Sheikh F A, Khil M S, Kim H Y, *Int. J. Appl. Ceram. Technol.,* Functionalization of Electrospun Titanium Oxide Nanofibers with silver nanoparticles: Strongly Effective Photocatalyst, 2010; **7**: E54-E63.

[12] Li X Z, Li F B, *Environ. Sci. Technol.* Study of Au/Au^{3+}-TiO_2 Photocatalysts toward Visible Photooxidation for Water and Wastewater Treatment, 2001; **35**: 2381-2387.

[13] Nikolajsen T, Leosson T, Salakutdinov K, Bozhevolnyi S, *Appl. Phys. Lett.* "Surface Plasmon polariton based modulators and switches operating at telecom wavelengths, 2004; **85**: 5833-5836.

[14] Ko K H, Lee Y C, Jung Y J, *J. Colloid Interface. Sci.* Enhanced efficiency of dye-sensitized TiO_2 solar cells (DSSC) by doping of metal ions 2005; **283**: 482-487.

[15] Sigmund W, Yuh J, Park H, Maneeratana V, Pyrgiotakis G, Daga A, Taylor J, Nino J. C, *J. Am. Ceram. Soc.,* Processing and structure relationships in electrospinning of ceramic fiber systems, 2006; **89**: 395-407.

[16] Barakat N A M, Kim H Y, “Metal oxides nanostructures and their applications”, American Scientific Publisher, 2009: **Ch.9**.

[17] B. Ding, C. K. Kim, H. Y. Kim, M. S. Khil, S. Park, Titanium dioxide nanofibers prepared by using electrospinning method, Fibers Polym., 2004, **5**, 105-109.

Cast iron cutting with nano TiN and multilayer TiN-CrN coated inserts

M Perucca[1], S Durante[1], U Semmler[2], C Rüger[2], G G Fuentes[3] and E Almandoz[3]

[1] DIAD, Buttigliera Alta, Italy
[2] IWU Fraunhofer-Institute for Machine Tools and Forming Technology, Chemnitz, Germany
[3] Center of Advanced Surface Engineering AIN, Spain

E-mail: ulrich.semmler@iwu.fraunhofer.de

Abstract. During the past decade great success has been achieved in the development of duplex and multilayer multi-functional surface systems. Among these surface systems outstanding properties have nanoscale multilayer coatings. Within the framework of the M3-2S project funded in the 7th European Framework Programme, several nanoscale multilayer coatings have been developed and investigated for experimental and industrial validation. This paper shows the performance of TiN and TiN/CrN nanoscale multilayer coatings on WC cutting inserts when machining GJL250 cast iron. The thin films have been deposited by cathodic arc evaporation in an industrial PVD system. The multilayer deposition characteristic and its properties are shown. The inserts have been investigated in systematic cutting experiments of cast iron bars on a turning machine specifically equipped for force measurements, accompanied by wear determination. Furthermore, equivalent experiments have been carried out on an industrial turning unit. Industrial validation criteria have been applied to assess the comparative performance of the coatings. The choice of the material and the machined parts is driven by an interest in automotive applications. The industrial tests show the need to further optimise the multi-scale modelling approach in order to reduce the lead time of the coating development as well as to improve simulation reliability.

1. Introduction

A large variety of different surface engineering systems (including coatings and surface modification), with thickness ranging from the nanometre scale to the millimetre scale, have been developed in past decades to combat surface-related degradation such as wear, corrosion, oxidation and fatigue. However, no single surface engineering system or coating can provide an overall solution. Significant progress has been made during the past decade in the development of duplex or multilayer multi-functional surface systems to meet the ever-increasing demands for high performance engineering components operating under severe working conditions, for tools for emerging manufacturing techniques and for the tribological applications of several alloys in the automotive industry: nanoscale multilayer or supper-lattice coatings with high hardness, good oxidation resistance and high wear resistance for high-speed and/or dry cutting conditions [1]; new self-lubricating coatings with good wear resistance for micro-forming and machining [2, 3]; duplex surface systems combining oxygen deep-case hardening and low friction wear resistant multilayer coatings [4]. Through the development of deep-case hardening and multilayer self-lubricating coatings, a wide exploitation of light alloys in automotive engines will provide added fuel efficiencies and reductions in CO_2 emission [5].

This paper addresses nanostructured and nanolayered coatings developed with a PVD process on tool inserts for turning of grey cast iron GJL250. The application is relevant for the automotive sector in which machining of grey cast iron plays an important role.

The paper is a collaboration of the three institutes DIAD (Italy), IWU (Germany) and AIN (Spain) within the 7th Framework Program in the M3-2S project consortium [6]. The experimental plan has been established in accordance with the modelling and the characterization of the coatings in this project. The turning machining tests on cast iron bars refer to typical automotive applications.

In this presented work AIN did the deposition of the coatings by cathodic arc evaporation in an industrial PVD system and investigated the coatings (chapter 2). The inserts have been investigated in systematic cutting experiments both at the IWU on a turning lathe, accompanied by force and wear measurements (chapter 3), and at the DIAD on an industrial turning unit (chapter 4). Based on experimental results numerical simulations of the same cutting processes were performed at the IWU in order to calculate the acting thermo-mechanical loads (chapter 5).

The properties of the coatings and their applicability have been tested under experimental conditions close to reality and, at the same time, under real industrial conditions. For both cutting series, experimental and industrial, the same cutting conditions, the same workpiece material, identical insert and insert holders, and the same coatings were chosen.

2. Coatings

2.1. Deposition

The three coatings have been produced by cathodic arc evaporation in a commercial METAPLAS MZR-323 PVD reactor. The reactor is equipped with two opposing columns, each of them hosting three circular cathodes (10 cm diameter) aligned vertically. Both columns face each other and leave an effective volume of 0.5 m^3 available for the substrates.

Table 1. Deposition parameters.

Sample	**Arc current [A]**	**DC Bias [V]**	**Total pressure [Pa]**	**Rotating speed**	**Adhesion layer**	
					Compo-sition	**Thickness [nm]**
TiN	(65Ti) 60	-200	1.0	3	Ti	200
CrN	(6Cr) 60	-100	1.2	3	Cr	200
Multilayer	(3Ti & 3Cr) 60	-100	1.1	5	Ti	290

Two different cathode configurations were used for the preparation of the coatings: six Ti cathodes (99.8% purity) resp. Cr cathodes (99.5% purity) for the preparation of monolayers in a work pressure of 1.0 resp. 1.2 Pa of high purity N2 gas. In the case of the TiN-CrN multilayers, three Ti sources were placed on one side of the chamber and three Cr sources on the opposite side. During deposition the work pressure was set to 1.1 Pa in high purity N2 gas together with high purity Ar.

In all cases steady DC bias was applied on the substrates during the deposition process. During the complete process the substrates rotated around the central vertical axis of the reactor. In all cases a metallic adhesion layer of approximately 200-290 nm was deposited.

2.2. Characterization

Glow discharge optical emission spectroscopy (GDOES) was used to obtain the chemical composition of the films. The GDOES analyses were carried out with a Jovin-Yvon JY 1000 RF optical spectrometer equipped with more than 40 channels and an optical monochromator. Results are shown in Figure 1.

Table 2. Hardness, reduced Young's modulus and thickness of the three coatings

	Hardness H [Gpa]	**Reduced Young's modulus E_r [Gpa]**	**Total thickness [µm]**
TiN	31.3 ± 7.0	356 ± 62	3.17
CrN	22.6 ± 2.7	263 ± 19	3.50
Multilayer	27.5 ± 2.0	252 ± 17	2.24

For each sample 60 points were measured in three different zones, each containing 20 indents. Table 2 shows the averaged values and their standard deviations which were calculated from the three zones.

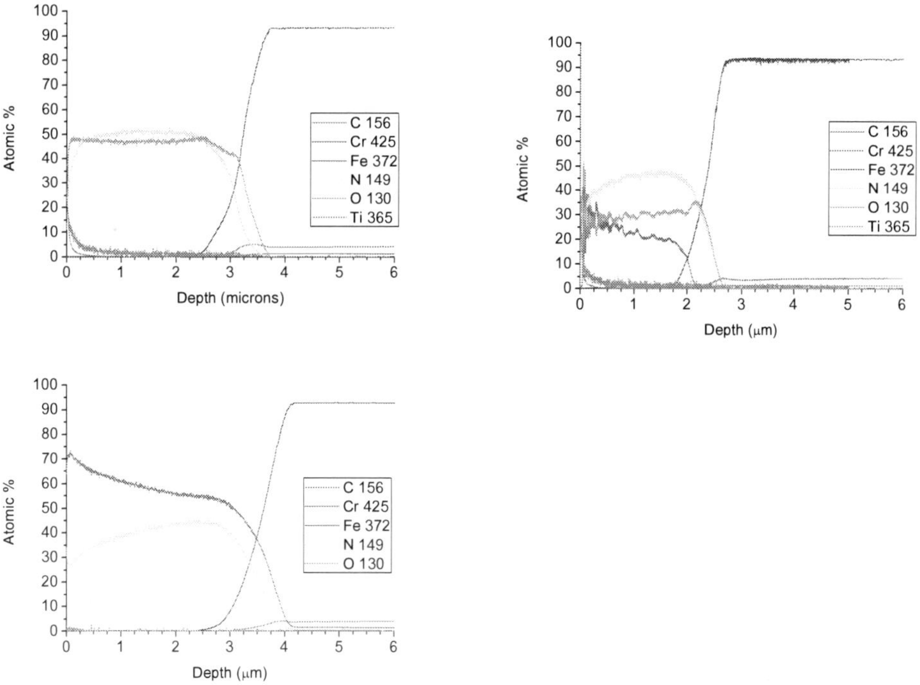

Figure 1. GD-OES analyses of the coatings: TiN (top left), CrN (top right) and multilayer.

A field emission scanning electron microscope (SEM) HITACHI S-4800 was used to inspect the cross-sections of the films and to measure the coating thickness. Hardness and reduced Young's modulus $(1/E_r=(1-\nu_{indenter}^2)/E_{indenter}+(1-\nu_{coating}^2)/E_{coating})$ of the coatings were determined by nanoindentation using a HYSITRON TI 950 TriboIndenter equipped with a Berkovich diamond indenter. The Oliver and Pharr method was used for the calculation of mechanical properties. The indentations were carried out at a constant strain rate of 2 mN/s up to a maximum load of 10 mN. In all cases the indentation depth was less than one tenth of the coating thickness, thus the substrate effect on the measured hardness can be neglected. The equipment had been previously calibrated using a fused silica standard sample with a hardness of 9.25 GPa and a Young's modulus of 69.6 GPa.

3. Cutting Experiments

The experiments were done in the laboratories of the Fraunhofer Institute in Chemnitz. The equipment allowed the on-line monitoring of cutting forces and the interruption at predefined cutting distances for the control of the developing wear. The conditions of the experiments are summarized in table 3.

The geometrical parameters of the process, of the tool and of the insert are as follows (Fig. 2): working cutting edge angle $\kappa = 45°$, cutting depth $a_p = 1.5$ mm and cutting edge angle of inclination $\lambda = 6$ °. The cutting edges are stabilized by a chamfer of 13° angle and 200 μm width. The cutting edge radius varies between 20 and 30 μm, depending on whether the insert is coated or not. The micro-geometry of the cutting edge has been measured by the structured-light 3D scanner system MikroCAD.

The parameter space of cutting conditions was limited by the following three combinations of cutting velocity v_c and feed f leading to equal cutting rate $\Delta V/\Delta t = 112{,}300$ mm^3/min: $v_c = 250$ m/min, f = 0.3 mm (CC1); $v_c = 375$ m/min, f = 0.2 mm (CC2); $v_c = 500$ m/min, f = 0.15 mm (CC3); all under flood cooling conditions in order to avoid too high of temperatures.

Table 3. Experimental conditions.

Workpiece material	Lamellar graphite cast iron GG25 (EN-GJL-250), HB 190 – 230
Workpiece size	Cylindrical blocks, initial diameter 150 mm, length 500 mm
Tool	Sandvik CoroTurn Holder CSDNN-4 with the inserts SNMG 1209408-QM
Machine tool	NC Lathe N20, Niles-Simmons
Force measurement	Kistler 3-component measuring system type 9129A

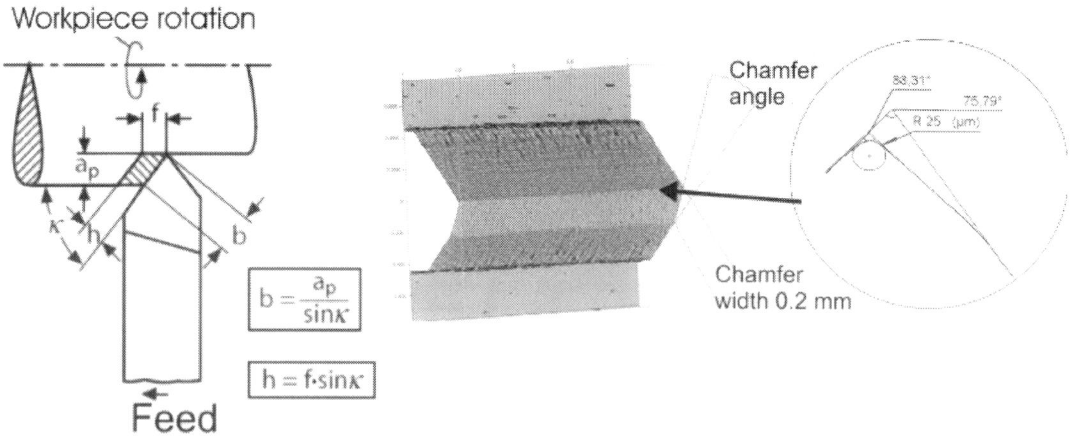

Figure 2. Cutting process parameters and cutting edge micro-geometry

In all cutting experiments a total feed of 400 mm was the goal. The total feed of 400 mm was calculated for the initial workpiece diameter of 150 mm, decreasing diameters led to proportional increasing total feeds. The total cut volume after 400 mm is 273.32 cm^3. All series were interrupted after 10, 40, 100 and 200 mm total feed for wear measurements. Some experiments had to be interrupted after 200 mm because the wear was too high.

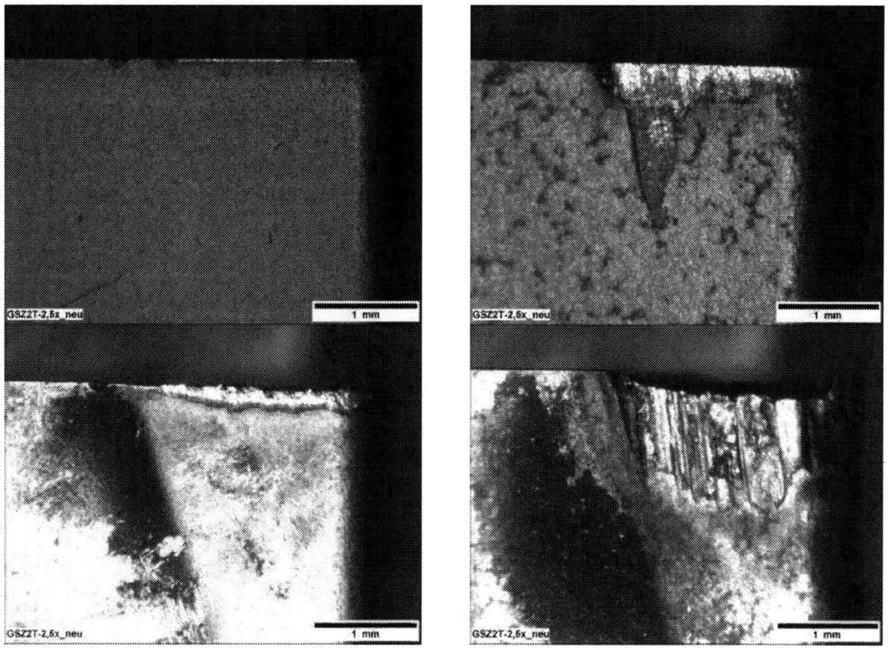

Figure 3. Land wear of the TiN coated (left) and of the uncoated inserts (right) after 40 and 200 mm total feed for CC2 (v_c = 375 m/min, f = 0.2 mm).

Compared with uncoated WC inserts and with CrN coatings, the TiN and TiN/CrN coatings had significantly better results in all cases, with TiN being slightly better than the multilayer coatings. The following Fig. 3 compares the land wear of TiN coated and uncoated inserts under medium cutting conditions CC2 (v_c = 375 m/min, f= 0.2 mm) after 40 and 200 mm total feed.

Fig. 4 compares the development of the land wear width and of the forces under the same cutting condition CC2.

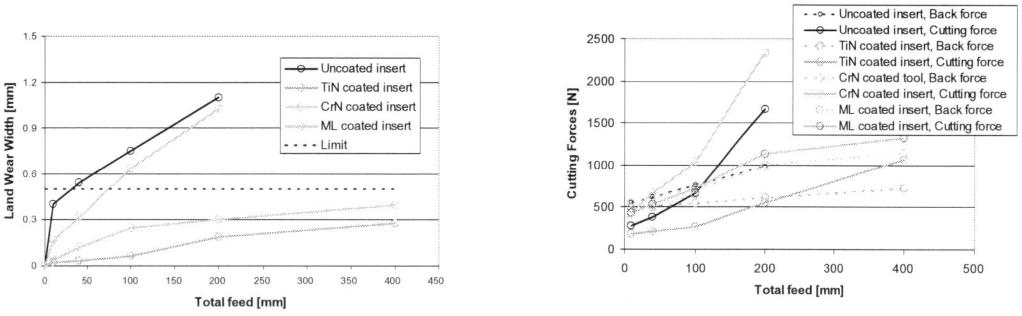

Figure 4. Land wear width (left) and cutting forces (right) for the cutting conditions: f = 0.2 mm, v_c = 375 m/min.

The better performance of the TiN and TiN/CrN coated inserts shown at the medium cutting condition CC2 becomes also clear for the other series CC1 and CC3, with the wear being smaller for the lower cutting velocity v_c = 250 m/min (CC1).

3. Industrial Validation

Industrial validation has been carried out using identical WC inserts (table 3) from the same coating processes as reported in chapter 1. The cutting conditions CC1 – CC3 are also identical with the experiments, as well as the insert tool holders, the cooling conditions of the processes and the machined cast iron GJL250 as bars of 150 mm diameter.

The industrial turning tests have been realized on an industrial turning unit model URSUS 320 with CMC numerical control. The total initial feed of 400 mm, defined for the initial bar diameter of 150 mm, was progressively increased with decreasing bar diameter. By measuring the change of bar diameter along the bar length during the cutting process, the land wear development of the inserts can be measured. This was done at the positions 0 mm, 200 mm, 300 mm and 400 mm from the beginning of the bar (see figure 6) defined for the initial bar diameter of 150 mm. These positions increase proportionally when the diameter decreases. After each pass, this conical surface had to be brought back into cylindrical form using spare sharp insert edges. Uncoated inserts and all coatings have been tested under all cutting conditions (see previous chapter) with a minimum of three repetitions for a higher confidence level.

Figure 5. Low surface quality and worn insert after machining with CrN coated insert.

Some of the tests exhibited considerably large wear with observable low surface quality and cutting edge destruction (figure 5). Other coatings such as TiN and the multilayer coatings allowed a substantially increased cutting edge life (figure 6). This was indirectly determined by the cutting depth changes indicating the land wear development during the cutting pass.

The uncoated inserts show a significantly higher cutting depth reduction than the best case of nanocomposite TiN coatings. Less severe cutting condition provide, on average, better results in terms of diameter tolerances.

Concerning industrial validation tests it can be concluded that

- the results of experimental tests and industrial validation are in accordance,
- the diameter change in machining grey cast iron is minimal for TiN coating in comparison with the other insert coatings, mostly evident for the severest cutting conditions $v_c = 500$m/min; f = 0.15mm,
- under all cutting conditions all coatings are damaged along the cutting edge after the turning tests, whereby uncoated, multilayer coated and CrN inserts collapsed earlier than the TiN

ones. (TiN coatings did not show sparking over the total feed. On the other side the CrN coated inserts showed distinct sparking already for small total feed length.)

The observed diameter decrease is not acceptable for industrial machining conditions. In all cases the quality of the machined surface was insufficient towards the end of the cutting feed. The selected turning conditions are extreme in comparison with the standard industrial turning conditions for grey cast iron, being two times less than the mildest conditions chosen in the tests.

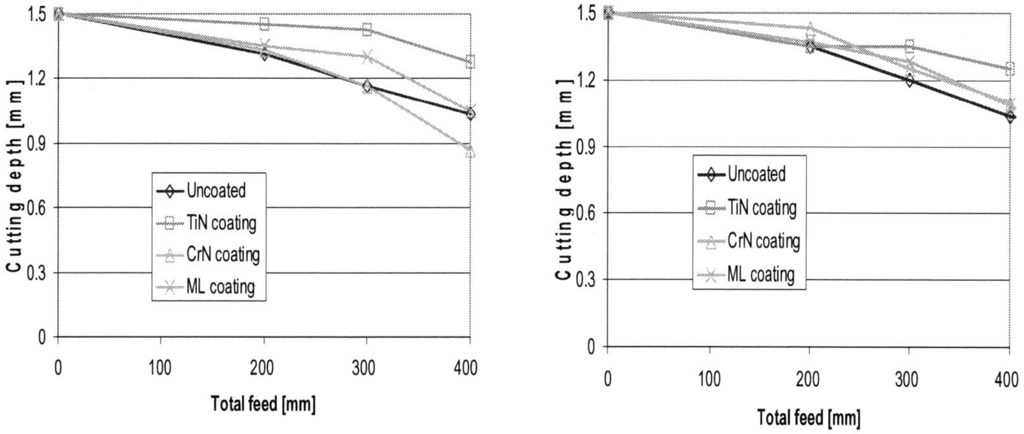

Figure 6. Measured cutting depth as a function of total feed (averaged values) for the cutting condition CC3 ($v_c = 500$ m/min, f = 0.15 mm, left) and CC2 ($v_c = 375$ m/min, f = 0.2 mm, right).

4. Numerical Simulations

Numerical simulation of the elementary cutting process, i.e. the chip formation, makes it possible to calculate the acting thermo-mechanical loads along the cutting edge. By means of the Finite Element Program AdvantEdge^{TM} [7], the geometry of the cutting process and of the cutting edge has been modeled in 3D (for parameters, see Fig. 2). The included in AdvantEdge materials properties for cast iron, for the carbide tool material, and for the coatings have been used.

Table 3. Average measured and simulated cutting forces for CC1.

	Cutting force	**Back force**
measured [N]	664.2	363.9
simulated [N]	697.7	405.7
deviation [%]	5.0	11.5

The cutting forces have been calculated and compared with the measured cutting forces for the model validation (performed for unworn inserts, i.e. models with sharp cutting edge and real process forces at the beginning of cutting). The results show a good coincidence (table ...).

Fig. 7 shows the initial FE mesh and the chip form after 3 mm cutting length for the cutting condition CC1 with TiN coated insert (coating thickness 5 μm).

The magnitude and the distribution of the loads in the contact zone along the cutting edge can be calculated by means of the FE method (Fig. 8). Then the thermo-mechanical loads of the coatings and the wear of the inserts can be assessed based on these values.

The temperature and stress values in the contact zone lead to considerably high wear rates, as shown in the experiments and in the industrial tests.

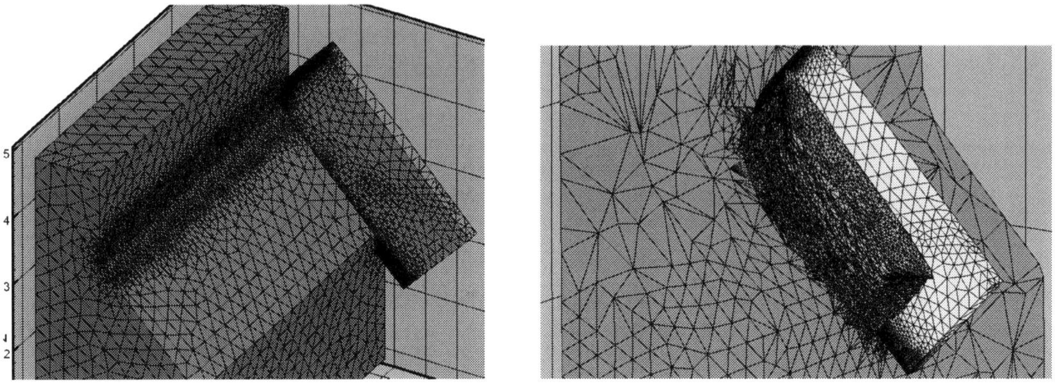

Figure 7. Initial FE mesh and the chip form after 3 mm cutting length for CC1.

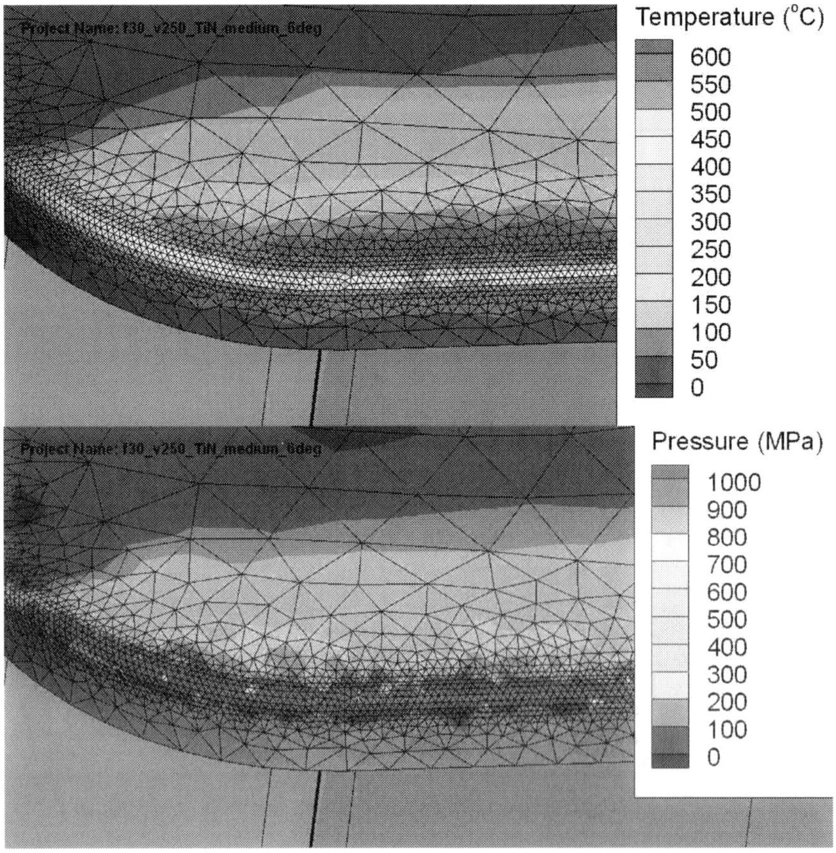

Figure 8. Distribution of temperature and surface pressure for CC1 Calculated by FEM

5. Summary and Conclusions

The nanostructured TiN coatings followed by multilayered coatings have the best performance for cutting cast iron GJL250 among the investigated coatings. The advantages of these coatings are more distinctive for higher cutting velocities. The investigations were carried out under laboratory and under industrial conditions. The results are in quite good accordance. However, for all investigated coatings and for all considered cutting conditions the tool live times are too short and the process uncertainties are too high. The coatings should be optimized and their qualification should be investigated for other cutting conditions such as dry cutting, rough cutting with large cutting depth and finishing cutting with small feed values.

Numerical simulations by means of the Finite Element Method show a good coincidence between measured and simulated cutting forces. As observed in the experiments by sparking and confirmed by the numerical simulations even for cutting under flood cooling the temperature effects become dominant and the coatings tend to delaminate. The simulation showed that on the rake face of the insert the temperatures are higher than 600 °C and the pressure reaches values over 1500 MPa along the cutting edge accompanied by high temperatures of over 500 °C, which is the reasons for the observed high wear of the cutting insert and for early coating destruction or delamination.

The modelling of coating systems under such conditions is a quite challenging task. In order to improve the coating performance in cast iron cutting the coatings should be investigated under high thermal and mechanical loads which are realistic for the considered cutting conditions.

6. Acknowledgement

The authors thank the European Committee for their support in the FP7 project "Multiscale Modelling for Multilayered Surface Systems (M3-2S)", Grant No: NMP3-SL-2008- 213600.

References

[1] C S Fox-Rabinnovich et al., Surface and Coatings Technology, 201(2006), 1852-1860.
[2] Y Qin et al. , MASMICRO, EU FP6 Project (500095-2)
[3] S H Yao, Y L Su, W H Kao and T H Liu, Materials Science & Engng, A392 (2005), 340-347.,
[4] H Dong, T Bell and A Mynott, Sports Engineering, 2, 213-219
[5] A M Merlo, Surface and Coatings Technology, 174-175 (2003), 21-26.
[6] www.m3-2s.bham.ac.uk
[7] Third Wave Systems, Inc., Minneapolis, USA, www.ThirdWaveSys.com

Al-Al_3Ti nanocomposite produced in situ by two-step hot-press sintering

B Nikfar, H Ghiabakloo, H R M Hosseini and AV Mohammadi

Department of Materials Science & Engineering, Sharif University of Technology, Tehran, Iran

Abstract. Aluminum reinforced with a large amount (up to about 55 vol.%) of Al_3Ti particles can be fabricated from Al-20Ti elemental nanometer-sized powder mixture via in-situ two step hot press sintering (TSS). For production of intermetallic reinforced in-situ composite, TSS can provide elevated temperature to facilitate the formation of intermetallic phase in situ and hot consolidation to form a fully dense solid. The first step sintering was employed at a higher temperature to obtain an initial high density, and the second step was held at a lower temperature by isothermal sintering for more time than the first one to increase bulk density without significant grain growth. The optimum TSS regime consisted of heating at 625°C for 5 min (1st step) and 570°C for 40 min (2nd step) with applied pressure of 470 MPa, resulting in the formation of near full dense microstructure (0.97 TD) with suppressed grain growth.

1. Introduction

Metal matrix composites (MMCs) are an important class of material for structural and electrical applications [1]. Particulate-reinforced MMCs are of particular interest because of their easy fabrication, low cost, and isotropic properties [2]. In conventionally processed powder metallurgy composites, the reinforcing particles are formed prior to their addition to the matrix metal. In this case, the scale of reinforcing phase is limited by the starting powder size, which is rarely below 1 μm [2]. It is widely recognized that the mechanical properties of MMCs are controlled by the size and volume fraction of the reinforcements as well as the nature of the matrix–reinforcement interface [1]. Superior mechanical properties can be achieved when fine and stable reinforcements with good interfacial bonding are dispersed uniformly in the matrix [2]. A possible alternative is to synthesize the reinforcement in situ in the metal matrix [3]. The advantages of in situ MMCs are that they have more homogeneous microstructures and are thermodynamically more stable. Moreover, they also have strong interfacial bonding between the reinforcements and the matrices [2]. Recently most researches have focused on production of composites and nanocomposites by using different sintering processes such as Hot Pressing, hot extrusion, and hot isostatic pressing (HIP) [4]. Hot pressing is widely used in the manufacturing process of high strength materials such as Al and Ti alloy and stainless steel [4-5]. However, grain growth occurs readily during the high temperature consolidation of nanocrystalline materials [6, 7], because nanocrystalline materials are thermally unstable. The full benefit of nanocrystalline materials may only be preserved if the consolidated specimen maintains its nano-sized grains. Two-Step Hot Pressing is a technique that in comparison to Single-Step hot pressing enables reaching higher densities while controlling grain size [8, 9]. Aluminum composites reinforced with trialuminide particles (Al_3Ni, Al_3Fe, Al_3Ti, Al_3Zr, etc.) possess high specific strength, high specific modulus and excellent properties both at ambient and elevated temperatures [9]. In comparison to most other aluminum-rich intermetallics, Al_3Ti with higher melting point (~1623 K), relatively low density (3.4 g/cm^3) and high Young's modulus (217

GPa; which are comparable with the properties of SiC) is the most considerable tri-aluminide for Al matrix composites [4, 9, 10]. Therefore, the presence of Al_3Ti intermetallic phase is very effective in increasing the strength and stiffness of aluminum composites and nanocomposites.

2. Experimental

The starting material used were aluminum fine powder Merck (99.9% purity and particle size of 45-150 μm) and titanium powder Merck (99.9% purity and average particle size of 10 μm). Titanium content of 20 wt. % with aluminum and Stearic acid (for lubricating the die) were mixed for 20 minutes using a mixer. Then 40 gr of the mixture was milled using a planetary ball mill for 12 hours in argon circumstance to prevent oxidation of the powder. The final powder after mechanical alloying would have two phases of Al and Al_3Ti. During the MA, samples of the mixture were taken out in argon box to be studied by SEM and EDS. Before hot pressing, the mechanical alloyed powder was degasified in a vacuum furnace at 350°c for 1 hour. The degasified powder was passed through mesh 200 and then was hot pressed using a die built by M2 tool steel with internal diameter of 18 mm. Two step hot pressing was performed to manufacture the bulk nanocomposite by in situ reaction. The used furnace was a cylindrical one with the coils on the perimeter with fireproof wool at the top and bottom. Hot press temperatures were chosen according to Al-Ti binary phase diagrams, which would cause the formation of intermetallic phase Al_3Ti [2]. Two first specimens were done under the constant pressure of 275 MPa with the first and the second step at 625°c for 15 min and 650°c for 5 min respectively for the first one and vice versa for the second one. The third specimen was hot pressed under 470 MPa with the first step at 625°c for 5 min and the second one at 570°c for 40 min. Applying force was performed by a Gotech general testing machine. Apparent density was performed according to ISO 39231:2008 standard using Archimedes method and theoretical density measurements according to ASTMD854-92e1 standard using pycnometry method. Micro hardness of the specimen were measured using Vickers hardness test applying 1 N of force for 15 sec. SEM and EDS analysis were done using TESCAN VGEA scanning electron microscopy and image analyzing was performed on SEM images.

3. Results and Discussion

3.1. Mechanical alloying

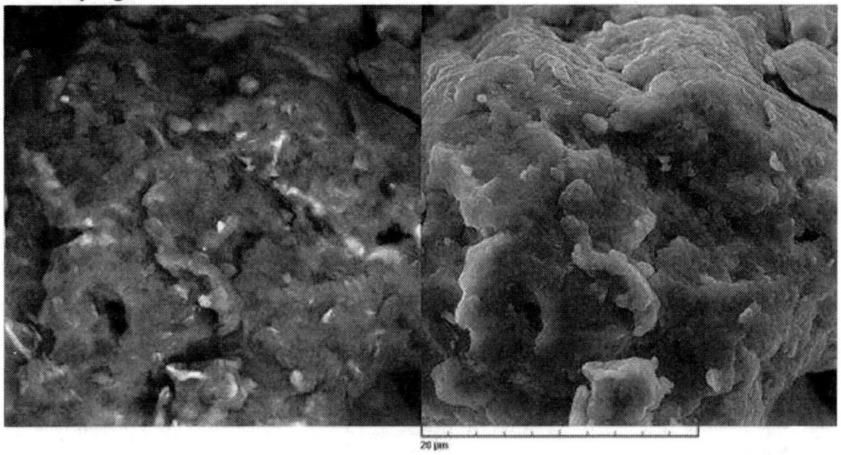

Figure 1. SEM picture of Al-20 wt.%Ti powder after 12 hours of MA

After 12 hours of mechanical alloying of the Al and Ti powders, the particles sizes were reached to under micron size as shown in Fig. 1. EDS test on the Al-20wt.%Ti milled powder after mechanical alloying shows Ti content of about 15 wt.% which means about 5 wt.% of Ti was entered in to Al network and supersaturated mechanical solution of Ti in Al was formed as shown in Fig. 2.

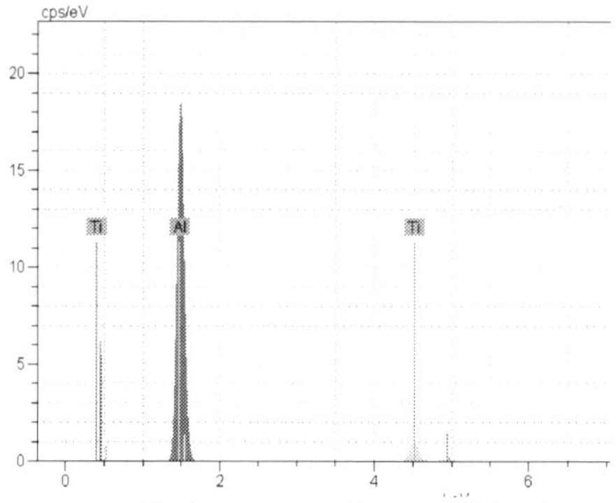

Element	Series	unn. C [wt.-%]	norm. C [wt.-%]	Atom. C [at.-%]
Aluminium	K series	79.32	84.08	90.36
Titanium	K series	15.02	15.92	9.64
	Total:	94.3 %		

Figutr 2. EDS test on Al-20 wt.%Ti powder after 12 hours of MA.

3.2. Two step hot press

Powders made by Al-20%Ti after 12 hours of mechanical alloying were two step hot pressed. The first specimen was hot pressed in the first step at 625°c for 15 min and in the second step at 650°c for 5 min using 275 MPa of pressure. The temperature, time and pressure condition of the second specimen was the same as the first one with reverse condition for the steps, to study the effect of setting the higher temperature for the first step and for the second one. The higher temperature of one of the steps is aimed for sintering and joining of the particles to increase strength and density and the lower temperature of the other step is to form intermetallic phase of Al_3Ti homogeneously. Setting the higher temperature for the first step or for the second one can be discussed. For this alloy, if sintering is performed in the second step, its time should be controlled exactly because additional time can dissolve the nucleated crystals in the first step and thus formation of coarse phases [11]. Densities of the first two specimens were measured 82% and 87% respectively using Archimedes method. Thus for the third specimen the higher temperature for the first step was set. The third specimen was hot pressed under the pressure of 470 MPa with the first step at 625°c for 5 min and the second step at 570°c for 40 min. Temperature, time and external pressure of the first and the second step of the two step hot press for the specimens are shown in Table 1. Apparent density of this specimen was measured 97% which is mostly because of the pressure increase related to the other specimens. Crystalline materials densify during sintering by mass flow processes of grain boundary diffusion, volume diffusion and plastic flow [12]. Densification occurs because the pores are annihilated at the interparticle boundaries, dislocations, phase boundaries or other microstructural grain boundaries [13]. For most materials, grain boundary diffusion is active during sintering because of its lower activation energy in comparison with that of volume diffusion. This is one reason that nanocrystalline materials where grain boundaries consists of 10–50 vol.% of materials can be successively consolidated in a short period of time at lower temperature [12]. In addition, applied pressure above the yield strength of compacting material increase the contact area between powders by plastic flow so that mass flow occurs more easily during the sintering process and thus aids densification [14]. Results of measured apparent density for the specimens are confirmed using pictures taken from the surface of them using optical microscope at 200X magnification as shown in Fig.3. Porosity (dark areas), according to the type of sintering employed in each case, is decreased from the first specimen (a) to the third one (c) which is not showing any porosity.

Micro hardness test is performed on different phases of the specimens. Hardness results of the first specimens are 221, 170, 221 and 469 Hv. With respect to Fig. 4, hardness of 221, 170 and 221 are related to the gray areas (light and dark) which contain Al (dark gray) and Al_3Ti (light gray). Since hardness of pure Al is 167 Hv, hardness of 170 Hv is related to areas with very low content of Al_3Ti and hardness of 221 Hv is related to areas with more content of Al_3Ti. Hardness of 469 Hv is related to the white areas and with respect to hardness of pure Ti which is 970 Hv, these areas consist of remained Ti and Al. Hardness of the second specimen are 297,237, 135 and 297 for the gray areas (Al+ Al_3Ti) that areas with less hardness (237 Hv) are with less content of Al_3Ti and areas with more hardness (297 Hv) are with more Al_3Ti. Hardness of 135 Hv is probably taken from the area with porosity. For the white areas, hardness test result is 729 Hv which is related to Ti. SEM micrograph of this specimen is shown in Fig. 5. Results of the hardness test on the third specimen are 221, 254, 143, 135 and 151 Hv for the gray areas which include Al (dark gray) and Al_3Ti (light gray). Hardness of 221 and 254 Hv are areas with more Al_3Ti and hardness of 143, 135 and 151 are related to areas with less Al_3Ti content and probably with porosity in around. Results for white areas are 274, 274, 279 and 255 Hv which is related to Ti that the less hardness related to pure Ti is because of existence of phases with lower hardness and porosity. Lower measured hardness for the third specimen related to the other ones is most probably because of homogeneous distribution of the porosities which is like thick grain boundaries as can be seen in Fig. 6. In this condition, any chosen point for hardness test is close to porosity.

Table 1. Sintering Conditions

Specimen	First Step		Second Step		External
	Temperature (°c)	Time (min)	Temperature (°c)	Time (min)	Pressure
First	625	15	650	5	275
Second	650	5	625	15	275
Third	625	5	570	40	470

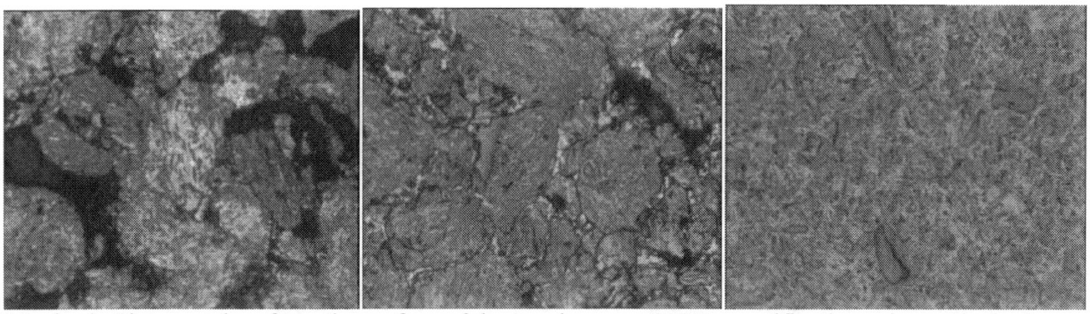

Fig. 3 (a, b, c): Pictures taken from the surface of the specimens at 200X magnification using optical microscope.

Also dispersion of white areas in this specimen causes that the indentor diameter is bigger than the white area and while measuring the hardness, the hardness of combined white and gray areas are measured. For measuring accurate hardness, it is better to perform nano hardness test on this specimen.

3.3. Formation of Al_3Ti

Aim of this project is producing Al- Al_3Ti nanocomposite by in situ reaction which means Al_3Ti intermetallic should be formed during sintering from the primary elements Al and Ti. Because of the formation importance of this intermetallic, formation of Al_3Ti is studied using EDS and photos of SEM. Result of EDS test for the first specimen is shown in Fig. 7. As can be seen, atomic percentages of Al and Ti are 79.98 and 20.02 respectively. With respect to the fact that EDS test is done on the light gray areas which is Al_3Ti and ratio between Al and Ti is 3 to 1, The atomic percentage from the EDS test shows more than 3 to 1 ratio which means the percentage of Al is measured more than actual that is probably because the probe diameter was bigger that the Al_3Ti particles size and thus Al of matrix was also detected. For the second specimen the EDS result is 3 to 1 for Al to Ti which means Al_3Ti is formed as shown in Fig. 8. According to this result the percentage of Ti is more than 33% which means the studied Al_3Ti phase for the EDS test is placed beside Ti phase. The EDS test for the third specimen shows the same percentage of 3 to 1 for Al to Ti as can be seen in Fig. 9.

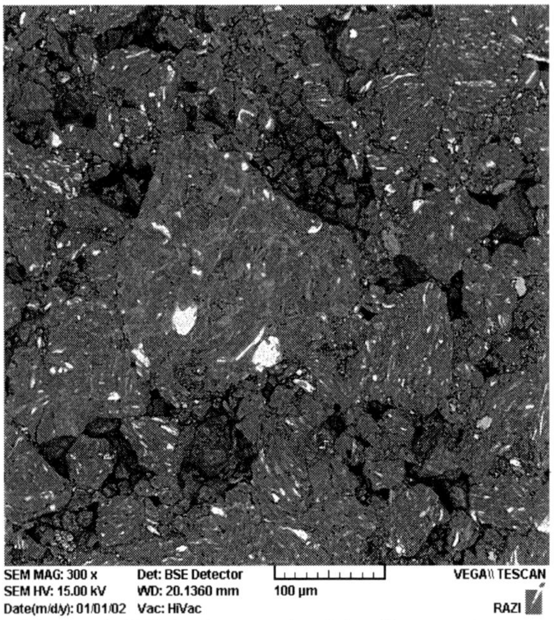

Figure 4. SEM micrograph of the first specimen.

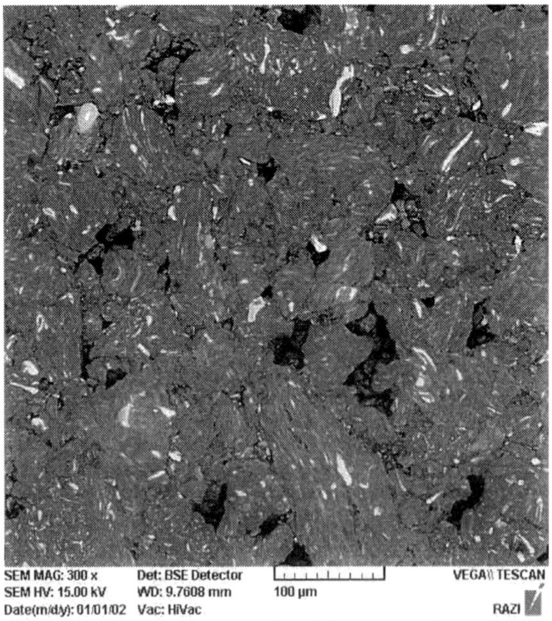

Figure 5. SEM micrograph of the second specimen.

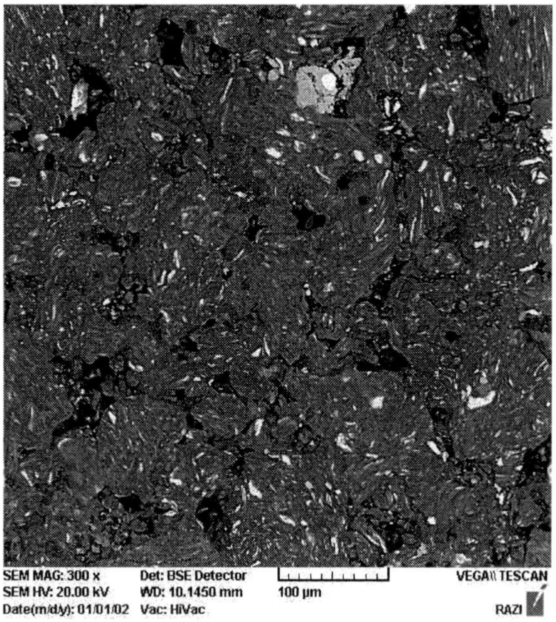

Figure 6. SEM micrograph of the third specimen.

SEM pictures of the specimens with 2000X magnification are shown in Fig. 10 and 11. In the pictures, the dark gray, light gray and white areas are related to the Al matrix, Al_3Ti phase and remained Ti respectively. Results of image analysis test on the SEM pictures are shown in Table 2.

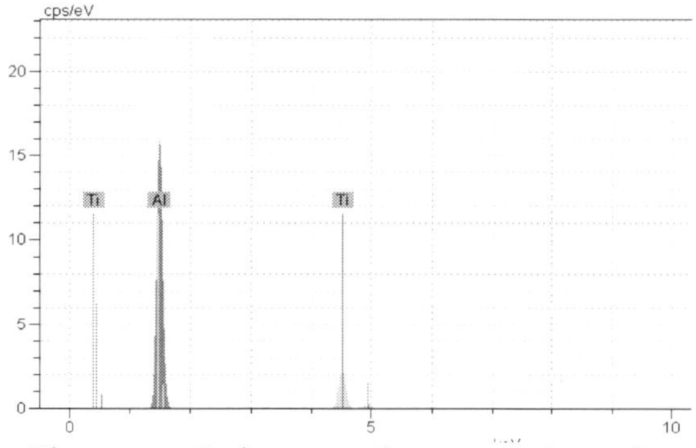

Element	Series	unn. C [wt.-%]	norm. C [wt.-%]	Atom. C [at.-%]
Aluminium	K series	77.92	69.25	79.98
Titanium	K series	34.60	30.75	20.02
	Total:	112.5 %		

Figure 7. EDS test on the first specimen.

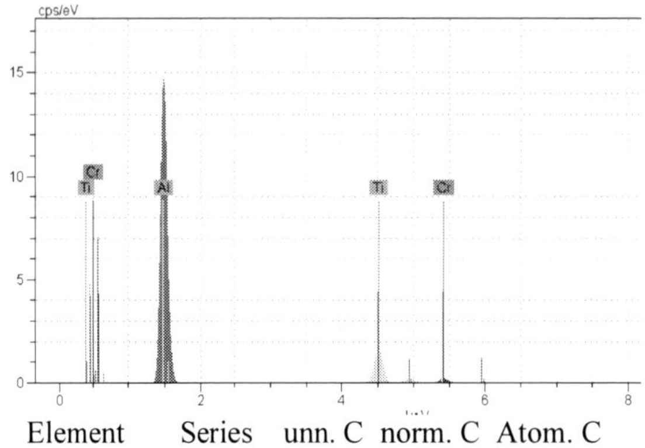

Element	Series	unn. C [wt.-%]	norm. C [wt.-%]	Atom. C [at.-%]
Aluminium	K series	38.37	57.79	71.09
Titanium	K series	23.70	35.70	24.75
Chromium	K series	4.33	6.52	4.16

Total: 66.4 %

Figure 8. EDS test on the second specimen.

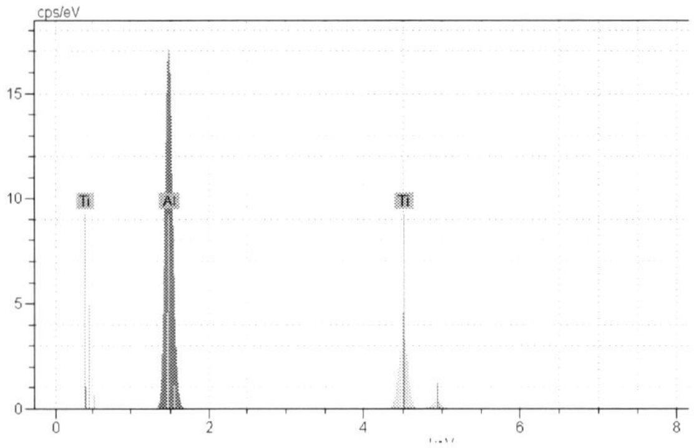

Element	Series	unn. C [wt.-%]	norm. C [wt.-%]	Atom. C [at.-%]
Aluminium	K series	67.23	65.92	77.44
Titanium	K series	34.76	34.08	22.56

Total: 102.0 %

Figure 9. EDS test on the third specimen.

As can be seen in Table 2, Al_3Ti percentages for the first specimens to the third one are 47.3, 46.3 and 55.5% respectively which is demonstrating that the lower temperature and the more time caused increase in formation of Al_3Ti in the third specimen. In the first specimen, amount of formed Al_3Ti is 1% more than that in the second one which can be neglected because of the oscillation in the furnace temperature and image analysis procedure (One picture represents whole the specimen). Therefore can be concluded that the first and the second specimens which were produced with the same temperature, time and pressure conditions but with reverse steps, have the same percentage of produced Al_3Ti and their difference is related to distribution and size of Al_3Ti particles. As can be seen in Fig. 10 and 11, formed Al_3Ti particles in the Al matrix in the second specimen are finer and have more homogeneous dispersion than in the first one. It is clearly seen that Al_3Ti particles in the first specimen are coarser and formation of these coarse areas caused having poor areas of Al_3Ti in around. Higher hardness for the second specimen related to the first one confirms this matter.

Table 2. Image analysis results on the SEM pictures of the specimens

	First specimen	Second specimen	Third specimen
Al matrix	48.9%	48.1%	40.5%
Al_3Ti	47.3%	46.3%	55.5%
Ti	3.8%	5.6%	4.0%

The most important effect of pressure on the specimens is on their density that with pressure increase from 275 MPa to 470 MPa in the third specimen, the apparent density is a lot higher that the first two specimens. But high pressure can be a factor for abnormal grain growth [12]. With respect to the lower temperature and more time during hot press for the third specimen, diffusion and thus grain growth of the Al matrix was decreased related to the others and the structure is finer as can be seen in Fig. 11.b.

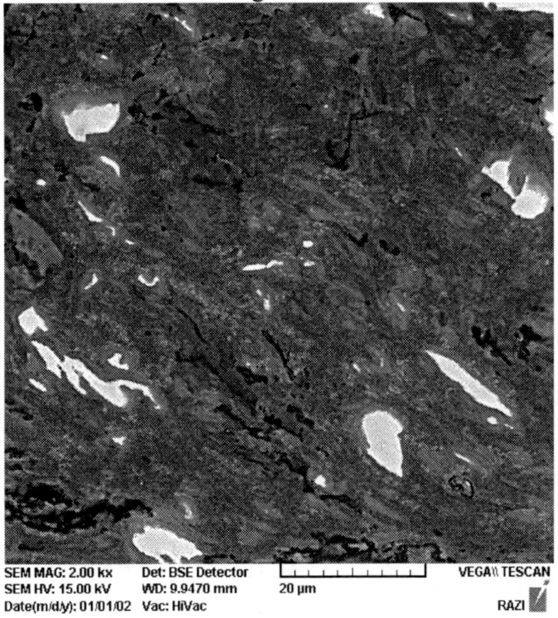

Figure 10. SEM micrographs of the first specimen at 2000X magnification.

Finer and homogeneously dispersed particles of Al_3Ti, Al matrix and Ti phase is seen in the figure. But external pressure increase in the third specimen has caused abnormal grain growth in some areas. With respect to the results of Table 2, percentage of remained Ti in the specimens one to three is 3.8, 5.6 and 4.0% respectively which is about one fifth of the preliminary content and shows that more Al_3Ti could be formed with giving more time to the specimens specially the third one which has finer and more homogeneously dispersed Ti which has more contact surface with Al matrix thus atomic exchange will be faster. Specially with respect to the lower temperature for the third specimen there is less concern about grain boundary diffusion which is the main reason of grain growth at high temperature. In the third specimen since external pressure was high, high density was gained but in some areas big porosities are seen which are of course smaller than in the first two specimens and has a reason that is well-known as Kirkendall's effect [15].

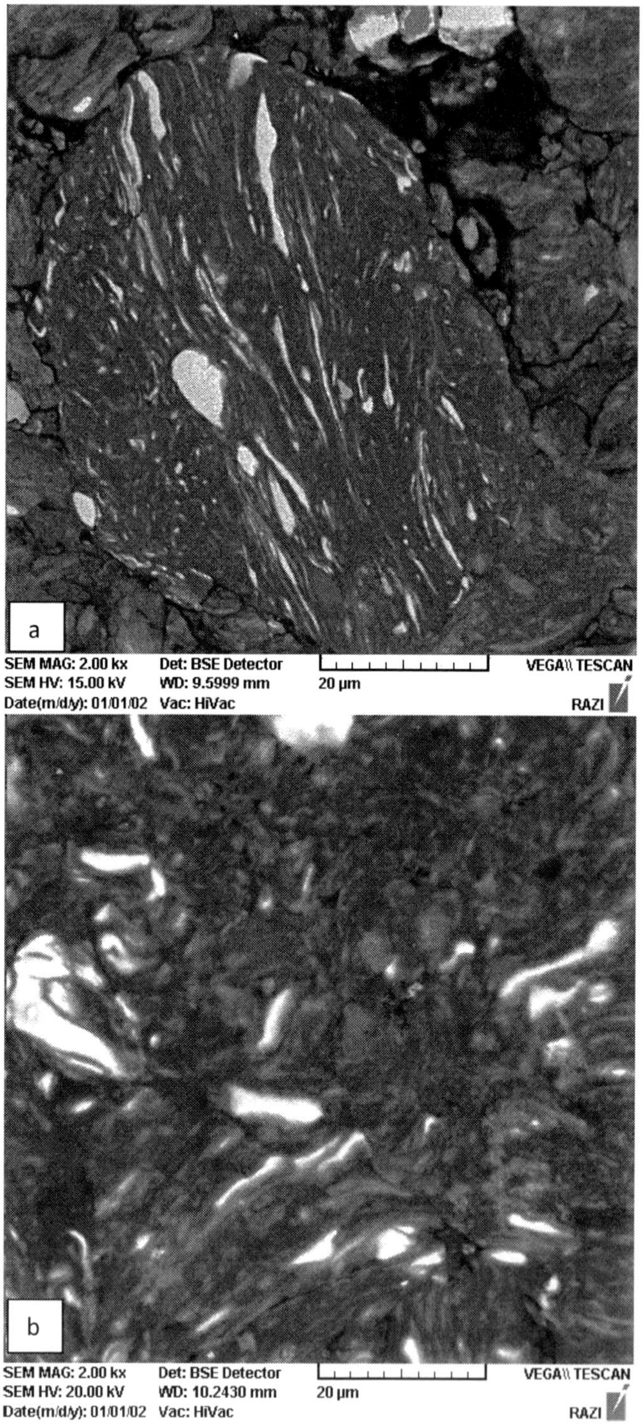

Figure 11. SEM micrographs of: a) the second and b) the third specimen at 2000X magnification.

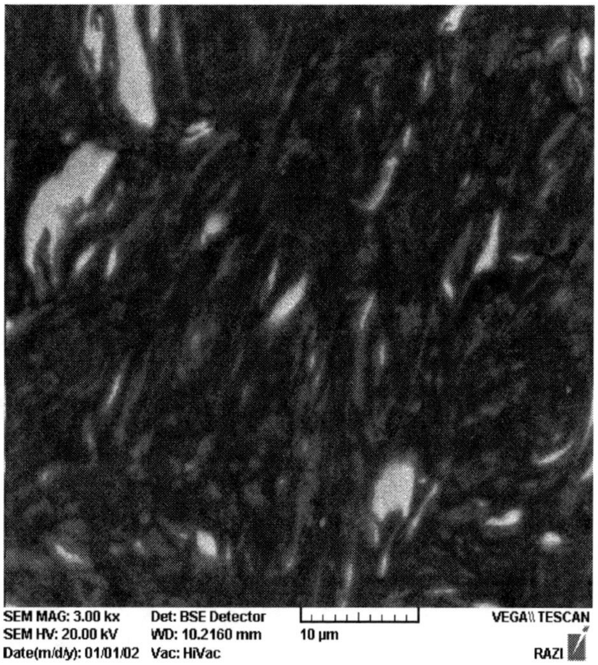

Figure 12. SEM micrograph of the third specimen at 3000X magnification.

Because of high external pressure, grains are plastic deformed and in some cases are rotating [12] that this deformations cause the formation of supersaturated vacancies. Then these vacancies gradually move towards the grain boundaries during sintering and cause growing of the cavities [15] which is one of the main reasons of low micro hardness results for the third specimen.

SEM micrograph of the third specimen at 3000X magnification is shown in Fig. 12. As can be seen in the picture, size of Al_3Ti phases are less than micron but for seeing the phase clearly more magnification is required.

4. Conclusions

Influence of temperature, time and external pressure of two step hot pressing on apparent density and microstructure of Al-Al_3Ti nanocomposite and also formation of Al_3Ti intermetallic was studied that using the results of densitometry, microhardness, EDS and SEM micrographs and analyzing these results we conclude the following:

- Densitometry results demonstrated that increasing the external pressure of pressing on the specimen while sintering, its density is increasing significantly so that with increasing the pressure from 275 MPa to 470 Mpa, the apparent density was increased from 82 and 87% to 97%. Pictures of optical and electron microscopes also confirmed the results.
- The higher temperature for the first step of sintering fines the microstructure and forms more homogeneously dispersed Al_3Ti reinforcement and therefore causes the increase of hardness and stiffness of the nanocomposite. Results of microhardness test demonstrated that for the first specimen which was hot pressed in the first step at 625°c for 15 min and in the second step at 650°c for 5 min, harness is about 121 to 170 Hv and for the second specimen with the first step at 650°c for 5 min and the second step at 625°c for 15 min, hardness is about 137 to 297 Hv. SEM micrographs also confirms this matter.
- In production of the Al- Al_3Ti nanocomposite, setting the higher temperature for the first step or for the second one does not influence the amount of produced Al_3Ti.
- Decreasing the temperature and increasing the time of sintering related to the condition with higher temperature and less time causes more formation of Al_3Ti intermetallic and fining of them and also reduction of grain growth rate of the Al matrix and the preliminary Ti. Analysis of the SEM micrographs demonstrated that in the third specimen with the first step at 625°c for 5 min and the second step at 570°c

for 40 min the volumetric percentage of formed Al_3Ti was 55.5% but for the second specimen was 46.3%. SEM micrographs confirm the finer microstructure of the third specimen.

5. References

[1] Clyne TW, Withers PJ. An introduction to metal matrix composites. Cambridge: Cambridge University Press; 1993.
[2] C.J. Hsu, C.Y. Chang, P.W. Kao , N.J. Ho, C.P. Chang. Al–Al3Ti nanocomposites produced in situ by friction stir. Acta Materialia 54 (2006) 5241–5249.
[3] Tjong SC, Ma ZY. Mater Sci Eng R 2000;29:49.
[4] M. Nofar, H.R. Madaah Hosseini, N. Kolagar-Daroonkolaie, Fabrication of high wear resistant Al/Al3Ti metal matrix composite by in situ hot press method, **Materials and Design**, 30 (2009) 280–286.
[5] K.T. Kim, H.C. Yang, Densification behavior of titanium alloy powder during hot pressing, **Materials Science and Engineering**, A313 (2001) 46–52.
[6] H. Hahn, J. Logas, R.S. Averback, J. Mater. Res. 5 (1990) 609.
[7] Kyoung Il Moon, Hee Sub Park, Kyung Sub Lee, Study of the microstructure of nanocrystalline Al–5 at.% Ti compacts prepared by reactive ball milling and ultra-high-pressure hot pressing, **Journal of Alloys and Compounds**, 325 (2001) 236–244.
[8] Chih-Jen Wang, Chi-Yuen Huang, Yu-Chun Wu, Two-step sintering of fine alumina–zirconia ceramics, **Ceramics International**, 35 (2009) 1467–1472.
[9] Karel Maca, Vaclav Pouchly, Pavel Zalud, Two-Step Sintering of oxide ceramics with various crystal structures, **Journal of the European Ceramic Society**, 30 (2010) 583–589.
[10] S.S. Nayaka, S.K. Pabi, B.S. Murty, Al–(L12)Al3Ti nanocomposites prepared by mechanical alloying: Synthesis and mechanical properties, **Journal of Alloys and Compounds**, 492 (2010) 128–133.
[11] J. Cintas, F.G. Cuevas, J.M. Montes and E.J. Herrera, Scripta Materialia 52 (2005), p. 344.
[12] Kyoung Il Moon, Hee Sub Park and Kyung Sub Lee, Journal of Alloys and Compounds 325 (2001).
[13] R.M. German (Ed.), Sintering Theory and Practice, Wiley, New York, 1996, p. 95.
[14] N.A. Fleck, L.T. Kuhn, R.M. McMeeking, J. Mech. Phys. Solids 40 (1992) 1139.
[15] V. Abbasi Chianeh, H.R. Madaah Hosseini and M. Nofar, Journal of Alloys and Compounds 473 (2009).

Polymer (PTFE) and shape memory alloy (NiTi) intercalated nano-biocomposites

S S Anjum, J Rao and J R Nicholls

School of Applied Sciences, Cranfield University, Bedfordshire, MK43 0AL, UK

Abstract. Engineering on a nano-scale has been undertaken to mimic a biomaterial by forming an intercalated nano-composite structure by PVD sputtering of a polymer with a nickel-titanium (NiTi) shape memory alloy (SMA). A PTFE polymer has been selected due to its elastic properties, low interactions with water, optimum surface energies, stability and chemical resistance. NiTi SMAs allow the coatings to be energy absorbent and thus suitable in load bearing situations. The coatings are aimed to constantly withstand variable adverse biological environments whilst maintaining their characteristics. The nano-intercalated structures have been characterised for their wettability, friction coefficients, chemical composition, and morphology. Intercalation of a polymer with energy-absorbing alloys uncovers a set of material systems that will offer characteristics such as self-healing of hierarchal tissue in the body. The reformation of PTFE following sputter deposition was confirmed by FTIR spectra. According to SEM analysis PTFE shows a promising surface interaction with NiTi, forming stable coatings. Surface interactions are evident by the hydrophobic behaviour of films as the composite's water contact angle is around 86°, which lies in-between that of PTFE and NiTi. The nano composite films are lubricious and have a measured CoF below 0.2 which does not vary with layer thickness.

1. Introduction

A biomaterial is intentionally designed to interact with biological systems with the capability to implement an appropriate host response in a specific application [1]. The physical, chemical, mechanical and biological properties of a biodegradable material vary with respect to time, application, thickness and interaction to tissue. Suitable biomaterials should be able to retain an appropriate level of such properties and should not induce a toxic or inflammatory response on implantation, whilst having a suitable shelf life and matching healing and regeneration processes [2]. For a biomaterial to be bio-functional it needs adhesion and self-growth, which is achieved by increasing the surface area and porosity in three-dimensional structures [3].

Sputtering has been used since 1970 for plasma assisted materials deposition. Advantages of using sputtering to prepare thin films include uniform coating over larger surface area, controlled deposition rate and efficient target consumption [4]. In physical vapour deposition sputtering of a polymer film, the target is subjected to energetic positive ion bombardment that causes the emission of volatile fragments from the target. These fragments take part in plasma polymerization process adjacent to the substrate, which results in the deposition of a plasma polymer on surfaces [4, 5].

The use of composite solutions have shown promise in past due to a combination of different desired characteristics and ability match to natural tissue. In this research polymer to smart memory alloys combination is selected. The main advances for the use of polymers include their long term biocompatibility, ease of processing, porosity hence bioactivity, possibility to do chemical, plasma and surface modifications with them such as altering functional group hydrophobic or hydrophilic behaviour, morphology and porosity etc. [6; 7]. PTFE has low surface energy [8] hence low surface tension [9], extreme chemical inertness {{46 Bártolo,Paulo. 2008}} [9], thermal resistant [8][9], and tribological properties [8][9]. Although metals and alloys have excellent mechanical properties, however they do not entirely satisfy the arduous demands required for biomedical systems such as, biocompatibility and biofunctionality. FeCrNi, CoCr and TiAlV are commonly accepted metallic implant materials, whereas NiTi offers excellent mechanical stability, corrosion resistance, biofunctionality and biocompatibility [10] [11].

A multilayer approach involving the PVD deposition of PTFE and NiTi has been implemented; by altering the individual layer thickness t produce intercalated nano-composite formed between the two materials. The fabrication of such a nanocomposite is aimed to construct a functional biomaterial. The intercalated coatings fabricated in this study are characterised in terms of their morphology, lubricity, surface roughness and hydrophobicity.

2. Methods

2.1. Reparation of PTFE and Ni Ti samples

Coatings were prepared at room temperature by PVD magnetron sputtering using a Leybold L560® coater equipped with two 3-inch magnetrons. The system is turbo-pumped to a base pressure of around 1x10-6mbar. Argon was used as working gas, with a baratron used to measure process pressures. The target–substrate distance is kept constant at 70 mm. Glass was the substrate used in this study. PTFE was coated using a radio frequency (13.56MHz) power supply at 100 Watts at a pressure of 25mb Ar, and NiTi was deposited using an Advanced Energy® Pulsed DC (f = 200kHz; rate = 1.6μs) power supply operating at 150 Watts at a pressure of 25mb Ar. The deposition rate of PTFE and NiTi was determined by using a Dectac surface profilometer and calculated to be 307nm/h for PTFE and 548nm/h for NiTi, respectively.

2.2. Formation of intercalated layered nano composite coating

Two sets of samples were prepared for investigation; those having 10 alternative layers of PTFE and NiTi (sample I) and sample II having 20 alternative layers of PTFE and NiTi. Table 1 summarises the film thickness and the thickness of the individual layers produced.

Table 1. details for intercalated nano composite sample I and sample II.

Sample I	**Sample II**
Total thickness = 1 μm = 1000nm	Total thickness = 1 μm = 1000nm
10 alternative layers of PTFE and NiTi	20 alternative layers of PTFE and NiTi
Individual layer thickness = 100 nm	Individual layer thickness = 50 nm

2.3. Composite coatings characterisation

The intercalated nanocomposite materials were characterised by Fourier Transform Infrared (FTIR) Spectroscopy using a JASCO – FTIR 6200 under absorption mode, a high resolution electron microscopy was undertaken using a FEI XL30 SFEG analytical SEM. Static water contact angle (WCA) and scratch tests with single pass scratch test was also undertaken. Scratch testing was performed using a tungsten carbide ball indenter under continuous progressive loads with forces in the range from 1 to 100N, at loading rate of 100 N/min with maximum friction force of 50 N.
Results and discussion

2.4. FTIR characterisation

Sputtering is energetic process and deposition of polymer material can lead to formation of complex chains. It is important to ensure that the composition of film is the same as that of the target material after PVD deposition [12]. Therefore, FTIR spectra (**Error! Reference source not found.**) of the PTFE were performed to determine the characteristics after RF sputtering. The FTIR spectrum of PTFE was obtained after subtracting glass slide component, as shown in

Figure **1**. The area of prime interest in the spectrum lies between 800cm−1 and 1800cm−1. The broad and complex absorption peaks absorption peaks of CFx (x: 1, 2) are fairly distinctive between 1012cm−1 and 1494cm−1 [13]. Absorption peaks at 893cm−1, 1102 and 1230cm−1 are for CF2 symmetric stretch and at 1430cm−1 are for CF2 asymmetric stretching vibration for the films [13][14][15]. Peaks in the region from 783 to 810cm-1 and especially at 719cm−1 (wagging CF2) indicate presence of and amorphous PTFE [14][15]. Absorption band at between 1598 and 1863cm−1 having distinct peak at 1719 belongs to C=C stretching vibrations [13]. Peaks around 777cm−1 show vibrations of CF3 groups [13][14][15]. The ratio between the area under the peak for CFx (x: 1, 2) and CF3 is 50:1 (2%), which indicates the polymerization has occured in the deposited fluorocarbon. The intensity of CF3 absorption peaks is low due to increase in chain cross-linking and decrease in abundance of CF3 assemblies after sputtering [13]. The presence of CF and CF3 within polymer chain shows branching within polymer chains and tertiary carbon structures around carbon skeleton. The lack of any OH stretching vibration peaks between 3000 to 3500cm− indicates the purity of PTFE film [13]. So PTFE thin film has successfully formed after PVD sputtering of a PTFE target.

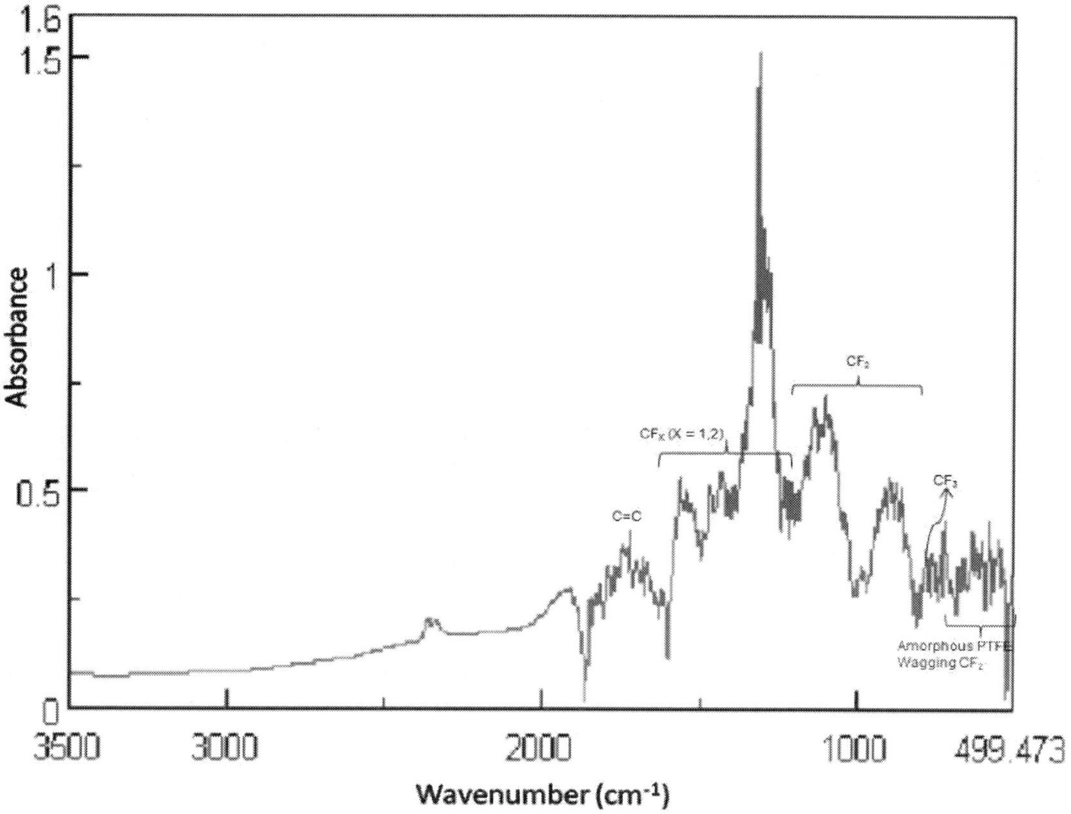

Figure 1. The resulting FTIR absorbance spectrum of a PTFE thin film after subtracting the blank glass substrate.

2.5. Water contact angle measurements

According to the literature, the WCA of PTFE varies between 112o and 170o [13]. Static water contact angle (WCA) measurements of a PTFE coating, (sample I) was found to be 113.7o (standard deviation: 0.115), which shows hydrophobic behaviour of the film. Hydrophobicity of the film has been reported to be altered during deposition, as it is highly dependent on chemical composition and surface morphology of deposited films which may vary with sputtering conditions such as working gas, working distance etc. [13][16]. The highly hydrophobic surface of PTFE demonstrates that a stable and compact surface thin film has been formed [16].

In comparison to PTFE, NiTi produces a low water contact angles of 62.57o (standard deviation: 0.115). When intercalated layers of PTFE and NiTi are designed, the resulting nano-composite surface forms a hybrid surface with WCAs of 86.9o (standard deviation: 0.1) and 85.6o (SD: 0.115) offering greater hydrophobicity then NiTi alone, but slightly lower than that of pure PTFE. PTFE has favourable surface interactions with NiTi to form stable layers which also has clear impact on hydrophobic behaviour of the NiTi surface as the top later of the composite. Altering the number of layers from 10 to 20 layers, and reducing the layer thickness from 100 to 50nm, does vary the WCA measurements significantly. Further experiments need to be undertaken in future, to study the pattern of water contact angle, which changes with changes in PTFE and NiTi layers thickness, and hence their surface interactions and hydrophobicity.

2.6. SEM characterisation

SEM images (x15000 magnification) of all samples show smooth and continuous surface morphology with minimal defects.

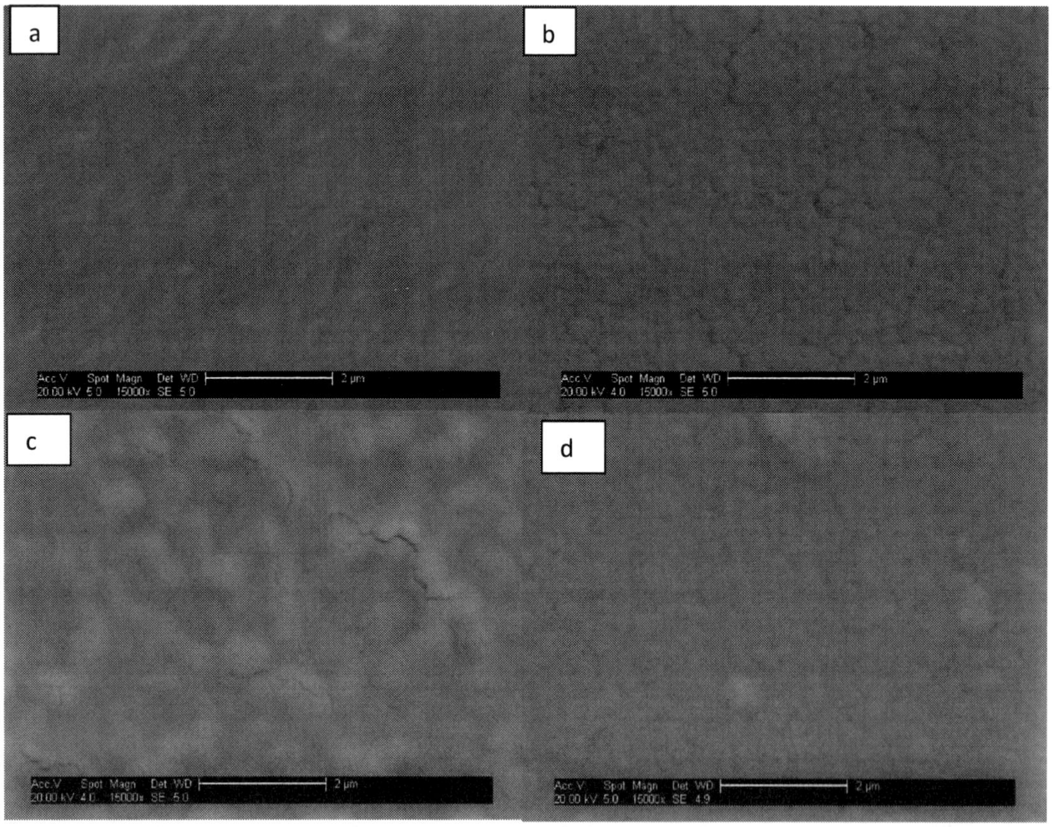

Figure 2. SEM images (x15000) for (a) PTFE (b) NiTi (c) Sample I: (PTFE-NiTi - 10 layers), (d) Sample II: (PTFE-NiTi - 20 layers).

Figure 2a shows the surface SEM of a PTFE deposited on a glass slide. There are no signs of irregularities, and the film compares well with other methods used to prepare nano-PTFE layers [16] [17]. Figure 2b shows the surface of NiTi thin film prepared on a glass slide. It is not possible to determine the parent phase of NiTi without XRD analysis, however, previous studies conducted by Rao et al. [11] show NiTi to be in martensite at room temperature. The film deposits with a columnar microstructure in zone1 of the Thornton diagram giving rise to NiTi furrows and channels around each column, which are clearly visible. The surface morphology of the layered nano-composites are shown in Figure 2c and Figure 2d, sample I (100nm individual layer thickness) and sample II (50nm individual layer thickness). There are no major of irregularities and few furrows effects and channels visible. Some cracking at column boundaries is observed. The continuous and stable surface of the nanocomposite in sample I and II confirm the ability of two materials to bind together. PTFE seems to lock in the column boundaries of the NiTi within the composite, the shape memory alloy acting as the matrix for PTFE within the top layer. This observation is also evident from the WCA measurements, as described above, as in the composite layers the WCA of NiTi increases from 62.5 to 86 after forming the nano composite even though NiTi is the top layer in composite. Figure 2c shows the sample I to be under more stress compared to sample II (Figure 2d). This is indicated by the opening of the column boundaries within the composite due to the thicker layers deposited. In turn, thinner layers in sample II (Figure 2d) show more similar morphological characteristics as the NiTi film (Figure 2c).

2.7. Friction coefficient analysis

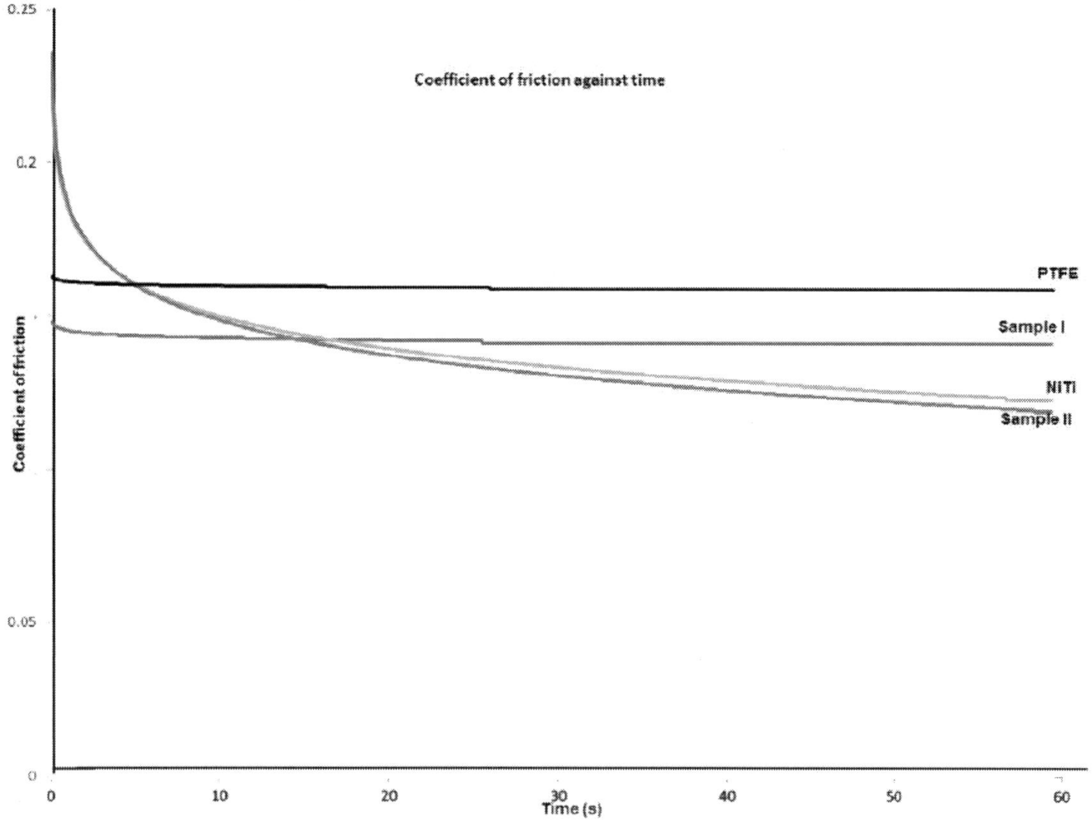

Figure 3. Trend lines of Coefficient of friction vs time graph (trend / regression type: logarithmic); Sample I: (PTFE-NiTi - 10 layers), Sample II: (PTFE-NiTi - 20 layers); measured by scratch testing using WC indenter under continuous progressive load (1-100N), loading rate 100 (N/min), max f: 50N.

Low friction coefficient coatings are considered to provide longer lifespan and excellent tribological properties [18]. Friction studies of PTFE and NiTi, prepared in this study, reveals that they have a low coefficients of friction, which is maintained for the intercalated nano composite samples all having values within the range of 0.1 to 0.2 as shown in Figure 3. PTFE and sample I (10 layers) maintain their low coefficient of friction throughout and do not show much variation in CoF with increase in load over time. The nano-composite with layer thickness of 50 nm has the lowest coefficient of friction and hence is more lubricious, compared to the nano-composite with layer thickness of 100 nm and the PTFE film alone, although it does not show much variation when compared to NiTi. The composite sample, which has a layer thickness of 50nm (sample II), has a similar friction coefficient to that of NiTi as shown in Figure 3. When the layer thickness is increased to 100nm (Sample I) the surface interactions alter and lead to larger asperities with increase in the friction coefficient. This observation together with the observation of cracks at column boundaries in sample I (Figure 2c) being under more stress compared to the sample II (Figure 2d) accounts for the increased observed CoF. Shape memory alloys are energy absorbing materials and this stress effect could be due to SMA conformation with WC indenter in sample II as it has a thicker layer (100nm) of NiTi on top of composite. The coefficients of friction for all four samples are comparable to those of commercially available lubrication materials, which are in the range of 0.1 to 0.18 according to the survey carried by American polywater [19; 20]. DLC (CoF: below 0.2) Diamond (CoF: 0.5-0.1), and tungsten carbide (CoF: 0.12) and Teflon (CoF: 0.04 – 0.1) are also considered as virtuous lubricating surfaces [19] [18], and CoF values of PTFE, NiTi and both nanocomposite sample all fall within a similar range of values and can be classified as lubricious. Further analysis of these coatings under loading conditions using a nano-indenter is underway to evaluate their load bearing capabilities.

3. Conclusions and future studies

Intercalated PTFE-NiTi thin films have been formed on glass substrates using PVD sputtering. It is archetype to accomplish suitable biomaterial properties in one framework. FTIR analysis shows that a polymerised PTFE forms after PVD sputtering. The layer thicknesses of nano-composite materials have been varied from 100nm to 50nm and their characteristics such as surface morphology, WCA and CoF evaluated. SEM images of all materials result in smooth and continuous surface morphology with minimal defects forming stable layers. XRD analysis will be carried out to further investigate the phases within SMA after forming nano composite with different layer thicknesses. The hybrid surface of the intercalated layers offered WCA of 86o, being more hydrophobic than NiTi, but slightly lower than that of pure PTFE. Further study will be carried out to understand how the WCA changes with layer thicknesses and hence their surface interactions and hydrophobicity.

All samples produced are lubricious, having low friction coefficients within the range of 0.1 to 0.2. Nano-indentation will be used in future to further investigate those coatings under loading conditions to evaluate their load bearing capabilities.

Reference

[1] Williams, D., Bassiouni, M. A. and Orooji, A. (2001), "Parallel Database Systems", in Wiley Encyclopedia of Electrical and Electronics Engineering, John Wiley & Sons, Inc., .

[2] Lloyd, A. W. (2002), "Interfacial bioengineering to enhance surface biocompatibility.", Medical device technology, vol. 13, no. 1, pp. 18-21.

[3] Gilson Khang, Moon Suk Kim, Hai Bang Lee (2007), "A manual for biomaterials / scaffold fabrication technology ", Manuals in Biomedical Research, vol. 4, pp. 1-12.

[4] Biederman, H., Stelmashuk, V., Kholodkov, I., Choukourov, A. and SlavÄ±´nskÃ¡, D. (2003), "RF sputtering of hydrocarbon polymers and their derivatives", Surface and Coatings Technology, vol. 174-175, no. 0, pp. 27-32.

[5] MFátima, V. S. and J.R. Nicholls (b) (2001), "A model for calculating the thickness profile of TiB2 and Al multilayer coatings produced by planar magnetron sputtering", Surface & Coatings Technology, vol. 142, pp. 934-938.

[6] Nair, L. S. and Laurencin, C. T. (2007), "Biodegradable polymers as biomaterials", Progress in Polymer Science, vol. 32, no. 8–9, pp. 762-798.

[7] Pişkin, E. (1995), "Biodegradable polymers as biomaterials.", Journal of biomaterials science.Polymer edition, vol. 6, no. 9, pp. 775-795.

[8] Bártolo, P. and Bidanda, B. (2008), Bio-materials and prototyping applications in medicine, Springer, New York.

[9] Liu, C., Gibbons, A. R., Arnell, R. D., Tong, J. and Ren, L. (2001), "Reactive depositon of aluminium oxide coatings on PTFE substrate", Surface and Coatings Technology, vol. 139, no. 2-3, pp. 111-117.

[10] Otsuka, K. (1999), Shape memory materials, Cambridge Univ. Press, Cambridge.

[11] Rao (□), J., Roberts, T., Lawson, K. and Nicholls, J. (2010), "Nickel titanium and nickel titanium hafnium shape memory alloy thin films", Surface & Coatings Technology, vol. 204, pp. 2331-2336.

[12] Bhushan, B., and Gupta, B. K. (1991), "Handbook of tribology : materials, coatings, and surface treatments", in McGraw-Hill, New York, pp. 13.56-13.57.

[13] Drábik, M., Polonskyi, O., Kylián, O., ?echvala, J., Artemenko, A., Gordeev, I., Choukourov, A., Slavínská, D., Matolínová, I. and Biederman, H. (2010), "Super-Hydrophobic Coatings Prepared by RF Magnetron Sputtering of PTFE", Plasma Processes and Polymers, vol. 7, no. 7, pp. 544-551.

[14] Bodas, D. S., Mandale, A. B. and Gangal, S. A. (2005), "Deposition of PTFE thin films by RF plasma sputtering on 〈1 0 0〉 silicon substrates", Applied Surface Science, vol. 245, no. 1–4, pp. 202-207.

[15] Selvakumar, N., Barshilia, H. C. and Rajam, K. S. (2010), "Effect of substrate roughness on the apparent surface free energy of sputter deposited superhydrophobic polytetrafluoroethylene coatings: A comparison of experimental data with different theoretical models", Journal of Applied Physics, vol. 108, no. 1, pp. 013505-013505-9.

[16] Kwong, H. Y., Wong, M. H., Wong, Y. W. and Wong, K. H. (2007), "Superhydrophobicity of polytetrafluoroethylene thin film fabricated by pulsed laser deposition", Applied Surface Science, vol. 253, no. 22, pp. 8841-8845.

[17] Jia, Z. and Yang, Y. (2012), "Self-lubricating properties of PTFE/serpentine nanocomposite against steel at different loads and sliding velocities", Composites Part B: Engineering, vol. 43, no. 4, pp. 2072-2078.

[18] Rao, J., Cruz, R., Lawson, K. J. and Nicholls, J. R. (2005), "Sputtered DLC-TiB2 multilayer films for tribological applications", Diamond and Related Materials, vol. 14, no. 11–12, pp. 1805-1809.

[19] , Carbide depot, technical resources for manufacturing professionals. Coefficient for static friction values (lubricated)(2004), available at: http://www.carbidedepot.com/formulas-frictioncoefficient.htm (accessed 05/02).

[20] Coefficient of friction measurement on polywater's friction table, a publication for engineers involved in cable installations(2007), , White paper ed., American polywater technical talk.

IOP Conf. Series: Materials Science and Engineering **40** (2012) 012007 doi:10.1088/1757-899X/40/1/012007

Thermoplastic polymer nanocomposites with montmorillonite-Lab vs industrial scale fabrication

M Giacomelli[1,*], K Pielichowski[2,] and A Leszczyńska[2]**

[1] Grado Zero Espace, via Nove, 2/A, Montelupo F. no, Italy
[2] Department of Chemistry and Technology of Polymers, Cracow University of Technology, ul. Warszawska 24, 31-155 Kraków, Poland

E-mail: * marco.giacomelli@gzespace.com
** kpielich@usk.pk.edu.pl

Abstract. Fabrication of polypropylene or polyamide-6/montmorillonite nanocomposites in a laboratory- and industrial-scale by melt processing has been described. Depending on the fabrication scale, different technological problems were observed that need to be taken into account during production of PP and PA-6-based nanocomposites with MMT which constitute a promising class of modern materials for engineering and packaging applications. Fabrication of PP and PA-6 nanocomposites with montmorillonite both on laboratory and industrial scale needs to be optimized in terms of processing conditions that include pre-mixing parameters, feeding rate, temperature of the barrel zone and extrusion die, screw speed and head pressure. Proper cleaning procedures need to be applied and the selection of the 'main' fraction enables to get nanostructured materials with enhanced properties.

1. Introduction

Substantial research efforts have been undertaken toward preparation (in the lab-scale) and characterization of polymer nanocomposites with layered silicates in the last two decades. It has been found that there is a number of factors which contribute to the structure of polymer-based clay nanocomposites and those factors include molecular weight of polymer, the type and concentration of organic modifier and the processing conditions in the polymer melt-direct intercalation or exfoliation. The latter group of factors includes extruder's screw diameter which governs the throughput of the processing equipment. However, most of the experiments are laboratory-scale and the question of scale transfer to industrial production scale remains not fully answered.

Thermoplastic polymer nanocomposites with montmorillonite were the first polymer nanomaterials studied [1-7] and they still gain considerable interest due to enhanced mechanical, thermal and barrier properties [8-13].

In our research performed in laboratory scale PP and PA-6 nanocomposites with montmorillonite were obtained by melt mixing in a lab-scale twin-screw extruder equipped with compounding screws, whereby industrial scale dispersion tests were done using large-scale extruders in the dedicated pilot-plant of an Italian masterbatch producer (VIBA SPA) and yielded several kilograms of the 'main' fraction of each nanocomposite sample.

2. Materials and techniques

Polypropylene (PP) homopolymer was obtained from Basell Orlen (Basell Polyolefins Group) under trade name Moplen HP500J. Polypropylene grafted with maleic anhydride (PP-g-MA) under trade name Polybond 3200 was applied as a compatibilizer for polypropylene nanocomposite material. Polyamide-6 (PA-6) was acquired from Zakłady Azotowe w Tarnowie-Mościcach, Poland, under trade name Tarnamid T30. Montmorillonite (MMT, Laviosa/Dellite 43B) was used, which has been surface-modified with N,N-dimethyl-N-benzyl-N-octadecyl ammonium.

PP and PA-6 nanocomposites with MMT were compounded on a Zamak twin-screw extruder (ZMK/116/10) with extruder screws' diameter 24 mm, length/diameter ratio 32 and consisting of 6 barrel zones. A conventional screw configuration was used for material compounding, equipped with two high-shearing zones in the second and fifth zone.

Industrial scale dispersion tests were performed using an Universal Aghetto turbomixer, having 1,500 rpm of the shaft (corresponding to a rotor speed of 1,884 m/min) in position 1 and 3,000 rpm of the shaft (corresponding to a rotor speed of 3,768 m/min) in position 2. Then a Maris 30 mm co-rotating twin screw extruder with 52 length/diameter ratio and up to 650 rpm of maximum speed, equipped with a spinneret, cooling water bath and a shearing and sieving machine (pelletizer system) was applied. The extruder was equipped with a vacuum pump used for the removal of water (such as the humidity hygroscopically absorbed by the feeding) and possible volatile substances.

3. Results and discussion

The flowsheet of the processing procedure in the laboratory scale is presented on Figure 1.

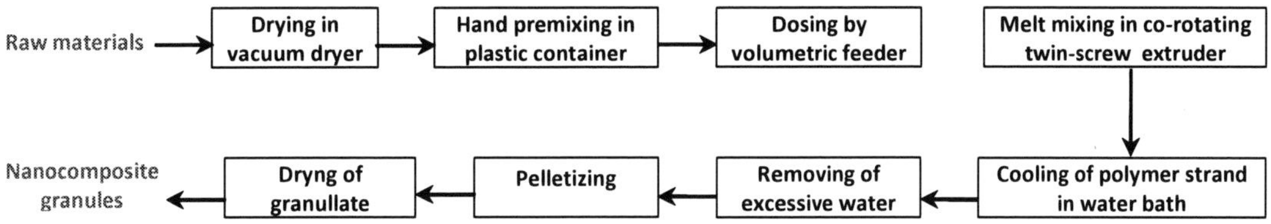

Figure 1. Flow-sheet of the laboratory-scale process applied for the production of thermoplastic polymer nanocomposites with montmorillonite.

In the laboratory scale polymers and fillers were dried for 2 h at 80°C in vacuum dryer prior to processing in extruder. The drying conditions of PA 6 were optimized in terms of moisture content and oxidation sensitivity of PA 6 revealed by yellowing. Organically modified montmorillonite (5 wt. %) and compatibilizer (PP-g-MA) used in polypropylene-based nanocomposite materials were premixed with polymer in plastic container by shaking for c.a. 3 minutes and then batched by main feeder to the first barrel zone. The filler revealed strong electrostatic adhesion to the walls of plastic container and feeding system. The remaining nanofiller was gathered after extrusion process and weighed. In the laboratory scale the nanofiller losses due to sticking to machine walls should not be neglected in the total mass of nanofiller especially when relatively low concentrations of additive (~ 1 wt %) are introduced to the polymer. Although the premixing of polymer and additives were effective in providing good distribution of nanofiller the wearing of initial barrel zone and screws elements suggested dosing the bentonite nanofiller to the polymer melt. However in laboratory scale it is impossible to apply feeder capable of dosing such a small amounts of powder with the desired precision.

The temperature of the main feeding zone was controlled by cooling system (the cooling medium was compressed air). Nanocomposites based on polypropylene were obtained after a set of optimization experiments at temperatures: 200, 200, 210, 210, 210 and 220°C of 1th-6th barrel zones respectively. The processing parameters of extrusion of PA6-based nanocomposites were: temperatures of barrel zones (1th – 6th) 250, 250, 260, 260, 270 and 270°C. The screw speed was set on 150 rpm for all prepared compositions. The extrusion die temperature was equal to the temperature of the last barrel zone. The cleaning procedure is needed for twin screw extruder when changing the composition of nanocomposite. If only the concentration of nanofiller was changed but the polymer and nanofiller were the same the removing of remaining nanocomposite from the extruder was done by extruding neat polymer for as long as the dwell time for set parameters. The initial part of extruded nanocomposite strand was also rejected since MMT content in material was lowered by the remaining 'cleaning' polymer. As the polymer was changed from polypropylene to polyamide, the extruding parts were dismantled and mechanically cleaned. The total mass of wasted material was not high but could be considered as significant in relation to the mass of nanocomposites produced.

During the preparation of dry premix and extrusion of both nanocomposites the exposition of operator to the nanofiller dust needs to be limited. During the extrusion of polyamide 6/MMT nanocomposites evolution of volatile gases with strong odour was observed. It is likely that slight decomposition of organically modified bentonite and/or polymer occurred though the colour of the extruded strand was light creamy or ivory and did not varied much from the neat polyamide monofilament.

Produced rod was cooled in a water tank. The excess of water was removed from a polymer strand by suction extractor. Then the pelleting machine was used to prepare polymeric granulate for further processing/structural investigations. The flowsheet of the industrial-scale process applied for the production of thermoplastic polymer nanocomposites with montmorillonite is presented in Figure 2.

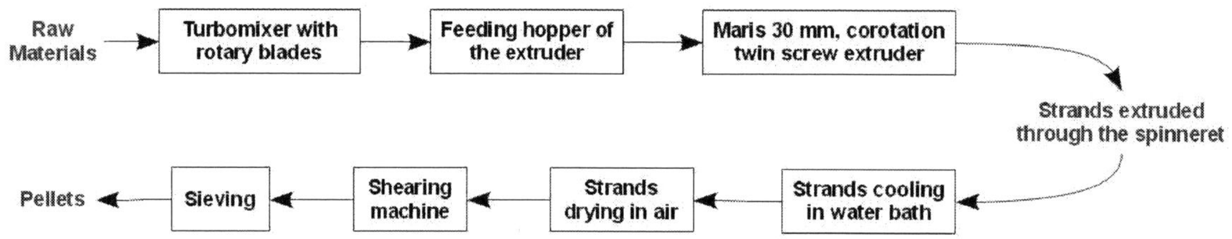

Figure 2. Flow-sheet of the industrial-scale process applied for the production of thermoplastic polymer nanocomposites with montmorillonite.

Raw materials were initially pre-mixed in a proper turbomixer; then the obtained mix was fed to the twin screw extruder and extruded through a spinneret; afterwards the obtained strands were cooled in a water bath and successively dried in air, cut at the desired length in the shearing machine, sieved and finally collected.

In order to warrant the highest homogeneity of composition in the finally obtained pellets, the first stage of the industrial dispersion tests consisted in the realization of a pre-mixing of each charge and the polymeric matrices, all in solid (powder) form.

This activity was performed by the use of a turbomixer with rotatory blades: the one chosen and used for these trials wasn't the industrial one because of the minimum quantities (20-30 kg) that need to be charged to this last one. Hence, a semi-industrial one was used: it was able to treat up to 4-5 kg of material each time, but for 'safety' reasons (to warrant minimum losses of material and impact in case of incurred problems in the mixing phase) it was decided to split each sample in two pre-mixing portions, put together at the end of this preliminary stage, in the bag used to feed the extruder.

Another important characteristic of the turbomixer used for this pre-mixing stage consists in being equipped with blades that are sharper than those used in the bigger one: this fact causes an easier and faster heating of the polymeric matrix, which (especially in case of PP) brings to a sticky behaviour that facilitates the process, minimizing volatility problems of the charges.

The turbomixer was loaded with the charges and, afterwards, the polymer; then, it was properly closed by the use of a plastic film (in order to maximize adhesion between cover and main body) and its own cover, properly blocked with three arresting hooks.

All six tests were performed with a double pre-mixing step, the first one at lower speed (1,500 rpm) and the second one at higher one (3,000 rpm): this choice was made in order to warrant the maximum homogeneity of the premix and, on the other hand, to subject the polymer to a small temperature stress that, improving binding between polymeric matrices and added charges, brings to the aforementioned advantages.

Before and during the pre-mixing phases, the whole extrusion system (composed by extruder, water bath, shearing and sieving machine) was started and brought in the predefined operating conditions (of speed, temperature, pressure, etc.): in order to do it and to properly clean and rinse the whole equipment, PP (homopolymer and characterized by a MFI of 12) was fed and processed for some hours. The same cleaning procedure, although with much lower times (20-30 minutes) was repeated to clean the equipment between the tests. While the extruder (and spinneret) was cleaned in the just described way, the feeding hopper and the shearing and sieving machine (processing materials in the solid state) had to be (also) manually cleaned to remove excess materials.

Once the equipment was ready for starting the extrusion of the prepared samples, the pre-mixed materials were inserted in the empty feeding hopper of the extruder; this system allowed to pour within the extruder a constant and predefined flow of material. Such feeding rate was kept constant during the processing of each polymer, although after the first test it was slightly increased in order to warrant the minimization of fluctuations in the extrudate production rate.

Once properly fed to the extruder, the materials were processed according to the predefined temperature profile: the choice of a (pilot plant) co-rotation twin screw extruder characterized by a high length/diameter ratio (52) allowed to warrant the best dispersion and highest homogenization

At the end of the extruder, the system was equipped with a spinneret that forced the formation of strands (Figure 3), successively cooled down in a (direct contact and closed circuit) water bath. The obtained strands were then dried in open air and subsequently fed to the shearing and sieving machine, that finally produces the desired pellets.

Figure 3. Spinneret and cooling water bath.

As easily understandable, in the beginning phase of the extrusion process (since the addition of the new feeding) the obtained material will not be the desired one, due to the presence of the residual 'cleaning polymer', and there will be a transient condition after which the system will be producing the aimed product. On the other hand, after the end of the material and since the addition of the 'cleaning polymer', there will be a similar condition in which the produced pellets' composition will be changing till the complete absence of the tested materials. Hence, for each test three different phases can be recognized: the 'Preliminary', the 'Main' and the 'Final' one.

In the common industrial production 'Preliminary' and 'Final' phases are usually treated as wastes, bringing to a loss of productivity around 3% (starting from 100 kg of raw materials 97 kg are usually collected as product and 3 kg are treated as production wastes).

In the case of these industrial scale dispersion tests, instead, the referred transitory phases brought to not negligible losses of material, mainly due to two main factors:

- The amount of material subject of these tests is very low for the type of equipment, which is studied and built up for the treatment of tens of kilograms of polymers; since the fact that starting-up and cleaning procedures are independent of the amount of material to be treated, the processing of few kilograms of material results in a high percentage of losses.
- The tests performed are still far from the optimized procedures used in industrial production and produced pellets are not characterized in function of the time, yet; hence, the most suitable starting-up time isn't defined, yet, and, as an obvious consequence, a margin must be left and it's not possible to obtain the maximum yield.

Anyhow, all fractions were collected and properly labelled: in this way not only any unnecessary loss of materials is avoided, but, at the same time, there will be the possibility to make analyses and evaluations about the load distribution in function of the processing time and, hence, on environmental and health issues related to materials that, in the industrial production, will be treated as wastes.

As aforementioned concerning the pre-mixing stage, all the industrial dispersion tests were accomplished without major problems, till the production of 2-3 kg of the 'main' fraction of each product. From the constant and careful observation of each test, the next remarks have to be highlighted:

PP+MMT - Strands (and pellets) were characterized by a brown colour (due to the presence of MMT), especially in the initial phase; later on, the colour became lighter (a sort of beige), with a good constancy throughout the whole test. Bubbles were present, again, but well-spaced and with medium dimensions (0,5-1 cm). Very low electrostatic behaviour was registered.

PA+MMT - In the initial phase of the extrusion, the obtained strands progressively turned from transparent and homogeneous aspect to an opaque and beige look, with total absence of bubbles and electrostatic behaviour. During the 'Main' phase, nearly in the middle of the process, strands produced by the spinneret presented a high degree of black dirt, most probably carbon traces; this fact was very limited in time (1-2 m of strands), but for obvious reasons it was decided to take this small section apart, for possible future analyses. Almost simultaneously, the tendency to a light temperature increase was registered in the third zone of the extruder, but the system reacted keeping the temperature profile on the foreseen values.

Hence, the only encountered problem was the discharge of black particles during the processing of the PA+MMT sample. It may arise from the lack of any filter within the extruder, but the cause of the production of these residues should be evaluated by the realization of extruding tests on bigger amounts of materials: it should be determined, indeed, whether these problems are due to the processing of one particular mix or it's ascribable to the fair wear and tear of the equipment.

4. Conclusions

Fabrication of PP and PA-6 nanocomposites with montmorillonite both on laboratory and industrial scale needs to be optimized in terms of processing conditions that include pre-mixing parameters, feeding rate, temperature of the barrel zone and extrusion die, screw speed and head pressure. Proper cleaning procedures need to be applied and the selection of the 'main' fraction enables to get nanostructured materials with enhanced properties [14,15]. For PP+MMT nanomaterials in form of strands and pellets brown colour (due to the presence of MMT) has been observed in the initial phase. Later on, the colour became lighter with a good constancy throughout the whole test. In the initial phase of the extrusion of PA+MMT composites the obtained strands progressively turned from transparent and homogeneous aspect to an opaque and beige look, with total absence of bubbles and electrostatic behaviour.

5. Acknowledgements

This work was funded by the European Commission (FP7 NEPHH Project-CP-FP; Project Reference No.228536-2).

6. References

[1] Fornes T D and Paul D R 2003 *Polymer* **44** 3945
[2] Paul M A *et al* 2003 *Polymer* **44** 443
[3] Utracki L A 2000 *Clay-Containing Polymeric Nanocomposites* (Shawbury: Rapra)
[4] Alexandre M and Dubois P 2000 *Mater. Sci. Eng.* **28** 1
[5] Kumar A P *et al* 2009 *Prog. Polym. Sci.* **34** 479
[6] Ray S S and Okamoto M 2003 *Prog. Polym. Sci.* **28** 1539
[7] Zhu L and Xanthos M 2004 *J Appl. Polym. Sci.* **93** 1891
[8] Dennis, H R et al 2001 Polymer 42 9513
[9] Paul D R and Robeson L M 2008 Polymer 49 3187
[10] Pielichowski K et al 2011 Mechanisms of thermal degradation of layered silicates modified with ammonium and other thermally stable salts, in Thermally Stable and Flame Retardant Polymer Nanocomposites ed V Mittal (Cambridge: Cambridge University Press) chapter 2 pp 29-63
[11] Leszczyńska A et al. 2007 Thermochim. Acta 453 75
[12] Leszczyńska A et al. 2007 Thermochim. Acta 454 1
[13] Pielichowski K et al 2010 Mechanisms of Thermal Stability Enhancement in Polymer Nanocomposites, in Optimization of Polymer Nanocomposite Properties ed. V. Mittal (Weinheim: Wiley-VCH Verlag) chapter 9 pp 195-210
[14] Sachse S et al. 2012 J Nanomater. in press.
[15] Gascon J A 2012 NEPHH Project Periodic Report

Elastic behavior of silica/poly(dimethylsiloxane) nanocomposites: nano-size effects

I A M Ibrahim,[1] A A F Zikry,[1] M A Sharaf,[1] J E Mark,[2] K Jacob,[3] I M Jasiuk,[4] and R Tannenbaumn[3]

[1] Department Chemistry, College of Science, Helwan University, 11795 Egypt.
[2] Department of Chemistry, University of Cincinnati, Cincinnati, OH 45221-0172, USA.
[3] School of Materials Science and Engineering, Georgia Institute of Technology, Atlanta, GA 30332-0245 USA.
[4] Department of Mechanical Science and Engineering, University of Illinois at Urbana-Champaign, Urbana, IL 61801 USA.

Abstract. Elastomeric materials require the incorporation of reinforcing fillers in order to improve their mechanical properties. The enhancement of properties is very much dependent on the size and any surface modification of the reinforcing agent. It is widely accepted that the reinforcement effects are primarily due to molecular interactions of the polymeric matrix and the filler inclusions and it involves both chemical and physical interactions. Herein, we have incorporated silica nano fillers (Stober silica) into poly (dimethylsiloxane) (PDMS) elastomeric networks. The mechanical and swelling properties of the networks were investigated as a function of filler sizes (50, 130, and 170 nm), volume fraction of the filler inclusions, and surface treatment of the particles by were analyzed to give the Mooney-Rivlin constants $2C_1$ and $2C_2$. These properties were found to be dependent on the size of the nano inclusions. Thus, a nano size phenomenon had been discerned and was one of the highlights of this investigation. This phenomenon was largely attributed to the high specific surface area of the nano fillers used that leads to significant increase in the interfacial interactions. Also, and as would be expected, the properties of the polymeric networks filled with unmodified particles were different than those obtained for the polymeric networks filled with surface-modified particles. This has been primarily attributed to changes in the surface properties, and as would be expected, the elastic properties of the networks were thus shown to be strongly dependent on type and concentration.

1. Introduction

Polysiloxanes (or "silicones") are semi-inorganic polymers with the repeat unit [-SiRR'-O-] (with various alkyl or aryl R and R' groups). These materials have many unusual properties, such as good thermal stability, unusually low viscosity-temperature coefficients, unusually high permeabilities, low surface tensions, good dielectric properties, good weather resistance, good release and lubrication properties, good biocompatibility, and excellent optical clarity [1-3]. The most important feature in the present context is the fact that they are generally elastomers and maintain their elastomeric properties down to very low temperatures.

Poly (dimethylsiloxane) (PDMS), [-Si (CH3)2O-], is by far the most important and most widely explored and utilized polysiloxane, both scientifically and commercially [1–4]. Its extremely low glass transition temperature Tg and high thermal stability over a wide range of temperatures make it high-performance material, but its elastomers in the unfilled state generally have poor mechanical strengths [2–5]. As a result, most applications require that PDMS elastomers be reinforced by fillers in order to improve their mechanical properties [6–8].

The reinforcement of PDMS elastomers by fillers has been the subject of numerous studies [9, 10]. Earlier work on filled PDMS focused on the effects of fillers on macroscopic properties several examples of which are given elsewhere [11, 12]. The role of molecular structure at the polymer–filler interface has been considered only in recent years. Notably, Litvinov et al. studied by NMR the structure and behavior of grafted PDMS chains on silica and found that the layer consisted of immobilized chain segments at the PDMS–silica interface but mobile chain portions beyond the interface [13]. Chains at the interface exhibited substantial decrease in heat capacity at Tg. Some of the PDMS chain units were immobilized as a result of physical adsorption at the silica surface, the amount being proportional to the number of silanol groups at the silica surface. Gussoni et al. also used NMR spectroscopy to investigate silica filled PDMS elastomers, and observed a decrease of PDMS segment mobility in the vicinity of the filler surface [14]. Berriot et al. employed H NMR experiments on filled elastomers and also observed a layer of immobilized segments at the particle surface [15]. In the case of chains chemically grafted to the silica surface, Berriot et al. observed immobile layers, which they described as a glassy shell around the filler surface. Cosgrove et al. used differential scanning calorimetry and NMR measurements, and showed that the reduction in the mobility of the PDMS chains corresponded to a shift of Tg to higher temperatures [16]. These shifts are relatively small. Nevertheless, they indicate the presence of a gradient of segmental mobility in going from the elastomer towards the silica surface.

In the present study, we report the incorporation of Stober silica into PDMS elastomers, and the investigation of their mechanical and swelling properties as a function of filler size, surface modification, and loading. The silica particles that were prepared under ambient conditions using sol-gel technique were then blended into the PDMS network precursor, prior to cross linking. The equilibrium stress–strain properties of the filled (unmodified and modified fillers) and unfilled PDMS elastomers were determined. The silica particle dispersion in the polymer network was characterized using transmission electron microscopy (TEM).

2. Experimental

2.1. Materials

The materials required for the synthesis of silica nanoparticles are described as follows. Tetraethyl orthosilicate (TEOS 98%) was obtained from the Fischer Company, ammonium hydroxide solution (31.5% NH_3) was used as received from the Aldrich Company, and absolute ethanol 99 % from ADWIC Company was used after distillation.

The materials required for preparation of organically-modified silica were: silane coupling agent (hexamethyldisilazane) obtained from Fluka and toluene from ADWIC Company.

The materials required for synthesis of PDMS networks were: α, ω-hydroxy-terminated PDMS network precursor of number average molecular weight Mn = 18,000 g/mol, obtained from UCT (Petrarch) company, and the catalyst bis (2-ethyl hexanoate) tin was obtained from the Aldrich Company. Dichloromethane was obtained from BDH, and all chemicals were directly used without any further purification.

2.2. Preparations of Silica Nanoparticles

Silica nanoparticles were prepared by hydrolysis and condensation of TEOS in ethanol, and in presence of ammonia as catalyst [17]. First, a solution containing appropriate quantities of absolute ethanol, ammonia, and deionized water were stirred for 5 min to ensure complete mixing. Then a chosen amount of TEOS in absolute ethanol was added to the above solution and the reaction permitted to proceed at ambient temperature for 24 hr. Thereafter the colloidal solution was separated by high-speed centrifuge, and the resulting silica particles were washed by absolute ethanol three times to remove undesirable particles, and they were then dried in an oven at 100 °C for 2 hr to prevent continuous reaction.

2.3. Preparation of Organically Modified Silica

After thermal treatment at 150°C for 2 hr to remove physisorbed water, a chosen amount of silica was put into a glass flask equipped containing the desired amount of toluene. The liquid reagent (10% hexamethyldisilazane in toluene) was then added and the mixture was stirred and refluxed at 110°C for 3 hr. After treatment, the solid was filtered and repeatedly washed with toluene and acetone in order to remove any unreacted silane coupling agent. It was finally dried at 150°C for 2hr.

2.4. Network Synthesis

The hydroxyl-terminated PDMS was end-linked into networks by reacting it with a stoichiometric amount of cross-linking agent, tetraethoxysilane (TEOS). The end-linking reaction was performed in solution to facilitate the mixing process and the samples were sonicated to obtain better dispersion of the silica particles. The hydroxyl-terminated PDMS was dissolved in dichloromethane. The desired amount of silica particles was dispersed in dichloromethane and sonicated for 15 min and was then added to the polymer solution. The mixtures were stirred for 1 hr to obtain a homogeneous dispersion. The solvent was removed before the cross-linking process, and then TEOS and the catalyst, (0.5wt%) bis (2-ethyl hexanoate) tin, were added to the dispersion with stirring. The cross-linking reactions were carried out in a Teflon mold at room temperature for 2 days. Each elastomeric network was then exhaustively extracted for five days with toluene in order to remove any soluble and/or extractable material. After deswelling using methanol as the non-solvent, the networks were dried to constant weight. The samples were removed and stored at room temperature in air- tight vials to prevent adsorption of moisture from the surroundings.

2.5. Transmission Electron Microscope Analysis

In order to investigate the dispersion of the silica nanoparticles in the respective elastomers, transmission electron microscope (TEM) analyses were performed. TEM analyses of the silica particles were carried out using a JEOL JEM 2010 unit on both the unmodified silica and organically-modified silica. To prepare samples for the TEM analysis, silica powder was dispersed in absolute ethanol and a drop of this silica colloidal mixture was placed on a copper grid coated with carbon. The solvent was evaporated at room temperature, leaving the silica particles on the grid.

The sample preparation was achieved was designed to produce thin sections having a thickness less than 1000 A°. They were cut from bulk pieces of the nanocomposite using either a glass or a diamond knife on an ultramicrotome at -125 to -130°C. Specimens were collected on standard copper grids, which were then coated with carbon to increase their conductivity.

2.6. FTIR Analysis

Fourier transform (FTIR) analyses were carried out on both unmodified silica and organically-modified silica using an ATI Mattson model Genesis series (USA) infrared spectrophotometer adopting KBr technique. For all samples, a small amount of silica was placed between the two electrolyte plates to obtain transmission spectra.

2.6.1. Equilibrium Stress-Strain Measurements. Portions of the unfilled elastomer and some filled with silica were then characterized by obtaining their elongation stress-strain isotherms at 298 K [11, 18, 19]. Of particular interest was the nominal stress given by

$$f^* = \frac{f}{A^*} \tag{1}$$

where f is the elastic force and A* is the undeformed cross-sectional area. The reduced stress or modulus is defined as,

$$[f^*] = \frac{f^*}{\alpha - \alpha^{-2}} \tag{2}$$

where α = L/Li is the elongation or length of the sample, L, relative to its initial (undeformed) length, Li. All measurements were made up to the rupture points of the samples.

*2.6.2. Equilibrium Swelling Measurements.*Values of the volume fraction v2m of polymer in the networks at swelling equilibrium for the unfilled and filled PDMS nanocomposite elastomeric networks were determined in toluene at 298K.

3. Results and Discussion

3.1. Transmission electron microscope analysis of silica particles

Figure 1 shows the TEM image of unmodified silica nanoparticles [~65nm]. The resulting particles in the product suspension were spherical in shape and nano-sized. The figure also shows the dispersion of the particles and the absence of any significant aggregation among particles. This is inherent in the method of preparation in alkaline medium [17].

Figure 2 shows the TEM image of silica nanoparticles whose surface has been modified with hexamethyldisilazane. The figure shows an appreciable dispersion of the nanoparticles. As would be expected, this is attributed to the fact that the particles possess organophilic character after treatment with hexamethyldisilazane which converted the majority of the surface OH groups to [-(CH3)3] groups. Undoubtedly, such hydrophobic groups result in electrostatic repulsions and in segregation of the particles.

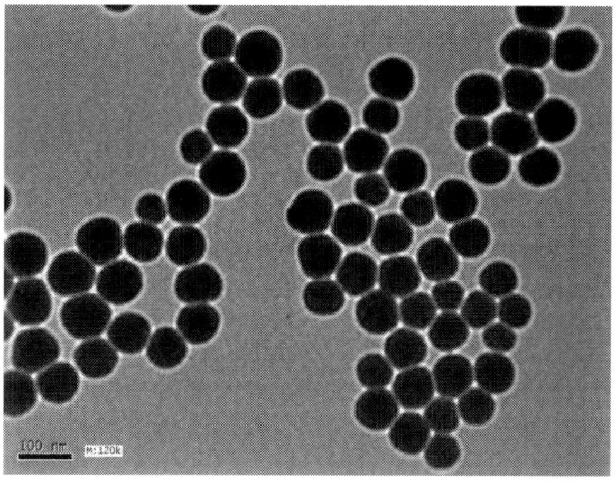

Figure 1. TEM image of unmodified silica nanoparticles (experimental conditions: TEOS: 0.18M; NH3: O.2M; H2O: 1M) [Particle size ~65nm].

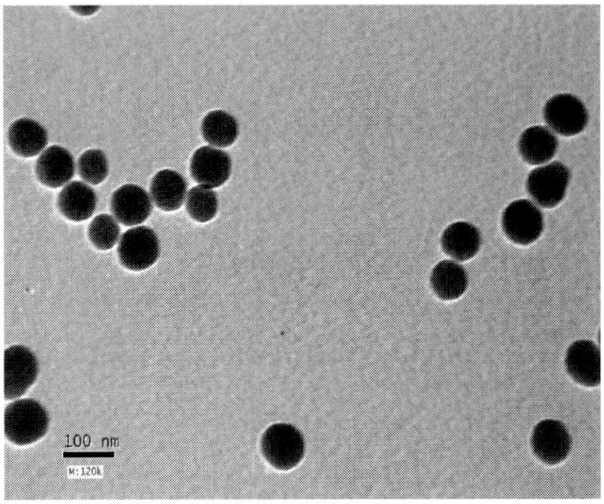

Figure 2. TEM image of organically modified silica nanoparticles (experimental conditions: TEOS: 0.18M; NH3: O.2M; H2O: 1M, 10% HMDS in toluene) [Particle size ~65nm].

3.2. FTIR Analysis is.

Figures 3 and 4 show the FTIR spectra for both untreated and nano silica particles treated by the silane coupling agent, respectively. For untreated nano silica particles, there is absorption band at around 3400 cm-1, which corresponds to Si–OH groups [20]. Also the peak at 1050-1250 cm-1 is due to an asymmetric stretching and vibration of Si-O-Si bonds [21]. After the surface treatment, a decrease in the intensity of Si-OH bands has been observed. Also the occurrence of absorption bands at around 2950 cm-1 and 1400 cm-1 are due to the stretching and bending of C-H bonds and the occurrence of peaks at around 847 cm-1 are due to the Si-C bonds [22] which confirm the conversion of the surface OH groups to [-(CH3)3] groups.

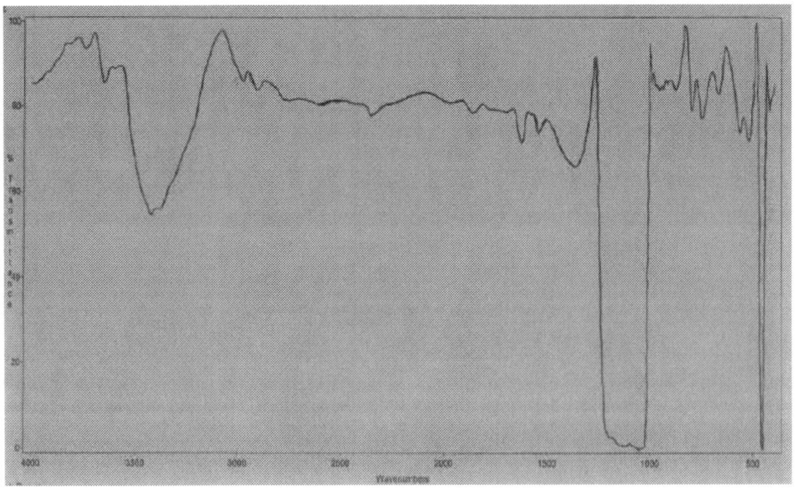

Figure 3. FTIR spectra of unmodified silica nanoparticles.

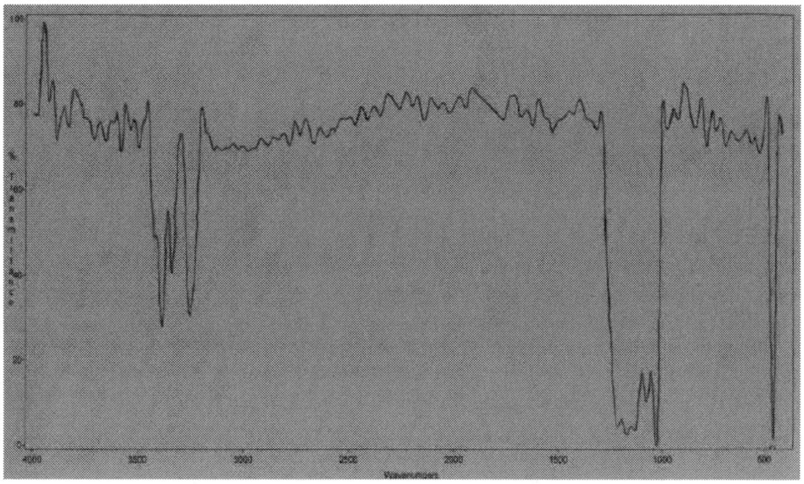

Figure 4. FTIR spectra of organically-modified silica nanoparticles.

3.3. Transmission electron microscope analysis of composites

The dispersion of the unmodified and organically-modified silica nanoparticles in the respective PDMS elastomeric networks was studied by TEM. A thin section of the elastomer had been frozen and cut using an ultramicrotome, and then analyzed. The data obtained by TEM analyses are presented in Figures 5 and 6. In the case of the unmodified silica nanoparticles, good dispersion of the particles in the PDMS networks was obtained (as shown in Figure 5). As expected, when the silica had been treated with the silane coupling agent, the particles were well dispersed in the PDMS networks, (as shown in Figure 6).

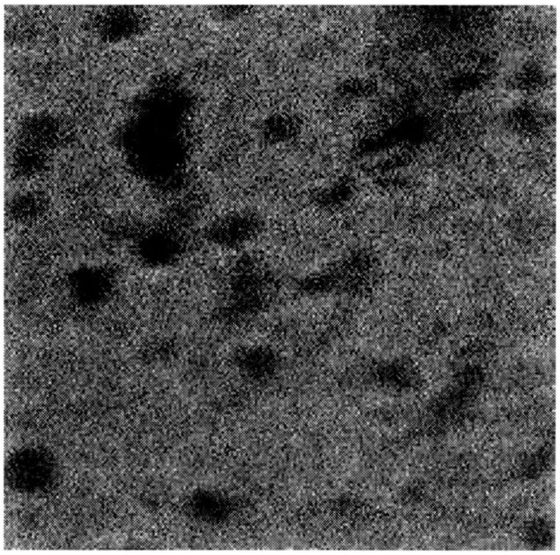

Figure 5. TEM of 5 wt% unmodified silica- PDMS nanocomposites [Particle size 50 nm].

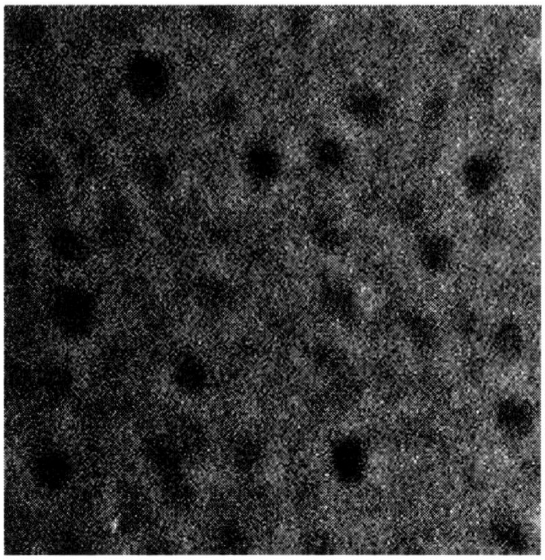

Figure 6. TEM of 5wt% organically modified silica PDMS nanocomposites [Particle size 50 nm].

3.4. Stress-strain measurement

Table 1. Stress-strain and equilibrium swelling characteristics of unfilled, unmodified, and modified nano-silica/PDMS elastomeric networks.

Size, nm	Weight % of silica	2C1 N mm-2	2C2 N mm-2	2C1+2C2 N mm-2	2C2 /2C1	v2m	Sol fraction wt%
Unfilled							
	0	0.0521	0.1016	0.154	1.95	0.203	5.0
Unmodified							
170	5	0.117	0.114	0.231	0.98	0.267	1.8
130	5	0.071	0.185	0.256	2.61	0.258	2.0
50	5	0.114	0.178	0.293	1.56	0.270	4.2
50	5	0.114	0.178	0.293	1.56	0.270	4.2
	10	0.086	0.234	0.320	2.73	0.263	5.2
	15	0.193	0.144	0.336	0.75	0.270	5.6
	20	0.201	0.182	0.383	0.91	0.279	6.8
Modified							
170	5	0.123	0.067	0.190	0.54	0.249	2.0
130	5	0.133	0.086	0.218	0.65	0.241	2.6
50	5	0.124	0.112	0.236	0.91	0.243	4.8

The equilibrium stress-strain isotherms obtained at 298 K for the unfilled PDMS network and those that had been filled with silica nanoparticles are represented as a function of the reduced modulus [f*] against α-1, in accordance with the Mooney-Rivlin equation [23-25]:

$$[f^*] = 2C1 + 2C2$$

From the respective isotherms, values for the constants 2C1 and 2C2 were calculated and the results are presented in Table 1. In the absence of a good molecular theory for filled elastomers [24, 26], the first Mooney-Rivlin term most likely represents an approximation of the high-deformation modulus. On the other hand, 2C2 and 2C2/2C1 are measures of the increase in the non-affineness of the deformation with an increase in elongation [24].

The values of the modulus for PDMS filled with unmodified and organically-modified silica nanoparticles are higher than the corresponding value of the unfilled PDMS, as is clearly demonstrated in Figures 7 and 8. Clearly, this was largely due to the reinforcement effect induced by the incorporation of the silica nanoparticles in the polymeric matrix.

The modulus or reduced stress values of the PDMS elastomeric networks were found to increase with a decrease in the size of both the modified and unmodified silica nanoparticles incorporated into the networks, as is illustrated in Figures 7 and 8. Again, this is primarily owing to the high specific surface area of the nano-fillers and subsequent significant increases in the interfacial adhesion between the filler and the polymer.

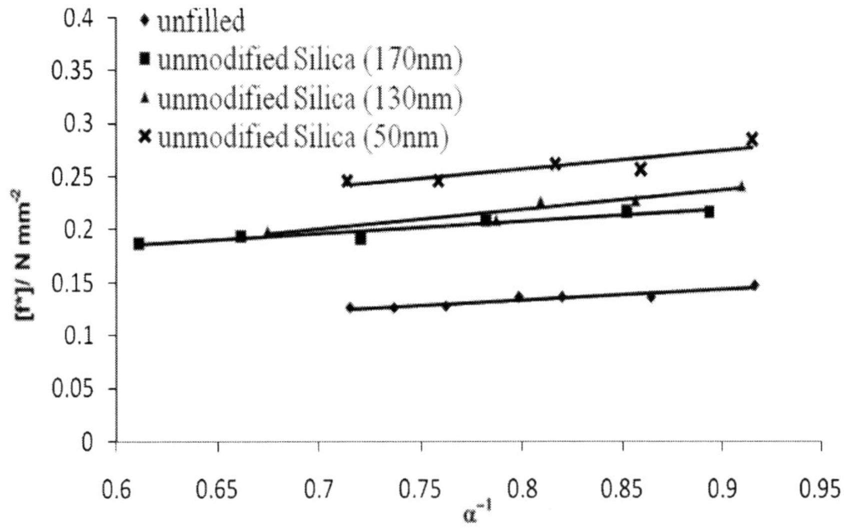

Figure 7. Equilibrium stress-strain isotherms for unfilled and 5 wt% unmodified silica filled PDMS elastomers at 298 K.

Approximate values of the modulus as expressed as the sum of 2C1 + 2C2 for the PDMS filled with organically modified silica nanoparticles are lower than those for the PDMS networks filled with unmodified silica, as is clearly seen in Figure 9 for networks filled with 50 nm inclusions. This behavior could be plausibly attributed to cross linking of some of the silanol OH groups present on the surface of silica with some end capped hydroxyl groups present on the PDMS oligomers. Intuitively, such cross links would result in junctions having higher functionality and consequently higher affine moduli. In addition, the higher values of the affine modulus are a reflection on the presence of a larger number of effective numbers of shorter network chains. This is a plausible result of H-bonding between O-atoms along the chain backbone and silanol groups present on the surface of silica. Again, the presence of such short network chains could offer a valid explanation of the higher values of the 2C2 constant that is known to serve as a measure of the nonaffiness of deformation observed for those networks incorporating unmodified nanoparticles. In networks having short chains, the junctions are not as firmly embedded in the matrix and so phantom deformations would be expected. The higher values of 2C2 for the networks incorporating unmodified silica serve as a clear indication of higher degree of nonaffiness of the deformations.

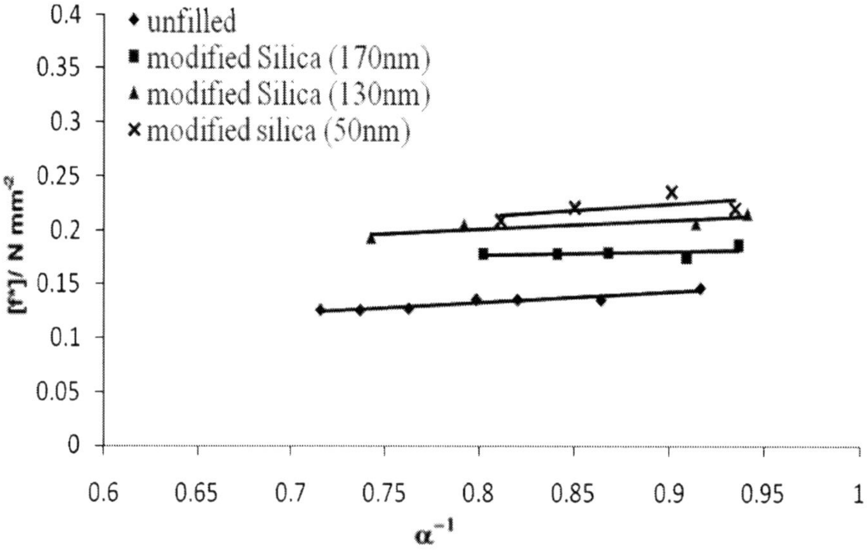

Figure 8. Equilibrium stress-strain isotherms for unfilled and 5 wt% modified silica filled PDMS elastomers at 298 K.

On the other hand, values of the large-deformation (phantom) modulus are higher for the networks incorporating modified silica. Also, their values of 2C2 are by far much smaller than those obtained for the unmodified case. Again, this is an indication that the deformations of networks incorporating modified nanoparticles tend to be more nearly affine in nature.

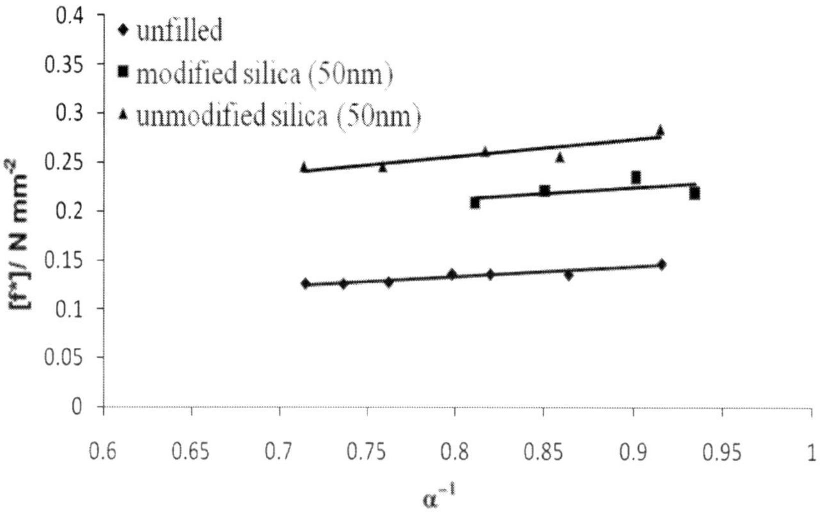

Figure 9. Equilibrium stress-strain isotherms for unfilled and 5 wt% unmodified and modified silica (50nm) filled PDMS elastomers at 298 K.

The modulus or reduced stress values of the PDMS elastomers were found to increase as a function of the amount of silica incorporated into the network, as is clearly seen in Figure 10. This could be explained in terms of an increase in the filler reinforcement and interfacial adhesion. Such increases are comparable with those observed for PDMS elastomers filled in-situ with silica [27].

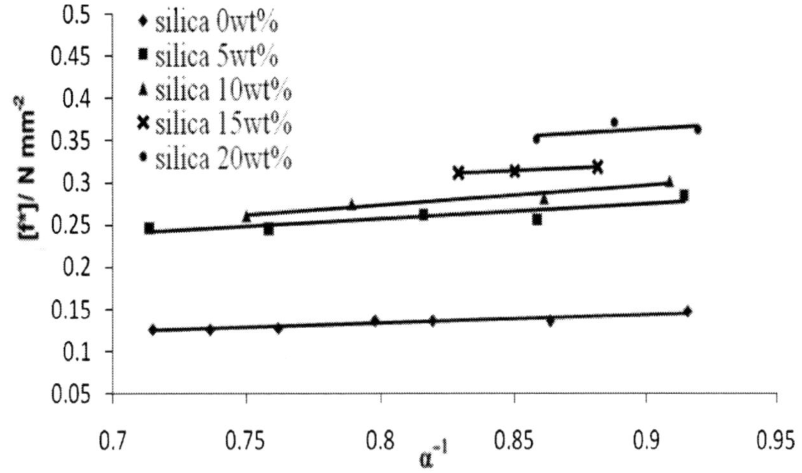

Figure 10. Equilibrium stress-strain isotherms for PDMS elastomers at 298 K unfilled and filled with 5, 10, 15 and 20 wt% unmodified silica (50 nm).

Values of the reduced force f* data of the PDMS networks filled with unmodified silica inclusions having 50 nm diameters is represented against the elongation are represented in Figure 11. In this way that the area under each curve corresponds to the energy, Er, of rupture [28,29]. The rupture energy serves as a measure of toughness of the polymeric network. However, this is the subject of another investigation.

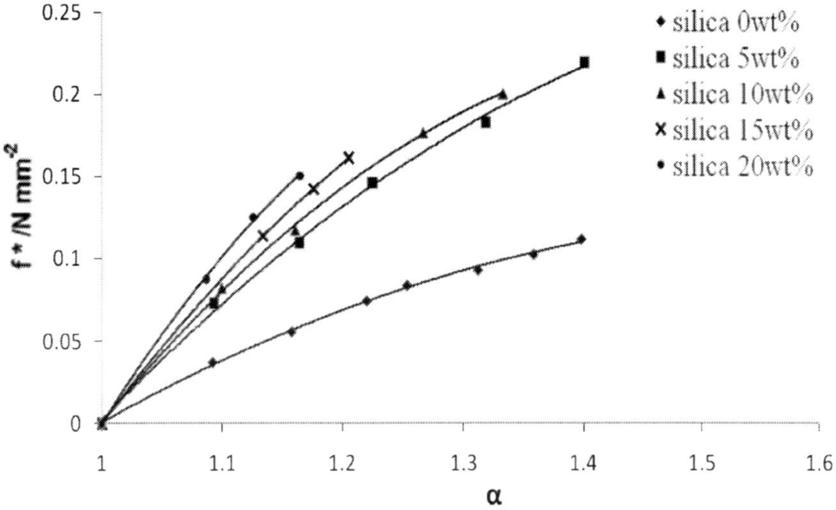

Figure 11. Equilibrium stress-strain isotherms for PDMS elastomers at 298 K unfilled and filled with 5, 10, 15 and 20 wt% unmodified silica (50 nm).

3.5. Reinforcement mechanisms from equilibrium swelling data.

The volume fraction of polymer υ2m at equilibrium swelling in toluene at 298 K is tabulated in Table 1. The unfilled network exhibited the highest degree of swelling, as would be expected.

It is possible to interpret equilibrium swelling measurements obtained on unfilled elastomers to estimate the degree of adhesion between elastomer and filler particles. According to Kraus [30],

$$\frac{V_{r_o}}{V_{r_f}} = 1 - m\left(\frac{\phi_f}{1-\phi_f}\right)$$

with

V_{r_o} being the volume of the polymer in the unfilled network, V_{r_f} being the volume of polymer in the filled network, and □f is the volume fraction of the filler. Kraus [30] has reported that a plot of V_{r_o} / V_{r_f} vs. filler content, □f, corrected for the non-swelling filler, is indicative of filler reinforcement. He has also shown that solvent uptake is inversely proportional to the filler reinforcing strength which strongly depends on filler surface area and agglomerate structure. Strongly interacting fillers reduce swelling, while non-reinforcing fillers result in solvent uptake which may exceed that of the unfilled polymer [31]. This behavior has been attributed to the formation of "bound polymer", which is polymer in close proximity to the filler which is either physisorbed or chemisorbed and therefore restricts swelling. Non-reinforcing fillers prevent the formation of bound polymer, allowing large vacuoles to form during solvent exposure, resulting in increased solvent uptake [31]. From the theory, if V_{r_o} / V_{r_f} decreases with increasing filler loading then the filler is adhering to the polymer. This is clearly seen for the PDMS-silica hybrid investigated here. Similar behavior has been reported for PDMS-silicate nanocomposites [31], PDMS filled with in situ precipitated silica and titania [32], and carbon black filled styrene-butadiene rubber (SBR) [31]. On the other hand, styrene-butadiene rubber (SBR) filled with kaolin showed an increase in swelling with an increase in filler loading [33].

Sol fractions obtained from swelling of the networks with toluene are shown in Table 1. The sol fraction did not exceed 7 wt % of the dry network, indicating that the preparation method was efficient.

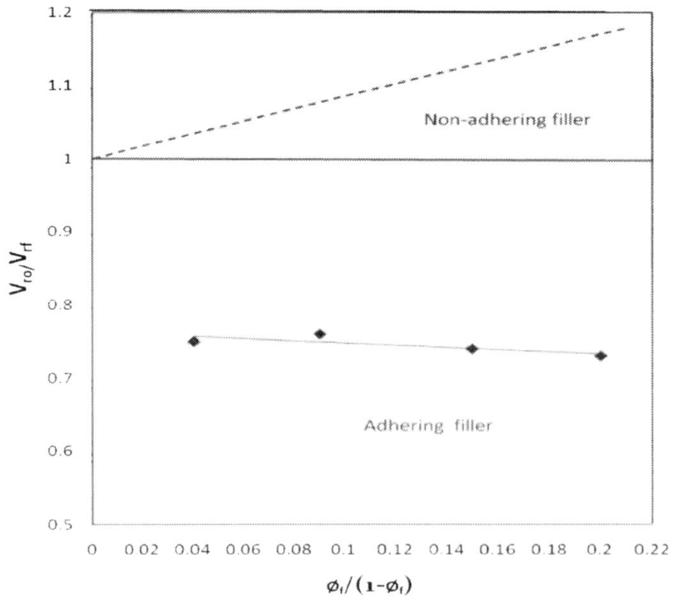

Figure 12. Kraus plots to illustrate the unmodified filler interacting with the matrix polymer at 298 K, according to filler (50 nm) fractions 5, 10, 15, and 20 wt%.

Figure 12 shows changes in values of V_{r_o} / V_{r_f} against the volume fraction of untreated silica (50 nm), where V_{r_o} / V_{r_f} decreasing with increasing filler loading which indicates adherence to the polymer.

Figure 13 shows change of V_{r_o} / V_{r_f} against the particle size 170, 130 and 50 nm of unmodified and modified silica, where unmodified and modified filler are both found to be adhering to the polymer, according to the Kraus theory. Also, the results clearly demonstrate that the unmodified nano filler inclusions tend to be more adhering than the modified one, as has already been discussed.

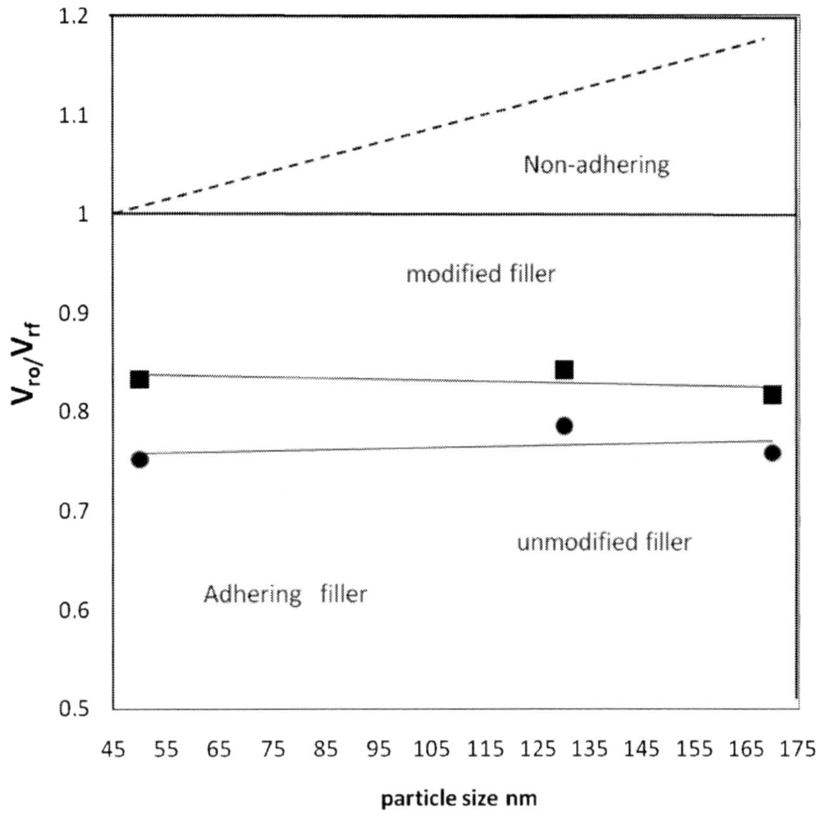

Figure 13. Kraus plots to illustrate 5wt% unmodified and modified filler (170, 130 and 50nm)-network polymer interactions at 298 K: (•) for unmodified filler and (■) for modified filler.

4. Conclusion

In particular, the equilibrium PDMS stress-strain behavior has been studied and improvements in mechanical properties such as increases in modulus occurred with decrease in filler size. Values of the modulus for the silica/PDMS nanocomposite networks manifested increases in comparison to those for the unfilled networks, as would be expected. The modification of filler surface was accompanied by an expected decrease in modulus values when compared to the case of incorporation of unmodified nano silica. More importantly, the swelling data lend support to the fact that both the unmodified and modified filler were classified as adhering to the matrix. The unmodified showed improved adherence than the modified filler and this is plausibly owed to the formation of partial bonds between the hydroxyl groups the PDMS backbone oxygen atoms.

The transmission electron microscope (TEM) micrographs clearly indicated that surface modification improved the dispersion of the nano particle inclusions. In the case of unmodified silica, however, a tendency for agglomeration of the nano inclusions was obvious.

One of the highlights of this work is the observation of a size phenomenon with regard to the broad-spectrum dielectric properties of the nanocomposite networks. These properties showed a discernible dependence on the nanosize of the filler particles. Such results clearly demonstrate the paramount role played by the interface and its subsequent effects on the properties in nano systems.

5. Acknowledgments

MAS acknowledges the US Egypt Science and Technol. Joint Fund through grant (MA7-001-002) for support for this research. Partial support was also received through US-Egypt Grant INT-01-11334. It is also a pleasure to acknowledge the financial support provided JEM by the National Science Foundation through Grant DMR-0803454 (Polymers Program, Division of Materials Research).

References

[1] A. J. Barry, H. N. Beck, "Inorganic Polymers", Academic Press, New York (1968).
[2] J. E. Mark, H. R. Allcock, R. West, "Inorganic Polymers", Prentice Hall, Englewood Cliffs, NJ, (1992).
[3] S. J. Clarson, J. A. Semlyen, "Siloxane Polymers", Prentice Hall, Englewood Cliffs, NJ, (1993).
[4] W. Noll, "The Chemistry and Technology of Silicones", Academic Press, New York (1968).
[5] J. E. Mark, M. Kato, J. Ko, J. Polym. Sci., Polym. Symp., 54, 217 (1976).
[6] K. E. Polmanteer, C. W. Lentz, Rubber Chem. Technol., 48, 795 (1975).
[7] E. M. Dannenberg, Rubber Chem. Technol., 48, 410 (1975).
[8] M. P. Wagner, Rubber Chem. Technol., 49, 703 (1976).
[9] W. Q. Yuan, J. E. Mark, Macromol. Chem. Phys., 200, 206 (1999).
[10] J. M. Breiner, J. E. Mark, Polymer, 39, 5483 (1998).
[11] J. E. Mark, B. Erman, Rubberlike Elasticity a Molecular Primer. New York: Wiley Interscience; p. 61(1988).
[12] B. Erman, J. E. Mark, Structures and Properties of Rubberlike Networks, Oxford University Press, (1997).
[13] W. M. Litvinov, H. Barthel, J. Weis, Macromolecules, 35, 4356 (2002).
[14] M. Gussoni, F. M. Grecp, A. Mapell, E. Vezzoli, P. Ranucci, P. Ferruti Macromolecules, 35, 1722 (2002).
[15] J. Berriot, F. Lequeux, L. Monnerie, H. Montes, D. Long, P. Sotta, J. Non-Cryst. Solids, 307–310, 719 (2002).
[16] T. Cosgrove, C. Roberts, T. Garasani, G. R. Schmidt, V. G. Gordon, Langmuir, 18, 10080 (2002).
[17] W. Stober, A. Fink and E. Bohn, Journal of Colloid and Interface Science, 26, 62 (1968).
[18] L. R. G. Treloar, The Physics of Rubber Elasticity, 3rd edition. Oxford: Clarendon Press; (1975).
[19] B. Erman and J. E. Mark, Ann. Rev. Phys. Chem., 40, 351 (1989).
[20] B. Y. Park, J. E. Mark, Polymer (Korea), 13, 597 (1989).
[21] J. K. Hong, H.R. Kim, H.H. Park, Thin Solid Films, 332, 449 (1998).
[22] N. Hering, K. Schriber, R. Riedel, O. Lichtenberger, J. Wolters-odorf, Appl. Organometal. Chem., 15, 879 (2001).
[23] L. R. G. Treloar, The physics of rubber elasticity, 3rd edition. Oxford: Clarendon Press; (1975).
[24] J. E. Mark, B. Erman, Rubberlike Elasticity: A Molecular Primer, New York: Wiley; (1988).
[25] P. J. Flory, Principles of polymer chemistry, Cornell University Press; New York: Ithaca; (1953).

[26] T. Coradin, J. Livage, Colloids Surf. B, 21, 329 (2001); T. Coradin, O. Durupthy, J. Livage, Langmuir, 18, 2331 (2002).
[27] S. B. Wang, J. E. Mark J.E. Polym. Bull., 17, 271 (1987).
[28] L. R. G. Treloar, The Physics of Rubber Elasticity. 3rd ed. Oxford: Clarendon Press; (1975).
[29] J. E. Mark, B. Erman, Rubberlike Elasticity: A Molecular Primer, New York: Wiley; (1988).
[30] G. Kraus, Reinforcement of Elastomers, Interscience; NewYork; (1965).
[31] S. D. Burnside, E. P. Giannelis, Chem. Mater., 7, 1597 (1995).
[32] J. Wen, J. E. Mark Rubber Chem. Technol., 67:806 (1994).
[33] G. Kraus Appl Polym Sci, 7, 861 (1963).

Advanced materials based on polymer blends/polymer blend nanocomposites

A V Shikaleska[1] and F P Pavlovska[2]

[1] Research Associate, Integrated Business Institute, Skopje, R. Macedonia
[2] Professor, Institute for Testing Materials and Development of New Technologies, Skopje, R. Macedonia

E-mail: aneta@fbe.edu.mk

Abstract. Processability, morphology, mechanical properties and rheological behavior of poly(vinylchloride) (PVC)/poly(ethylmethacrylate) (PEMA) blends and PVC/PEMA/montmorillonite (MMT) composites, prepared by melt processing in a brabender mixer, were studied. Samples were characterized using SEM, mechanical testing, DMTA and a parallel plate rheometer.

Plastograms show that there is noticeable drop of fusion times and increase in melt viscosity torque of both, polymer blend and polymer blend nanocomposite, in comparison with those of neat PVC. SEM images show that homogenous dispersions are obtained. Tensile tests indicate that PVC/PEMA and PVC/PEMA/MMT samples have greater tensile strength and elastic modulus and lower elongation compared to PVC. When solid viscoelastic properties are considered (DMTA), slightly higher storage moduli are obtained whereas more prominent increase of storage modulus is observed when nanoclay particles are added in a PVC/PEMA matrix. From the calculated area of tandelta peak of all tested samples, nanocomposites exhibit the lowest damping behavior. Oscillatory measurements in a molten state were used for determining the frequency dependencies of storage G' and loss G" moduli. It was found that G" curves of neat PVC lie above those of G' suggesting that PVC behaves like viscoelastic liquid. Similar results, but with significantly higher values of G' and G" over the whole frequency range for PVC/PEMA blends were obtained. Steady shear measurements show that the presence of PEMA and nanoclay particles increases the shear stress and shear viscosity of neat PVC. In order to define the rheological equations of state the three material functions were determined. According to these functions ($\eta(\dot{\gamma}), \psi_1(\dot{\gamma}), \psi_2(\dot{\gamma})$) all samples exhibit shear thinning behavior and the curves obey the power law equation.

As rheological behaviour was found to be strongly dependent on blend's micro and macro structure and it is one of the main factors defining the end properties, attempt was made to use this material functions to define the structure-properties relationship of polymer blends/polymer blend nanocomposites.

1. Introduction

Over the last two decades polymer blend technologies have attracted an increased amount of attention predominantly due to two major reasons: firstly, the ability to manufacture materials with desired properties that are not easy to achieve with neat polymers and secondly, the possibility to obtain such materials with reduced research and development expenses. Undoubtedly the strategy of polymer blending offers a wide range of materials with enhanced mechanical properties and fracture resistance, chemical resistance, barrier, thermal and other properties [1,2]. However, nowadays the rapid growth of nanoscience opens a new opportunity to use polymer blends as matrices for incorporation of nano-sized particles. Incorporated nanoparticles might cause changes at the polymer-polymer interface and depending on the interactions between them and each of the polymer components in the blend different spatial arrangement of nanocomposites can be obtained [3-4]. As a result, nanocomposites can exhibit remarkably improved properties which are almost impossible to reach by polymer combination in blends. Understanding the rheological behaviour of nanocomposites is another important issue that has to be borne in mind. On the one hand, rheological properties are strongly dependent on material's micro and macro structure and on the other hand rheological behaviour is one of the main factors that determines the processability of the material and defines the rate of production, its quality and end properties[5-7].

In that context, the aim of this work is to find out what kind of changes will cause the montmorillonite nanoparticles incorporated in the blend of PVC and PEMA and how those changes will affect the processability characteristics, rheological behaviour and mechanical properties of the resulting nanocomposite.

2. Experimental

2.1. Materials

The materials used in this study, PVC, PEMA and organically modified nanoclay Dellite 67G were commercially available products in a powder state. Some characteristics of the pure polymers are listed in Table 1.

Table 1. Materials characteristics

Polymer	Mw	Tg (°C)	Density (kg/m^3)	Solubility parameter ($MPa^{1/2}$)	Supplier
PVC	79 000*	80*	1,30*	21,46	OHIS Skopje
PEMA	350 000	63*	1,19*	22,69	SP[2]

*these data were experimentally obtained by the authors [8-10].

In order to prevent PVC degradation and to assist passage of the material through processing equipment, 2wt. % of thermal stabilizer (Irgastab 17 MOK) and 1wt. % of lubricant (WAX-E) were added.

The nanoclay used in this investigation Dellite 67G is montmorillonite modified with a high content of quaternary ammonium salt. The bulk density reported by the supplier is 0.45 g/cm3 and particle size after dispersion 500 nm.

2.2. Mixing

The blending of materials in a molten state was performed in a Haake Rheomix 600 internal mixer at 170 °C and 30 rpm for 5 minutes. PVC/PEMA blends with 90/10 weight ratio as well as PVC/PEMA blends with 5 wt% of organic montmorillonite nanoclay were prepared. In order to exclude the effect of different thermo-mechanical history and thus to evaluate accurately the PEMA and nanoclay effects, neat PVC was also treated under the same conditions as the PVC/PEMA and PVC/PEMA/MMT blends.

3. Characterizations

3.1. Processability

The processability measurements were performed in a Haake Rheomix 600 internal mixer with digital data acquisition. Torque-time curves (fusion curves), melt temperature and fusion time of all investigated samples were determined.

3.2. Morphological analysis
Investigated samples were fractured under liquid nitrogen and coated with a thin golden layer. Micrographs of cross sections of fractured surfaces were obtained using SEM JEOL JSM 5610.

3.3. Mechanical tests
Tensile mechanical properties (Young's modulus, tensile strength and elongation at maximum load, tensile strength and elongation at break) for neat polymers, PVC/PEMA blends and PVC/PEMA/MMT nanocomposites were carried out in a mechanical tester LLOYD LR10K. Experiments were performed at a crosshead speed of 10mm/min using specimens with a dumbbell shape and thickness of 2mm. Test specimens were previously prepared by injection molding; they were preheated at temperature of 180 °C for 4 minutes and then molded under pressure of 7 bar while the temperature in the mold was 50-60°C. For each material ten samples were prepared and tested. The reported tensile properties are averages of at least six individual tests.

3.4. Dynamic Mechanical Thermal Analysis (DMTA)
DMTA technique was used to evaluate the viscoelastic properties (storage modulus, loss modus and $\tan\delta$) of the material in a solid state. Dynamic mechanical properties of samples were measured using TA Instruments DMA 2980 analyzer. Temperature scans from 30°C to 130°C at constant frequency of 1 Hz and heating rate of 2 °C/min were carried out in a bending mode. Rectangular test specimens for DMA experiments were prepared by injection moulding as previously described.

3.5. Parallel plate rheometry
ARG2000 rheometer with parallel plate geometry was used to determine rheological properties under dynamic shear and steady shear conditions. The test specimens with disk geometry were obtained by compression moulding after preheating at 170°C for 5 min, then being subjected to a pressure for 2 min. and rapidly cooled to room temperature. In order to reduce degradation at high temperatures the measurements were carried out under nitrogen atmosphere.
The small amplitude oscillatory measurements performed with each sample included:
- dynamic strain sweeps at 200°C and angular frequency of 10 rad/s, in order to determine the region of linear viscoelasticity;
- dynamic frequency sweeps at 200°C and angular frequency from 0,1 to 100 rad/s at a strain of 1%.

Steady shear experiments were carried out at temperature of 200 °C and shear rate from 0,1 to 10 s^{-1}.

3.6. Capillary rheometry
The flow behaviour of the specimens under steady state shearing conditions was studied also using capillary rheometer Goettfert Rheo-Tester 1000. Measurements were conducted at temperature of 180°C and shear rate ranging from 10 to 1000 s^{-1}.

4. Results and discussion
The experimental results were processed and analyzed according to the aim of this investigation and they are as follows:

4.1. Processability
The fusion curves given in Figure 1 were used to evaluate the processability of PVC, PVC/PEMA blend and PVC/PEMA/MMT nanocomposite through fusion time, fusion energy and melt viscosity torque (MVT). The values of these processability parameters are presented in Table 2.

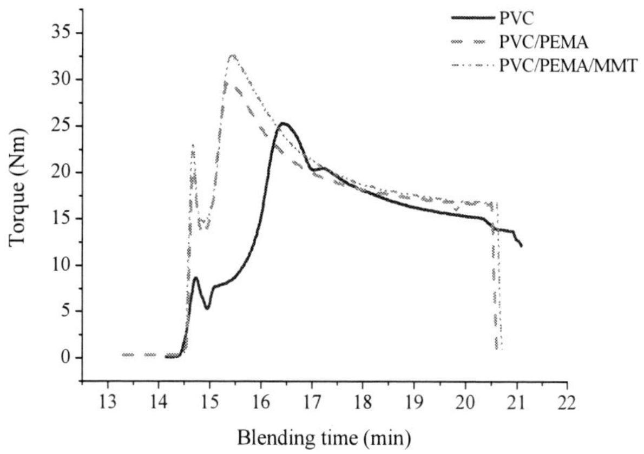

Figure 1. Fusion curves of PVC, PVC/PEMA and PVC/PEMA/MMT at 170 °C.

Table 2. Processability of PVC, PVC/PEMA and PVC/PEMA/MMT at 170 °C.

	Fusion time [min]	MVT [Nm]	Energy of fusion [Nm]
PVC	1.45	26	103
PVC/PEMA	0.37	30	118
PVC/PEMA/MMT	0.39	33	125

The fusion curves and values in Table 2 indicate that fusion times of PVC/PEMA blend and PVC/PEMA/MMT nanocomposite are significantly lower (about 4 times) compared to fusion time of neat PVC. MVT and energy of fusion of polymer blend and polymer blend nanocomposite are slightly higher (up to 20 %) than those of neat PVC. It is obvious that these changes are due to the presence of the second polymer component in the blend, while the effect of nanoclay on the processability parameters of this blend is not so noticeable. These observations imply that the nanocomposite can be processed using the same operations and processing conditions as those for polymer blends.

4.2. Morphology

SEM micrographs of tested samples are given in Figure 2. It can be seen that in all cases the components are uniformly distributed and homogenous morphologies are obtained. The fracture surfaces of samples have the characteristics of ductile failure of the material with high extent of plastic deformation.

a) b) c)

Figure 2. SEM photomicrographs for a) neat PVC b) PVC/PEMA and c) PVC/PEMA/MMT.

4.3. Tensile properties

Tensile properties are presented in Figures 3, 4 and 5.

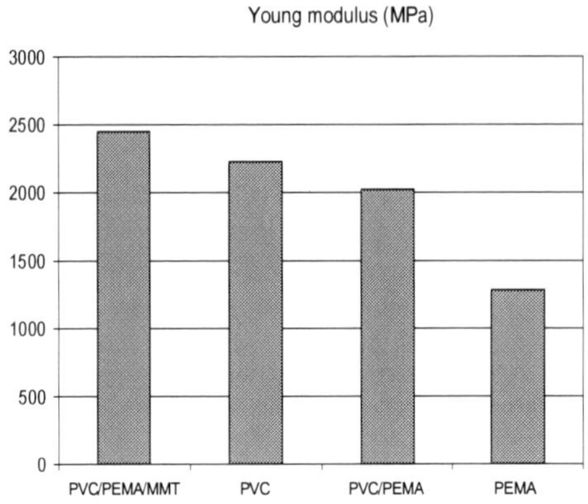

Figure 3. Young's modulus of PVC/PEMA, PVC/PEMA, PVC/PEMA/MMT and pure polymer

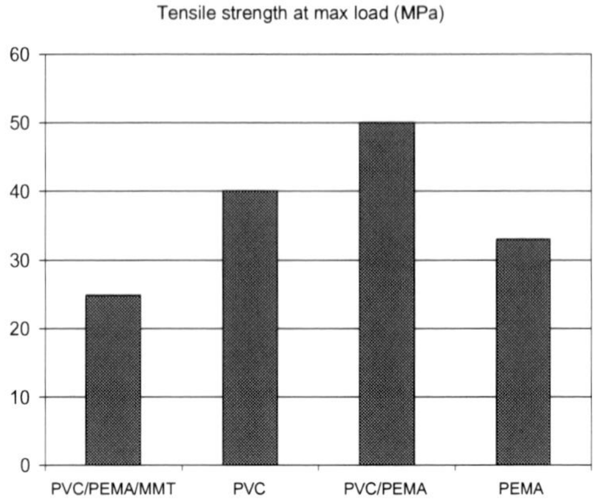

Figure 4. Tensile strength at break of PVC/PEMA/MMT and pure polymers

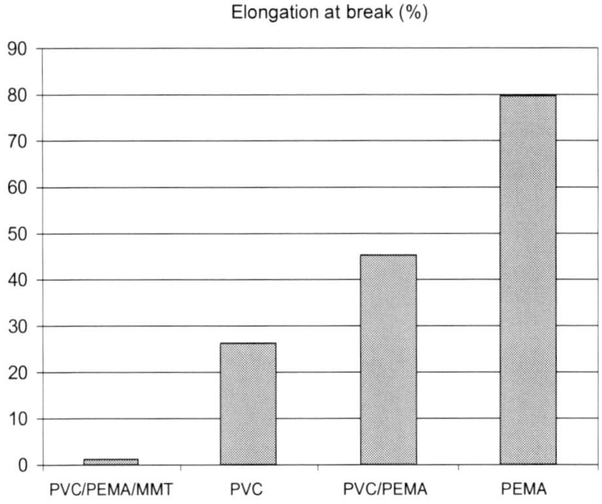

Figure 5. Elongation at break of PVC/PEMA, PVC/PEMA/MMT and pure polymers.

The results show that the ultimate mechanical properties of the PVC/PEMA 90/10 blend are significantly influenced by the presence of PEMA and nanoclay particles. It can be observed that a PVC/PEMA blend has a lower Young's modulus and higher tensile strength and elongation at break than neat PVC. As expected, by incorporation of nanoclay particles into PVC/PEMA matrix materials with higher Young's modulus and lower tensile strength and elongation at break are obtained. These measurements indicate that the nanoclay addition has more pronounced effect on the tensile strength and elongation at break than on the modulus of elasticity.

4.4. DMTA

The storage modulus E' as a function of temperature at 1 Hz is presented in Figure 6. It provides information about the viscoelastic behaviour and stiffness of investigated samples.

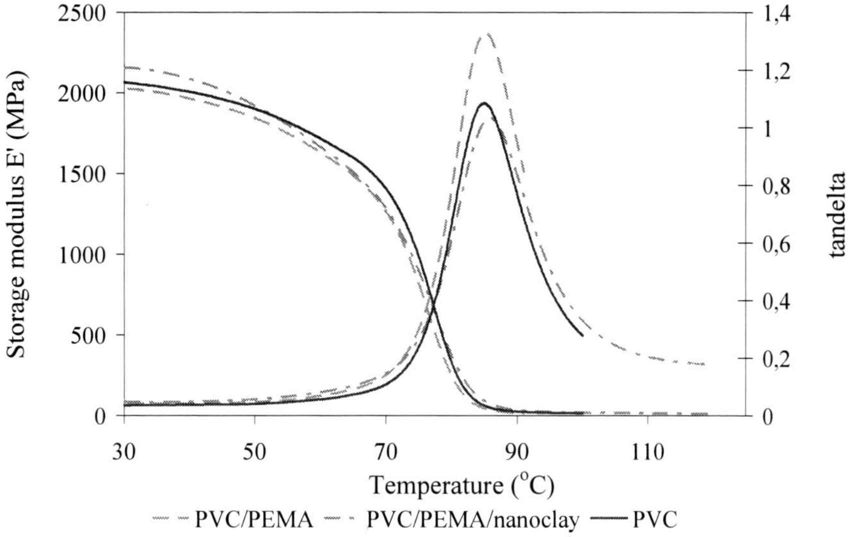

Figure 6. $E' = f(T)$ and $\tan\delta = f(T)$ curves for PVC, PVC/PEMA and PVC/PEMA/MMT.=

It can be seen that at temperatures up to 60°C the $E' = f(T)$ curve of the nanocomposite is above that of pure PVC and PVC/PEMA blend which indicates that nanoclay particles reinforce the PVC/PEMA matrix. As a result stiffer material is obtained. These data are in consistency with those obtained from tensile measurements. The areas of $\tan\delta = f(T)$ curves were used for assessing the damping behaviour of tested samples. It was found that nanocomposites exhibit the lowest damping behaviour. Also, from the maximum of these curves the glass transition temperatures (Tg's) were determined. It was shown that Tg of the nanocomposite is slightly higher then that of the polymer matrix. This is expected as the addition of nanoclay particles may restrict the movement of polymer segments which will result in an increase of the Tg.

4.5 Dynamic oscillatory and steady shear rheology

Oscillatory measurements in a molten state were used for determining the frequency dependencies of storage modulus G', loss modulus G" and $\tan\delta$ (Figure 7a).

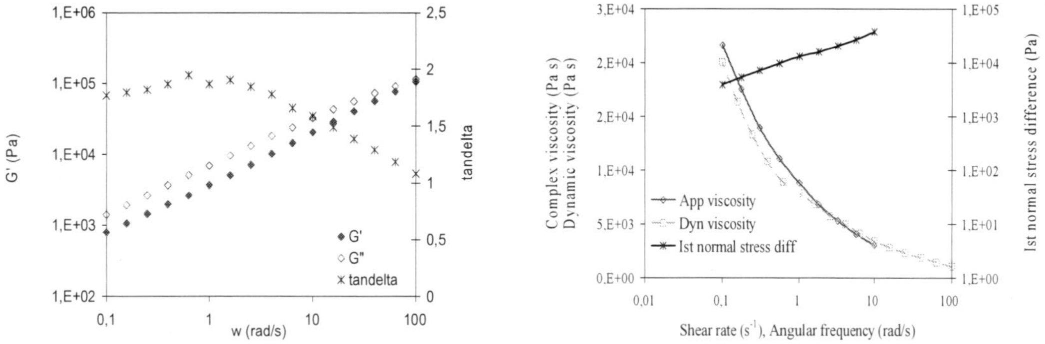

a) $G' = f(\omega)$ $G'' = f(\omega)$ and $\tan\delta = f(\omega)$ b) $\eta_{app} = f(\gamma)$, $\eta' = f(\omega)$ and $(\tau_{11} - \tau_{22}) = f(\gamma)$

Figure 7. Rheological properties of neat PVC at 200 °C

It was found that G" curves of neat PVC lie above those of G' suggesting that PVC behaves like viscoelastic liquid. Similar results, but with significantly higher values of G' and G" over the whole frequency range for PVC/PEMA blends were obtained. Using the data from steady and dynamic shear measurements (figure 7b) the shear dependent viscosity function $\eta_a(\dot{\gamma})$ and first normal stress coefficient function $\psi_1(\dot{\gamma})$ were determined and they are as follows (Table 3):

Table 3. Shear dependent viscosity and first normal coefficient functions at 200 °C

	Steady shear		Dynamic shear
	$\eta_a(\dot{\gamma})$	$\psi_1(\dot{\gamma})$	$\eta'(\omega)$
PVC	$\eta_a = 0.009\dot{\gamma}^{-0.42}$	$\psi_1 = 0.012\dot{\gamma}^{-1.52}$	$\eta' = 0.008\dot{\gamma}^{-0.4}$
PVC/PEMA	$\eta_a = 0.014\dot{\gamma}^{-0.6}$	$\psi_1 = 0.025\dot{\gamma}^{-1.78}$	$\eta' = 0.012\dot{\gamma}^{-0.47}$

The shear stress and apparent viscosity as a function of shear rate were determined also using a capillary rheometer. These dependencies are presented in Figure 8.

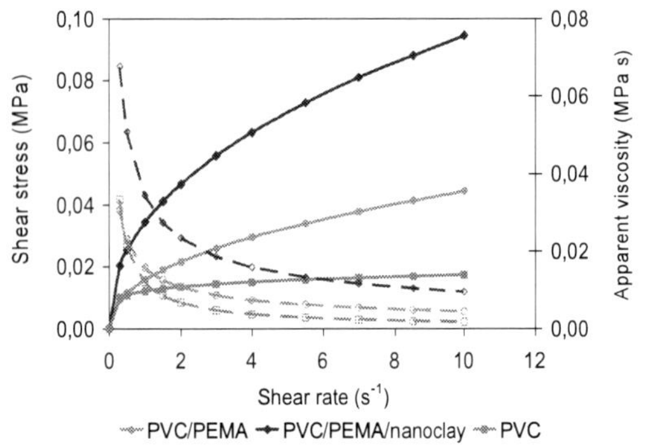

Table 4. Flow characteristics of PVC, PVC/PEMA and PVC/PEMA/MMT at 180 °C

	n	K (MPa s^n)
PVC	0.16	0.012
PVC/PEMA	0.45	0.016
PVC/PEMA/MMT	0.44	0.034

Figure 8. Flow and viscosity curves of PVC, PVC/PEMA and PVC/PEMA/MMT at 180 °C.

The flow and viscosity curves show pseudoplastic behaviour. The smallest values for shear stress and apparent viscosity were obtained for the neat PVC. Again, it was found that the presence of PEMA and nanoclay strongly affects the rheological properties the material. Also, it was observed that the flow and viscosity curves of PVC/PEMA blends and PVC/PEMA/MMT composites have the same shape and follow the same power law as does the pure PVC (Figure 8) with *n* up to 0.45 (Table 4). It implies that the components of the polymer blend and polymer blend nanocomposite form an isotropic medium with no phase separation and are rheologically compatible. As the measurements were performed in the range of linear viscoelasticity the second normal stress coefficient function $\psi_2(\gamma)$ for all investigated samples is approximately zero ($\psi_2 \approx 0$).

5. Conclusion remarks

PVC/PEMA blends and PVC/PEMA/MMT nanocomposites were prepared by melt mixing. It was found that modifying PVC with small amounts of PEMA and MMT significant improvements in processability, mechanical and rheological properties were obtained.

The processability parameters (as fusion time, MVT and energy of fusion) are more influenced by the presence the second polymer component than of the nanoclay. On the other hand, the nanocomposite exhibited higher storage modulus than the polymer blend and neat PVC. The storage modulus enhancement is possibly due to the stiffness of MMT particles and reduced volume of the polymer matrix. The addition of the organic MMT increased Young's modulus and the tensile strength but decreased the ductility of the nanocomposite. The calculated material functions show that PVC/PEMA and PVC/PEMA/clay samples are viscoelastic with pseudoplastic character of the apparent viscosity and the first normal stress coefficient.

The results of this investigation show that polymer blends used as a matrix offer wide opportunities for creating novel nanocomposites for diverse applications.

6. Acknowledgements

The authors gratefully acknowledge prof. d-r Paula Moldenaers and the Laboratory for Applied Rheology and Polymer Processing, K.U. Leuven, prof. d-r Marek Kozlowski and the Materials Recycling Centre of Excellence, Wroclaw University of Technology, d-r Donatella Duraccio, d-r Clara Silvestre and d-r Sossio Cimmino from ICTP-CNR (NA) Italy and the European Community's for assigning STSM in the framework of COST action FA0904.

Reference

[1] Robeson L M 2007 *Polymer Blends: A Comprehensive Review* (Hanser Verlag)
[2] Paul D R and Bucknall C B 2000 *Polymer blends: Formulation (*John Wiley & Sons, Inc.)
[3] Suprakas S R and Masami O 2003 *Prog. Polym. Sci.* **28** 1539–1641
[4] Rozenberg B A Tenne R 2008 *Prog. Polym. Sci.* **33** 40–112
[5] Awad W H *et al* 2009 Polymer **50** 1857–1867
[6] Xu, W Ge M Pan W-P 2004 *Journal of Thermal Analysis and Calorimetry* 78
[7] Yu Z-Z *et al* 204 *Polymer International* 53 (8) 1093-1098
[8] Popovska-Pavlovska 1998 F Final project report *Morphological, rheological, thermal and mechanical properties of thermoplastic polymer blends* (Ministry of Science of R. Macedonia)
[9] Popovska-Pavlovska F *et al* 1996 Polimery **41** 11-12.
[10] Popovska-Pavlovska F and Raka Lj 2004 *J. of Pol. Sci.: Part B* 42 267-279

IOP Conf. Series: Materials Science and Engineering **40** (2012) 012010 doi:10.1088/1757-899X/40/1/012010

Energy absorption characteristics of nano-composite conical structures

F Silva, S Sachse and J Njuguna*

Centre for Automotive Technology, School of Applied Sciences, Cranfield University, Bedfordshire, MK43 0AL, UK.

Email: * Corresponding Author:j.njuguna@cranfield.ac.uk

Abstract. The effect of the filler material on the energy absorption capabilities of polyamide 6 composite structures is studied in details in the present paper. The axial dynamic and quasi-static collapse of conical structures was conducted using a high energy drop tower, as well as Instron 5500R electro-mechanical testing machine. The impact event was recorded using a high-speed camera and the fracture surface was investigated using scanning electron microscopy (SEM). The obtained results indicate an important influence of filler material on the energy absorption capabilities of the polymer composites. A significant increase in specific energy absorption (SEA) is observed in polyamide 6 (PA6) reinforced with nano-silica particles (SiO_2) and glass-spheres (GS), whereas addition of montmorillonite (MMT) did not change the SEA parameter.

1. Introduction

Thermoplastic polymers, such as polyamide 6 (PA6), are widely used in the automotive industry because of their good mechanical performances, processing properties and low cost. However, their application as structural materials is limited due to their low impact resistance and energy absorption capabilities [1]. Incorporation of glass-fibres and various nano-sized fillers such as: nano-particles, carbon nano-tubes, and clay nano-plates; can be an appropriate solution to that problem [2; 3]. In case of light-weight structures, made of polymer composites, the most widely used nano-reinforcements are silica based particles, due to their good mechanical properties and high thermal stability [4].

For the purpose of measuring the energy absorption in composite structures, tube crashing experiments are the most prevailing. The ability of a composite structure to absorb energy was found to be highly dependent on the mode of fracture. Materials which fail in a progressive manner, with extensive delamination and fragmentation, are able to absorb much higher energies than those which tend to fail in a brittle manner. Mamalis et al. [5] studied polyester cones, cylinders and tubes, reinforced with random orientated glass fibres, divided failure of the samples into four different modes: progressive crashing with micro-fragmentation (Mode I), brittle fracture with catastrophic failure (Mode II and III, depending on the crack form), progressive folding and hinging, similar to metallic tubes (Mode IV). Each of these modes is characterized by different energy absorption capabilities.

Numerous researches have been conducted to study the influence of nano-particles on the mechanical behaviour of polymer composites and main factors influencing their enhancing capabilities were outlined. This includes key parameters such as: shape [6] and size [7] of the nano-fillers, matrix and reinforcement material [8; 9], interfacial strength and interphase characteristics [10], as well as volume fraction [11] and quality of dispersion within the matrix [12]. However, there is a lack of crash experiments conducted on nano-composites presented in the literature [2]. Energy absorption capabilities of nano- composites have been mainly characterized by means of compression [13], flexural [14] and Charpy or Izod impact testing [15]. That is why the relation between mechanical properties of nano-filled materials and energy absorption characteristics of nanocomposite structure is not fully understood.

2. Experimental

2.1. Materials

Polyamide 6 (PA6) (Tarnamid T-30 from Azoty Tarnow Poland), was used as a matrix material. As nano-filler two different types of material were used: silica-particles (SiO_2) (AEROSIL 200 from Degussa) and montmorillonite (MMT) (Dellite 43Bfrom Laviosa). Additionally, two different glass-reinforced composite materials, supplied by MACOMASS Verkaufs AG Germany, were used to prepare nano and glass-reinforced composite samples: glass-fibre (GF) reinforced polyamide 6 (MM-PA I 1F30) and glass-spheres (GS) reinforced polyamide 6 (MM-PA I 1K30).

Preparation of nano and glass reinforced polymer composites was conducted in three main steps: preparation of nano-composite granulate, mixing and extrusion of nano and glass reinforced composite granulate and injection moulding of the structural cones. In the first step nano-reinforcement and polymeric matrix, all in solid (powder) form, were premixed before extrusion, in order to warrant the highest homogeneity of the composition. Subsequently, the premixed materials were fed into the twin-screw extruder. In the second step, nano-composite granulates and glass-fibre reinforced polymers were mixed in the extruder. As a result eight different composite materials were prepared as shown in Table 1. In the third step, crash cones were produced using injection moulding machine (Engel ES200/60 HL ST).

Table 1. PA6composites.

Name	PA/GF	PA/GF/GS	PA/GF/SiO_2	PA/GF/MMT
Matrix	PA	PA	PA	PA
1st filler [wt%]	GF [30%]	GF [30%]	GF [30%]	GF [30%]
2nd filler [wt%]	-	GS [12%]	SiO_2 [2%]	MMT [2 %]

2.2. Methods

Quasi-static compression testing of the crash cones was conducted using Instron 5500R universal machine, at a crosshead speed of 0.1mm/s. The load was measured using a 100kN load cell. Impact tests of the crash cones were carried out on a high energy capacity drop tower machine at the velocity of 6.2m/s. The impactor mass of 54kg was constant in all experiments, giving an overall impact energy of 1050J. The load was measured using a 200kN load cell, placed underneath the sample. In order to measure the displacement of the falling mass, the linear variable differential transformer (LVDT) displacement transducer was used, with precision of 0.01mm and a maximum displacement speed of 10m/s.

The fracture surface of the impacted cones was exanimated with FEI XL30 field emission scanning electron microscope (SEM). The operating voltage was in the range of 10-20 kV and the specimens were gold sputtered to minimise charging of the sample.

3. Results and discussion

3.1. Crashing behaviour

Crashing behaviour and energy absorption characteristic of the composite structures were studied by means of quasi-static compression and dynamic impact testing. The results obtained are listed in

Tables 2 and 3. Analysing these results it can be seen that crushing characteristic under dynamic load are different from those subjected to the quasi-static compressive load. All materials tested under dynamic load absorbed similar amount of impact energy. However, a big difference can be seen in the specific energy absorption (SEA) parameter. This discrepancy was caused by the fact that each material failed with a different crushing length. For PA/GF/GS composite the crushing length was much smaller than for the other PA6 composites. As the specific energy absorption is a function of a crushed mass the SEA parameter was the highest in these materials, which absorbed the energy at a small crushing length.

Regarding the loads induced during the impact, the mean crushing load was much closer to the initial peak in case of PA/GF/SiO_2 composite, which had a direct influence on the amount of energy absorbed by the structure. In case of the other PA6 composites, the mean crushing load was significantly smaller than the initial peak indicating weaker energy absorption capabilities (see Figure

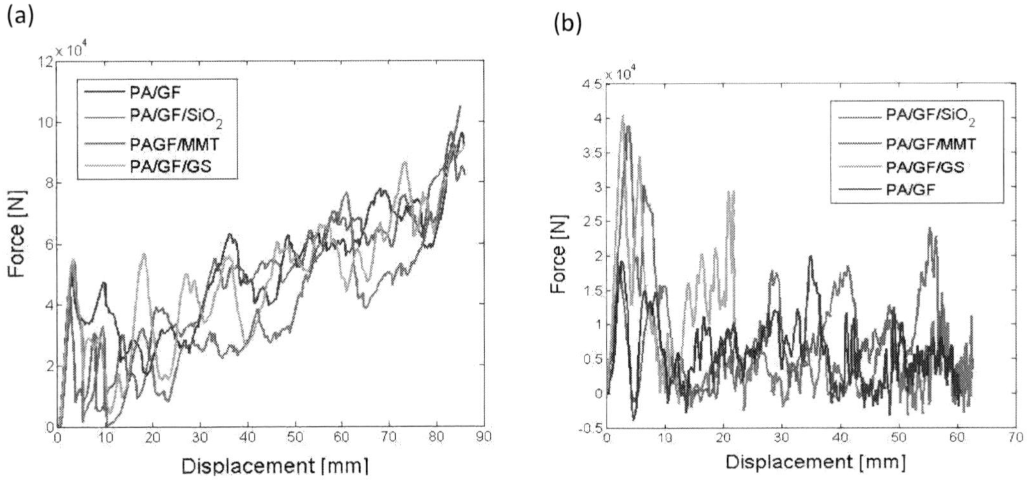

Figure 1. Load-displacement curves (a) static (b) dynamic.

Table 2. Quasi-static crashing characteristics.

Material	Crash length [mm]	Collapse mode	Initial peak [kN]	Mean crashing load [kN]	Energy absorbed [kJ]	SEA [kJ/kg]	Change in SEA [%]
PA/GF	86	II	47.66	50.44	4.33	58.1	
PA/GF/SiO_2	86	II	44.61	45.66	4.15	54.5	-6.1
PA/GF/MMT	86	II	54.59	40.65	3.23	42.9	-26.2
PA/GF/GS	86	II	55.10	45.74	4.11	51.7	-11.0

Table 3. Dynamic crashing characteristics

Material	Crash length [mm]	Collapse mode	Initial peak [kN]	Mean crashing load [kN]	Energy absorbed [kJ]	SEA [kJ/kg]	Change in SEA [%]
PA/GF	60.5	I	19.99	5.64	0.35	7.7	-
PA/GF/SiO_2	57.56	II	26.51	8.98	0.43	9.8	27.0
PA/GF/MMT	62.61	I	38.82	4.48	0.37	7.7	0.1
PA/GF/GS	22.03	II	40.42	15.58	0.32	22.3	188.5

IOP Conf. Series: Materials Science and Engineering **40** (2012) 012010 doi:10.1088/1757-899X/40/1/012010

The loads induced during the impact are directly correlated with the fracture mode and propagation of the cracks. The following main fracture modes could be identified and classified: (i) Mode I - Brittle fracture with large fragmentation. This fracture mode corresponds to unstable and catastrophic failure of the sample. Its characteristic part is formation of large debris due to the propagation of axial cracks. These cracks become initiated at the early stage of the impact event and cause a significant decrease in post-failure strength and stability of the structure. This mode indicates weak energy absorption and was observed in PA/GF and PA/GF/MMT composites tested under dynamic load (see Figure 2a and 2c). (ii) Mode II - Brittle fracture with progressive crashing and medium fragmentation. In this mode propagation of the axial cracks, initiated at the early stage of the impact event, stops quickly after the formation. Therefore, the size of the generated debris is significantly smaller than the debris size observed in Mode I. Additionally, a delamination effect was observed, as a separation of the composite layers. That is why the structure does not suffer catastrophic failure, indicating relatively good energy absorption, compared to Mode I. This mode was observed in PA/GF/SiO_2 and PA/GF/GS composites tested under the dynamic load (see Figure 2b and 2d), as well as in all PA based materials tested under the quasi-static load (see Figure 3).

Relating the energy absorption characteristic with the crashing characteristics, it can be seen that the materials which fail in a progressive manner, with small local cracks induced (Mode II), are able to absorb much higher energies than those with large continuous cracks (Mode I). This is caused by the fact that the fracture mode has got direct influence on the crushing parameters such as: crushing length, value of the peak loads and mean crashing load. The crushing length of the structure increases if the large cracks and debris become initiated. Additionally, the post-failure strength of the material is also reduced, what was recorded as a decrease in mean crushing load. As a result, the specific energy absorption of the material, which depends on these two parameters, was decreased as well.

(a) (b) (c) (d)

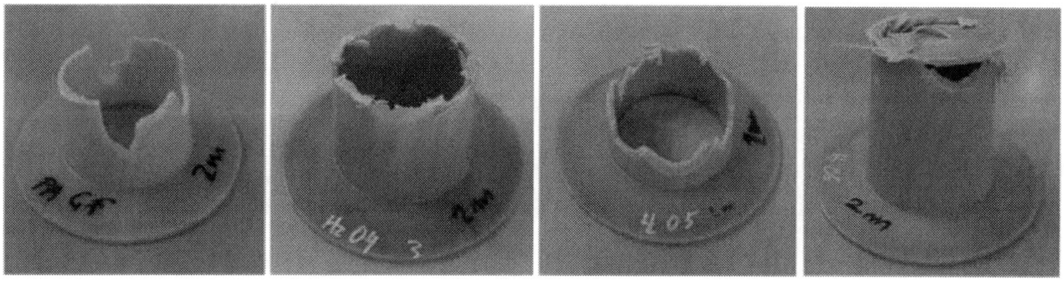

Figure 2: Dynamic collapse mode of PA6 composites (a) PA/GF (b) PA/GF/SiO_2 (c) PA/GF/MMT (d) PA/GF/GS

(a) (b) (c) (d)

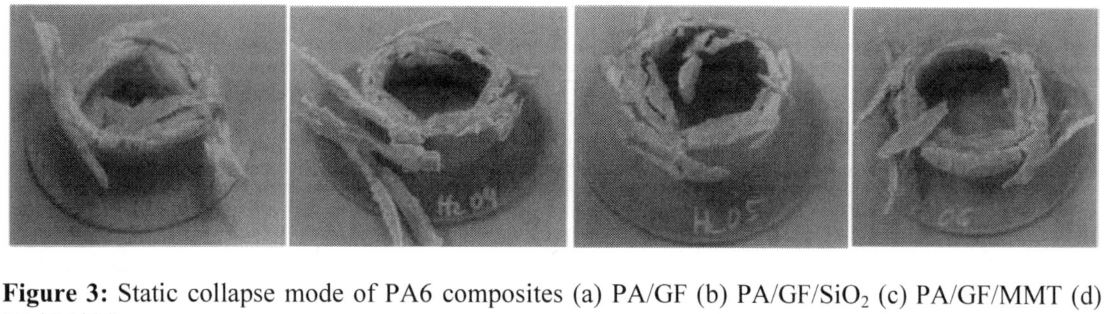

Figure 3: Static collapse mode of PA6 composites (a) PA/GF (b) PA/GF/SiO_2 (c) PA/GF/MMT (d) PA/GF/GS.

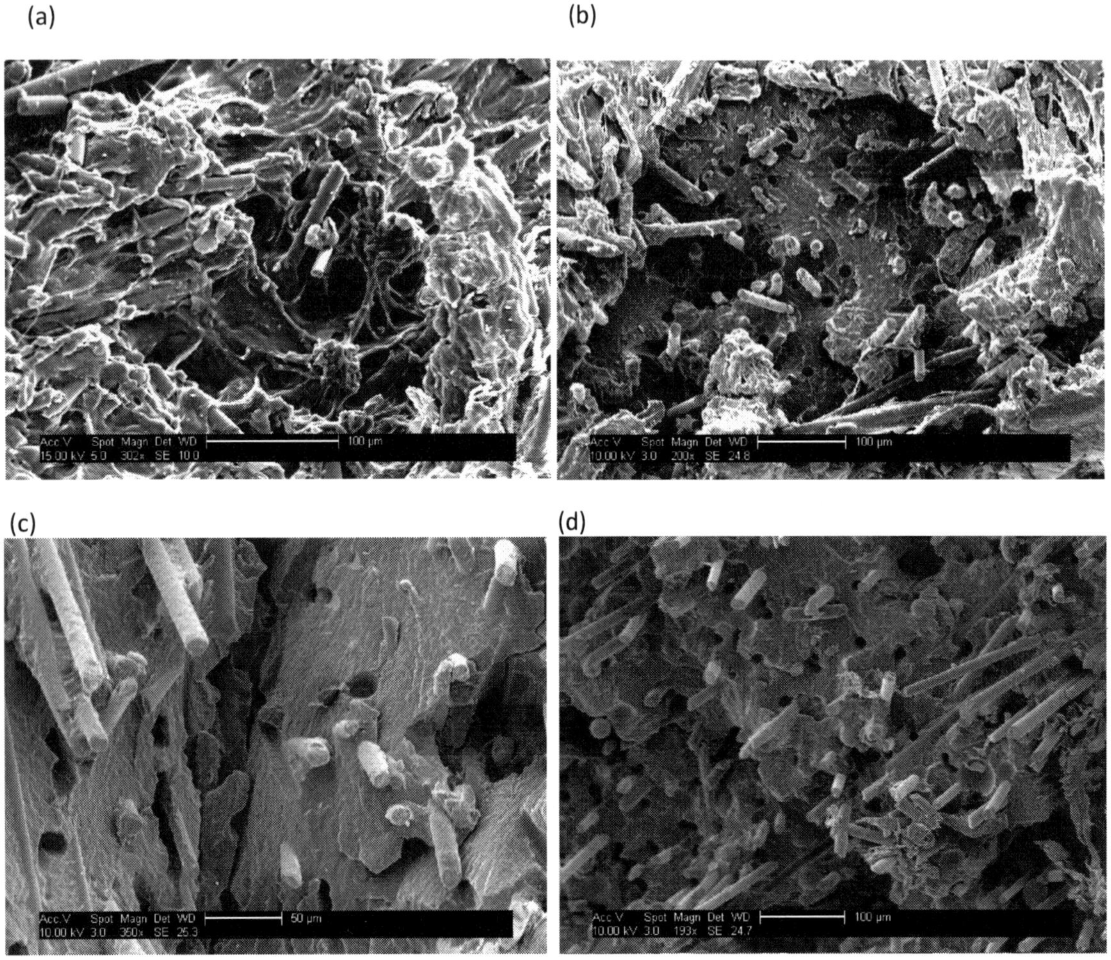

Figure 4: PA composites: (a) neat, (b) SiO_2, (C) MMT and (d) GS.

Analysing the influence of the secondary filler on the energy absorption of polymer composites, a significant differences could be observed. The SEA parameter increased in SiO_2 and GS reinforced composites, whereas it decreased in MMT filled ones. In PA/GF composite the axial cracks were initiated at a relatively low load, indicating low impact resistance of the material. The cracks propagated quickly along the height of the cone, leading to catastrophic failure of the structure (Mode I) and low energy absorption. Incorporation of SiO_2 particles did not increase the impact strength of the material but it changed the fracture behaviour. This was observed as a transition from fracture Mode I to Mode II. This change was caused by the significant reduction of the material brittleness, which was observed as an increase in elongation to break. As a result, the strain induced in the structure did not initiate severe cracks, as the material below the crush zone did not reach the failure strain. Moreover, an extensive delamination was observed, increasing the energy absorption capability of the material.

The opposite behaviour was observed in PA/GF/MMT composite. In this case the impact strength of the material was increased, but at the cost of reduced ductility. That is why the nano-composite became even more brittle than neat PA/GF composite. Hence, the strain in radial direction reached the maximum allowable limit and the axial cracks propagated along the height of the structure, leading to a complete failure of the structure. As a result the energy absorption capability of the material remained on the same level as for the neat PA/GF, in spite of the increase in properties such as: impact strength, tensile strength and stiffness.

The biggest increase in the SEA parameter was found in the PA/GF/GS composite. Similar to the SiO_2 reinforced PA/GF, the fracture mode has changed from Mode I to Mode II, after the addition of the secondary reinforcement, but the toughening mechanism was different. In this instance, the properties such as: stiffness, impact and tensile strength were improved, but with reduced elongation to brake, analogously to PA/GF/MMT composite. However, the increase in stiffness was much more significant, and additionally, the material was subjected to delamination effect. As a result the radial stress did not initiate any axial cracks, due to the high resistance of the material and propagation of the interlaminar cracks. That is why the crashing length of the cone was importantly reduced increasing the value of the SEA parameter.

3.2. SEM Analysis

The fracture surface of the crash cones tested under dynamic load was examined using SEM. In all PA6 composites the fracture was dominated by matrix and fibre cracking. The glass reinforcement was covered with polymer residuals, which was a sign of good interfacial adhesion. Moreover, there was a visible difference in the fracture mode between various PA6 composites. In neat PA/GF and PA/GF/SiO_2 composites the plastic deformation of the matrix was the most evident and the fibres pull out and debonding was of little meaning. An extensive plastic deformation was clearly visible in PA/GF/SiO_2 as a non-smooth texture and characteristic deformation paths. Contrary, in PA/GF/MMT and PA/GF/GS composites, the plastic deformation of the matrix was reduced, due to the transition to more brittle failure. There were also visible signs of fibre pull out and debonding.

4. Conclusion

It has been shown that addition of secondary reinforcement into glass-fibre reinforced polymer composites can have a significant influence on the energy absorption capabilities of the material. The carried out experiments showed that by changing the secondary filler material it is possible to change the micro-mechanism of the crash and therefore control the energy absorption characteristics of the composite. The following general remarks could be drawn, regarding the energy absorption of polymer composites: (i) Secondary reinforcement in PA6 composites leads to an increase in energy absorption capabilities of the structure. (ii) The transition from brittle to ductile fracture mode was clearly demonstrated as a main reason for the increased energy absorption capabilities. (iii) Two different toughening mechanisms were observed. First, due to the increase in elongation to brake. Second, due to the increase in the material impact strength and stiffness.

5. Acknowledgments

This work was funded by the European Commission (FP7 Project-CP-FP; Project Reference: 228536-2). The authors also gratefully acknowledge the EPSRC for testing equipment provided, as well as Grado Zero Espace SRL and Laviosa Chimica Mineraria SPA for the preparation of the materials.

References

[1] Yuan, Q. and Misra, R. Impact fracture behavior of clay-reinforced polypropylene nanocomposites. *Polymer,* vol. 47, no. 12, pp. 4421-4433 (2006).

[2] Sun, L., Gibson, R. F., Gordaninejad, F. and Suhr, J. Energy absorption capability of nanocomposites: A review. *Composites Science and Technology,* vol. 69, no. 14, pp. 2392-2409 (2009).

[3] Njuguna, J., Michałowski, S., Pielichowski, K., Kayvantash, K. and Walton, A. C. Fabrication, characterization and low-velocity impact testing of hybrid sandwich composites with polyurethane/layered silicate foam cores. *Polymer Composites,* vol. , no. , pp. (2011).

[4] McWilliams, A. Nanocomposites, nanoparticles, nanoclays, and nanotubes. *NANO21C, BCC Research, Norwalk, CT,* vol. , no. , pp. (2006).

[5] Mamalis, A. G. *Crashworthiness of composite thin-walled structural components,* th ed, Technomic Pub. Co., Lancaster, Pa. (1998).
[6] Crosby, A. J. and Lee, J. Y. Polymer nanocomposites: the "nano" effect on mechanical properties. *Polymer reviews,* vol. 47, no. 2, pp. 217-229 (2007).
[7] Ng, C., Schadler, L. and Siegel, R. Synthesis and mechanical properties of TiO2-epoxy nanocomposites. *Nanostructured Materials,* vol. 12, no. 1-4, pp. 507-510 (1999).
[8] Lin, J. C. Investigation of impact behavior of various silica-reinforced polymeric matrix nanocomposites. *Composite Structures,* vol. 84, no. 2, pp. 125-131 (2008).
[9] Deshmane, C., Yuan, Q., Perkins, R. and Misra, R. On striking variation in impact toughness of polyethylene-clay and polypropylene-clay nanocomposite systems: The effect of clay-polymer interaction. *Materials Science and Engineering: A,* vol. 458, no. 1-2, pp. 150-157 (2007).
[10] ZHANG, M., ZENG, H., ZHANG, L. and LIN, G. Fracture characteristics of discontinuous carbon fibre-reinforced PPS and PES-C composites. *Polymers & Polymer Composites,* vol. 1, no. 5, pp. 357-365 (1993).
[11] Zhang, H., Zhang, Z., Friedrich, K. and Eger, C. Property improvements of in situ epoxy nanocomposites with reduced interparticle distance at high nanosilica content. *Acta Materialia,* vol. 54, no. 7, pp. 1833-1842 (2006).
[12] Wetzel, B., Haupert, F. and Qiu Zhang, M. Epoxy nanocomposites with high mechanical and tribological performance. *Composites Science and Technology,* vol. 63, no. 14, pp. 2055-2067 (2003).
[13] Guo, Y. and Li, Y. Quasi-static/dynamic response of SiO2-epoxy nanocomposites. *Materials Science and Engineering: A,* vol. 458, no. 1-2, pp. 330-335 (2007).
[14] Han, J. and Cho, K. Nanoparticle-induced enhancement in fracture toughness of highly loaded epoxy composites over a wide temperature range. *Journal of Materials Science,* vol. 41, no. 13, pp. 4239-4245 (2006).
[15] Viana, J. Polymeric materials for impact and energy dissipation. *Plastics, Rubber and Composites, 35,* vol. 6, no. 7, pp. 260-267 (2006).

IOP Conf. Series: Materials Science and Engineering **40** (2012) 012011 doi:10.1088/1757-899X/40/1/012011

Dielectric behavior of Silica/Poly(dimethylsiloxane) nanocomposites. nano size effects

I A M Ibrahim[1], A A F Zikry[1], M A Sharaf[1], J E Mark[2], K Jacob[3], I M Jasiuk[4] and R Tannenbaumn[3]

[1]Department of Chemistry, College of Science, Helwan University, 11795 Egypt.
[2]Department of Chemistry, University of Cincinnati, Cincinnati, OH 45221-0172, USA.
[3]School of Materials Science and Engineering, Georgia Institute of Technology, Atlanta, GA 30332-0245 USA.
[4]Department of Mechanical Science and Engineering, University of Illinois at Urbana-Champaign, Urbana, IL 61801 USA.

Abstract. The enhancement of properties of elastomeric composite materials is very much dependent on the size and the surface modification of the reinforcing filler inclusions. It is well accepted that the reinforcement effects are primarily due to molecular interactions of the polymeric matrix and the filler inclusions and it involves both chemical and physical interactions.

In the present study, we have incorporated silica nano fillers (Stober silica) into poly(dimethylsiloxane) (PDMS) elastomeric networks. The dielectric properties of the networks were investigated as a function of filler nano filler size, volume fraction, and surface treatment by hexamethyldisilazane. The broad-spectrum dielectric properties (in particular, the dielectric constant, the dielectric loss, and tan δ) were characterized. These properties were found to be dependent on the size of the nano inclusions; thus, the results clearly showed a nano size phenomenon that was a highlight of the present research. Certainly, that can be largely attributed to the high specific surface area of the nano fillers, which significantly leads to a pronounced increase in interfacial interactions. Also, and as would be expected, the dielectric properties of the polymeric networks filled with unmodified particles were different from those for the polymeric networks filled with surface-modified particles. Again, this is mainly attributed to changes in the surface properties. The expected dependence of properties of the nanocomposite networks on the nature of the filler and its concentration has thus been demonstrated.

1. Introduction

Polysiloxanes (or "silicones") are semi-inorganic polymers with the repeat unit [-S'RR'-O-] (with various alkyl or aryl R and R'- groups). These materials have many unusual properties, such as good thermal stability, unusually low viscosity-temperature coefficients, unusually high permeabilities, low surface tensions, good dielectric properties, good weather resistance, good release and lubrication properties, good biocompatibility, and good optical clarity [1-3]. The most important feature in the present context is the fact that they generally exhibit rubberlike elasticity, and maintain their elastomeric properties down to very low temperatures.

Poly(dimethylsiloxane) (PDMS), [-Si(CH_3)$_2$O-], is by far the most important and most widely explored and utilized polysiloxane, both scientifically and commercially [1–4]. Its extremely low glass transition temperature and high thermal stability over a wide range of temperatures makes it a high-performance material, but its elastomers in the unfilled state generally have poor mechanical strength [2–5]. As a result, most applications require that PDMS elastomers be reinforced by fillers, in order to improve their mechanical properties [6–8].

The reinforcement of PDMS elastomers by fillers has been the subject of numerous studies [9, 10]. Earlier work on filled PDMS concentrated on the effects of fillers on macroscopic properties and involved several polysiloxanes [11, 12]. The role of molecular structure at the polymer–filler interface, however, on the filler effect has been considered only recently. Notably, Litvinov et al. used NMR to study the structure and behavior of PDMS chains grafted onto silica. They found immobilized chain segments at the PDMS–silica interface, but mobile chain portions that were located beyond the interface [13]. Chains at the interface were found to have substantial decreases in heat capacity at the glass transition temperature T_g. Thus, some of the PDMS chain units were immobilized as a result of physical adsorption at the silica surface, the amount being proportional to the number of silanol groups at the silica surface. Gussoni et al. also used NMR spectroscopy to investigate silica-filled PDMS elastomers, and observed a decrease of segmental mobility of the PDMS chains in the vicinity of the filler surface [14]. Berriot et al. employed H NMR experiments on filled elastomers and observed a layer of immobilized segments at the particle surface [15]. In the case of chains chemically grafted to the silica surface. Berriot et al. observed immobile layers, which they described as a glassy shell around the filler surface. Cosgrove et al. used differential scanning calorimetry and NMR to show that the reduction in the mobility of the PDMS chains corresponded to a shift in T_g to higher temperatures [16]. These shifts are relatively small but they, nevertheless indicate the presence of a gradient of segmental mobility in going from the elastomer towards the silica surface.

In the present report, we describe the incorporation of Stober silica into PDMS elastomers and the values of their dielectric properties as a function of filler size, modification, and concentration. Silica particles that had been prepared under ambient conditions using sol-gel techniques were then blended into the PDMS networks precursor, prior to their crosslinking. The dielectric constant, dielectric loss, and tan δ related to this loss were determined for the filled (both unmodified and modified) and unfilled PDMS elastomers. The silica particle dispersion in the polymer network was characterized using transmission electron microscopy (TEM).

2. Experimental

Materials

The materials required for the synthesis of silica nanoparticles are described as follows. Tetraethyl orthosilicate (TEOS 98%) was obtained from the Fischer Company, Ammonium hydroxide solution 31.5% (NH_3) was used as received, from the Aldrich Company, and absolute ethanol (99 %) from the ADWIC Company was used after distillation.

The materials required for preparation of the organically modified silica are as follows. The silane coupling agent (hexamethyldisilazane) was obtained from Fluka company and toluene from the ADWIC Company.

The materials required for synthesis of the PDMS networks are as follows. α, ω-hydroxy-terminated PDMS network precursor of average number molecular weight M_n = 18,000 g/mol, was obtained from the UCT (Petrarch) Company. The catalyst bis (2-ethyl hexanoate) tin used to synthesize the PDMS networks was obtained from the Aldrich Company. Dichloromethane was obtained from the BDH Company, and all the chemicals were directly used without any further purification.

2.1. Preparations of Silica Nanoparticles

Silica nanoparticles were prepared by hydrolysis and condensation of TEOS in ethanol, and in the presence of ammonia as catalyst [17]. First, solutions containing appropriate quantities of absolute ethanol, ammonia, and deionized water were stirred for 5 min to ensure complete mixing. Then the proper amount of TEOS in absolute ethanol was added to the above solution and the reaction permitted to proceed at ambient temperature for 24 hr. The resulting colloidal solution was separated into components by high-speed centrifuge, and the silica particles obtained were washed in absolute ethanol three times to remove undesirable particles, The silica particles were then dried in an oven at 100 °C for 2 hr to prevent continuing reactions.

2.2. Preparation of Organically-Modified Silica

After thermal treatment at 150°C for 2 hr to remove physisorbed water, a chosen amount of silica was put into a glass flask containing the desired amount of toluene. The liquid reagent (10% hexamethyldisilazane in toluene) was then added and the mixture was stirred and refluxed at 110°C for 3 hr. After treatment, the solid was filtered and repeatedly washed with toluene and acetone in order to remove any unreacted silane coupling agent. It was finally dried at 150°C for 2h.

2.3. Network Synthesis

The hydroxyl-terminated PDMS was end-linked into networks by reacting it with a stoichiometric amount of crosslinking agent, tetraethoxysilane (TEOS). The end-linking reaction was performed in solution to facilitate the mixing process and the samples were sonicated to obtain better dispersion of the silica particles. The hydroxyl-terminated PDMS was dissolved in dichloromethane. The desired amount of silica particles was dispersed in dichloromethane and sonicated for 15 min and was then added to the polymer solution. The mixtures were stirred for 1 hr to obtain a homogeneous dispersion. The solvent was removed before the cross-linking process, and then TEOS and the catalyst, (0.5wt%) bis (2-ethyl hexanoate) tin, were added to the dispersion with stirring. The cross-linking reactions were carried out in a Teflon mold at room temperature for 2 days. Each elastomer was then exhaustively extracted for five days with toluene in order to remove any soluble and/or extractable material. After deswelling using methanol as the non-solvent the networks were dried to constant weight. The samples were removed and stored at room temperature in air- tight vials to prevent adsorption of moisture from the surroundings.

2.4 Transmission Electron Microscope Analysis

In order to investigate the dispersion of the silica nanoparticles in the respective elastomers, the transmission electron microscope (TEM) analyses were performed. TEM analyses of the silica particles were carried out using a JEOL JEM 2010 unit on both the unmodified silica and organically-modified silica. To prepare samples for the TEM analysis, silica powder was dispersed in absolute ethanol and a drop of this silica colloidal mixture was placed on a copper grid coated with carbon. The solvent was evaporated at room temperature, leaving the silica particles on the grid.

The sample preparation was designed to produce thin sections having a thickness less than 1000 A°. They were cut from bulk pieces of the nanocomposite using either a glass or a diamond knife on an ultramicrotome at -125 to -130°C. Specimens were collected on standard copper grids, which were then coated with carbon to increase their conductivity.

2.5. FTIR Analysis

Fourier transform (FTIR) analyses were carried out on both unmodified silica and organically-modified silica using an ATI Mattson model Genesis series (USA) infrared spectrophotometer adopting KBr technique. For all samples, a small amount of silica was placed between the two electrolyte plates to obtain transmission spectra.

2.6. Dielectric Relaxation Measurements

Before each measurement, the cell was cleaned by washing it several times with redistilled water, and then rinsed repeatedly with redistilled ethyl alcohol and ether. Then it was dried by passing a stream of dried filtered air over it.

Values of the dielectric constant and dielectric loss were measured using a HIOKI (3532) LCR type which has a frequency band from 42 to 5 MHz with high resolution. This device is an impedance meter, which uses a touch panel that simplifies operation.

The dielectric constant ε' and dielectric loss ε" were calculated from the equations [18,19],

$$\varepsilon' = \frac{d \times C}{A \times 8.85 \times 10^{-12}} \tag{1}$$

$$\varepsilon'' = \varepsilon' \times D \tag{2}$$

where d is the thickness of the sample in meters, and A is its area in m^2. C is the capacitance of the sample in farads, and D = tan δ is the loss coefficient as obtained directly from the device.

3. Results and Discussion

3.1. Transmission electron microscopic analyses of the silica particles

Figure 1 shows the TEM image of unmodified silica nanoparticles [sizes of ~65nm]. The resulting particles in the product suspension were spherical in shape and nano-sized. The figure also shows the dispersion of the particles and the absence of any significant aggregation among particles. This is inherent in the method of preparation in alkaline medium [17].

Figure 2 shows the TEM image of silica nanoparticles whose surface has been modified with hexamethyldisilazane. The figure shows the desired dispersion of the nanoparticles. As would be expected, this is attributed to the fact that the particles posses organophilic character after treatment with hexamethyldisilazane where the majority of the surface OH groups are converted to [$-(CH_3)_3$] groups. Almost certainly, such hydrophobic groups resulted in electrostatic repulsions and in dispersions of the particles.

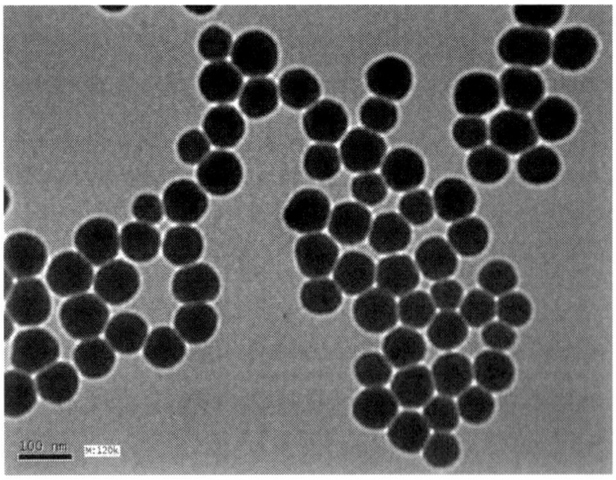

Figure 1. TEM image of unmodified silica nanoparticles (experimental conditions: TEOS: 0.18 M; NH_3: O.2 M; H_2O: 1 M). [Particle size ~ 65 nm].

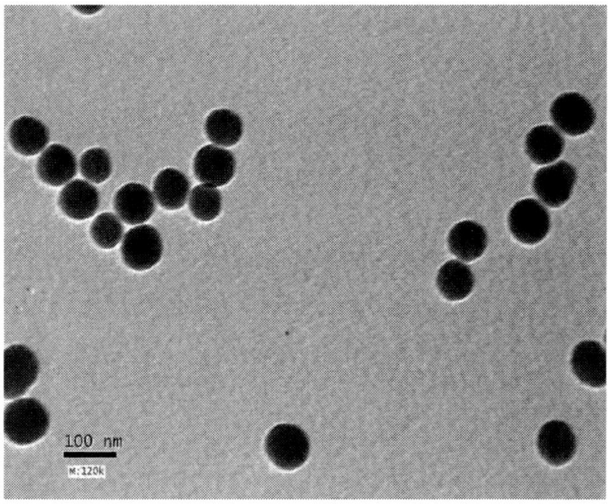

Figure 2. TEM image of organically modified silica nanoparticles (experimental conditions: TEOS: 0.18 M; NH_3: O.2 M; H_2O: 1 M, 10% HMDS in toluene). [Particle size ~ 65 nm].

3.2. FTIR Analysis

Figures 3 and 4 show the FTIR spectra of both untreated silica particles and those treated by the coupling agent. For untreated nano silica particle, there is absorption band around 3400 cm^{-1}, which corresponds to Si–OH groups [20]. Also the peak at 1050 - 1250 cm^{-1} is due to an asymmetric stretching and vibration of Si-O-Si bonds [21]. After the surface treatment, a decrease in the intensity of the Si-OH bands has been observed. Also , the occurrence of absorption bands around 2950 cm^{-1} and 1400 cm^{-1} are due to the stretching and bending of C-H bonds, and the occurrence of peaks around 847 cm^{-1} are due to the Si-C bonds [22] which confirm the conversion of the surface OH groups to [-$(CH_3)_3$] groups.

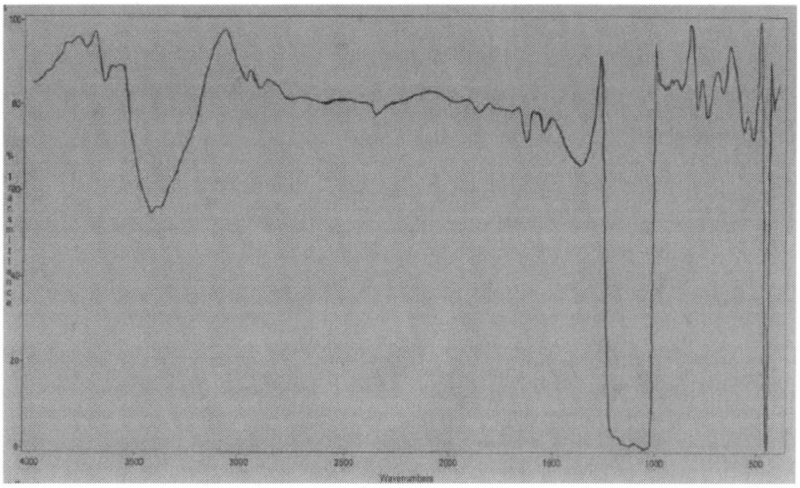

Figure 3. FTIR spectra of unmodified silica nanoparticles.

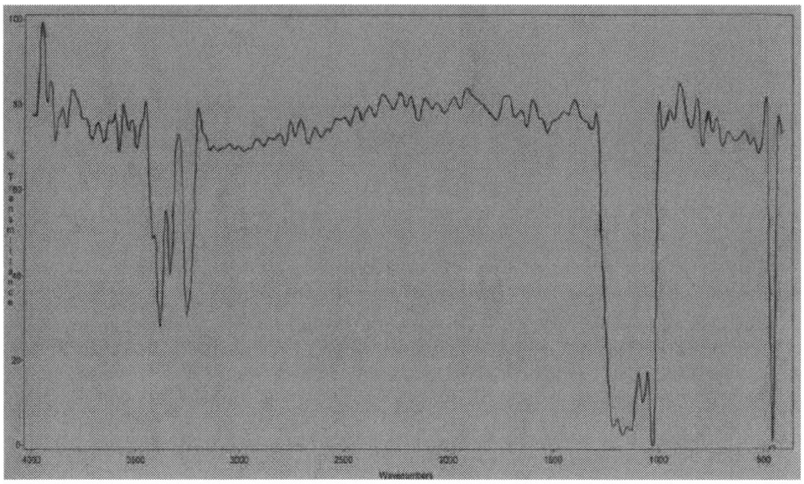

Figure 4. FTIR spectra of organically-modified silica nanoparticles.

3.3. Transmission electron microscope analysis of composites

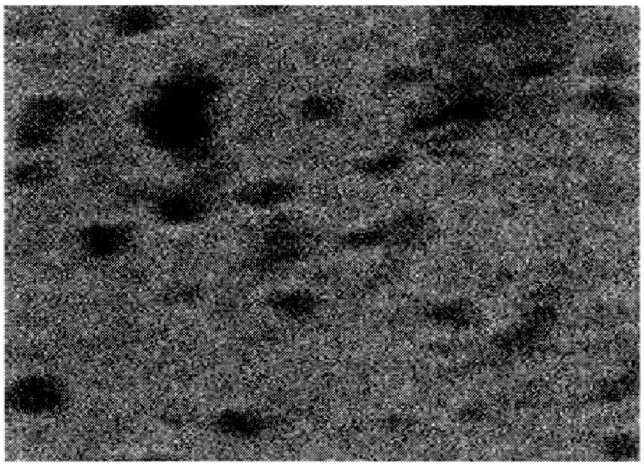

Figure 5. TEM of 5wt% unmodified silica PDMS nanocomposites *par*ticle size around 50 nm].

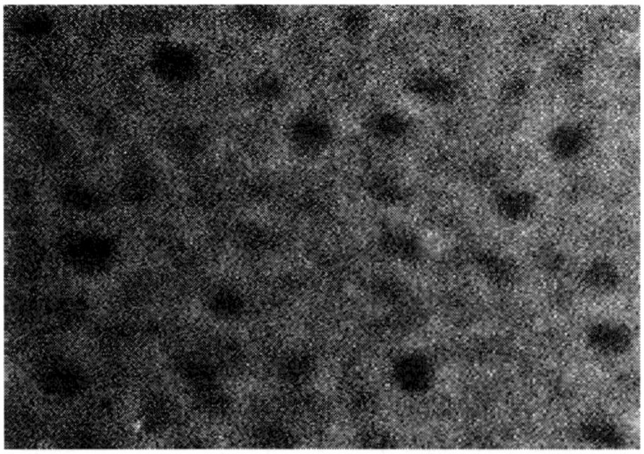

Figure 6. TEM of 5 wt% organically-modified silica-PDMS nanocomposites [Particle size around 50 nm].

The dispersion of the unmodified and organically-modified silica nanoparticles in the respective PDMS elastomeric networks was studied by TEM. A thin section of the elastomer had been frozen and cut using an ultramicrotome, and then analyzed. The data obtained by TEM analyses are presented in Figures 5 and 6. In the case of the unmodified silica nanoparticles, good dispersion of the particles in the PDMS networks was obtained (as shown in Figure 5). As expected, when the silica had been treated with the silane coupling agent, the particles were well dispersed in the PDMS networks, (as shown in Figure 6).

3.4. Dielectric properties

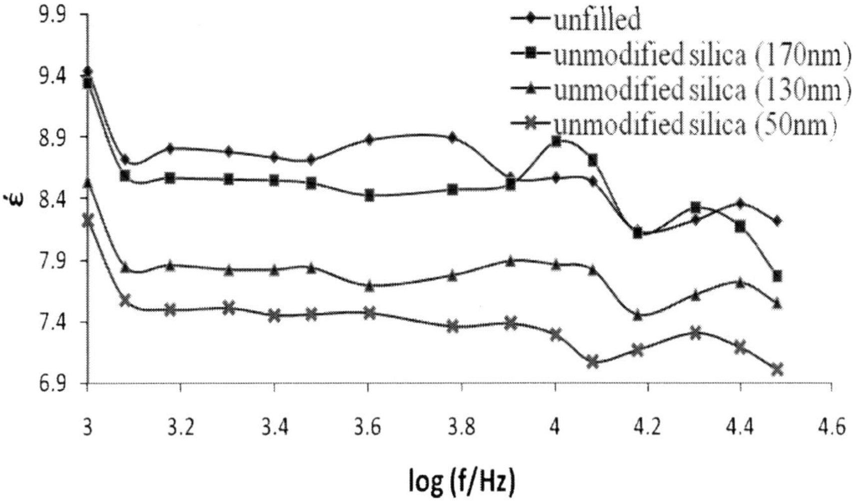

Figure 7. Dielectric constant of unfilled and 5wt% unmodified silica filled PDMS elastomers as a function of frequency at 298 K

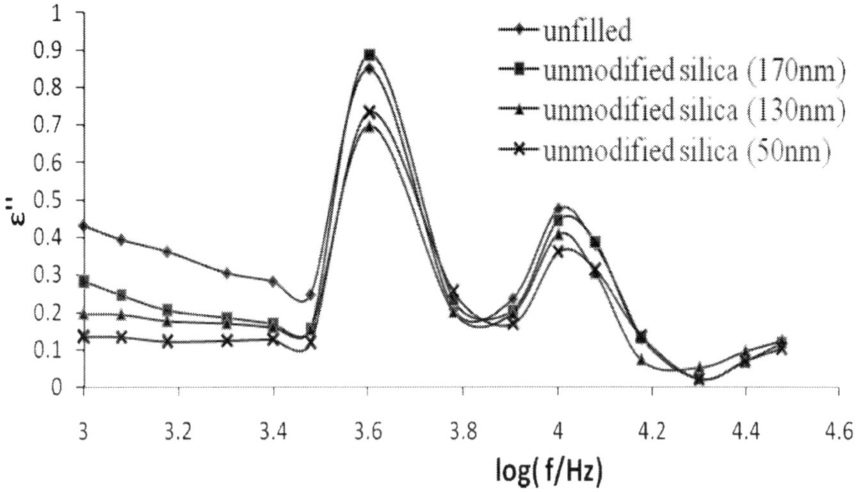

Figure 8. Dielectric loss of unfilled and 5wt% unmodified silica filled PDMS elastomers as a function of frequency at 298 K

This section describes the results of the broad-band spectrum dielectric measurements. Values of the dielectric constant έ and the dielectric loss ε" for unfilled PDMS and unmodified and modified silica/PDMS nanocomposite networks were obtained. In figure 7, the dependence of dielectric constant έ on the frequency for unmodified silica/PDMS nanocomposite networks is shown. The measurements covered a frequency range of $1x10^3 - 3x10^5$ Hz, at room temperature. It is to be noted here, that values of the dielectric constant of the nanocomposites are smaller than those for the corresponding unfilled network. This reduction in the dielectric constant values can be attributed to restrictions in the mobility of the polymer chains. As would be expected, the introduction of the nanosized particulate fillers resulted in a limited ability of the polarizable groups or segments in the nanocomposites to orient rapidly enough to keep up with the oscillations of an alternating electric field [23].

The dielectric constant values were found to decrease with decrease in the size of both the modified and unmodified silica nanoparticles. This expected behavior is due to an inadvertent increase in the interfacial area and accompanying increase in the degree of interaction between the filler and the polymer. This leads to an additional restriction on the mobility of the polar groups (as shown in figure 7).

It is evident from these figures that the dielectric constant at some value of the frequency decreases with increasing frequency. This may be due to the fact that at high frequencies, the polar groups in the polymer are not mobile at these short time scales and do not orient readily to provide a discernible change in polarization. As the frequency is decreased, the time scales increase and the most mobile groups begin to orient, thus an increase in the dielectric constant, ε', value was observed. This behaviour has also been observed in the study of the dielectric behavior of dimethacrylate networks [24].

One should note that values of the dielectric loss and the dielectric constant, in some instances, did not increase with an increase in frequency. This behavior could be interpreted in terms of the fact that the dipoles would cooperatively have sufficient rotational energy to permit passage over the energy barrier. Clearly, this would result in an apparent contribution to the orientation polarization [25].

Figure 8 show the dependence of dielectric loss ε" on the frequency for the unfilled polymer network and for the unmodified silica/PDMS nanocomposite networks. The magnitude of the dielectric loss is an important material parameter in the construction of capacitors. Ideally, the ε" value in a capacitor should be as low as possible [26]. It can be noted that the dielectric loss values of the nanocomposites are lower than those for the corresponding unfilled networks. This reduction in the dielectric loss values is attributed to the restriction in the mobility of the polymer chains subsequent to the introduction of the nanosized filler inclusions. Also, the dielectric loss values are shown to decrease with decrease in the size of both the unmodified and modified silica nano particles. Again, this would be expected owing to an increase in the interfacial area and subsequent increase in the degree of interaction between the filler and the polymeric matrix chains. Clearly, such interactions will result in restrictions on the mobility of the polar groups. In addition, a major relaxation peak is observed and had been generally attributed to the normal-mode relaxation behavior of the PDMS [27, 28]. Also, an additional minor relaxation peaks was observed at a relatively higher frequency that could be primarily attributed to the presence of dangling chains. It is noteworthy here that the nanocomposite networks had undergone exhaustive solvent extraction procedures. Nonetheless, there is a slight possibility of the presence of some unreacted trapped chains that could plausibly contribute to this minor relaxation, as well. Figures 7 and 8 clearly manifest a size phenomenon that affects the normal relaxation-mode behavior of networks having different nanosize silica particles, at the same volume fraction, for the untreated particles. Certainly, this nano size effect phenomenon is a clear indication of the role of the nano-size of the particles on the interfacial behavior in nanocomposites. More restrictions are clearly apparent on the mobility of the polar groups with a size decrease.

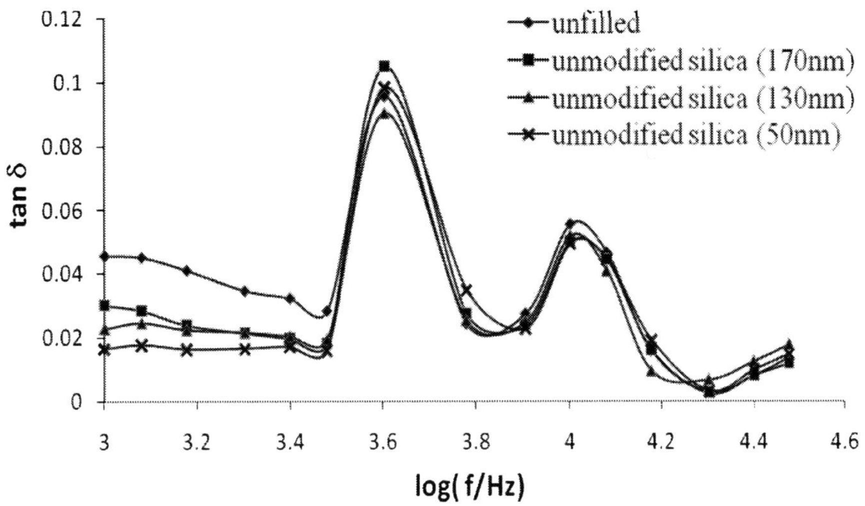

Figure 9. Tan δ for unfilled and 5 wt% unmodified silica-filled PDMS elastomers as a function of frequency at 298 K

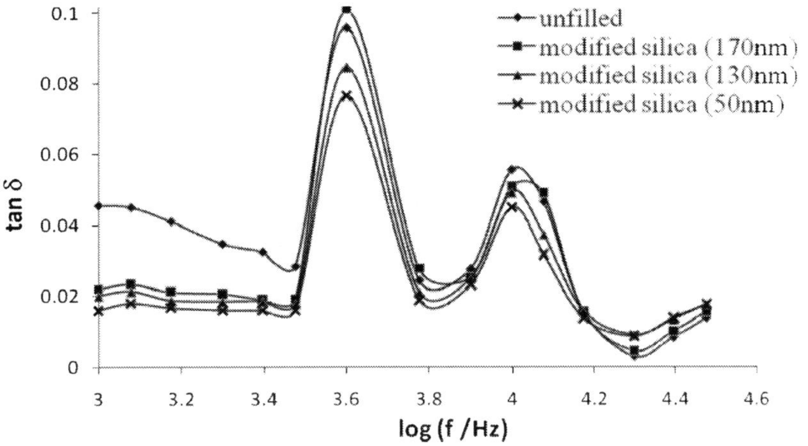

Figure 10. Tan δ for unfilled and 5wt% modified silica filled PDMS elastomers as a function of frequency at 298 K.

Similar trends in the dielectric constants ε' and ε'' was also manifested for the surface-modified nano particles.

Figures 9 and 10 show the dependence of tan δ on the frequency for the unfilled polymer and the unmodified and modified silica/PDMS nanocomposites. This provides further confirmation of the preceding interpretations of behavior observed for the dielectric loss.

In Figure 11, it is clearly apparent that values of tan δ for the surface-modified silica/PDMS nanocomposites networks are smaller than those for the case of the unmodified nano silica particles. It is clear that such behavior is due to decrease in the percentage of the polar hydroxyl groups resulting from surface modification. It is worth noting that the behavior of tan δ is size dependent, even at the same volume fraction of filler.

Also, similar trends and interpretations to those cited above apply to the dielectric loss and dielectric constant values for the PDMS nanocomposite networks filled with unmodified and surface-modified nano silica (as shown in Figure 11). Again, it is worth noting here that the behavior of the dielectric constant, dielectric loss, and tan δ is size dependent, even at the same volume fraction of filler.

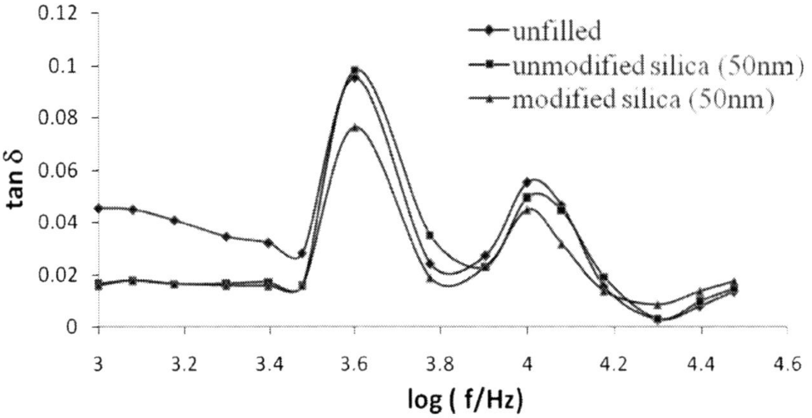

Figure 11. Tanδ of unfilled and 5 wt% unmodified and modified silica (50nm) filled PDMS elastomer as a function of frequency at 298 K

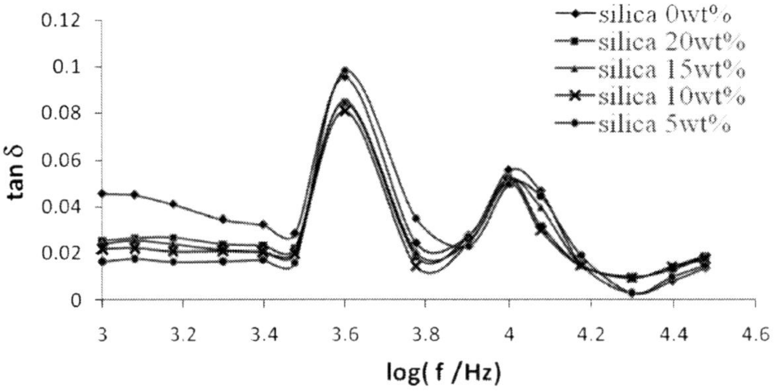

Figure 12. Tan δ of unfilled and 5, 10, 15 and 20 wt% unmodified silica (50nm) filled PDMS elastomers as a function of frequency at 298 K

In Figure 12 the dependence of tan δ for the unmodified silica/PDMS nanocomposites on frequency is illustrated for various contents of unmodified silica. Again, it noteworthy that an increase in the content of unmodified silica will lead to an inconspicuous increase in the dielectric loss and tan δ owing to an increase in the number of polar OH groups.

Siimilar trends in the dielectric constants ε' and ε'' have been reported for silica/PDMS nanocomposites having different volume fractions of the particles.

4. Conclusion

Values of the dielectric constant and the dielectric loss for silica/PDMS nanocomposite networks exhibited a decrease, as would be expected, in comparison to those for the unfilled polymer. Again, as would be expected, the dielectric properties, namely the dielectric constant and dielectric loss, were lower for the networks incorporating surface-modified silica in comparison to those reported in the case of incorporation of unmodified silica. Two peaks were observed in the broad-spectrum dielectric loss measurements. Both are attributed to the normal-mode relaxations, with the major one attributed to nanocomposite network chains and the minor one to dangling and uncross-inked chains. This has been also confirmed by the tan δ measurements.

The transmission electron microscope (TEM) micrographs clearly indicated that surface modification improved the dispersion of the nano particles. In the case of unmodified silica, however, a tendency for agglomeration of the nano particles was obvious.

One of the highlights of this work is the observation of a size phenomenon with regard to the broad-spectrum dielectric properties of the nanocomposite networks. These properties showed a discernible dependence on the nanosize of the filler particles. Such results clearly demonstrate the paramount role played by the interface and its subsequent effects on the properties in nano systems.

5. Acknowledgments

MAS acknowledges the US Egypt Science and Technol. Joint Fund through grant (MA7-001-002) for support of this research. Partial support was also received through US-Egypt Grant INT-01-11334. It is also a pleasure to acknowledge the financial support provided JEM by the National Science Foundation through Grant DMR-0803454 (Polymers Program, Division of Materials Research).

References

[1] A. J. Barry, H. N. Beck, "Inorganic Polymers", Academic Press, New York (1968).
[2] J. E. Mark, H. R. Allcock, R. West, "Inorganic Polymers", Prentice Hall, Englewood Cliffs, NJ, (1992).
[3] S. J. Clarson, J. A. Semlyen, "Siloxane Polymers", Prentice Hall, Englewood Cliffs, NJ, (1993).
[4] W. Noll, "The Chemistry and Technology of Silicones", Academic Press, New York (1968).
[5] J. E. Mark, M. Kato, J. Ko, J. Polym. Sci., Polym. Symp., 54, 217 (1976).
[6] K. E. Polmanteer, C. W. Lentz, Rubber Chem. Technol., 48, 795 (1975).
[7] E. M. Dannenberg, Rubber Chem. Technol., 48, 410 (1975).
[8] M. P. Wagner, Rubber Chem. Technol. 49, 703 (1976).
[9] W. Q. Yuan, J. E. Mark, Macromol. Chem. Phys., 200, 206 (1999).
[10] J. M. Breiner, J. E. Mark, Polymer, 39, 5483 (1998).
[11] J. E. Mark and B. Erman, "Rubberlike Elasticity. A Molecular Primer", New York: Wiley Interscience; p. 61(1988).
[12] B. Erman, J. E. Mark, "Structures and Properties of Rubberlike Networks", Oxford University Press, (1997).
[13] W. M. Litvinov, H. Barthel, J. Weis, Macromolecules, 35, 4356 (2002).
[14] M. Gussoni, F. Grecp, M. Mapell, A. Vezzoli, E. Ranucci, P. Ferruti P, Macromolecules, 35, 1722 (2002).
[15] J. Berriot, F. Lequeux, L. Monnerie, H. Montes, D. Long, P. Sotta, J. Non-Cryst. Solids, 307–310, 719 (2002).
[16] T. Cosgrove, C. Roberts, T. Garasani, G. R. Schmidt, V. G. Gordon. Langmuir;18, 10080 (2002).
[17] W. Stober, A. Fink and E. Bohn, Journal of Colloid and Interface Science, 26, 62 (1968).
[18] F. Kremer and A. Schönhal, "Broadband Dielectric Spectroscopy", Springer-Verlag Berlin Heidelberg; p. 1 (2003).
[19] A. R. Blythe and David Bloor "Electrical Properties Of Polymers", 2nd edition, Cambridge University Press, New York; p. 29 (2005).
[20] B. Y. Park, J. E. Mark, polymer (Korea), 13, 597 (1989).
[21] J. K. Hong, H. R. Kim, H. H. Park, Thin solid Films, 332, 449 (1998).
[22] N. Hering, K. Schriber, R. Riedel, O. Lichtenberger, J. Wolters-Odorf, Appl. Organometal. Chem., 15, 879 (2001).
[23] J. Huang, P. Lim, P. Pallathaka, K. Zeng, C. He, Acta Materialia, 53, 2395 (2005).

[24] R. Anandkumar, R. Kannurpatti, C. N. Bowman, Macromolecules, 31, 3311 (1998).
[25] A. A. F. Zikry, International Journal of Polymeric Materials, 57, 383 (2008).
[26] R. Popielarz, C. K.Chiang, R. Nozaki, J. Obrzut, Macromolecules, 34, 5910 (2001).
[27] J. P. Lewicki, M. Patel, P. Morrell, J. Liggat, J. Murphy, R. Pethrick, Sci. Technol. Adv. Mater., 902, 4403 (2008).
[28] J. P. Lewicki, R. S. Maxwell, M. Patel, J L. Herberg, A. C. Swain, J. Liggat, R. A. Pethrick, Macromolecules, 41, 9179 (2008).

IOP Conf. Series: Materials Science and Engineering **40** (2012) 012012 doi:10.1088/1757-899X/40/1/012012

Physical characteristics of nanoparticles emitted during drilling of silica based polyamide 6 nanocomposites

Sophia Sachse[1], Francesco Silva[1], Adeel Irfan[2], Huijun Zhu[2], Krzysztof Pielichowski [3], Agnieszka Leszczynska [3], Maria Blazquez [4], Olga Kazmina [5], Oleksandr Kuzmenko [6] and James Njuguna[1*]

[1] Department of Environmental Science and Technology, Cranfield University, Bedfordshire MK43 0AL, UK
[2] Cranfield Health, Cranfield University, Bedfordshire MK43 0AL, UK
[3] Department of Technology of Polymers, Cracow University of Technology, ul. Warszawska 24, 31-155 Krakow, Poland
[4] Ekotek Ingenieri y Consultoria Medioambiental, Erandio, Bizkaia, 48950 Spain
[5] Department of Silicate Technology and Nanotechnology, Tomsk Polytechnic University,634050 Tomsk, Russia
[6] Palladin Institute of Biochemistry, National Academy of Sciences of Ukraine (NASU), Kiev, 01601 Ukraine

E-mail: j.njuguna@cranfield.ac.uk (*Corresponded Author:)

Abstract.
During the past decade, polymer nanocomposites have emerged as a novel and rapidly developing class of materials and attracted considerable investment in research and development worldwide. However, there is currently a lack of information available in the literature on the emission rates of particles from these material. In this study, real-time characterization of the size distribution and number concentration of sub-micrometer-sized particles (5.6-512 nm) emitted from polyamide 6 nanocomposites during mechanical drilling was made. For the first time, four different silica based filler of commonly use were assessed. Further, the respective emission rates were determined based on the particle population and the time. The measurements showed that the particle emission rates ranged from 1.16E+07 (min^{-1}) to 1.03E+09 (min^{-1}) and that the peak diameters varied from 29.6 to 75.1 nm. Airborne particles in the nanometer range (11.1-46.8 nm), in the ultrafine range (51.3-101.1 nm) and in the accumulation mode range (111.9-521 nm) accounted for 34.1% to 76.6%, 8.3% to 47% and 4.1% to 24.2% of the total emission rates, respectively, depending on the type of filler. Additionally, deposited particles were sampled and characterized, to explore any possible correlation between deposited and airborne particles. The result clearly showed that with increasing airborne particle concentration the deposit particle concentration decreased and vice verse.

1. Introduction

Polymer nano-composites have proven huge potential in both mechanical and thermal application J. et al. [2010], Njuguna et al. [2011]. The global consumption of nanocomposites is estimated to exceed 214,081 metric tons and 1.38 billion USD, by 2014BCCResearch [2010]. Of which the consumption of clay-based polymer nanocomposites will increase to 181,094 metric tons and 692.3 million USD by 2014. For example, one of the leading automotive manufacturers is using 300000 kg of nanoclay composites annually for various automotive exterior part/panel applications at presentBCCResearch [2010]. However, only little attention has been payed to the end user contact with nanocomposites by science and regulators. Despite the worst case scenario of a nanocomposite being the encounter with a do it yourself (DIY) worker C. et al. [2010]. The DIY-worker's personal protection in general consist of an low-cost filter mask, which might be inadequate for protection against nanoparticle particles inhalation. In the car tuning scene, a common way to reduce the weight of a car is the drilling of small holes. Further other car modifications e.g. integration of a new stereo, system or body kit are imaginable and will involve drilling into those nanocomposites Scott J. [2006]. However, during drilling, shear forces may detach free nanofillers as ultrafine airborne particles. Due to their size, these particles are able to remain airborne for a long period M. [2004] and are suspected to enter the human cells more easily via respiratory, dermal or oral absorptionA. et al. [2011]. The high biological activity can be explained by the large surface-to-volume ratio; hence many types of nanoparticles have shown toxic impact A. et al. [2011], L. [2004], G. [2010], D. et al. [2011]. In the current work, airborne particles emitted by silica filled polyamide 6 nanocomposites by mechanical drilling were investigated in a controlled environment. The particle number concentration and the size distribution were measured continuously in real time using a particle sizer. In addition, the particles emission rates were evaluated as this information could be useful in assessing the resulting human exposure.

2. Methodology

2.1. Materials

Polyamide 6 (Tarnamid T30, Azoty Tarnow, Poland) was used as matrix material for all composites. As fillers the following materials were utilised (5wt.%); montmorillonite layered nanosilicates (Laviosa/Dellite 43B), nanosilica (Degussa/AEROSIL 200), foam-glass-crystal-materials (Produced by Tomsk Polytechnic University) and glass fibres (Taiwan glass Industry Co./ 473H).

2.2. Nanocomposite manufacturing

A twin-screw extruder (ZMK/116/10, ZAMAK - Cable Machinery Plant, Poland) was utilized to compound the PA6 matrix and 5 wt.% of montmorillonite (MMT), nanosilica (SiO_2) and foam-glass-crystal-materials (FGC). Extruder screws diameter was 24 mm, had a length/diameter ratio of 32 and consisted of 6 barrel zones. A conventional screw configuration for material compounding was. The screws were equipped with two high-shearing zones in a second and fifth zone. Single-screw extruder (ZMK/117/10, ZAMAK Zaklad Maszyn, Poland) with a length/diameter ratio of 25 and a screw diameter of 25 mm was used for manufacturing of the PA6-glass fiber (PA6/GF) granulate. The single-screw extruder provided effective homogenization of classic fillers without changing the geometry of glass fibres. The panels were prepared by compression moulding technique. The mould temperature was kept at 250°C over the compression time of 5 min. Due to the high dimensions (100x100x10 mm^3) of the panels, cooling of polymer melt was completed in the mould in order to avoid warping.Additional unreinforced PA6 panels have been manufactured in the same way as the reinforced ones, as reference material

2.3. Measurement of number concentration and size distribution

To characterize the physical properties of particles generated during drilling, composite panels were fixed in a chamber, with dimensions 68.58cm (width) x 33.02cm(depth) x 55.88cm(height) as illustrated in Fig. 1. The particle emissions were measured using a Condensation Particle Counter "CPC" 5.403 with Classifier "Vienna"-DMA 5.5-U (SMPS+C, Grimm Aerosol,

Germany). SMPS+C measured sub-micrometer particles generated during drilling process over a particle size range of 5.6-1083 nm and a particle size resolution of 32 channels in total. An angle drill (Makita BDA351Z 18V LXT Angle Drill) was used for drilling with a maximum speed of 1800 min^{-1}, adapted with a conventional drill bit of 10mm diameter. Prior to measurements, the chamber was purged with laboratory air for about 20 min. Each sampling cycle comprised a 60 min background air monitoring in the chamber, 14 min. of active drilling, and a 60 min postdrilling period. The experiment was repeated 3 times for each material composition. A total of about 2000 to 3000 data sets were collected for each particular sequence.

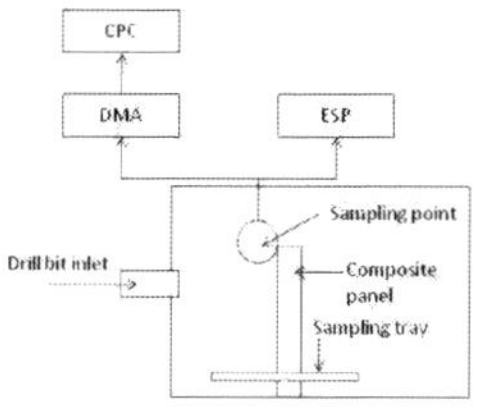

Figure 1: Apparatus and setup for the chamber experiments

2.4. Determination of emission rates

The emission rates of particles of different sizes were determined from the number concentration and size distribution data based on a single compartment population balance model. Liu et al. W. et al. [2003] used a mass balance model to estimate emission rates of particles based on mass concentrations. As reported by See et al. S.W. et al. [2007], the mass balance model calculation can also be used to estimate the particle emission rate based on particle population. Basic assumptions for this model were made as followed: a) particles are assumed to be perfect spheres with a constant density; b) background concentration is zero; b) particle concentrations are homogeneous within the chamber; and d) emission rate and decay rate of the particles remain constant throughout the entire period of generation.

The respective removal rates k_x (min^{-1}) of a particle size x were evaluated using the following equation:

$$k_x = \frac{ln(C_x/C_{max;x})}{T-t}; \quad \text{for } t \geq T \tag{1}$$

where C_x (cm^{-3}) represents the number concentration of particles of size x at time t (min), while $C_{max;x}$ (cm^{-3}) represents the maximum concentration of particles of size x which at the time T (min). The background concentration of particles, averaged over the 60 min pre-burning interval, was subtracted from the concentration readings to compute C_x (cm^{-3}) and $C_{max;x}$.

After calculating the removal rates k_x, the emission rate of particles of size x, P_x (min^{-1}) in the chamber volume V (1,26540x10^5 cm^{-3}), was calculated from:

$$P_x = \frac{VC_xk_x}{1-e^{k_xt}}; \quad \text{for } 0 \leq t \leq T \tag{2}$$

3. Results and Discussion

3.1. Sequential alteration of number concentration and size distribution

The sequential alteration of the total number concentration of particles for a typical sampling cycle is shown in Fig. 2. It consists of a 60-min background measurement ($t \geq T$), the 15min drilling period($0 \leq t \leq T$), and a 120-min post-drilling period ($t \geq T$). Figure 2, shows that the total number concentration of particles was essentially constant with an average of 1000 particles/cm^{-3} before the drilling process. As soon as drilling started inside the experimental chamber, the number concentration increased rapidly. The maximum number concentration, C_{max}, was reached at time T. Subsequently, the particle decay was observed with the concentration falling back to the original background level approximately after a post drilling period of 2h. In significant research studies I.K et al. [2009], K. et al. [2009], W.-C. [1997], R. et al. [2007] the influence of the machining engines as "background noises" on the results has been reported. As the drilling was conducting in a controlled environment, the influence of the machining engine could be reduced to a minimum, as shown in figure 2. Highest C_{max} could be measured for the nanosilica filled composites. Wohlleben et al. W. et al. [2011] reported that, while comparing nanocomposites with their reference materials without nanofillers, the differences are insignificant in the number concentrations of aerosols during normal abrasion use. This statement can not be substantiated, as the results strongly suggest that the integration of nanofillers changes the material behavior and thus the amount of particle emitted during drilling.

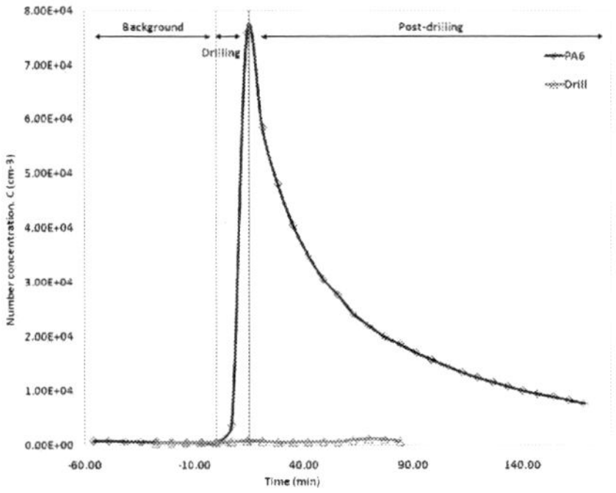

Figure 2: Sequential alteration of number concentration of a typical sampling cycle

Diffusion, gravitational deposition, convection, impaction, and coagulation are some of the complex processes which influence formation and removal of particles in the chamber. Removal of small particles is primary dominated by diffusion while larger particles are mainly affected by gravitation. During diffusion, small particles collide with one another and form larger particles Grassian V. H. [2008]. This process is known as coagulation and strongly depended on particle size and concentration. While the coagulation rate of a simple monodisperse particle population can be calculated using equation 3 Hinds [1982],

$$\frac{dN}{dt} = KN^2 \tag{3}$$

where N is particle number concentration and K is the coagulation coefficient. Based on this equation, the coagulation rate is directly proportional to the diffusion coefficient and particle size and therefore decreases with particle size. An indication of the coagulation rate can be identified by the half-life of an individual particle, which is introduced into an atmosphere embracing a defined concentration of such particles Grassian V. H. [2008]. However, according to Hinds Hinds [1982] coagulation can be neglected for laboratory experiments if the particle concentration is less the $1x10^{12}m^{-3}$. As the maximum concentration measured in this study did not exceed $1.7x10^{11}m^{-3}$, coagulation was neglected.

The particle size distributions of drilling into different material systems is shown in Figure 3. The plots reflected particle size distributions at T, the time particle concentration reached it maximum and drilling was terminated. The plots presented the normalized distribution with $\Delta N/\Delta \ln d_p$ versus the particle diameter d_p, where ΔN is the concentration of particles within a specified size interval and $\Delta \ln d_p$ is the difference in the natural logarithm of the largest and smallest particle sizes of that interval. From visual inspection of the graph, it could be noted that from different composite distinct modal diameters were obtained and that they appear to be dependent of the filler used. A major difference in peaks could be observed between the different fillers. The peaks varied from 29.6 nm for the nanosilica composite to 75.1nm for neat PA6. This showed that integration of nanofillers changes the physical properties of the emitted particles. This results contradict with the results obtained in some recent studies I.K et al. [2009], W. et al. [2011], M. et al. [2009]. For example, Wohlleben et al. W. et al. [2011] reported, while comparing nanocomposites with their reference materials without nanofillers, that both the differences in the number concentrations and in the actual size distribution of aerosols during normal abrasion use, were insignificant. However, these studies dealt with abrasion and sanding of surfaces and composites. The difference in mechanical processing could explain the alterations in the obtained results.

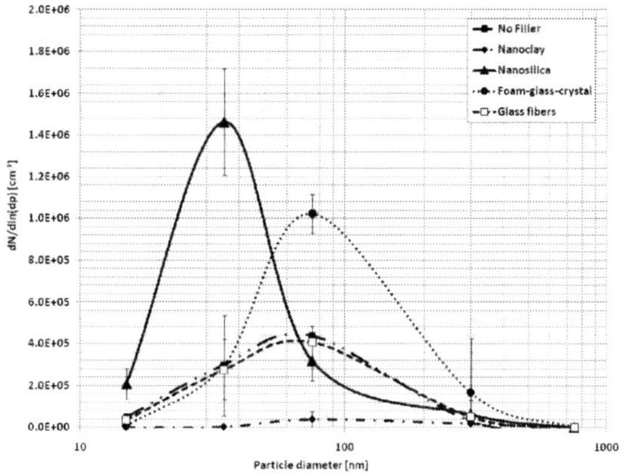

Figure 3: Normalized particle size distributions at time T

To support the results obtained by SMPS+C particles were sampled via an electrostatic precipitator (ESP, Grimm Aerosols, Germany) and subjected to JFEI XL30 field emission scanning electron microscope. The scanning electron micrographs of the particles generated during drilling of different composites are represented in Figure 4. Micrographs of the particles generated from PA6/FGC (Fig. 4a)., PA6/SiO_2 (Fig. 4b)and the neat PA6 (Fig. 4e) composites showed very similar structures of coagulated particle. Smaller generated particles (20-50nm), coagulated to larger particles 150nm, which were then sampled by ESP over a time frame of 2h. According to the literature Grassian V. H. [2008], the half time of particles in the size range of 10-200nm in the measured concentration would be between 16-83 min, therefore coagulates of particles are natural for such a long sampling period. The SEM image of the generated particles obtained from drilling of the glass fiber panels (Fig. 4d), showed agglomeration of larger particles. Furthermore the particles seem to be glass fibers debris, as the structure is very similar to structure seen in the bulk panels. Visual investigation of the particles obtained from drilling of the nanoclay composites (Fig. 4c)were scattered over the sampling plate and no coagulation could be found. This could be explained by the low particle concentration that was measured for this composite.

3.2. Emission rates of different composite materials

The emission rates of size-distributed particles from different composites were determinate using Eq. 2, after taking into account the removal rates of particles of different sizes from Eq. 1. The stacked column charts of Figure 5 shows the number of particles that were emitted per minute

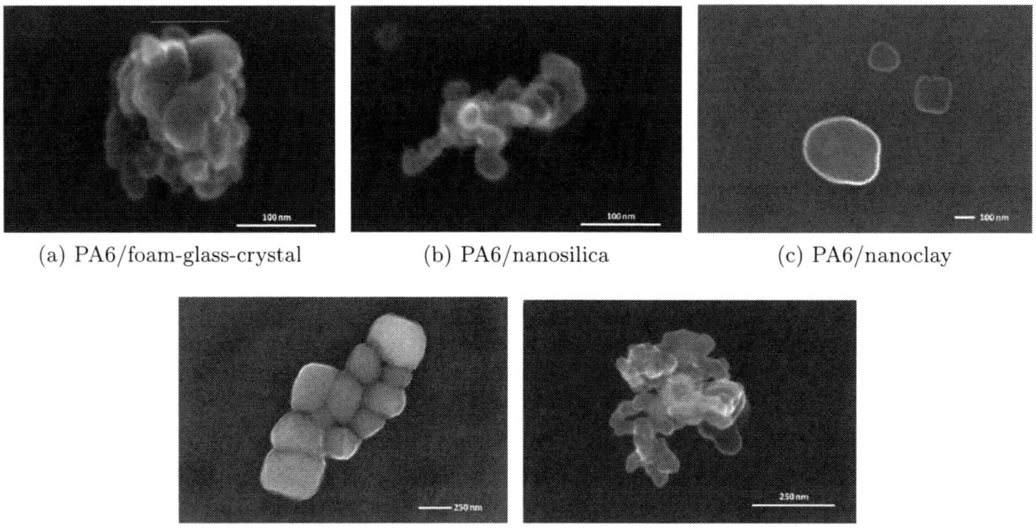

(a) PA6/foam-glass-crystal (b) PA6/nanosilica (c) PA6/nanoclay

(d) PA6/glassfibre (e) PA6

Figure 4: SEM micrographs of generated particles collected by ESP

during drilling of the composite materials. The error bars represent the standard deviation associated with the total emission rate of 11.1-521nm particles resulting from the three replicate measurements. Table ?? represents the total emission rates for the different composite materials. Lowest emission rates could be calculated for the nanoclay filled composites, followed by the neat PA6 panel. Highest emission was calculated for the nanosilica filled material. However, while comparing the percentage of the individual size intervals on the total emission, it could be seen that for all materials (beside nanoclay) the majority of particles, 45-59%, were in the size range of 22.6-42.6 nm (Figure 5). Further the un-filled material emitted the largest quantity (25%) of particles larger than 100nm. The emission rate increased by 56 times for the nanosilica filler and between 20-45 times for the glass fibre and foam-glass crystal filler. Interestingly, integration of nanoclay into the PA6 matrix reduced particle emission during drilling by 1.5 times. It is likely that the presence of nanoclay in some way retains the formation of high quantity of airborne particles. This could be caused due to modification of the nanoclay prior integration of the matrix, however other reason could be plausible e.g. shape of the nanoclay filler and integration mechanism into the matrix (intercalation/exfoliation) A. et al. [2007a,b].

4. Conclusion

Physical characterization of the number concentration and size distribution of sub-micron particles from 5.6 to 512nm was carried out, for the first time, for different silica based nanocomposites. Two kind of nanosilica fillers (nanoclay, nanocilica) and two kind of macro silica filler (glass fibres and foam-glass crystal) were successfully integrated into a PA6 matrix. Based on size-resolved particulate emission rates from various silica based composites during mechanical drilling process were evaluated. In general, nano and ultrafine airborne particles were emitted from all investigated materials, even the non reinforced polymer. However emission increased by 56 times for the nanosilica filler and between 20-45 times for the glass fiber and foam-glass crystal filler. Interestingly, integration of nanoclay into the PA6 matrix reduced particle emission during drilling by 1.5 times. However, the characterisation of deposited particles showed exactly the opposite particle behavior, as with decreasing airborne particle concentration the deposit particle concentration increased and vice verse. This study has certainly shown that the integration of

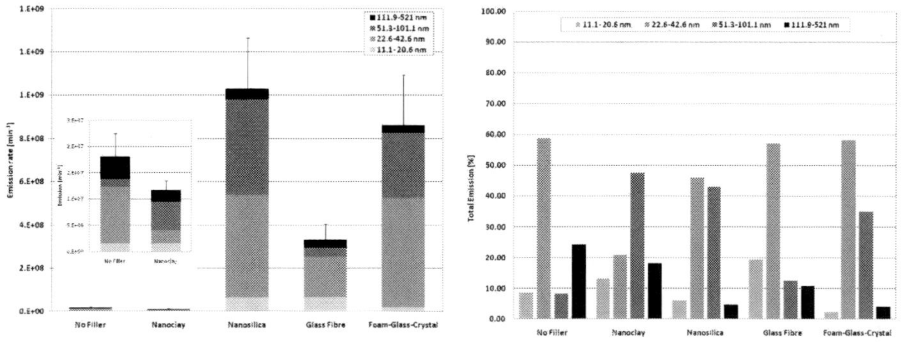

(a) Emission rates (b) Particle influence on emission rate

Figure 5: Emission rates of particles for different composites

nanofillers into polymeric matrix, changes the behaviour of the composite material in terms of particle emissions. Therefore, future work will involve a series of comprehensive chemical characterization of size-distributed particles released during drilling of nanocomposites to make a better assessment of public health outcomes due to exposure.

5. Acknowledgment
This work was funded by the European Commission (FP7 NEPHH Project- CP-FP; Project ReferenceNo. 228536 - 2).

References

Njuguna J., Sachse S., Ifran A., Zhu H., Michalowski S., and Pielichowski K. Nanoparticles released from structural nano-enhanced products following mechanical loading at low velocity impacts: A case study on structural polyurethane nanoreinforced foams. In *4th International Seminar on Modern Polymeric Materials for Environmental Applications*, volume 4, pages 247–254, 2010.

J. Njuguna, Silva F., and Sachse S. *Nanocomposites for Vehicle Structural Applications, Nanofibers - Production, Properties and Functional Applications*, chapter 19, pages 402–434. InTech, 2011.

BCCResearch. Nanotechnology - nanocomposites, nanoparticles, nanoclays, and nanotubes. Technical Report NAN021D, BCC Research, 2010.

Som C., Berges M., Chaudhry Q., Dusinska M., Fernandes T.F., Olsen S.I., and Nowack B. The importance of life cycle concepts for the development of safe nanoproducts. *Toxicology.*, 269: 160–9, 2010.

and Scott J. *How to Rebuild and Modify Your Muscle Car.* MotorBooks International, 2006.

Harrison R. M. Review: Key pollutants - airborne particles. *Sci. Total Environ.*, 334:3–8, 2004.

Sotiriou G. A., Diaz E., Long M. S., Godleski J., Brain J., Pratsinis S. E., and Demokritou. P. A novel platform for pulmonary and cardiovascular toxicological characterization of inhaled engineered nanomaterials. *Nanotoxicology*, 0(0):1–11, 2011.

Dreher K. L. Health and environmental impact of nanotechnology: Toxicological assessment of manufactured nanoparticles. *Toxicol. Sci.*, 77:3–5, 2004.

Oberdrster G. Safety assessment for nanotechnology and nanomedicine: concepts of nanotoxicology. *Journal of Internal Medicine*, 267(1):89–105, 2010. ISSN 1365-2796.

Maynard A. D., Warheit D. B., and Philbert M. A. The new toxicology of sophisticated materials: Nanotoxicology and beyond. *Toxicological Sciences*, 120(suppl 1):S109–S129, 2011.

Liu W., Zhang J., Hashim J.H., Jalaludin J., Hashim Z., and Goldstein B. D. Mosquito coil emissions and health implications. *Environ. Health Perspect.*, 111:1454–1460, 2003.

See S.W., Balasubramanian R., and Joshi U.M. Physical characteristics of nanoparticles emitted from incense smoke. *Science and Technology of Advanced Materials*, 8(1-2):25, 2007.

Koponen I.K, Jensen K.A., and Schneider T. Sanding dust from nanoparticle-containing paints: Physical characterisation. *Journal of Physics: Conference Series*, 151(1):012048, 2009.

Jensen K., Koponen I., Clausen P., and Schneider T. Dustiness behaviour of loose and compacted bentonite and organoclay powders: What is the difference in exposure risk? *Journal of Nanoparticle Research*, 11:133–146, 2009. ISSN 1388-0764.

Chen W.-C. Some experimental investigations in the drilling of carbon fiber-reinforced plastic (cfrp) composite laminates. *International Journal of Machine Tools and Manufacture*, 37(8): 1097–1108, 1997. cited By (since 1996) 71.

Khettabi R., Songmene V., and Masounave J. Effect of tool lead angle and chip formation mode on dust emission in dry cutting. *Journal of Materials Processing Technology*, 194(1-3):100 – 109, 2007. ISSN 0924-0136.

Wohlleben W., Brill S., Meier M. W., Mertler M., Cox G., Hirth S., von Vacano B., Strauss V., Treumann S., Wiench K., Ma-Hock L., and Landsiedel R. On the lifecycle of nanocomposites: Comparing released fragments and their in-vivo hazards from three release mechanisms and four nanocomposites. *Small*, pages 2384–2395, 2011. ISSN 1613-6829.

and Grassian V. H. *Nanoscience and nanotechnology: environmental and health impacts.* John Wiley & Sons, 2008.

W. C. Hinds. *Aerosol Technology: Properties, Behavior, and Measurement of Airborne Particles.* New York:Wiley, 1982.

Vorbau M., Hillemann L., and Stintz M. Method for the characterization of the abrasion induced nanoparticle release into air from surface coatings. *J. Aerosol Sci.*, 40:209–217, 2009.

Leszczynska A., Njuguna J., Pielichowski K., and Banerjee J.R. Thermal stability of polymer/montmorillonite nanocomposites: Part i: Factors influencing thermal stability and mechanisms of thermal stability improvement. *Themochimica Acta*, 453:75–96., 2007a.

Leszczynska A., Njuguna J., Pielichowski K., and Banerjee J.R. Polymer/clay (montmorillonite) nanocomposites with improved thermal properties: Part ii: A review study on thermal stability of montmorillonite nanocomposites based on different polymeric matrixes. *Themochimica Acta*, 454:1–22., 2007b.

Ecological assessment of nano-enabled supercapacitors for automotive applications

M Weil[1, 2], H Dura[1], B Shimon[1, 2], M Baumann[1], B Zimmermann[1], S Ziemann[1], C Lei[3], F Markoulidis[3], T Lekakou[3] and M Decker[1]

[1] Karlsruhe Institute of Technology (KIT), ITAS, Germany
[2] Helmholtz Institute Ulm for Electrochemical Energy Storage, Germany
[3] University of Surrey, United Kingdom

E-mail: marcel.weil@kit.edu

Abstract. New materials on nano scale have the potential to overcome existing technical barriers and are one of the most promising key technologies to enable the decoupling of economic growth and resource consumption. Developing these innovative materials for industrial applications means facing a complex quality profile, which includes among others technical, economic, and ecological aspects. So far the two latter aspects are not sufficiently included in technology development, especially from a life cycle point of view. Supercapacitors are considered a promising option for electric energy storage in hybrid and full electric cars. In comparison with presently used lithium based electro chemical storage systems supercapacitors possess a high specific power, but a relatively low specific energy. Therefore, the goal of ongoing research is to develop a new generation of supercapacitors with high specific power and high specific energy. To reach this goal particularly nano materials are developed and tested on cell level. In the presented study the ecological implications (regarding known environmental effects) of carbon based nano materials are analysed using Life Cycle Assessment (LCA). Major attention is paid to efficiency gains of nano particle production due to scaling up of such processes from laboratory to industrial production scales. Furthermore, a developed approach will be displayed, how to assess the environmental impact of nano materials on an automotive system level over the whole life cycle.

1. Introduction

Electromobility is considered as one of the most promising technologies for replacing fossil fuelled mobility. Depending on the electricity generation mixture of a region, electric cars can have already today an environmental advantage against gasoline cars. With a growing share in regenerative production of electric energy this advantage will substantially increase, particularly when synergy effects of the progressively merging energy and transport networks will appear. The key element of an electric vehicle is the energy storage system. At present for hybrid and full electric vehicles commonly nickel metal hydride and lithium based batteries are used. If electro mobility is to succeed there are a number of obstacles to overcome including great improvements in energy density, peak power, life time (calendar and cycle stability), operation temperature and costs. Furthermore, the energy storage systems should rely on available resources, be easily recyclable and should have an overall good environmental performance.

Supercapacitors are a very promising candidate as an alternative energy storage system. It has the advantage of high power density, high cycle stability, high calendar life time, a broad temperature range for operation and relies only on good available raw materials like carbon. However, they are facing a number of problems including low energy density and high cost. In Figure 1.1 the performance regarding power and energy density of capacitors, supercapacitors and batteries are shown.

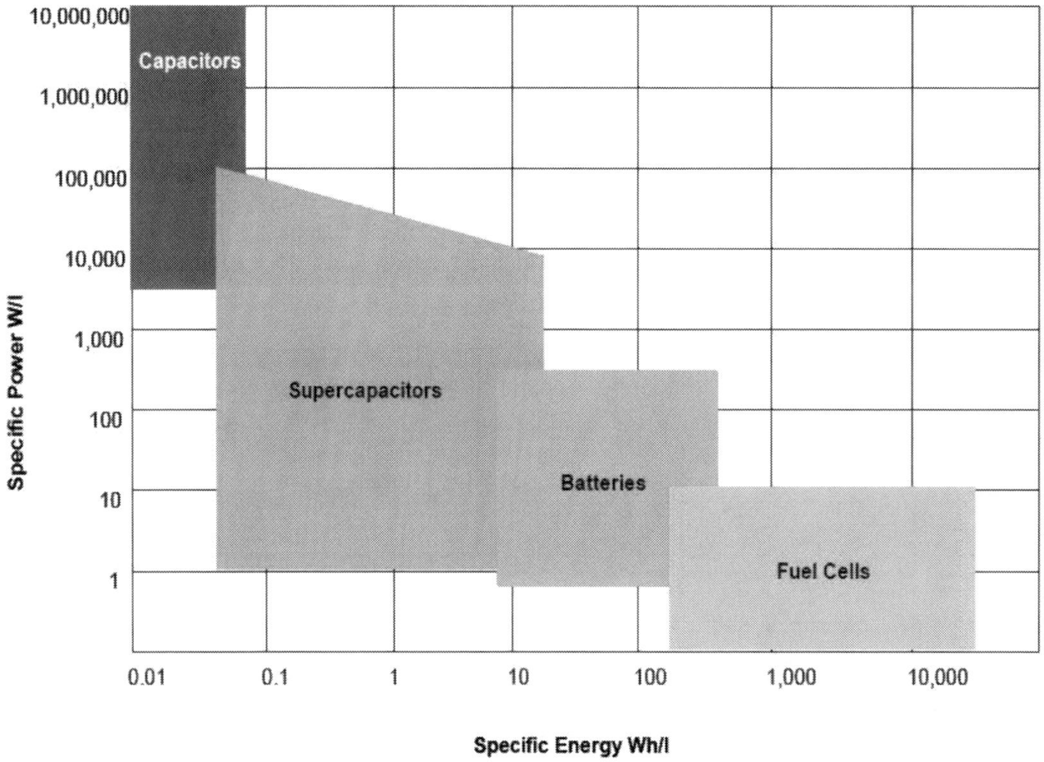

Figure 1.1. Specific energy and specific power of various energy storage systems [1].

2. Improved Supercapacitors by Nano Materials

2.1 Performance implications by the use of nano materials

The structure of a super capacitor is related to the structure of a battery encompassing current collectors, high surface area electrodes, an electrolyte and separator. However, the energy is not stored chemically but electrostatically in an electric double layer (EDL) forming at the interface between the electrodes and the electrolyte as depicted in Figure 2.1 [2]. Thus, the amount of energy stored is proportional to the surface area of the electrode [3], the higher the surface area, the greater the specific energy. Since carbonaceous nano materials offer great surface areas at low specific weight and volume, their application for supercapacitor electrodes is increasingly being investigated. As already shown in Figure 2.1 supercapacitors have a very good specific power, but need great improvement in specific energy in order to be more applicable for automotive energy storage.

To overcome the stated problem the EU project "Autosupercap" focuses on the development of supercapacitors with high power and high energy density while reducing weight and costs. To achieve such an improved performance it is planned to assess different kinds of carbon based nano elements for the production of the electrodes [4].

Often research largely focuses on the development of high area electrodes. This may be achieved using carbonaceous nano materials, as these offer high surface to volume and weight ratio.

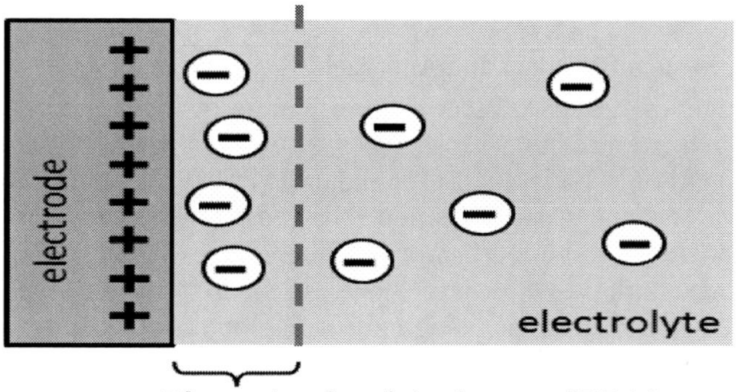

Figure 2.1: Simplified drawing of the electric double layer.

The energy and power performance of the capacitor as well as the lifetime are strongly influenced by the internal resistance. The internal resistance is the product of the carbon material itself, the carbon particle-particle contact, and the carbon-current collector contact [4].

Normally carbon based aerogels and powders are used for the synthesis of electrodes. A major improvement of the electrode resistance can be achieved by using conducting additives, such as black carbon or carbon nanotubes [5].

In both ways the intelligent use of nano materials may improve the technological performance of the supercapacitor.

2.2 Environmental implications of nano materials

High performance carbonaceous nano materials may be produced by three methods: arc discharge, laser ablation and chemical vapour deposition (CVD). The arc discharge method and laser ablation are based on vaporization of carbon feedstock. In the arc method two carbon rods are placed in inert gas or liquid nitrogen. A high temperature discharge between the two rods causes a vaporization of the surface of one of the carbon rods while deposition takes place on the other. In laser ablation vaporization of graphite rods is attained using a dual pulsed laser and a subsequent thermal treatment. Due to the underlying principal of evaporation of the carbon source, these two production approaches appear to be difficult for up scaling to an industrial level [6].

Chemical Vapour Deposition (CVD) on the other hand has the greatest potential for large scale production. In a batch process, plates containing a catalyst are placed on the bottom of a reaction chamber, which is filled with a synthetic gas mixture. Upon heating, the gas reacts with the catalyst forming nanotubes on the catalyst surface [7]. A further development of the CVD process uses a fluidized bed reactor and allows the continuous and mass production of CNTs [8].The production of nano materials is very often an energy and resource intensive process, which may partially explains also the high price of these products. Using such materials increases the efforts for the production of the supercapacitor. To reach a net environmental and economic gain the efforts for the production have to be at least compensated if not significantly outweighed by a better environmental performance during the use phase and at the end of life treatment of a nano enabled supercapacitor.

In this regard an environmental assessment of the entire life cycle is a key element. But until today only few LCAs exist on the production of nano materials [9]. Furthermore, there is a need for additional models that describe adequately the paths of release and effects of nano materials throughout their life cycle [10]. This problem is further enhanced by the high variability of nano materials even within one material group, i.e. a multi walled carbon nano tube (MWCNT) is not necessarily comparable to another MWCNT. The associated environmental impact depends on several factors including processes of synthesis, purification and dispersion as well as functionalization. That means the real environmental impacts of such materials are neither well assessed nor well understood. Therefore, the first step is to investigate the environmental impact of commonly used nano materials for electrode production.

2.3 LCA Methodology, System Boundary and Approach

LCA offers a methodological framework for a consistent assessment and comparison of different technological systems with respect to various types of environmental impacts. According to the framework set forth by DIN EN ISO 14040 [11] and 14044 [12] a LCA is comprised of four work packages starting with a goal and scope definition followed by a life cycle inventory (LCI) and life cycle impact assessment (LCIA). The fourth work package, the evaluation is included as an iterative process which is performed during all three prior stages. In the first step the objective of the study is to be stated and further details explained including information on how the study will be performed, the functional unit, system boundaries as well as allocation methods and impact categories used. The LCI is purely descriptive documenting all input and output flows into and out of the earlier defined system. In a subsequent step, the LCIA is performed in order to determine environmental repercussions on the basis of the flow network.

For the presented work the cumulated energy demand (CED) is displayed and used as a screening indicator for environmental repercussions as it depicts the direct and indirect energy used e.g. from use of raw materials.

The considered system boundaries of the investigated nano materials are from "cradle to gate", that means all pre chains of the production process are included. General problems of LCA investigations for nano materials are complex and described in [7].

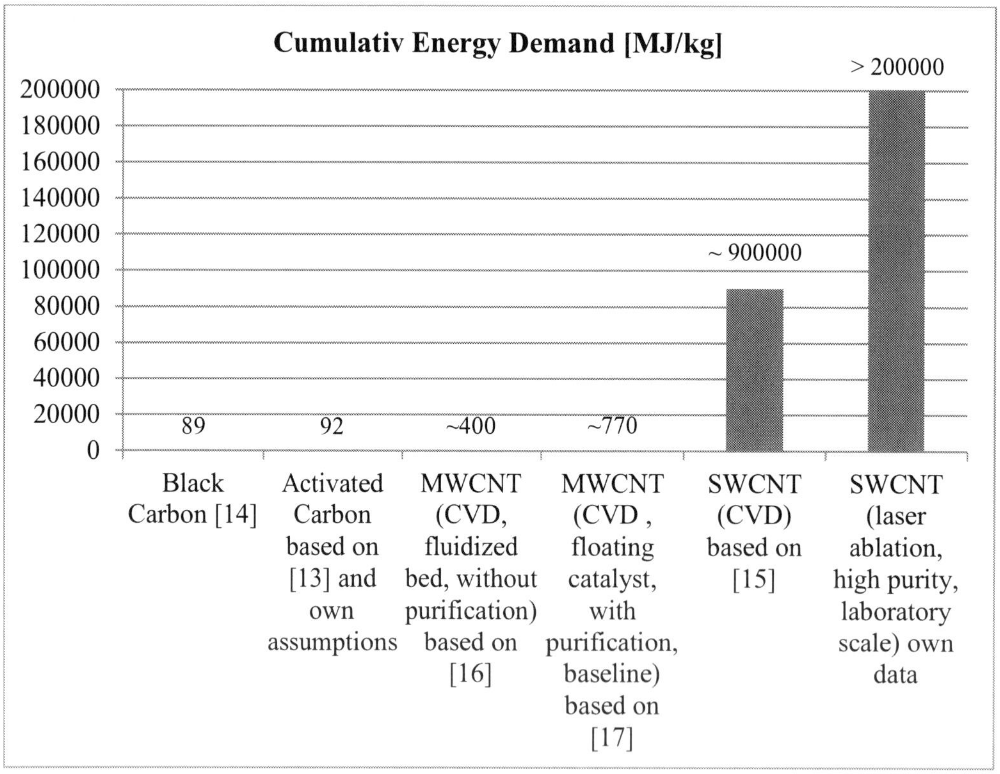

Figure 4.1. Comparison of different carbon based materials for electrode production regarding the cumulative energy demand (CED) [13-17].

2.4 LCA Results of Carbonaceous Materials on Nano and Micro Scale

The CED values for different carbonaceous materials and production procedures are shown in Figure 4.1. Most striking is the extremely high energy intensity of SWCNT production by laser ablation creating an energy demand of more than 200.000 MJ/kg, more than twice as much as SWCNT produced by CVD. MWCNT cause a much smaller CED, although this also strongly depends on the production method applied. Although significant reductions may be achieved with the fluidized bed and CVD production processes, in comparison with the production of black carbon and activated carbon both types of the carbon nano materials create a relatively high CED, from 4 fold larger (MWCNT fluidized bed) up to approximately 4 orders of magnitude larger (SWCNT laser ablation). Problematic is also the circumstance that MWCNT produced by a fluidized bed reactor tend to have high impurities and agglomerations of CNTs, which might cause in addition for some application fields further efforts for purification and dispersion.

2.5 Assessment of nano materials on automotive system level

In order to assess the economic and environmental effects of the use of nano materials in new developed supercapacitor the whole life cycle in a use case has to be considered.
For this the life cycle models regarding an automotive application have to cover the following sub-models:

- Supercapacitor cell
    - Supercapacitor bank
        - Whole powertrain (including thermal and electric management system)
            - Electric vehicle
                - Use phase of electric vehicle (driving cycles)
                    - After use phase of electric vehicle

Only by considering the whole life cycle the net value in respect of economic and ecological aspects can be evaluated. General statements are not possible, due to the strong dependency of the application field. This kind of prospective evaluation has to be conducted for each single application of supercapacitors.

3. Summary and Outlook

The next generation of supercapacitors which are under development uses especially nano materials to improve the technical performance. It could be shown, that the production of carboniferous materials on nano scale can have a significant environmental impact. The environmental impact depends on several factors, but an upscale of the synthesis process from laboratory to industrial scale exhibit a very promising improvement. Unfortunately the scale up process can have an important influence on the quality of the nano materials. The quality losses may cause a higher use of the nano materials or cause further efforts for the improvement of the quality, e.g. by additional dispersion or purification processes. To evaluate the possible environmental and economic saving due to the gained higher technical performance, the supercapacitor has to be investigated on a system level over the entire live cycle. This will be conducted in the future for a simulated automotive application with measured data from a test rig.

4. Acknowledgments

This research was funded by the European Commission FP7 project "AUTOSUPERCAP".
We would like to express our sincere gratitude also to Regina Fischer and Prof. Dr. Manfred Kappes from the KIT, Division of Physical Chemistry of Microscopic Systems, Germany.

Reference

[1] Bertoldi, Odile, und Sébastien Berger. 2009. *WP2_5Energy_3Thermoelectricity.pdf*. European Commission, ObservatoryNANO.
[2] Winter, Martin, und Ralph J Brodd. 2004. „What are batteries, fuel cells, and supercapacitors?" *Chemical Reviews* 104 (10): 4245–4269.

[3] Diederich, L., E. Barborini, P. Piseri, A. Podesta, P. Milani, A. Schneuwly, und R. Gallay. 1999. „Supercapacitors based on nanostructured carbon electrodes grown by cluster-beam deposition“. *Applied Physics Letters* 75 (17) (Oktober): 2662 –2664. doi:10.1063/1.125111.
[4] Lei, Chunhong, Foivos Markoulidis, Peter Wilson, und Lekakou Constantina. 2012. „Reduction of the Internal Resistance of Carbon Electrodes for an Electric Double-layer Capacitor (EDLC)“. In OH, USA.
[5] Snook, Graeme A., Pon Kao, und Adam S. Best. 2011. „Conducting-polymer-based supercapacitor devices and electrodes“. *Journal of Power Sources* 196 (1): 1–12. doi:10.1016/j.jpowsour.2010.06.084.
[6] Wilson, Michael. 2002. *Nanotechnology: Basic Science and Emerging Technologies*. Chapman & Hall/CRC.
[7] Weil, Marcel. 2012. *System Analysis in the Early Phase of Technology Development - Responsible Development and Production of Carbon Nanotube* Paper. P. 301-312 - In: Decker, M., Grunwald, A., Knapp M. (Hrsg.): Der Systemblick auf Innovation – Technikfolgenabschätzung in der Technikgestaltung. Berlin: Edition Sigma 2012
[8] See, Chee Howe, und Andrew T Harris. 2007. „A Review of Carbon Nanotube Synthesis via Fluidized-Bed Chemical Vapor Deposition“. Industrial & Engineering Chemistry Research 46 (4): 997–1012.
[9] Hischier, Roland, und Tobias Walser. 2012. „Life cycle assessment of engineered nanomaterials: State of the art and strategies to overcome existing gaps“. *Science of The Total Environment* (0). doi:10.1016/j.scitotenv.2012.03.001.
[10] Fries, René, Sabine Greßler, und Myrtill Simkó. 2011. „Kohlenstoff-Nanoröhrchen (Carbon Nanotubes) - Teil ll: Risiken und Regulierungen“. *nano trust dossiers* (024) (Mai).
[11] DIN-ISO 14040. 2006. „Environmental management Life Cycle Assessment - Requirements and Guidelines 14040“.
[12] DIN-ISO 14040. 2006. „Environmental management Life Cycle Assessment - Requirements and Guidelines 14040“.
[13] Noijuntira, I. and Kittisupakorn, P. 2009. Life Cycle Assessment for the Activated Carbon Production by Coconut Shells and Palm-Oil Shells. (Thailand, Jul. 2009)
[14] Ecoinvent Life Cycle Inventory Database. Swiss Centre for Life Cycle Inventories. www.ecoinvent.org.
[15] Healy, ML, Dahlben, LJ, Isaacs, JA 2008. Environmental Assessment of Singel-Walled Carbon Nanotube Processes. Journal of Industrial Ecology 12, 3: 376-393
[16] Steinfeld, Michael und von Gleich, Arrnin 2008. Entlastungseffekte für die Umwelt durch nanotechnische Verfahren und Produkte. Umweltbundesamt (UBA)
[17] Kushnir, D Saden, BA 2008. Energy Requirements of Carbon Nanoparticles Production. Journal of Industrial Ecology 12, 3: 360-375

IOP Conf. Series: Materials Science and Engineering **40** (2012) 012014 doi:10.1088/1757-899X/40/1/012014

Life cycle assessment of the application of nanoclays in wire coating

ATellaetxe[1,*], M Blázquez[2], A Arteche[3], A Egizabal[4], V Ermini[5], J Rose[6], P Chaurand[7] and I Unzueta[2]

[1] EKOTEK, Ingeniería y Consultoría Medio Ambiental
[2] INKOA SISTEMAS
[3] Fundación Tecnalia Research and Innovat[illegible]
[4] Fundación Tecnalia Research and Innovat[illegible] Ciber BBN
[5] LAVIOSA Chimica Mineraria
[6] CEREGE, UMR 7330 CNRS Aix Marseille Univ.

*Corresponding Author – atellaetxe@ekotek.es

ABSTRACT. A life cycle assessment (LCA) is carried out to compare nanoclay-reinforced polymer wire coatings with conventional ones. While the conventional wire coatings contain standard halogen free retardants, in reinforced coatings, montmorillonite (nanoclay) is incorporated into electric cable linings as a rheological agent for an increased resistance to fire. In addition, a reduced load of standard halogen free retardants is obtained. The synergistic effect of the montmorillonite on traditional flame retardant additives (by the formation of a three-dimensional char network) can lead to a revolution in wire production. The application of nanoclays contributes also to anti-dripping effect and flexibility increase [1]. Some producers have already started commercializing wire with nanotechnology-based coating; in the short term the use of nanoclay in wire coating production will probably reach a significant market share replacing traditional formulations.

The main aim of this study is to compare the environmental impacts along the life cycle of a traditional wire coating (mineral flame retardants like ATH or MDH in a polymer matrix) with the nanoclay-reinforced wire coating, where the montmorillonite replaces a low percentage of the mineral flame retardant. The system boundaries of the study include the following unit processes: nanoclay production, thermoplastic material and mineral flame retardants production, cable coating manufacturing by extrusion and different end of life scenarios (recycling, incineration and landfill disposal). Whereas nanoreinforced composites have shown and increased fire retardance, the addition of nanomaterials seems to have no significant relevance in the environmental assessment. However, the lack of nano-specific characterization factors for nanomaterials and emission rates associated to the different life cycle stages –mainly in the extrusion and use phase, where accidental combustions can take place- still remains a challenge for realistic life cycle assessment modelling.

1. INTRODUCTION

Nanocomposites are a new field of research and development in which great hopes have been placed for significant social and environmental benefits linked to a greater economic development.

Wire coating represents a niche market of high value for thermoplastic composites. However, the demand for such applications suffered a tough fall of 23% between 2008 and 2009, mainly due to the drop in construction sector. Nevertheless, the use of polymers for cable insulation and coating has largely recovered, reaching 1.4 million tones in the EU in 2011 [1]

Nanocomposites are a new class of polymer systems. Dr. Günter Beyer observed new extraordinary properties after dispersing modified layered silicates as fillers at a nanometre-level within a polymer matrix. Through nanocomposite formation the thermal stability and flame retardancy of polymers were improved. Nanocomposites avoid the disadvantages of these traditional systems (very high loadings of the filler, reservations about general use of brominated compounds, expensive intumescent systems…) and show significant improvements while the filler content is just 2-10% [1,2]. As an example, a new hybrid nano-material derived from synthetic platy silicates was used with magnesium hydroxide to prepared polyolefin-based halogen free flame retardant (HFFR) wire and cable compounds. While providing improved flame retardance, the nanocomposites provide comparable tensile and strength. Investigations on ethylene vinyl acetate (EVA)- based nanocomposites demonstrate that a reduction of 47% for the peak of heat release can be achieved by only 5wt% on an organoclay within the polymer matrix can reduce peak heat release rate by 47% and increase the peak decomposition temperature by 40°C [2,3]

Nanoclays, montmorillonite (MMT) specifically, are actually being incorporated into electric cables linings as a rheological agent for an increased resistance to fire. In this application the organoclays are used in combination with mineral flame retardants like aluminium trihydroxide ATH ($Al(OH)_3$) or magnesium dihydroxide MDH ($Mg(OH)_2$) in a polymer matrix. In fire, these mineral flame retardants decompose, absorbing energy, releasing water (reducing fire intensity and diluting fire gases), and causing charring (fire barrier). Organoclays act as a synergist to flame retardants by the formation of a three-dimensional char network while the combustion of the cable [3]. This three-dimensional network promotes a strong carbon crust, behaving like an insulation material. Dripping of molten compound and the time until the burning stops is reduced.

Thereby, nanoclays application in wire coating has a huge growing potential. The alternative product, or baseline needed for LCA comparative work, is the conventional formulation including traditional halogen free flame retardants (HFFR) in a given concentration. A comparative Life Cycle Assessment between traditional product versus novel nanotechnology incorporating product for cable coating has been performed.

2. MATERIALS AND METHODS

Material compositions as recommended by LAVIOSA were selected for materials manufacturing. According to LAVIOSA for a UL94 V0 in EVA sheating, a formulation containing 55% of inorganic fire resistant (ATH or MDH) and 5% of MMT is advised. These recommendations were literally followed.

Montmorillonite was provided by LAVIOSA. Dellite 67G was selected as synergistic agent as it contains an excess (43-49%) of dimethyl dihydrogenated tallow ammonium, improving the dispersion of the nanoclay inside polymer matrix if compared with other grades. Dellite 67G's interlayer distance is 34,328 Å, while its particle size (dry) is 7-9 μm and 1 x 500 μm after dispersion.

ATH with a mass median diameter of 1.7-2.1 microns was purchased from Albemarle–Martinswerk. Albemarle–Martinswerk. MDH was purchased from Albemarle–Martinswerk with an approximate mass median diameter of 1.5 microns. Ethylene Vinyl Acetate – EVA- was Escorene 00328, acquired from ExxonMobil Chemical.

Weight percentages were defined as illustrated in table below:

Table 1 - Product systems

Non nano enabled coatings	Nano enabled coatings
40% wt EVA + 60% wt ATH	40% wt EVA + 55% wt ATH + 5% wt MMT
40% wt EVA + 60% wt $Mg(OH)_2$	40% wt EVA + 55% wt $Mg(OH)_2$ + 5% wt MMT

Reference materials included a currently used thermoplastic material (EVA - ethylene vinyl acetate) for cable coating production and two traditional fire retardants: MDH and ATH. As it can be noticed, the addition of MMT to the systems implies a slight reduction of the quantity of the traditional inorganic Flame Retardants.

Extrusions were performed by VIBA, in Italy. Universal Aghetto turbomixer was used for the necessary preliminary pulverization prior extrusion. Maris 30mm corotating twin screw extruder with 52 length/diameter ratio and up to 650 rpm of maximum speed, equipped with a spinneret, cooling water bath and a shearing and sieving machine (pelletizer system) was used to perform extrusions for the 4 specimens manufacturing. By means of extrusions MMT is exfoliated and dispersed into the polymer; strands are then cooled in a water tank and dried. Strands are then cut and sieved.

Injections were performed by means of Battenfeld CD 75 machine. Injection was a necessary step for the further verification of the performance of the UL94, and this process was not included in LCA study.

2.1 CHARACTERIZATION OF MANUFACTURED MATERIALS AS HFFR

Complementing this LCA study, the materials arising from the industrial dispersions were tested against their expected functionality as HFFR by means of the standard protocol, UL 94. The intention was to confirm the increased performance and associated potential replacement of traditional formulations by novel formulations enabled by Nanotechnology in terms of fire resistance for their use in cable lining industry.

The standard protocol, UL 94 was selected as for the characterization of the properties of the materials in terms of increased Fire Resistance. During this experiment, a specimen (13 mm x 125 mm) is supported in a vertical position and a flame is applied to the bottom of the specimen. The flame is applied for ten seconds and then removed until flaming stops at which time the flame is reapplied for another ten seconds and then removed. Two sets of five specimens were tested. The first set is comprised by materials not processes/stored after the injection whereas the second set includes materials that have been "aged" by keeping them at 70ºC during one week.

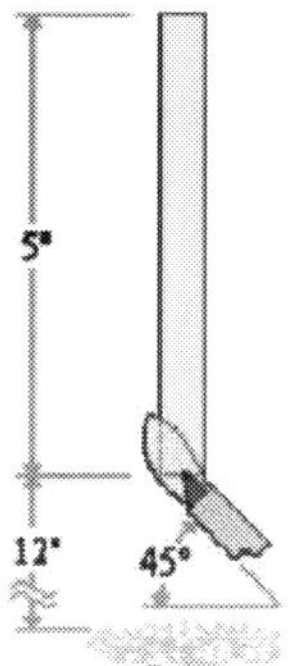

Figure 1 – UL94 vertical flame test (IEC 60695-11-10 Test Standard)

Table 2 – Results of fire resistance tests.

	Composition	Specimen	Result
1	40% wt EVA + 60% wt ATH	Not aged	Unclassifiable
		Aged	V-1
2	40% wt EVA + 55% wt	Not aged	**V-0**

	ATH + 5% wt MMT	Aged	**V-0**
3	40% wt EVA + 60% wt $Mg(OH)_2$	Not aged	V-1
		Aged	Unclassifiable
4	40% wt EVA + 55% wt $Mg(OH)_2$ + 5% wt MMT	Not aged	**V-0**
		Aged	**V-0**

UL94 establishes the following flammability rating: V0, V1 or V2.The lower the number the better the flammability rating is. As shown in Table 2, nanoenabled samples (samples 2 and 4) have a better fire resistance -V0 in both cases- than non nanoenabled samples (samples 1 and 3) which have revealed V1 or unclassifiable. This fact confirms the potential of MMT including formulations to replace traditional formulations, and therefore the need to evaluate these materials in terms of associated environmental impact.

2.2 LCA APPROACH – GOAL AND SCOPE DEFINITION

Life cycle assessment has been proposed to be a key to such comprehensive environmental assessments [2]. LCA is an established tool for identifying environmental "hot spots" and for comparing products that provide the same services. However, there are little published inventory data of nanomaterials and nanoenabled products [2,3,4]. Furthermore, to date only a very small number of studies that deals with the environmental impacts of nanomaterials and their applications is available even if reports by UK Royal Science Society, Michelson, Feigenbaum et al [2] point out the need for LCA studies for nanomaterials. For what specifically refers to nanocomposites, in general, applications for the automotive industry have been evaluated [2,3].

The ISO 14040: *Environmental Management—Lifecycle Assessment—Principles and Framework*, has defined the four major components of an LCA as:

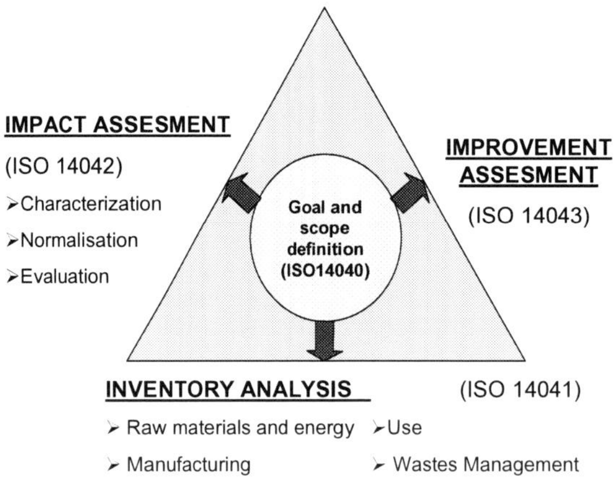

Figure 2 - Structure of LCA study.

1. **Goal and scope definition**;
2. **Life Cycle Inventory** - data collection and calculation of an inventory of materials, energy and emissions related to the system being studied;
3. **Life Cycle Impact Assessment** - analysis of data to evaluate contributions to various environmental impact categories; and impact assessment; and
4. **Interpretation of results** - where data are analysed in the context of the methodology, scope and study goals and where the quality of any study conclusions is assessed.

According to that document, that methodology has been performed beginning with the creation of an inventory of the inputs and outputs of most processes that occur during the whole life cycle of a product. Afterwards, all the data are analysed and modelled using a professional tool following the ISO 14040 series recommendations [3,4,5]

SimaPro is the chosen software to carry out this LCA study. The environmental impacts have been determined with the EcoIndicator 99 methodology that considers damages to three categories: Human Health, to Ecosystems and to Natural Resources.

Functional unit and system boundaries

Product systems in an LCA must be evaluated on a functionally equivalent basis to provide a reference for relating process inputs and outputs to the inventory and impact assessment across alternatives. According the LCA framework "**1 kg of polymeric pellets**" has been defined as functional unit or base of calculation.

Figure 3 shows the main stages in the lifecycle of a coated cable and the **system boundaries**. Stages in dash box have been excluded from the study as it is below explained.

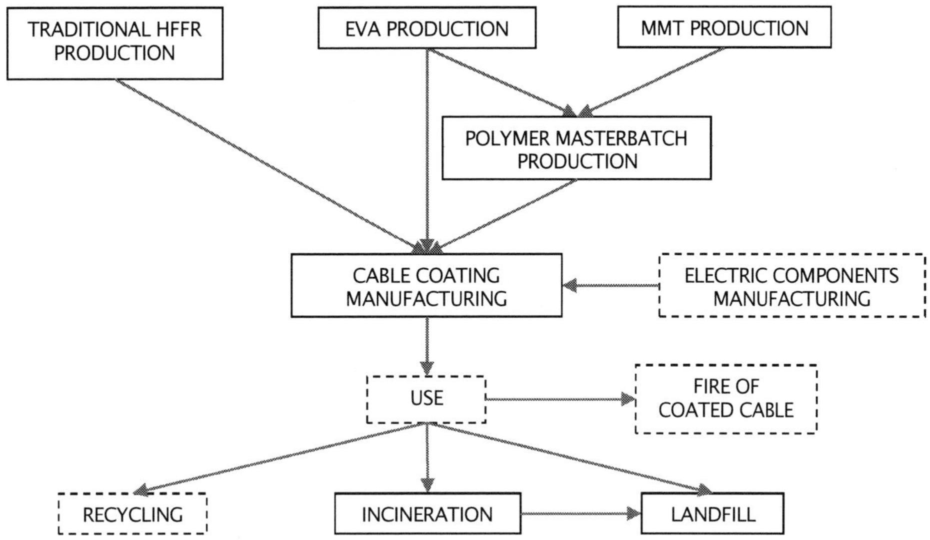

Figure 3 – System boundaries

- Being the main goal of present assessment to analyze the differences in the environmental impacts between using conventional versus nano-enabled materials, the processes that have equal inputs and outputs are simplified (as equal environmental impacts are expected).
- Impacts related to the infrastructure needed to support manufacturing facilities are excluded.
- Use stage of the final product, in which coated cable is installed, is excluded because the choice of flame retardant has not effect on either the energy demand of the electronic product or the lifetime of the wire.
- The manufacture of electric components is not included in the assessment since it is not expected that electric components will modify as a result of changing the sheath composition.

- For the primary data collected from organoclay and masterbatch manufacturers, transport is not included in the scope. The differences in transport among the different kinds of flame retardants in the associated life-cycle stages (i.e., manufacturing, use, and end of life are not expected to be significant. Therefore, excluding transport from primary data collection is not expected to adversely affect the study results.

2.3 INVENTORY ANALYSIS

Data inventory is composed of not only secondary data (data compiled from leading databases) but also primary data obtained through an exhaustive data collection (process description and inputs/outputs quantification) as well as through some specific tests related to: leaching tests.

Raw materials LCI

The first stage in the life-cycle of a product corresponds to the raw material extraction. Here the extraction of different chemicals which enter in the polymer composition has been placed. For this stage, SimaPro's Eco-invent database has been used, which considers a set of processes, since the extraction of minerals/chemicals to the transportation to different processing plants in order to treat and manufacture the product in cause. Therefore, the productions of EVA and ATH have been modelled according to the following Ecoinvent inventories: *Ethylene vinyl acetate copolymer, at plant* and *Aluminium hydroxide, at plant*.

$Mg(OH)_2$ can be produced by the hydration of magnesium oxide (MgO), which is usually produced industrially from the calcination of the mineral magnesite ($MgCO_3$). No LCA data for an actual production of MDH have been found despite extensive consultations to different suppliers and numerous bibliographic searches on technical papers and patents. However, as it has been explained, the production of $Mg(OH)_2$ could be considered very similar to that of MgO. Inventory for magnesium oxide from the Ecoinvent database has therefore been used as an approximation of production data for MDH (*Magnesium oxide, at plant*).

MMT production

Since the process of nanoclay manufacturing is not available in Ecoinvent Database, specific inventory data have been compiled with inputs from LAVIOSA in order to model the cradle to gate cycle of MMT. Nanoclay manufacturing process includes several unit processes as shown in next figure:

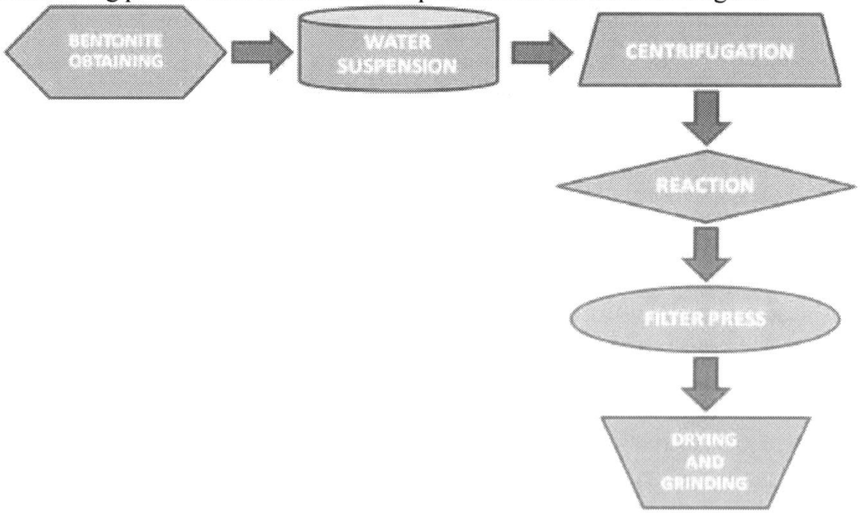

Figure 4 – MMT production

Extrusion processes
The extrusion processes involved in the life cycle studied were defined as "Extrusion, plastic pipes, RER U", from Ecoinvent database.

It is important to be noticed that life cycles of nanoenabled wire coating (system 2 and system 4) include two extrusion processes (firstly, one extrusion to polymer masterbatch production containing 40% wt MMT load and secondly, the extrusion to cable coating manufacturing with a 5% wt MMT load), while the life cycles of traditional cable coating include just the second extrusion (cable coating manufacturing). According to primary data provided by suppliers, an efficiency of 97% was assumed for all the extrusion processes.

End of life scenario
In the final stage, the end-of-life for used cable, one scenario has been considered including two different waste management routes: landfill and incineration.

In landfilling route, it has been assumed that the 12 % of final product in the end of its life will be put on a municipal landfill. In the incineration route, it has been considered that the 88% of used cable, after its lifetime, is sent to an incineration plant in order to produce energy. The data used for these two routes in this stage have been taken from SimaPro's Eco-invent database.

Regarding end of life scenarios, experimental tests were accomplished in order to support and adapt secondary data (inventory available in SimaPro) to the products systems.

Leaching tests of the four studied materials were carried out according to the European Decision 2003/33/CE to classify waste as hazardous or nonhazardous waste. After leaching tests, the obtained results of the chemical analysis of eluates allow to determine the character of waste (hazardous or non–hazardous) according to the European Decision 2003/33/CE. In this sense, Al, Mg, Si and COD have been the selected parameters to be analyzed in each sample.

In these tests, material or waste are exposed to a leaching solution in a laboratory container under a prescribed set of conditions and the concentrations of metals in the solution are measured after a specified time of exposure. According to the EN- 12457 part 4 demineralized water or equivalent ($5<pH<7.5$) has been used as leaching agent to simulate conditions in a municipal waste landfill. 100 g of material was required, which was size-reduced prior to leaching (squares about 1cm side size). Leaching took place at a 10:1 liquid solid ratio in a rotary extractor at 10 rpm for 24hours. The leachates were then filtered (0.45 μm) and analyzed for the chemicals elements of concern according to following standardized methods:

- Metals: Atomic Emission Spectrometry with Inductively Coupled Plasma (ICP-AES) according to Standard Methods S.M. 3120-B.
- Dissolved Organic Carbon: Internal method ref: TECNALIA 37-O-049 Catalytic oxidation and determination of organic carbon by infrared detector.

Figure 5 - Tecnalia's laboratory apparatus for Leaching tests EN-12457

The results of the analysis carried out on eluates obtained of leaching tests are shown in the table below:

Table 3 – Chemical analysis results

	System 1		System 2		System 3		System 4	
pH	4,53		4,9		-		8,83	
Conductivity (μS/cm)	10,6		4,5		-		13,1	
	mg/l	mg/Kg dm	mg/l	mg/Kg dm	mg/l	mg/Kg dm	mg/l	mg/Kg dm
Al	0,32	3,2	0,06	0,6	<0,05	<0,5	0,11	1,1
Mg	<0,05	<0,5	<0,05	<0,5	1.57	15.7	1,03	10,3
Si	<0,05	<0,5	<0,05	<0,5	<0,05	<0,5	<0,05	<0,5
COD	2,2	22	1,3	13	1,2	12	0,9	9

In accordance with the results obtained, no significant difference is expected between the performance of studied nanoenabled and non nanoenabled wire coatings. According to the Council Decision of 19 December 2002 establishing criteria and procedures for the acceptance of waste at landfills pursuant to Article 16 of and Annex II to Directive 1999/31/EC, these materials may be acceptable at landfills for inert waste.

Additional work is ongoing since experiment involving incineration of nanocomposites, under controlled conditions, are currently being performed. Since an incineration is an intrinsic part of the life cycle of HFFR in cable coatings (mainly in the use and end of life phases), emissions to air will be assessed.

3. RESULTS AND DISCUSSION

The ISO framework for LCA (ISO 14040:2006; ISO 14044:2006) is fully applicable to LCAs involving nanomaterials and nanoproducts. In some phases and steps, however, a number of issues need to be addressed in more detail [3].

In this sense, there is an urgent need of obtaining datasets about the most important nanotechnological processes and publishing these datasets in well-know LCA databases (inventory analysis). In parallel, there is also the need to develop a methodological approach for handling the assessment of nanomaterials by identifying what need to be measured as outputs, and how the impacts can be assessed in an impact assessment [3,4].

Furthermore, the European Commission as well as a number of authors have pointed out that for some specific materials, such as nanomaterials, no characterisation factors are available [3].

The method selected for carrying out the impact assessment is Ecoindicator 99 (EI 99H/H). Standard Eco-indicators are numbers that express the total potential environmental load of a product or process (dimensionless figures, Ecopoints, Pt). Eco-indicator 99 methodology assesses the seriousness of environmental impact on three damage categories:

1. *Damage to Human Health*, expressed as the number of year life lost and the number of years lived disabled. It includes the number and duration of diseases, and life years lost due to premature death from environmental causes. The effects we include are: climate change, ozone layer depletion, carcinogenic effects, respiratory effects and ionising (nuclear) radiation. These are combined as Disability Adjusted Life Years (DALYs), an index that is also used by the Worldbank and WHO.
2. Damage to Ecosystem Quality, expressed as the loss of species over a certain area, during a certain time. Damage is expressed in Potentially Disappeared Fraction (PDF)$\cdot m^2 \cdot year/m^2$ or m^2a. The effects we include are: ecotoxicity, acidification, eutrophication and land-use.
3. *Damage to Resources*, expressed as the surplus energy needed for future extractions of minerals and fossil fuels. The damage of resources will be experienced by future generations, as they will have to use more effort to extract remaining resources. This extra effort is expressed as "surplus energy" (MJ surplus)The depletion of agricultural and bulk resources as sand and gravel is considered under land use

As can be seen below, it is noteworthy that there are not statistically significant differences between nanoenabled systems and non nanoenabled system. The impacts of nanoenabled systems (system 2 and system 4) related to "Resp. organics", "Radiation", "Ozone layer", "Acidification/Eutrophization", "Land use" and "Fossil fuels" are slightly higher than the impacts of non nanoenabled systems (system 1 and system 3). On the other hand, environmental loads of nanoenabled systems related to "Carcinogens", "Resp. Inorganics", "Climate Change", "Ecotoxicity" and "Minerals" are a bit lower than the impacts of non nanoenabled systems. As it has been mentioned at the beginning of this section, databases and characterisation factors should be optimized and adapted to nanomaterials. As soon as new resources become available these impact assessments will be reviewed.

It is worth highlighting that the ozone layer impact is quite higher in systems incorporating ATH as traditional flame retardant, than in systems using $Mg(OH)_2$. However, all the systems studied have a very low contribution to the ozone layer depletion.

Table 4 – Characterisation results according EI99

Impact category	Unit	SYSTEM 1 EVA+ATH	SYSTEM 2 EVA+ATH+MMT	SYSTEM 3 EVA+Mg(OH)2	SYSTEM 4 EVA+Mg(OH)2+MMT
Carcinogens	DALY	1,25 E-06	1,16 E-06	2,22 E-06	2,04 E-06
Resp. organics	DALY	2,08 E-09	2,13 E-09	1,86 E-09	1,93 E-09
Resp. inorganics	DALY	1,38 E-06	1,36 E-06	1,41 E-06	1,39 E-06
Climate change	DALY	5,69 E-07	5,47 E-07	6,19 E-07	5,93 E-07
Radiation	DALY	8,80 E-09	9,42 E-09	8,41 E-09	9,05 E-09
Ozone layer	DALY	1,08 E-10	1,14 E-10	6,49 E-11	7,45 E-11
Ecotoxicity	PAF*m2yr	1,7721	1,6594	1,8904	1,7679
Acidification/ Eutrophication	PDF*m2yr	0,0268	0,02728	0,0211	0,0220
Land use	PDF*m2yr	0,02649	0,02823	0,0230	0,0250
Minerals	MJ surplus	0,6058	0,5618	0,0417	0,0447
Fossil fuels	MJ surplus	4,8332	5,0112	4,2774	4,5018

(figures in blue show that nanoenabled systems have lower environmental impact, while figures in orange show that nanoenabled systems have higher environmental impact)

As it can be noticed in next figures (Figure 6 and Figure 7) the addition of MMT does not entail a significant increase of the environmental load. In fact, in some impact categories such as land use or fossil fuels the figures related to nanoenabled systems (system 2 and system 4) are slightly higher than in non nanoenabled systems (system 1 and system 3). On the other hand, in other impact categories, for example climate change, carcinogens and ecotoxicity, nanoenabled systems have a bit less load than non nanoenabled systems.

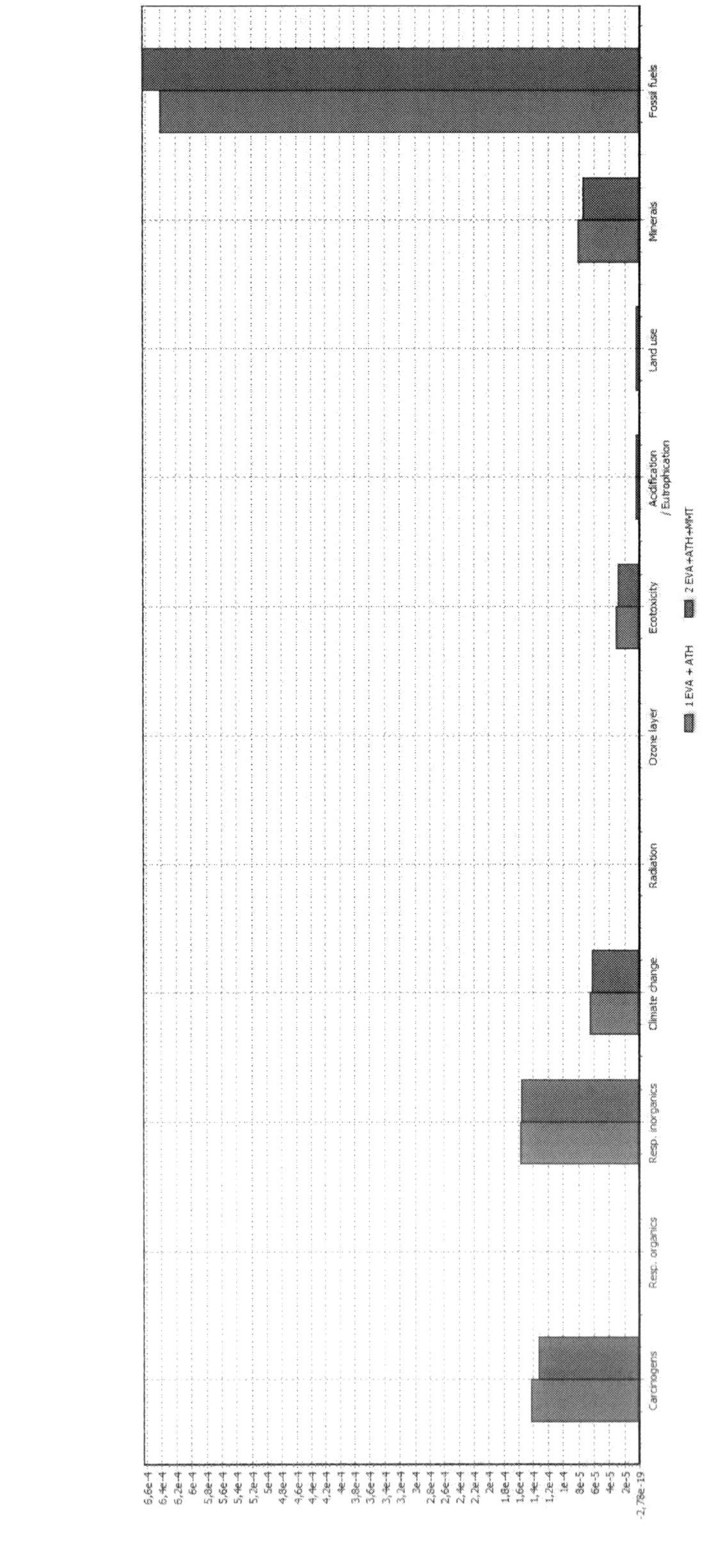

Figure 6 – System 1 and system 2 (Normalisation)

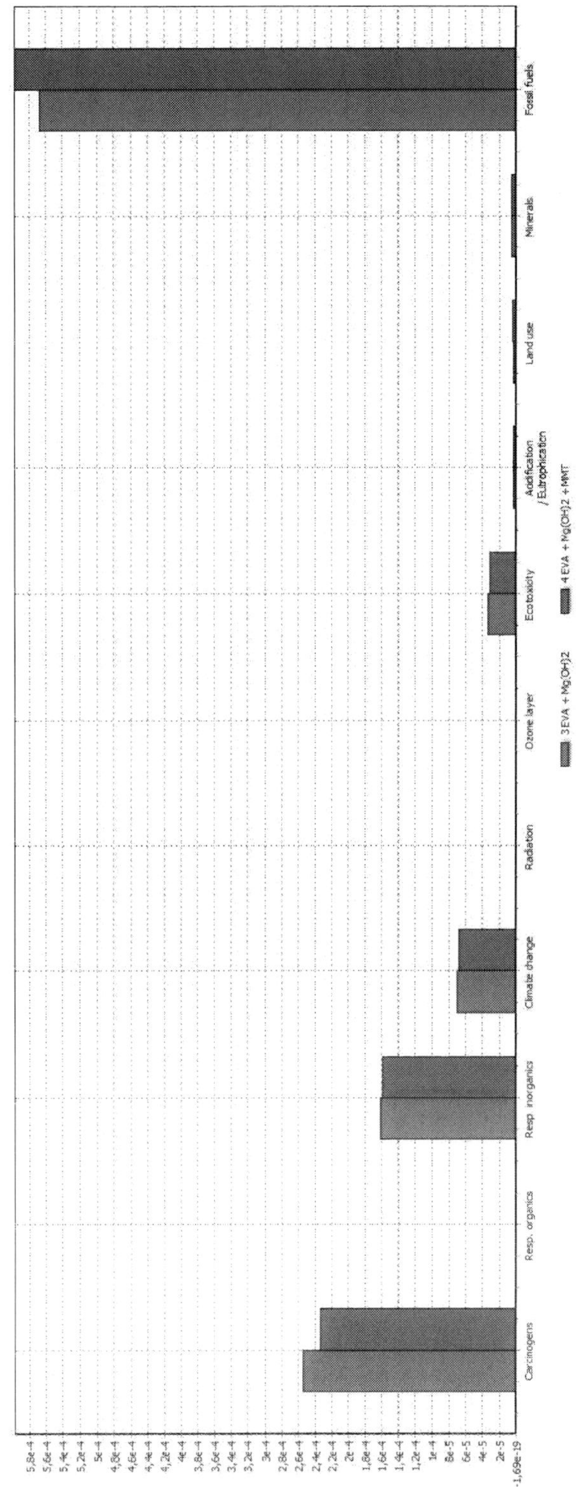

Figure 7 - System 3 and system 4 (Normalisation)

Table 5 shows the total environmental load of the studied materials in Ecopoints (Pt.). The environmental impacts are similar and very low in the four studied systems. In both comparisons (systems with ATH and systems with MDH) non nanoenabled systems (system 1 and system 3) turn out to be a bit more environmentally friendly than nanoenabled systems (system 2 and system 4).

Table 5 – Scoring by impact categories according to EI99 methodology

Impact category	Unit	SYSTEM 1 EVA+ATH	SYSTEM 2 EVA+ATH+MMT	SYSTEM 3 EVA+Mg(OH)2	SYSTEM 4 EVA+Mg(OH)2+MMT
TOTAL	**Pt**	**0,3426**	0,3428	**0,3340**	0,3349
Carcinogens	Pt	0,0431	0,0399	0,0761	0,0701
Resp. organics	Pt	7,13 E-05	7,30 E-05	6,37 E-05	6,60 E-05
Resp. inorganics	Pt	0,0474	0,0467	0,0484	0,0477
Climate change	Pt	0,0195	0,0187	0,0212	0,0203
Radiation	Pt	0,0003	0,0003	0,0003	0,0003
Ozone layer	Pt	3,72 E-06	3,92 E-06	2,22 E-06	2,55 E-06
Ecotoxicity	Pt	0,0124	0,0116	0,0132	0,0123
Acidification/ Eutrophication	Pt	0,0019	0,0019	0,0015	0,0015
Land use	Pt	0,0019	0,0019	0,0016	0,0017
Minerals	Pt	0,0241	0,0223	0,0017	0,0017
Fossil fuels	Pt	0,1921	0,1991	0,1700	0,1789

(figures in blue show that nanoenabled systems have lower environmental impact, while figures in orange show that nanoenabled systems have higher environmental impact)

It is worth pointing out that, in all cases, the higher contribution to the environmental burden corresponds to the "fossil fuels" category, ranging between the 50% up to 58% of the total score. Therefore, the "Damage to Resources" is the more affected damage category. Hence, a great chance to minimize the environmental load of any of these systems would be to reduce the fossil fuels consumption throughout the life cycle. So, one of the recommendations from this study would be the promotion of the replacement of fossil fuels for renewable energy sources in order to halving the potential environmental burden.

4. CONCLUSIONS:

Leaching tests results allowed to conclude that every studied materials could be accepted in municipal landfill and the presence of those materials as waste did not significantly changed the environment in a controlled landfill.

During the course of this study, previously described limitations have been found, mainly related to nanomaterial's data deficiency in inventory databases and the lack of characterization factors for them. In spite of these issues the study has been carried out as precisely as possible. Nevertheless, it is expected that databases and impact assessment methods will be adapted to nanomaterials in the short to medium term.

As already explained, the environmental impacts are similar and very low in the four studied systems.

Systems including ATH in their composition have a slightly higher impact than MDH systems. Although, it was a bit higher in case of systems including ATH, environmental loads were nearly equals.

The standard protocol UL 94 confirmed an increased fire retardance of the nanoreinforced EVA matrixes. However, the addition of nanomaterials seems to have no significant relevance in the environmental assessment taking into account the above mentioned considerations related to the lack of characterization factors for nanomaterials. In fact, nanoaenabled systems have a slightly higher impact than non nanoenabled systems.

Finally, it is worth highlighting that the "Damage to Resources" category is the main affected in all studied materials and specifically "fossil fuels" category presents the higher contribution to the total environmental burden score (50% - 58%). The total or partial replacement of the fossil fuel use in the whole life-cycle of the product will reduce considerably the potential environmental impact.

The LCA is still under development and efforts are underway to include the results of incineration tests in order to consider more accurately such aspects as emissions and slags treatment at landfill.

5. ACKNOWLEDGMENTS:

The research leading to these results has received funding from the European Community's Seventh Framework Programme (FP7/2007-2013) under Grant Agreement N°228536. The contents of this publication are the sole responsibility of the authors and can in no way be taken to reflect the views of the European Union.

Reference

[1] Pirngadi, P.; Ritcher, S. New Developments in the Application of Nanoclays for HFFR Cable Compounds. 2006

[2] Compounding World, Desenredando la industria de alambres y cables. Julio/Agosto 2011

[3] Lee, J.H. et al. Flame Retardancy and Mechanical Property of Polypropylene/Nylon Nanocomposite Reinforced with Montmorillonite. Polymer (Korea), Vol 27, No 6 576-582. – 2003 -

[4] Zanetti, M.; Camino, G.; Canavese, D.; Morgan, A.B.; Lamelas, F.J.; Wilkie, C.A. Fire Retardant Halogen-Antimony-Clay Synergism in Polypropylene Layered Silicate Nanocomposites. Chem. Mater. 2002, 14, 189-193

[5] Cogen, J.M.; Lin, T.S.; Morgan, A.B.; Garcés, J.M. Novel Synthetic Nanocomposite Materials and Their Application in Polyolefin-Based Wire and Cable Compounds. The Dow Chemical Company

Assessment of nanoparticle release and associated health effect of polymer-silicon composites

Huijun Zhu[1*], Adeel Irfan1, Sophia Sachse[2] and James Njuguna[2]
[1]. Cranfield Health, [2]. Centre for Automotive Technology, Cranfield University, UK

* Correspondence author: h.zhu@cranfield.ac.uk

Abstract. Little information is currently available on possible release of nanomaterials or/and nanoparticles (NP) from conventional and novel products and associated health effect. This study aimed to assess the possible release of NP during the application stage of conventional and nanoproducts. NP release was monitored during physical processing of polymer-silicon composites, and the toxicity of both the released NP and the raw silica nanomaterials that were used as fillers in the nanocomposites was assessed in vitro using human lung epithelial A549 cells. This study suggests that 1) NP can be released from the conventional and novel polymer-silicon composites under certain application scenario; 2) the level of NP release from polymer composites could be altered by different reinforcement materials; e.g. nanostructured MMT could reduce the release while SiO2 NP could increase the release; 3) working with polymer composites under certain conditions could risk inhalation of high level of polymer NP; 4) raw nanomaterials appeared to be toxic in the chosen in vitro system. Further study of the effect of novel filler materials on NP release from final polymer products and the effect of released NP on environment and human health will inform design of safe materials and minimization of negative impact on the environment and human health. Introduction.

Keywords: Polymer-silicon composite, nanoparticles, A549 cells.

1. Introduction

The fast advancement in the nanotechnology and manufacturing of engineering nanomaterials has inevitably caused concerns over their impact on environment and human health. Little information is currently available on possible release of nanomaterials or/and nanoparticles (NP) from conventional and novel products and associated health effect. The NEPHH project funded under the EU FP7 programme is aimed to assess the environment and health impact of polymer-silicon nanocomposites in comparison to conventional polymer composites from life cycle perspective. As part of the NEPHH project, this study focused on assessing the possible release of NP during the application stage of conventional and nanoproducts. NP release was monitored during physical processing of polymer-silicon composites, and the toxicity of both the released NP and the raw silica nanomaterials that were used as fillers in nanocomposites was assessed in vitro using human lung epithelial A549 cells.

2. Materials and methods

The components of some of the polymer composites tested in the NEPHH project are listed in table 1. The polyamide 6 (PA6)- and Polypropylene (PP)-composites were used as matrixes which were reinforced with 5% of glass fibres (GF, Taiwan Glass Industry Corporation), SiO2 NP (Aerosil 200 or Aerosil 974, both 12 nm, amorphous, fumed, Evonik Degussa Polska Sp. z. o. o), organically modified montmorillonite (MMT, Dellite 43B and Dellite 72T, Laviosa Chimica Mineraria S.p.A), or foam glass crystal (FGC, provided by Tomsk Polytechnic University, Russia). The PA6 and PP neat polymers were also included as reference materials (REF). Among the reinforcement materials SiO2 NP and MMT are nanomaterials and the rest are microsized materials. The testing specimens with dimension of 100x100x10 mm^3 were prepared by the compression moulding technique with a temperature of 250°C over the compression time of 5 min [1]. These specimens were subjected to drilling in a testing chamber. The release of NP from PA6 and PP group composites was comparatively quantified using a scanning mobility particle sizer (SMPS) during the 28 minutes of drilling. NP in the dust of the testing chamber were extracted using a two-step of filtration method.

The toxicity of the dust NP and the raw SiO2 NP at concentrations of 25 µg/ml, 50 µg/ml and 100 µg/ml was assessed in human lung epithelial A549 cells by membrane integrity (LDH) and viability (MTT) assays. Silica NP of size 7 nm (Si 7) and H2O2 (200 µM) were used as positive controls in these assays. In order to correlate the toxicity of NP with their size, all the testing NP were characterized for dispersion pattern and size distribution in cell culture medium by dynamic light scattering and SEM assays before toxicity studies.

Table 1. Components the polymer materials tested in the NEPHH project

Polymers	Reinforcement materials					
	Dellite 43B MMT Nanostructured	**Dellite 72T MMT Nanostructure**	**SAEROSIL® 200 SiO2 NP Hydrophilic**	**AEROSIL® 974 SiO2 NP Hydrophobic**	**FGC Microsize particles**	**GF Microsize particles**
PA6	PA-MMT		PA-SiO2		PA-FGC	PA-GF
PP		PP-MMT		PP-SiO2	PP-FGC	PP-GF

3. Result and discussion

It was shown in figure 1 that under drilling, the PA6 group materials generated more airborne NP than the PP group (compare A with B) with the polymeric-silica nanocomposites generating the most. Interestingly, the polymer-MMT composite produced the lowest amount of NP, suggesting that the use of MMT as fillers could prevent the release of NP from final polymer products under certain application conditions. The level of NP release from PA6 group materials is in the order of PA6-SiO2>PA6-FGC>PA6-GF>PA6-REF>PA6-MMT. The concentration of NP in the air reduced subsequently after reaching peak concentration at 500000 NP/cm^3 or higher, depending on the type of specimens. The airborne NP were also examined by scanning electron microscopy, suggesting that the released NP formed aggregates and were of polymer but not filler material origins.

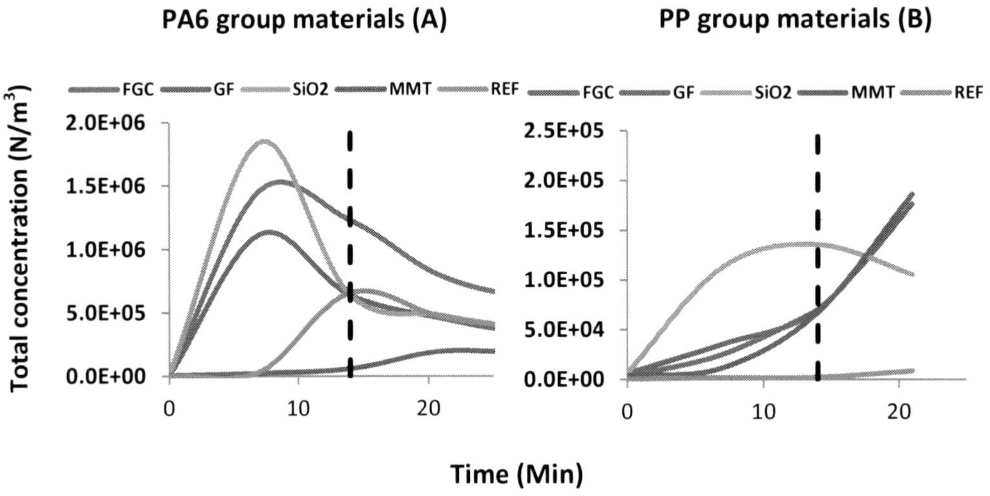

Figure 1. Monitoring of airborne NP using SMPS during 28 minutes of drilling. The concentration of NP generated from PA6 group materials shown in A; the concentration of NP generated from PP group materials shown in B. The lines represent NP released from polymer composites with reinforcement material of foam glass crystal (FGC), glass fiber (GF), silica NP (SiO2), montmorillonite (MMT), or without reinforcement material (REF).

At concentrations up to 100 µg/ml, all the dust NP disperse well in culture medium with an average size less than 100 nm. In contrast, all the raw silica NP at 10-100 µg/ml exhibit a polydispersion pattern with particle size increasing in a concentration dependent manner. Figure 2 illustrates a concentration dependent dispersion pattern of Aerosil 200 in culture medium. Other silica NP studied also showed very similar behaviour in the culture medium.

IOP Conf. Series: Materials Science and Engineering **40** (2012) 012015 doi:10.1088/1757-899X/40/1/012015

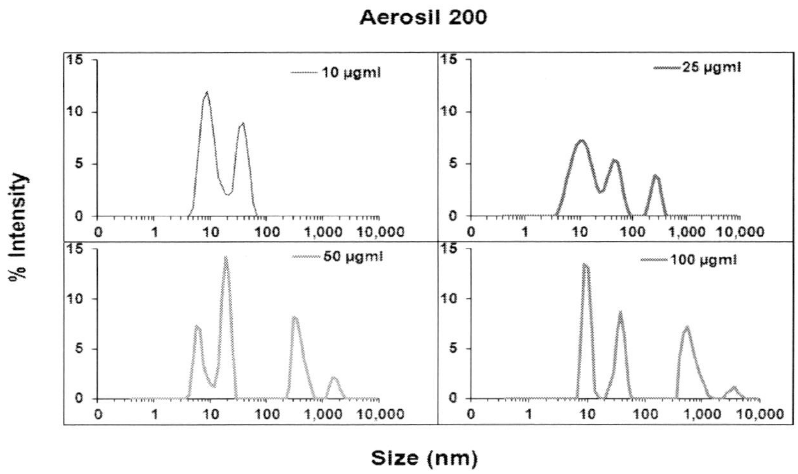

Figure 2. Concentration dependent dispersion pattern of silica NP in culture medium. Aerosil 200 was dispersed in culture medium at concentrations as indicated. The dispersion pattern and size distribution were determined by DLS assay.

The cytotoxicity in dust NP was not detectable as assessed by the LDH assay in A549 cells at the concentrations tested for up to 48 h (figure 3 A and B) and was only detected by the MTT assay at 72 h (Figure 3 C and D). In contract, the toxicity of raw SiO2 NP was detected by the LDH assay as early as 12 h (figure 4 A) and by the MTT assay at 24 h, 48h and 72 h (figure 4 B).

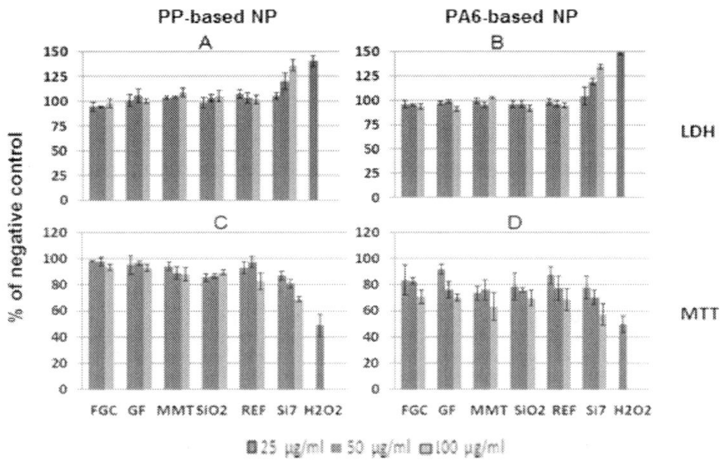

Figure 3. Cellular effect of NP from PP (A) and PA6 (B) materials. Cell membrane integrity (LDH) assay at 48 hour (A and B); cell viability assay at 72 hour (C and D).

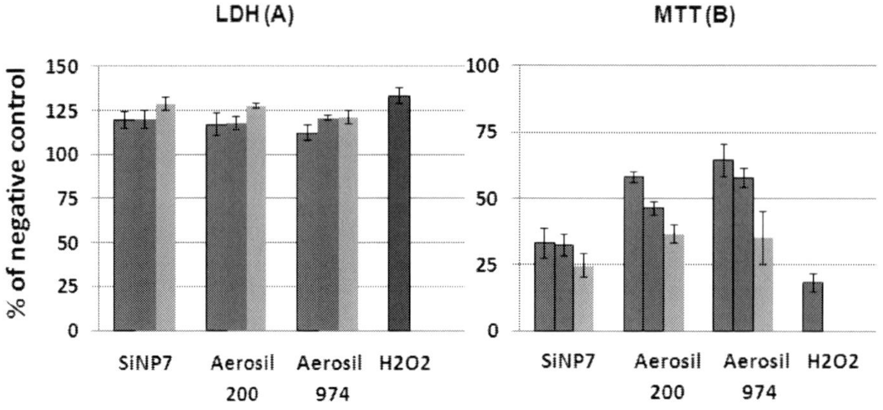

Figure 4. Raw SiO2 NP toxicity assessment. Cell membrane integrity (LDH) assay at 12 hours (A); cell viability (MTT) assay at 72 hours (B). ▨25 μg/ml ■ 50 μg/ml ▨100 μg/ml

4. Conclusion

Our study is consistent with previous work demonstrating that polymeric products can generate NP under certain application scenarios [2,3]. This study further suggests that 1) the level of NP release from polymer composites could be altered by different reinforcement materials; e.g. nanostructured MMT could reduce the release while SiO2 NP could increase the release; 2) working with polymer composites under certain conditions could risk inhalation of high level of polymer NP, although which exhibit low toxicity; 3) raw nanomaterials appeared to be toxic in the chosen in vitro system. Further study of the effect of novel filler materials on NP release from final polymer products and the effect of released NP on environment and human health will inform design of safe materials and minimization of negative impact.

Acknowledgement

This research was financially supported by the European Commission's FP7 programme under the project NEPHH (CP-FP 228536-2). We wish to thank Professor Krzysztof Pielichowski (Department of Chemistry and Technology of Polymers, Cracow University of Technology, Poland) for providing all the polymer specimens, and Professor Olga V Kazmina (Department of Silicate Technology and Nanotechnology, Tomsk Polytechnic University, Russia) for providing FGC.

Reference

[1] Sachse, S., Irfan, A., Zhu, H & Njuguna, J. 2011**7**:5-7,8,9.
[2] Roubicek V, Raclavska H, Juchelkova D, Filip P. Wear 2008; **265**167-75.
[3]Vorbau M, Hillemann L, Stintz M. J Aerosol Sci 2009; **40** 209-17.

IOP Conf. Series: Materials Science and Engineering **40** (2012) 012016 doi:10.1088/1757-899X/40/1/012016

Expression of circadian gens in different rat tissues is sensitive marker of in vivo silver nanoparticles action

D O MINCHENKO[1,2], O P YAVOROVSKY[2] , T O ZINCHENKO[1,2], S V KOMISARENKO[1] and O H MINCHENKO[1,*]

[1]Department of Molecular Biology, Palladin Institute of Biochemistry National Academy of Science of Ukraine, 9 Leontovycha St., Kyiv 01601, Ukraine;
[2]National O.O. Bohomolets Medical University, Kyiv, Ukraine;

* - corresponding author; address: 9 Leontovycha St., Kyiv 01601, Ukraine; e-mail: ominchenko@yahoo.com

Abstract. Circadian factors PER1, PER2, ARNTL and CLOCK are important molecular components of biological clock system and play a fundamental role in the metabolism at both the behavioral and molecular levels and potentially have great importance for understanding metabolic health and disease, because disturbance the circadian processes lead to developing of different pathology. The antibacterial effect of silver nanoparticles has resulted in their extensive application in health, electronics, home products, and for water disinfection, but little is yet known about their toxicity. These nanoparticles induce blood-brain barrier destruction, astrocyte swelling, cause degeneration of neurons and impair neurodevelopment as well as embryonic development. We studied the expression of genes encoded the key molecular components of circadian clock system in different rat organs after intratracheally instilled silver nanoparticles which quite rapidly translocate from the lungs into the blood stream and accumulate in different tissues. We have shown that silver nanoparticles significantly affect the expression levels of PER1, PER2, ARNTL and CLOCK mRNA in different rat tissues in time-dependent and tissue-specific manner. High level of PER1, ARNTL and CLOCK mRNA expression was observed in the lung on the 1st, 3rd and 14th day after treatment of rats with silver nanoparticles. At the same time, the expression level of PER1 mRNA in the brain and liver increases predominantly on the 1st and 14th day but decreases in the testis. Significant increase of the expression level of PER2 and ARNTL mRNA was detected only in the brain of treated by silver nanoparticles rats. Besides that, intratracheally instilled silver nanoparticles significantly reduced the expression levels of CLOCK mRNA in the brain, heart and kidney. No significant changes in the expression level of PER2 mRNA were found in the lung, liver, heart and testis, except kidney where this mRNA expression decreases on the 3rd and 14th day after treatment of rats with silver nanoparticles. It was also shown that expression level of PFKFB4, a key enzyme of glycolysis regulation, gradually reduces in the brain from 1st to 14th day being up to 4 fold less on 14th day after treatment of animals with silver nanoparticles. Thus, the intratracheally instilled silver nanoparticles significantly affect the expression of PER1, PER2, ARNTL, and CLOCK genes which are an important molecular component of circadian clock system. This is because a disruption of the circadian processes leads to a development of various pathologic processes. The results of this study clearly demonstrate that circadian genes could be a sensitive test for detection of silver nanoparticles toxic action and suggest that more caution is needed in biomedical applications of silver nanoparticles as well as higher level of safety in silver nanoparticles production industry.

1. INTRODUCTION

Most of physiological and metabolic processes in organism are controlled by regulatory factor network, which includes a lot of protein kinases, protein phosphatases and transcription factors [1–3]. These factors are key regulators of metabolism both in normal and different pathological conditions, its circadian character of regulation [4, 5]. Moreover, the mammalian molecular clockwork controls rhythmic expression of its own input pathway components and feedback repression is also required for circadian clock function [6–10]. The circadian rhythms oscillations are generated by endogenous circadian clocks, present in virtually all cells where they control key biological processes. The circadian factors PER1 (period 1), PER2 (period 2), ARNTL (aryl hydrocarbon receptor nuclear translocator-like), and CLOCK (circadian locomotor output circuits kaput) are important molecular components of biological clock system and play a fundamental role in the metabolism at both the behavioral and molecular levels and potentially have great importance for understanding metabolic health and disease, because disturbance the circadian processes lead to developing of different pathology [11–16].

Casein kinase-1epsilon and casein kinase-1delta are also important molecular components of circadian clock system and play essential roles in the mammalian circadian clock and in the metabolism at both the behavioral and molecular levels; potentially have great importance for understanding metabolic health and disease [17]. It was shown that casein kinase-1epsilon phosphorylates circadian regulatory proteins ARNTL, cryptochromes, PER1, PER2 and PER3 and that SCFbeta-TRCP (SKP1-CULL-F-box protein - transducin repeat-containing protein) controls clock-dependent transcription via degradation of the PER1 protein, down-regulates phospho-AKT via PTEN, following genotoxic stress-induced apoptosis in hematopoietic cells, regulates YAP (YES kinase-associated protein) stability through SCFbeta-TRCP and controls many other processes [18–24].

The antibacterial effect of silver nanoparticles has resulted in their extensive application in health, electronic, home products, and for water disinfection, but little is yet known about their toxicity [25 – 27]. Silver nanoparticles induce blood-brain barrier destruction, astrocyte swelling, caused neuronal degeneration and impair neurodevelopment and embryonic development, as well as induced oxidative stress and apoptosis [28 – 35]. Intratracheally instilled silver nanoparticles quite rapidly translocate from the lung into the blood stream and accumulate in different tissues; however, translocation pathway of the intratracheally instilled ultrafine particles from the lung into the blood circulation is not clear yet [36, 37]. Moreover, results of *in vitro* experiments clearly demonstrated that silver nanoparticles crossed the blood-brain barrier and accumulated inside brain microvascular endothelial cells, while the silver microparticles did not [38]. It is possible that silver nanoparticles pass the blood-brain barrier mainly by transcytosis of capillary endothelial cells. To see the possible effects of the silver nanoparticles on the organism, several key regulatory genes were chosen; among them were genes encoded the PER1, PER2, ARNTL and CLOCK, sensitive markers of cellular stress. Previously we have shown that circadian genes are sensitive to the action of ecotoxicant methyl tertial butyl ether as well as that SNF1/AMP-activated protein kinase (SNARK) and casein kinase-1epsilon are sensitive markers of silver nanoparticles action [39, 40].

This study was undertaken to examine the effect of the intratracheally instilled silver nanoparticles on the expression of genes encoded the key molecular components of circadian clock system (PER1, PER2, ARNTL and CLOCK) in rat brain, lung, liver, heart, kidney and testis.

2. MATERIALS AND METHODS

Male Wistar rats (an outbred strain of albino rats belonging to the species *Rattus norvegicus*; 230–240 g) were intratracheally instilled by 30% silver nanoparticles (28-30 nm) in sodium chloride matrix aerosol in dose 0.05 mg/kg body weight and the expression of the SNARK and casein kinase-1epsilon genes were analyzed in the lung, liver, brain, heart, kidney and testis on the 1st, 3rd or 14th day after treatment using quantitative polymerase chain reaction (qPCR). Silver nanoparticles were prepared in the Laboratory No. 84 of the E.O. Paton Electric Welding Institute of National Academy of Sciences of Ukraine.

Total RNA was extracted using Trizol reagent (Invitrogen, USA), precipitated with equal volume of 2-propanol, collected by centrifugation, washed twice with 75 % ethanol and dissolved in ribonuclease free water as described previously [41]. QuaniTect Reverse Transcription Kit (QIAGEN, Germany) was used for complementary DNA (cDNA) synthesis according to manufacturer protocol. Quantitative PCR was performed on "Mx3000P QPCR" (Stratagene, USA), using SYBRGreen Mix (AB gene, Great Britain) and specific for each gene primers:

for period-1 (PER1) – 5'– TCTCTTCTCAGAACTGGATG –3' (forward) and 5'– GGAAGCCTCTCATTAGACTGC –3' (reverse); nucleotide sequences 3699 – 3718 and 3983 – 3963 of rat PER1 (GenBank accession number NM_001034125);

for period-2 (PER2) – 5'– CTGGGAAGATCCTGTACATC –3' (forward) and 5'– GCTGGTAGCGAATCTCATTC –3' (reverse); nucleotide sequences 700 – 719 and 960 – 941 of rat PER2 (GenBank accession number NM_031678);

for aryl hydrocarbon receptor nuclear translocator-like (ARNTL; BMAL1) – 5'– TGACCCTCATGGAAGGTTAG –3' (forward) and 5'– AATCCATCTGCTGCCCTGAG –3' (reverse); nucleotide sequences 753 – 772 and 1042 – 1061 of rat ARNTL (GenBank accession number NM_024362);

for circadian locomotor output circuits kaput (CLOCK) – 5'– TGCACAGTCAGATGCTAGTG –3' (forward) and 5'– TGATCCACAAGATCAGATGG –3' (reverse); nucleotide sequences 264 – 283 and 455 – 436 of rat CLOCK (GenBank accession number NM_021856);

for 6-phosphofructo-2-kinase/fructose-2,6-bisphosphatase-4 (PFKFB4) – 5'– GCAGTGGAAGGTCCTCAACG –3' (forward) and 5'– GTCAAGGAAGTAGGCCAG –3' (reverse); nucleotide sequences 991 – 1010 та 1246 – 1229 (GenBank accession number NM_019333);

for glyceraldehyde-3-phosphate dehydrogenase (GAPDH) – 5'– ATCTTCCAGGAGCGAGATCC –3' (forward) and 5'– TCATGGATGACCTTGGCCAG –3' (reverse); nucleotide sequences 291 – 310 та 562 – 543 (GenBank accession number X02231);

for –actin (ACTB) – 5'– GGACTTCGAGCAAGAGATGG –3' (forward) and – 5'– AGCACTGTGTTGGCGTACAG –3' (reverse); nucleotide sequences 704 – 723 та 937 – 918 (GenBank accession number X00351). Primers were received from Sigma (USA).

Quantitative PCR reaction was performed in triplicate. Analysis of quantitative PCR was performed using special computer program "Differential expression calculator" and statistical analysis – in Excel program.

Expression levels of 6-phosphofructo-2-kinase/fructose-2,6-bisphosphatase-4 (PFKFB4), glyceraldehyde-3-phosphate dehydrogenase (GAPDH) and –actin were also analyzed by regular PCR using same primers and "MasterCycler Personal" (Eppendorf, Germany). Amplification products were analyzed by electrophoresis in 2% agarose gel, identified with Sight DNA Stain (EUROMEDEA) and analyzed in Quantity One BioRad System (USA).

3. RESULTS AND DISCUSSION

The quantitative polymerase chain reaction analysis of the expression of period 1 (PER1) mRNA shown that this mRNA level significantly increases in the brain, liver and lung after treatment of rats with silver nanoparticles, being more intense on the 1st and 14th day in the brain but in the lung – on the 1st, 3rd and 14th day (Fig. 1 and 2). We have also shown that PER1 mRNA expression level significantly decreases in the testis on the 1st day and was at the same level on the 3rd and 14th day after treatment of rats with silver nanoparticles as compared to control animals (Fig. 1).

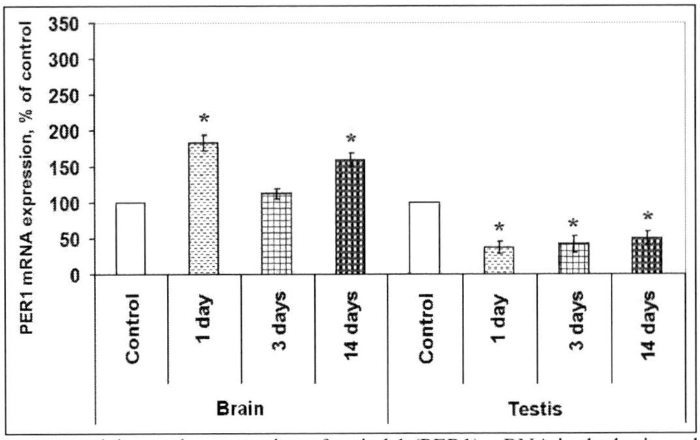

Figure 1. Effect of silver nanoparticles on the expression of period 1 (PER1) mRNA in the brain and testis in 1, 3 or 14 days after treatment. Values of PER1 mRNA expressions were normalized to □-actin mRNA expressions measured by qPCR; n = 3; * - P < 0.05.

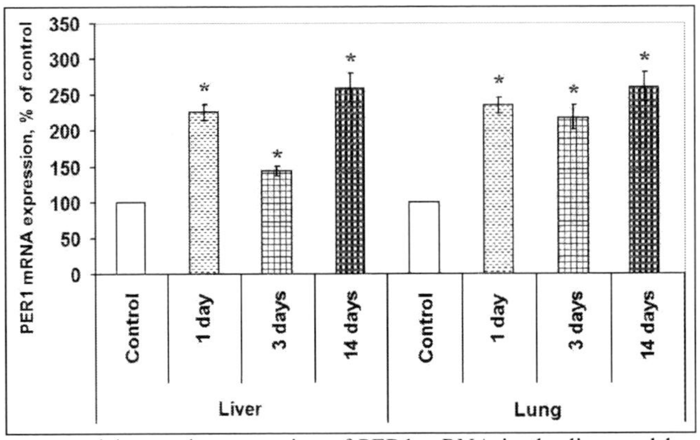

Figure 2. Effect of silver nanoparticles on the expression of PER1 mRNA in the liver and lung in 1, 3 or 14 days after treatment. Values of PER1 mRNA expressions were normalized to □-actin mRNA expressions measured by qPCR; n = 3; * - P < 0.05.

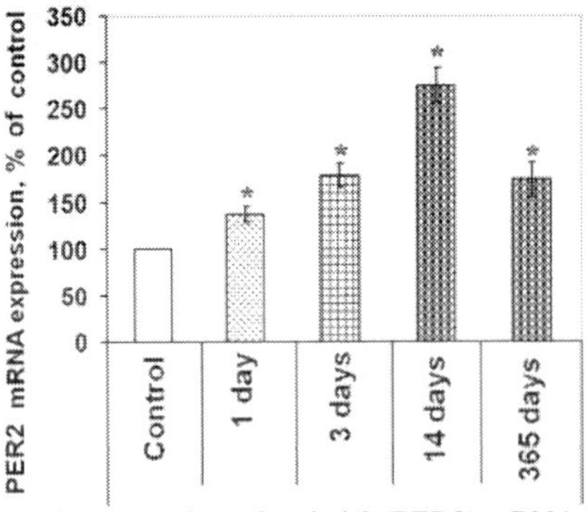

Figure 3. Effect of silver nanoparticles on the expression of period 2 (PER2) mRNA in the brain in 1, 3, 14 or 365 days after treatment. Values of PER2 mRNA expressions were normalized to □-actin mRNA expressions measured by qPCR; n = 3; * - P < 0.05.

As shown in Fig. 3, the expression level of PER2 mRNA is gradually increased in the brain on the 1st, 3rd and 14th day after treatment of rats with silver nanoparticles being still increased even in 1 year after treatment. No significant changes in PER2 mRNA level were observed after the treatment of rats with these nanoparticles in other tested tissues (lung, liver, heart and testis) except kidney where was shown a decreased level of this mRNA expression (Fig. 4 and 5).

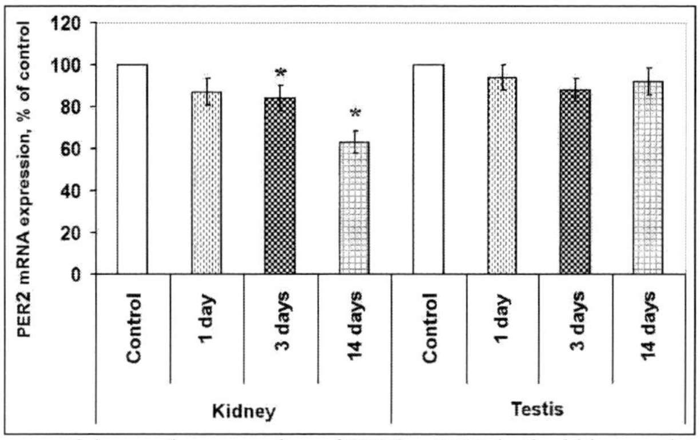

Figure 4. Effect of silver nanoparticles on the expression of PER2 mRNA in the kidney and testis in 1, 3 or 14 days after treatment. Values of PER2 mRNA expressions were normalized to β-actin mRNA expressions measured by qPCR; $n = 3$; * - $P < 0.05$.

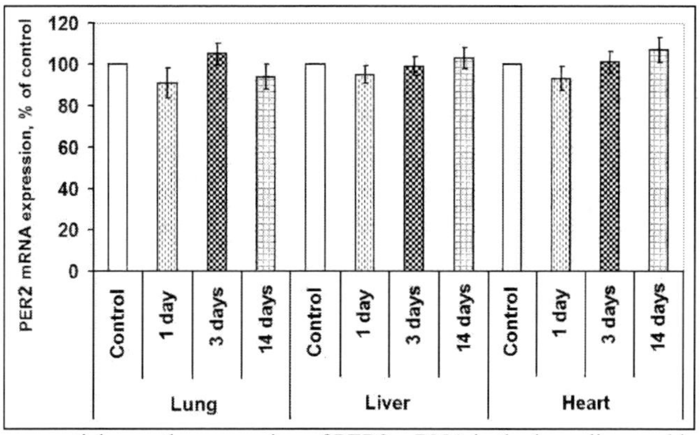

Figure 5. Effect of silver nanoparticles on the expression of PER2 mRNA in the lung, liver and heart in 1, 3 or 14 days after treatment. Values of PER2 mRNA expressions were normalized to β-actin mRNA expressions measured by qPCR; $n = 3$; * - $P < 0.05$.

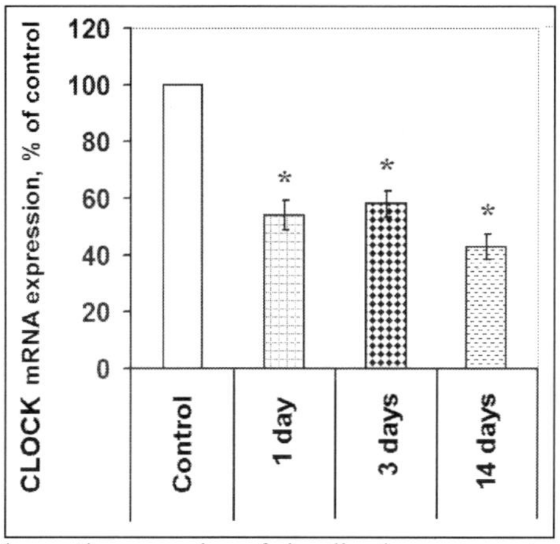

Figure 6. Effect of silver nanoparticles on the expression of circadian locomotor output circuits kaput (CLOCK) mRNA in the brain in 1, 3 or 14 days after treatment. Values of CLOCK mRNA expressions were normalized to β-actin mRNA expression measured by qPCR; $n = 3$; * - $P < 0.05$.

We have also demonstrate that silver nanoparticles significantly changes the expression of circadian locomotor output circuits kaput (CLOCK) mRNA in different rat organs on the 1st, 3rd and

14th day after treatment of rats (Fig. 6, 7 and 8). This effect of silver nanoparticles strongly depends on time after the treatment of animals with nanoparticles in a tissue-specific manner.

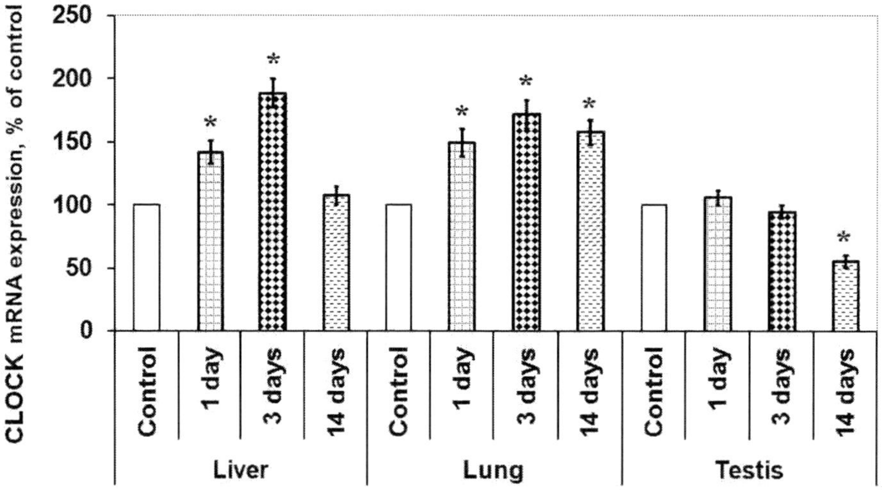

Figure 7. Effect of silver nanoparticles on the expression of CLOCK mRNA in the liver, lung and testis in 1, 3 or 14 days after treatment. Values of CLOCK mRNA expressions were normalized to β-actin mRNA expression measured by qPCR; $n = 3$; * - $P < 0.05$.

Thus, the expression level of CLOCK mRNA significantly decreased in the brain, kidney and heart in 1, 3 or 14 days after treatment (Fig. 6 and 8). At the same time, CLOCK mRNA expression level decreases in the brain and kidney almost two fold on the 1st day and was in the brain at the same level and in kidney increases on the 3rd and 14th day after treatment of rats with silver nanoparticles as compared to control animals. In contrast, the expression level of CLOCK mRNA in the heart slightly decreases on the 1st day and much stronger effect was observed on the 3rd and 14th day after treatment of rats with silver nanoparticles (Fig. 8).

As shown in Fig. 9, the expression level of aryl hydrocarbon receptor nuclear translocator-like (ARNTL) mRNA in the lung and brain also increases after intratracheally treatment of rats with silver nanoparticles being much more significant on the 14th day as compared to the 1st and 3rd day after treatment.

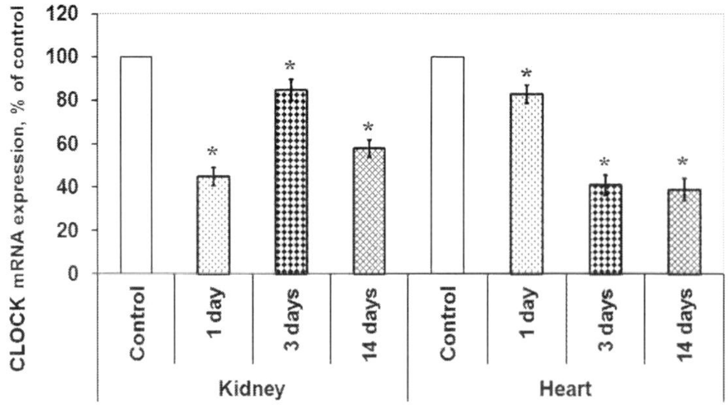

Figure 8. Effect of silver nanoparticles on the expression of CLOCK mRNA in the kidney and heart in 1, 3 or 14 days after treatment. Values of CLOCK mRNA expressions were normalized to β-actin mRNA expression measured by qPCR; $n = 3$; * - $P < 0.05$.

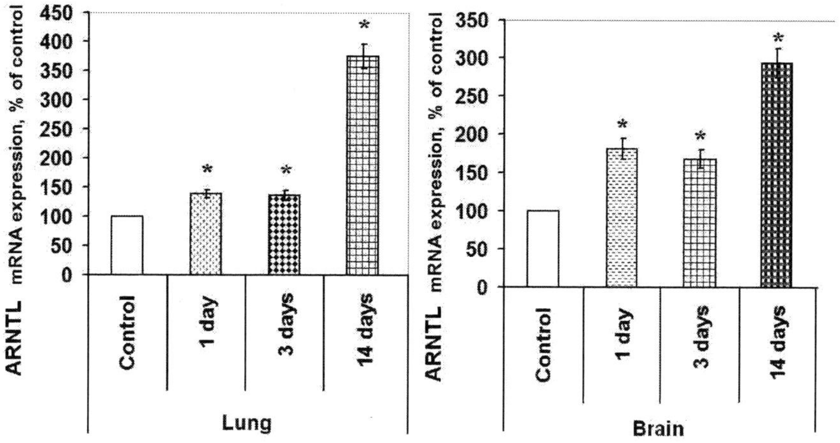

Figure 9. Effect of silver nanoparticles on the expression of aryl hydrocarbon receptor nuclear translocator-like (ARNTL or BMAL1) mRNA in the lung and brain in 1, 3 or 14 days after treatment. Values of ARNTL mRNA expressions were normalized to β-actin mRNA expression measured by qPCR; $n = 3$; * - $P < 0.05$.

We also demonstrate that silver nanoparticles significantly changes the expression level of 6-phosphofructo-2-kinase/fructose-2,6-bisphosphatase-4 (PFKFB4) and glyceraldehyde-3-phosphate dehydrogenase (GAPDH) mRNA in the brain (Fig. 9). Thus, the expression level of PFKFB4 in the brain does not change significantly on the 1st day and decreases in 2 and almost 4 fold on the 3rd and 14th day after intratracheally treatment of rats with silver nanoparticles. At the same time, the expression level of glyceraldehyde-3-phosphate dehydrogenase (GAPDH) mRNA in the brain increases on the 1st day and does not change significantly on the 3rd and 14th day under silver nanoparticles action.

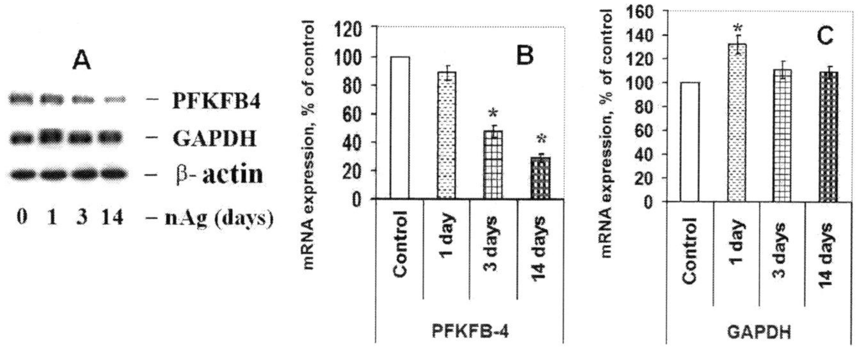

Figure 10. Effect of silver nanoparticles on the expression of 6-phosphofructo-2-kinase/fructose-2,6-bisphosphatase-4 (PFKFB4) and glyceraldehyde-3-phosphate dehydrogenase (GAPDH) mRNA in the brain in 1, 3 or 14 days after treatment. Values of PFKFB4 and GAPDH mRNA expressions were normalized to β-actin mRNA expression measured by PCR (A) or qPCR (B and C); $n = 3$; * - $P < 0.05$.

Thus, intratracheally instilled silver nanoparticles significantly change the expression of circadian genes encoded important regulatory factors in different rat organs, not only in the lung tissue. Moreover, this effect of silver nanoparticles strongly depends on time after the treatment of animals with these nanoparticles in a tissue-specific manner. These results demonstrate strong induction or suppression of the expression of different circadian factors mRNA in most vital organs of rats intratracheally treated with silver nanoparticles.

Results of this investigation clarify some molecular mechanisms of silver nanoparticles toxic action on organism after its intratracheally instillation. It is possible that silver nanoparticles as artificial chemical structures can quite rapidly translocate from the lung into the blood stream, accumulate in different tissues and initiate the stress conditions leads to unfolded protein response and activation of signaling systems of endoplasmic reticulum stress; however, translocation pathway of the intratracheally instilled ultrafine particles from the lung into the blood circulation is not clear yet [36, 37]. It is interested to note that the silver microparticles could not cross the blood-brain barrier and accumulated inside brain microvascular endothelial cells [38].

We suggest that silver nanoparticles affect the expression of circadian gene expression via unfolded protein response which controls the protein folding capacity of the endoplasmic reticulum through different signalling systems. Central to this signaling pathway is the endoplasmic reticulum-resident bifunctional transmembrane kinase/endoribonuclease ERN1 (endoplasmic reticulum to nuclei-1 or inositol requiring enzyme-1; IRE-1) [42 – 44]. The endoribonuclease domain of this enzyme initiates a non-conventional mRNA splicing reaction, leading to the production of a transcription factor XBP-1 that controls unfolded protein response target genes [45 – 49]. The mRNA splicing reaction is an obligatory step of IRE-1 signaling. Endoplasmic reticulum stress induced inositol requiring enzyme-1 signalling via XBP-1 control diverse cell type- and condition-specific transcriptional regulatory networks. XBP-1 binding sites were identified into promoter region of PER1 and PER2 [49]. This data support the possibility of PER1 and PER2 gene regulation via endoplasmic reticulum stress responsible mechanism. However, the cellular mechanism for survival under stress conditions is complex and further investigation of the mechanism by which silver nanoparticles affects circadian gene expression as well as biologic significance of silver nanoparticles induced alteration in these genes function are needed. The stage is now set for the elucidation of the molecular mechanisms responsible for these important regulatory factors responses to silver nanoparticles action.

The results of this investigation clearly demonstrate that silver nanoparticles have a significant effect on important regulatory mechanisms which control metabolic processes in different tissues via circadian gene expression, which can be considered as sensitive marker for the silver nanoparticles toxic action. These results suggest that more caution is needed in biomedical applications of silver nanoparticles as well as higher level of safety in silver nanoparticles production industry.

REFERENCES

[1] Virshup D.M., Eide E.J., Forger D.B., Gallego M., Harnish E.V. Reversible Protein Phosphorylation Regulates Circadian Rhythms. *Cold Spring Harb. Symp. Quant. Biol.,* **2007**, *72*, 413-420.

[2] Gonze D., Goldbeter A. Circadian rhythms and molecular noise. *Chaos,* **2006**, *16(2)*, 026110.

[3] Harmer S.L., Panda S., Kay S.A. Molecular bases of circadian rhythms. *Annu. Rev. Cell. Dev. Biol.*, **2001**, *17*, 215-253.

[4] Tsinkalovsky O., Smaaland R., Rosenlund B., Sothern R.B., Hirt A., Steine S., Badiee A., Abrahamsen J.F., Eiken H.G., Laerum O.D. Circadian variations in clock gene expression of human bone marrow CD34+ cells. *J. Biol. Rhythms.*, **2007**, *22(2)*, 140-150.

[5] Kovac J., Husse J., Oster H. A time to fast, a time to feast: the crosstalk between metabolism and the circadian clock. *Mol. Cells*, **2009**, *28(2)*, 75-80.

[6] Pfeffer M., Muller C.M., Mordel J., Meissl H., Ansari N., Deller T., Korf H.W., von Gall C. The mammalian molecular clockwork controls rhythmic expression of its own input pathway components. *J. Neurosci.*, **2009**, *29(19)*, 6114-6123.

[7] Motzkus D., Loumi S., Cadenas C., Vinson C., Forssmann W.G., Maronde E. Activation of human period-1 by PKA or CLOCK/BMAL1 is conferred by separate signal transduction pathways. *Chronobiol. Int.*, **2007**, *24(50)*, 783-792.

[8] Chen R., Schirmer A., Lee Y., Lee H., Kumar V., Yoo S.H., Takahashi J.S., Lee C. Rhythmic PER abundance definesa critical nodal point for negative feedback within the circadian clock mechanism. *Mol. Cell.,* **2009**, *36(3)*, 417-430.

[9] Sato T.K., Yamada R.G., Ukai H., Baggs J.E., Miraglia L.J., Kobayashi T.J., Welsh D.K., Kay S.A., Ueda H.R., Hogenesch J.B. Feedback repression is required for mammalian circadian clock function. *Nat. Genet.*, **2006**, *38(3)*, 312-319.

[10] Sasaki M., Yoshitane H., Du N.H., Okano T., Fukada Y. Preferential inhibition of BMAL2-CLOCK activity by PER2 reemphasizes its negative role and a positive role of BMAL2 in the circadian transcription. *J. Biol. Chem.*, **2009**, *284(37)*, 25149-25159.

[11] Turek F.W., Joshu C., Kohsaka A., Lin E., Ivanova G., McDearmon E., Laposky A., Losee-Olson S., Easton A., Jensen D.R., Eckel R.H., Takahashi J.S., Bass J. Obesity and metabolic syndrome in circadian Clock mutant mice. // Science. – 2005. – 308, N *Science*, **2005**, *308(5724)*, 1043-1045.

[12] You S., Wood P.A., Xiong Y., Kobayashi M., Du-Quiton J., Hrushesky W.J. Daily coordination of cancer growth and circadian clock gene expression. *Breast Cancer Res. Treat.*, **2005**, *91(1)*, 47-60.

[13] Zhu Y., Stevens R.G., Hoffman A.E., Fitzgerald L.M., Kwon E.M., Ostrander E.A., Davis S., Zheng T., Stanford J.L. Testing the circadian gene hypothesis in prostate cancer: a population-based case-control study. *Cancer Res.*, **2009**, *69(24)*, 9315-9322.

[14] Taniguchi H., Fernandez A.F., Setien F., Ropero S., Ballestar E., Villanueva A., Yamamoto H., Imai K., Shinomura Y., Esteller M. Epigenetic inactivation of the circadian clock gene BMAL1 in hematologic malignancies. *Cancer Res.*, **2009**, *69(21)*, 8447-8454.

[15] Woon P.Y., Kaisaki P.J., Braganca J., Bihoreau M.T., Levy J.C., Farrall M., Gauguier D. Aryl hydrocarbon receptor nuclear translocator-like (BMAL1) is associated with susceptibility to hypertension and type 2 diabetes. *Proc. Natl. Acad. Sci. U.S.A.*, **2007**, *104(36)*, 14412-14417.

[16] Laposky A.D., Bass J., Kohsaka A., Turek F.W. Sleep and circadian rhythms: Key components in the regulation of energy metabolism. *FEBS Lett.*, **2008**, *582(1)*, 142-151.

[17] Lee H., Chen R., Lee Y., Yoo S., Lee C. Essential roles of CKIdelta and CKIepsilon in the mammalian circadian clock. *Proc. Natl. Acad. Sci. U.S.A.*, **2009**, *106(50)*, 21359-21364.

[18] Eide E.J., Vielhaber E.L., Hinz W.A., Virshup D.M. The circadian regulatory proteins BMAL1 and cryptochromes are substrates of casein kinase Iepsilon. *J. Biol. Chem.*, **2002**, *277(19)*, 17248-17254.

[19] Miyazaki K., Nagase T., Mesaki M., Narukawa J., Ohara O., Ishida N. Phosphorylation of clock protein PER1 regulates its circadian degradation in normal human fibroblasts. *Biochem. J.*, **2004**, *380(Pt. 1)*, 95-103.

[20] Shirogane T., Jin J., Ang X.L., Harper J.W. SCFbeta-TRCP controls clock-dependent transcription via casein kinase 1-dependent degradation of the mammalian period-1 (Per1) protein. *J. Biol. Chem.*, **2005**, *280(29)*, 26863-26872.

[21] Gao Z.H., Seeling J.M., Hill V., Yochum A., Virshup D.M. Casein kinase I phosphorylates and destabilizes the beta-catenin degradation complex. *Proc. Natl. Acad. Sci. U.S.A.*, **2002**, *99(3)*, 1182-1187.

[22] Okamura A., Iwata N., Tamekane A., Yakushijin K., Nishikawa S., Hamaguchi M., Fukui C., Yamamoto K., Matsui T. Casein kinase Iepsilon down-regulates phospho-Akt via PTEN, following genotoxic stress-induced apoptosis in hematopoietic cells. *Life Sci.*, **2006**, *78(14)*, 1624-1629.

[23] Zhao B., Li L., Tumaneng K., Wang C.Y., Guan K.L. A coordinated phosphorylation by Lats and CK1 regulates YAP stability through SCF(beta-TRCP). *Genes Dev.*, **2010**, *24(1)*, 72-85.

[24] Eide E.J., Woolf M.F., Kang H., Woolf P., Hurst W., Camacho F., Vielhaber E.L., Giovanni A., Virshup D.M. Control of mammalian circadian rhythm by CKIepsilon-regulated proteasome-mediated PER2 degradation. *Mol. Cell. Biol.*, **2005**, *25(7)*, 2795-2807.

[25] Sahoo S.K., Parveen S., Panda J.J. The present and future of nanotechnology in human health care. *Nanomedicine: Nanotechnology, Biology and Medicine*, **2007**, *3(1)*, 20-31.

[26] Nangmenyi G., Yue Z., Mehrabi S., Mintz E., Economy J. Synthesis and characterization of silver-nanoparticle-impregnated fiberglass and utility in water disinfection. *Nanotechnol.*, **2009**, *20(49)*, 495705.

[27] Balbus J.M., Maynard A.D., Colvin V.L., Castranova V., Daston G.P., Denison R.A., Dreher K.L., Goering P.L., Goldberg A.M., Kulinowski K.M., Monteiro-Riviere N.A., Oberdörster

G., Omenn G.S., Pinkerton K.E., Ramos K.S., Rest K.M., Sass J.B., Silbergeld E.K., Wong B.A. Meeting report: hazard assessment for nanoparticles--report from an interdisciplinary workshop. *Environ Health Perspect.*, **2007**, *115(11)*, 1654-1659.

[28] Griffitt R.J., Hyndman K., Denslow N.D., Barber D.S. Comparison of molecular and histological changes in zebrafish gills exposed to metallic nanoparticles. *Toxicol. Sciences*, **2009**, *107(2)*, 404-415.

[29] Chen D., Xi T., Bai J. Biological effects induced by nanosilver particles: in vivo study. *Biomed. Mater.*, **2007**, *2(3)*, S126-S128.

[30] Lubick N. Nanosilver toxicity: ions, nanoparticles--or both? *Environ. Sci. Technol.*, **2008**, *42(23)*, 8617.

[31] Ji J.H., Jung J.H., Kim S.S., Yoon J.U., Park J.D., Choi B.S., Chung Y.H., Kwon I.H., Jeong J., Han B.S., Shin J.H., Sung J.H., Song K.S., Yu I.J. Twenty-eight-day inhalation toxicity study of silver nanoparticles in Sprague-Dawley rats. *Inhalat. Toxicol.*, **2007**, *19(10)*, 857-871.

[32] Ahamed M., Posgai R., Gorey T.J., Nielsen M., Hussain S.M., Rowe J.J. Silver nanoparticles induced heat shock protein 70, oxidative stress and apoptosis in Drosophila melanogaster. *Toxicol. Appl. Pharmacol.*, **2010**, *242(3)*, 263-269.

[33] Ringwood A.H., McCarthy M., Bates T.C., Carroll D.L. The effects of silver nanoparticles on oyster embryos. *Mar. Environ Res.* **2009**, *69*, Suppl., S49-S51.

[34] Powers C.M., Wrench N., Ryde I.T., Smith A.M., Seidler F.J., Slotkin T.A. Silver impairs neurodevelopment: studies in PC12 cells. *Environ Health Perspect.*, **2010**, *118(1)*, 73-79.

[35] Choi J.E., Kim S., Ahn J.H., Youn P., Kang J.S., Park K., Yi J., Ryu D.Y. Induction of oxidative stress and apoptosis by silver nanoparticles in the liver of adult zebrafish. *Aquat Toxicol.* **2009**, *100(2)*, 151-159.

[36] Tang J., Xiong L., Wang S., Wang J., Liu L., Li J., Yuan F., Xi T. Distribution, translocation and accumulation of silver nanoparticles in rats. *J. Nanosci. Nanotechnol.*, **2009**, *9(8)*, 4924-4932.

[37] Shimada A., Kawamura N., Okajima M., Kaewamatawong T., Inoue H., Morita T. Translocation pathway of the intratracheally instilled ultrafine particles from the lung into the blood circulation in the mouse. *Toxicol. Pathol.*, **2006**, *34(7)*, 949-957.

[38] Tang J, Xiong L, Zhou G, Wang S, Wang J, Liu L, Li J, Yuan F, Lu S, Wan Z, Chou L, Xi T. Silver nanoparticles crossing through and distribution in the blood-brain barrier in vitro. J *Nanosci. Nanotechnol.* **2010**, *10(10),* 6313-6317.

[39] Minchenko D.O., Yavorovsky O.P., Zavgorodny I.V., Paustovsky Y.O., Tsuchihara K., Esumi H., Minchenko O.H. Experimental data on disorders of expression of circadian genes in liver and lungs as a marker of methyl-tretbutyl ether toxic effect on the body. *Ukr. J. Occupational Health*, **2008**, *3(15),* 20-26.

[40] Minchenko D., Bozhko I., Zinchenko T., Yavorovsky O., Minchenko O. Expression of SNF1/AMP-activated protein kinase and casein kinase-1epsilon in different rat tissues is a sensitive marker of in vivo silver nanoparticles action. *Materialwissenschaft und Werkstofftechnik*, **2011**, *42(2)*, 21–25.

[41] Minchenko O.H., Opentanova I.L., Minchenko D.O., Ogura T., Esumi H. Hypoxia induces transcription of 6-phosphofructo-2-kinase/fructose-2,6-bisphosphatase 4 gene via hypoxia-inducible factor-1alpha activation. *FEBS Lett.*, **2004**, *576(1-2)*, 14-20.

[42] Moenner M., Pluquet O., Bouchecareilh M., Chevet E. Integrated endoplasmic reticulum stress responses in cancer. *Cancer Res.*, **2007**, *67(22),* 10631–10634.

[43] Saito A., Ochiai K., Kondo S., Tsumagari K., Murakami T., Cavener, D.R., Imaizumi K. Endoplasmic reticulum stress response mediated by the PERK-eIF2-ATF4 pathway is involved in osteoblast differentiation induced by BMP2. *J. Biol. Chem.*, **2011**, *286(6)*, 4809–4818.

[44] Fels D.R., Koumenis C. The PERK/eIF2a/ATF4 module of the UPR in hypoxia resistance and tumor growth. *Cancer Biology and Therapy*, **2006**, *5(7)*, 723–728.

[45] Aragón T., van Anken E., Pincus D., Serafimova I.M., Korennykh A.V., Rubio C.A., Walter P. Messenger RNA targeting to endoplasmic reticulum stress signalling sites. *Nature*, **2009**, *457(7230)*, 736-740.
[46] Korennykh A.V., Egea P.F., Korostelev A.A., Finer-Moore J., Zhang C., Shokat K.M., Stroud R.M., Walter P. The unfolded protein response signals through high-order assembly of Ire1. *Nature*, **2009**, *457(7230)*, 687-693.
[47] Korennykh A.V., Egea P.F., Korostelev A.A., Finer-Moore J., Stroud R.M., Zhang C., Shokat K.M., Walter P. Cofactor-mediated conformational control in the bifunctional kinase/RNase Ire1. *BMC Biol.* **2011**, *9*, 48.
[48] Korennykh A.V., Korostelev A.A., Egea P.F., Finer-Moore J., Stroud R.M., Zhang C., Shokat K.M., Walter P. Structural and functional basis for RNA cleavage by Ire1. *BMC Biol.* **2011**, *9*, 47.
[49] Acosta-Alvear D., Zhou Y., Blais A., Tsikitis M., Lents N.H., Arias C., Lennon C.J., Kluger Y., Dynlacht B.D. XBP1 controls diverse cell type- and condition-specific transcriptional regulatory networks. *Molecular Cell*, **2007**, *27(1)*, 53-66.

Enhancement of thermal conductivity of materials using different forms of natural graphite

S Desai[1] and J Njuguna[2]

[1] Tata Group, UK
[2]Cranfield University, UK

Abstract. The quest for improvement of thermal conductivity in every area including aerospace structures is gaining momentum as the modern day technology is embedded with electronics which generate considerable amounts of heat energy. In some other applications, the thermal energy must be absorbed or released at a very fast rate. Natural graphite is highly anisotropic and has excellent thermal and electrical conductivity in a-direction hence; it is one of the potential candidates to enhance thermal conductivity. This paper encompasses various preparation routes used to enhance thermal conductivity in materials/composites using graphite flakes in its natural and treated forms along with characterization results of thermal conductivity and/ or electrical resistivity where available. Addition of graphite has shown to improve thermal conductivity in some materials by 30-130 times whereas one method of preparation described here shows composites to have thermal conductivity as high as 750 W/m K.

1. Introduction

In recent years, the number of applications requiring more efficient and light weight thermal management, e.g. advanced aircraft, car navigation systems, high density electronic equipments, etc., has considerably increased. As each new generation of electronics equipment exhibits higher performance and more and more power into ever smaller packages, this has aggravated the problem associated with heat dissipation in the electronic industry and thus produced a need for improved thermal interface materials in modern chip packaging (Aiping Yu. et al, 2007).

Along this line of thinking, there exists increasingly new thermal management materials applications and these are also being considered as potential interface and attachment techniques.

Interfaces between materials have a significant impact on the thermal impedance of electronic systems and in practice they can be the dominant factor in achieving effective thermal transfer. The interface materials and processes in question are the methods used to join an electronic device to the thermal transfer medium (e.g. substrate, heat pipe, heat sink), including coatings and bonding techniques. In this respect they may need to perform the tasks of attachment, stress/strain relief and thermal transfer. Therefore, these industries are focusing attention on alternate materials to replace conventional heat sinks and heat spreaders.

Although miniaturization has built focus of thermally conducting materials, thermal energy storage (TES) for free-cooling applications also face a major challenge due to relatively low heat transfer rates / thermal conductivity of the materials usually used to fabricate these systems commonly known as phase change materials (PCM). Various methods such as metal matrix structures and finned tubes have been widely used (Zalba B. et al, 2003) and methods of using water as heat transfer medium are proposed to enhance the heat transfer in these latent heat thermal storage devices. Cabeza L. F. et al (2002) and Fukai J. et al (2000) and Ply X. et al (2001) proposed the use of graphite-compound material to enhance thermal conductivity by having the PCM embedded inside a graphite matrix.

In recent year, several research studies have been carried out to improve the thermal conductivity of the polymeric composites by using different fillers in the form of fibers and particles. The researchers at Oak Ridge National Laboratory (ORNL) identified the potential of carbon foams for enhancing heat transfer. Although pyrolitic graphite was well known for its high thermal conductivity, the processing cost has always led to have its applications only in niche areas. However, with the use of graphite foams as heat exchangers, natural graphite (NG) has become a popular material in thermal enhancement and thermal management research.

Mariner J. T. et al (1999) invented electrically conductive composites by combining thermally treated graphite flakes with a polymeric binder. The thermal conductivity of these composites was found to be greater than 100 W/m/K.

Prieto R. et al (2008) prepared different composites with different reinforcements for Al-Si and Ag-Si alloys. The fillers used were a combination of different range of proportions of graphite flakes and carbon fibers, graphite flakes and SiC, diamond, graphite particles, carbon fibers and graphite foam. They found that the graphite flake/ metal composites exhibited superior thermal properties (in-plane thermal conductivity ranging from 350 W/m K to 548 W/m K).

GrafTech International Ltd developed natural graphite/epoxy laminate material (eGraf^{TM}) for plasma display panels (PDP) and 'Spreadershield^{TM}' as heat-spreader in notebooks, digital cameras, etc.. The in-plane thermal conductivity of this laminate was reported to be ~370 W/m K, which is 77% higher than aluminum and comparable to that of copper whereas eGraf 1200 is reported to have an in-plane thermal conductivity of 120 W/m K (Technical Bulletin 268M) Following advances in the material development, an in-plane thermal conductivity of approximately 400 W/m K (i.e. approximately equal to that of copper) is achieved in this composite (Norley J., 2003). Spreadshield^{TM} heat spreaders developed by Graftech are known to have thermal conductivity upto 1500 W/m K.

This paper will focus on summarizing natural flake graphite and its forms used in enhancing thermal conductivity. In the end, we will briefly describe method of using graphene to prepare composites.

Natural graphite (NG) being well known as highly anisotropic forms of carbon with low density (d ≈ 1.94 g/cc) and is abundantly available in nature. The carbon atoms in graphite are hexagonally arranged in a planar condensed ring system with 1.42 Å and the layers are stacked over one another (Mantell C., 1968) with large spacing (3.35 Å). These layers are held together by weak van-der-Waals forces whereas the atoms within the plan are held by covalent bonds.

Single crystal graphite is one of the stiffest materials in nature (elastic modulus of over 1 TPa) with lower density (Zheng W. G. et al, 2002; Chen G. H. et al, 2001).

Natural graphite has good thermal conductivity (k≈2000W/m/K in a-direction) and is highly temperature dependent (it can be higher than the quoted value (Kelly B.T., 1981)]). The electrical conductivity of natural graphite is 10^4 S/cm at room temperature (Usuki A, Kojima Y., Kawassumi M., Okada A. and Fukushima Y., 1993; Giannelis E.P., 1996 and Lebaron P.C., Wang Z. and Pinnavaia T., 1999). The superior thermal property along two directions makes it one of the potential candidates to combat thermal management problems. Natural graphite has high chemical and thermal stability over a range of temperatures with a very high melting point, high thermal and electrical conductivity and is resistant to attack by most chemical reagents. Hence, graphite has attracted researchers to enhance thermal conductivity in various applications.

1.1 Possible uses of Graphite in different forms

Flake graphite can be directly used in its natural form in composites or it can be intercalated with different chemical species to form nanolayers which can then be used to form composites. These nanolayers can be obtained by taking advantage of this weak bonding between flakes by inserting various chemical species to form graphite intercalation compounds (GICs) (Guohua Chen G. H. et al, 2003). Such intercalated graphite is referred as expanded graphite/ exfoliated graphite/ worm like graphite/ garland graphite in the literature. However, expanded/ exfoliated graphite is more popular.

Pyrolytic graphite, graphite fibres, reinforced carbon and polymer matrix composites heat-treated to temperatures (~3000 ^{0}C), graphite foams, etc. are all forms of graphite that can be categorised as "synthetic graphite" or "artificial graphite". In spite of meeting the required thermal conductivity, synthetic graphite has found its application only in high performance applications due to its high processing costs.

1.1.1 Expanded graphite (EG) to enhance thermal conductivity

Natural graphite has no reactive ion groups on the surface layers hence; it is difficult to incorporate organic molecules or polymers directly into the interlayer of graphite through an ion exchange reaction to prepare the polymer/graphite composite. However, EG have worm-like appearance and the expansion along the c-direction can be as high as two orders of magnitude when intercalate is heated over a critical temperature. This form of graphite contains multipores ranging from 2 to 10 nm in large numbers, functional acids and OH groups that promote affinity of EG to both organic compounds and polymers. Hence, EG can absorb some monomers, initiators and polymers, and result in the formation of conductive/graphite nanocomposite (Zheng W., Wong S.C. and Sue H.J., 2002; Li J, Kim J. and Sham M., 2005). The layered structure in EG is similar to nanoclays but EG is very cost effective with superior mechanical, electrical and thermal properties as compared to carbon nanotubes (Fukushima H., 2003; Kalaitzidou K., 2006).

Lincoln V. F. and Claude Z. (1983) proposed dispersion of intercalated graphite in polymeric resin by conventional composite processing techniques in 1980s. Aylsworth J. W. (1916) developed and proposed EG as a reinforcement of polymer in 1910s. Research has been conducted on EG-reinforced polymer composites after these studies. More recently, research by the Drzal group (2007, 2009) has shown that exfoliated graphite nanoplatelets have the ability to provide a large enhancement in physical and chemical properties.

1.1.2 EG/Epoxy Composites

Debelak B. and Lafdi K. (2007) dispersed EG flakes in an epoxy resin and studied the thermal and electrical properties of the composites. In this study, natural graphite flakes of an average diameter of 500 μm and EPON Resin 862 was used as a polymer matrix. Exfoliated graphite was prepared by mixing nitric and sulfuric acid and natural graphite. After 24 h of reaction, intercalated graphite compound was obtained. The mixture was then filtered, washed with water and dried in an oven at low temperatures. The intercalated graphite compound was then subjected to sudden heat treatment temperature of 900 ^{0}C for rapid expansion. The expansion ratio was as high as 300 times (c-direction). During this study, three kinds of exfoliated graphite filled polymers, having different graphite particle sizes, were prepared. The graphite flakes were separated using 50, 100, and 150 mesh sieves and were labeled as large, medium, and small flakes respectively. Polymer materials containing different concentrations by weight of exfoliated graphite were prepared (0.1, 0.5, 1, 2, 4, 8, 12, 16, and 20%). Exfoliated graphite flakes were placed in a solvent and sheared in a homogenize rotating at a very high speed (13500 turns/minute). The blade of the homogenizer rotates and breaks the larger graphite flakes into even smaller nanosheets. Ultrasonication was used to assist and improve the dispersion of these graphite nanosheets. After the mixing process completed, the mixture was filled into a silicone rubber mold. The mold with the mixture was loaded in a hydraulic press to cure under a pressure of 10000 lb at 121 ^{0}C for 2 h and then at 177 ^{0}C for two more hours.

The electrical resistivity of the composites with varying EG concentration was studied for all three sizes of graphite flakes according to Debelak B. and Lafdi K. (2007). It should be noted that epoxy is not an electrically conducting polymer at room temperature (RT) and has a resistivity of 1.5^8 Ωcm in its dry state. Addition of EG was found to lower the electrical resistivity of the epoxy resin with a sharp transition of a polymer from electrical insulator to an electrical conductor. It was seen that the electrical resistivity of EG/epoxy composites decreased with increasing concentration of EG (Fig. 1). This is attributed to the Nanoscale dispersion of graphite nanosheets within the polymer and the formation of conducting networks. Hence, it is clear that the conductivity depends upon the factors such as the percentage of EG in the composite and its into the matrix. The larger graphite flake filled epoxy showed lowest resistivity for every concentration as compared to the medium and small flake filler. It is well reported that electrical conductivity of carbon composites depends upon the grain and crystallite size and hence, the large flake filler composites show higher conductivity as expected.

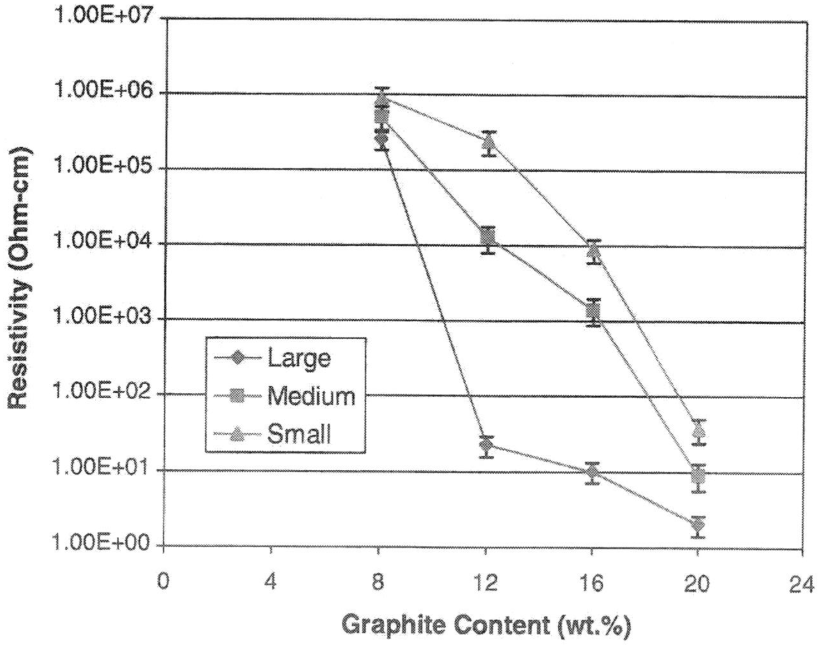

FIGURE 1 Electrical resistivity of exfoliated graphite filled polymer with different graphite sizes as a function of the graphite content (taken from Debelak B. and Lafdi K., 2007).

The electrical resistivity of the composites with varying EG concentration was studied for all three sizes of graphite flakes according to Debelak B. and Lafdi K. (2007). It should be noted that epoxy is not an electrically conducting polymer at room temperature (RT) and has a resistivity of 1.5^8 Ωcm in its dry state. Addition of EG was found to lower the electrical resistivity of the epoxy resin with a sharp transition of a polymer from electrical insulator to an electrical conductor. It was seen that the electrical resistivity of EG/epoxy composites decreased with increasing concentration of EG (Fig. 1). This is attributed to the Nanoscale dispersion of graphite nanosheets within the polymer and the formation of conducting networks. Hence, it is clear that the conductivity depends upon the factors such as the percentage of EG in the composite and its into the matrix. The larger graphite flake filled epoxy showed lowest resistivity for every concentration as compared to the medium and small flake filler. It is well reported that electrical conductivity of carbon composites depends upon the grain and crystallite size and hence, the large flake filler composites show higher conductivity as expected.

Fig. 2 shows thermal conductivity of composites prepared using the three sizes of exfoliated graphite filled epoxy composites with varying concentration of EG content. As seen with electrical resistivity, the thermal conductivity increases with increase in EG concentration. Superior transport properties are seen in composites with larger flake filler as compared to medium and small flake content for a given concentration level which again can be attributed to the larger grain and crystallite size of large flake graphite.

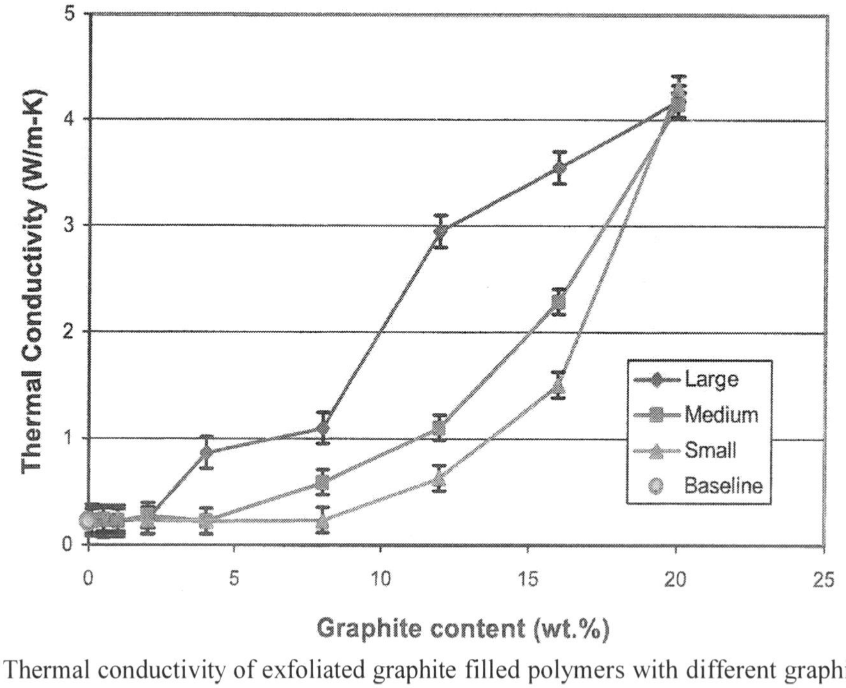

FIGURE 2 Thermal conductivity of exfoliated graphite filled polymers with different graphite flake sizes as a function of the graphite content (taken from Debelak B. and Lafdi K., 2007)

The authors conducted a comparative study of thermal conductivity of 20% large flake EG filled epoxy composite with polymer composites with carbon nanofiber and high heat-treated (HHT) nanofiber. The results showed (Fig. 3) EG filled composites to have superior thermal conductivity as compared to the nanofiber and HHT nanofibers (Matzek M. D., 2004). The superior property is due to the strong thermal conductive network produced by EG with its higher aspect ratio as compared to typical carbon nanofiber.

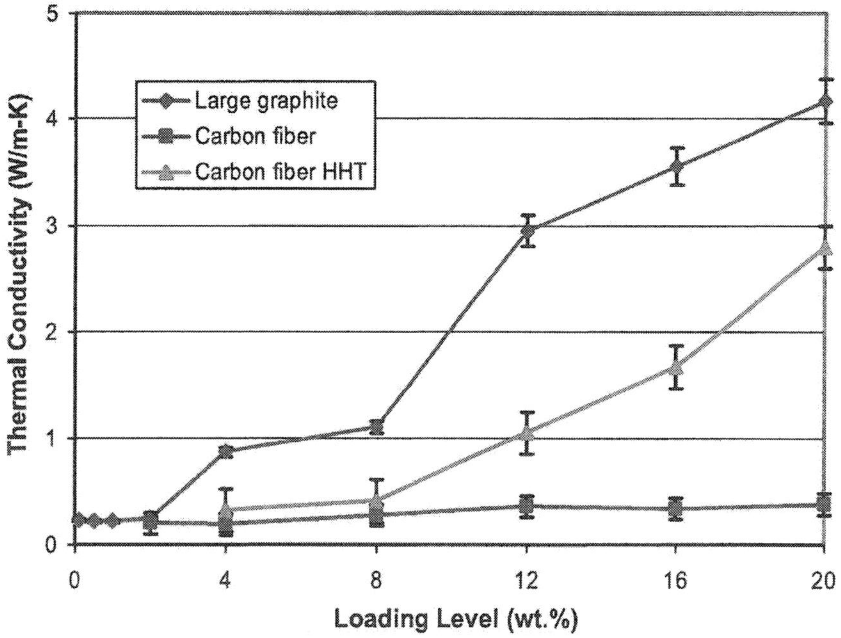

FIGURE 3 Thermal conductivity comparison of graphite flake filled polymers versus carbon nanofiber and HHT carbon nanofiber composites (taken from Debelak B. and Lafdi K., 2007)

Ganguli S. et al (2008) used EG supplied by Graftech International Ltd., Parma OH and prepared composites (EG/epoxy) using Epon 862 (an epoxy bisphenol F resin) and Epicure W (an aromatic amine curing agent). Epicure 537 (consisting of organic salts) was used as an accelerating agent to accelerate the curing reaction which in turn aided in 'locking in' the dispersed morphology. Flacktek Speedmixer was used to disperse the graphite nanoparticles in the epoxy. The high speed mixer (2750 rpm) consisted of a one direction spinning arm with the basket containing the mixture rotating in the opposite direction. This combined force in opposite planes enabled fast mixing. The composites were synthesized with different volume fraction of EG ranging from 2% - 20% (2, 4, 6, 8, 16 and 20%).

Ganguli S. (2008) prepared another set of samples with same volume fractions but with functionalized exfoliated graphite (FN/epoxy). The grafting reaction was carried out in the mixture of water and ethanol in the ratio of 25:75 by volume. 3 g of γ- APS (3-aminopropoxyltriethoxy silane) was first added into 1000 ml of the water/ethanol mixture. The mixture temperature was raised to 80 ^{0}C and then 10 g of exfoliated graphite was added. The grafting was achieved under shearing for 5 h at 80 ^{0}C. The reaction product was filtered and washed six times using a mixture of water/ethanol and freeze dried. The obtained product was ground and placed in sealed container for characterization.

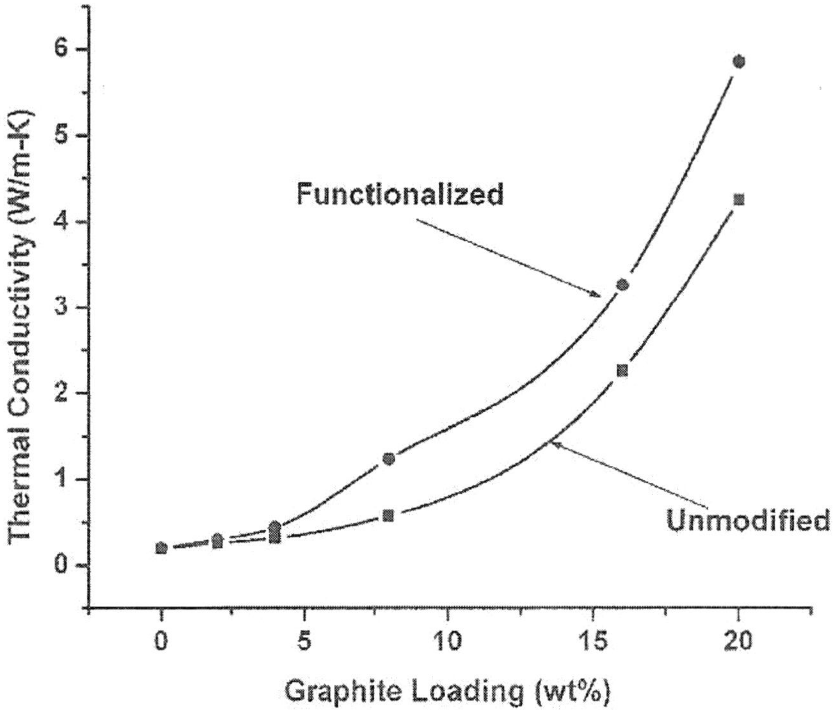

FIGURE 4 Thermal conductivity as a function of Graphite loading of EG/epoxy (unmodified) and FN/epoxy (functionalized) composites (taken from Ganguli S. et al, 2008)

Thermal conductivity of EG/epoxy and FN/epoxy composites was measured by the Netzsch laser flash diffusivity system LFA 457 with 5 mm thick POCO graphite specimen as reference heat capacity material. The thermal conductivity of neat Epon 862/W epoxy resin is around 0.2 W/m K. Figure 4 shows no significant change in thermal conductivity for composites with 2% and 4% of exfoliated graphite. At 8%, the thermal conductivity was seen to increase to 0.5 W/m K. As the filler concentration increased the conductivity was seen to increase. The composite with 20% filler concentration showed 19-fold enhancement in conductivity compared to neat resin.

Functionalized exfoliated graphite composites showed similar low conductivity for 2 and 4% but for higher loading of functionalized exfoliated graphite, the conductivity is seen to increase and was higher than EG/epoxy composites (Fig 4). The composite with 20 % of functionalized exfoliated graphite was measured to be 5.8 W/m K which is 28-fold improvement in comparison to neat resin. The mode of thermal conduction in amorphous polymers is primarily by phonons. In order to enhance thermal transport in filled amorphous polymeric systems, the acoustic impedance mismatch between the filler and the polymeric matrix has to be reduced. Functionalisation seems to improve this interfacial heat transfer between the graphite platelets and the epoxy matrix. This improvement is attributed to the covalent bonds formed between the graphite platelets and the epoxy matrix as a result of chemical functionlisation which reduce the acoustic mismatch impedance and enhance thermal conductivity.

Aiping Yu et al (2007) dispersed expanded graphite in acetone by high shear mixing for 30 min followed by bath sonication for 24 h and prepared GNP dispersions at a concentration of 2 mg/ml. Composites of GNP / epoxy were prepared by adding epoxy resin to the GNP suspension and subjecting it to high-shear mixing for 30 min. The solvent was then removed in vacuum oven at 50 ^{0}C and the curing agent (Diethyltoluenediamine, EPI-CURE W) was added in a ratio 100:26 (Epoxy: curing agent) by weight while the mixture was continuously stirred. The mixture containing the homogenously dispersed GNP was degassed and cured in vacuum in a stainless steel mold. The composite was initially cured at 100 ^{0}C for 2 h and then further cured at 150 ^{0}C for another 2 h to complete the curing cycle.

Composites with different loading levels of GNP ranging from 0.1 to 10 % by volume were prepared (Aiping Yu et al, 2007). The thermal conductivity of GNP/epoxy composite was compared with graphite macroplatelet (GMP)/epoxy composite. With 5.4% of GMP filler (GMP aspect ratio of length ≈ 30 μm, thickness ≈ 10 μm) at 30 ^{0}C was found to show a thermal conductivity of 0.54 W/m K which was comparably higher than pristine epoxy (K= 0.201 W/m K). Composites with ~ 5% volume of GNP (prepared with 800 ^{0}C exfoliating temperature) in composite showed thermal conductivity of ≈1.45 W/m K which compares very favorably with current available thermal interface materials which require about 10 times (i.e. 50 - 70 %) filler volume to achieve similar thermal conductivity.

The researchers studied GNP prepared from exfoliated graphite using different shock temperatures namely 200, 400 and 800 ^{0}C. The thermal conductivity was seen to enhance with both the volume fraction of the filler content in the composite and the shock treatment temperature (Fig. 5a). Further studies on GNP (prepared from 800 ^{0}C shock treatment temperatures) composites with 25 % of filler showed the thermal conductivity of these composites increased with temperature for a given volume fraction of the filler. The highest thermal conductivity achieved with these GNP composites with 25 % volume fraction of filler was 6.44 W/m K which corresponds to an enhancement of more than 100 % per 1 volume % loading (Fig. 5b). These composites when used in computer processor that usually operate at elevated temperatures could show thermal conductivity as high as 6.87 W/m K

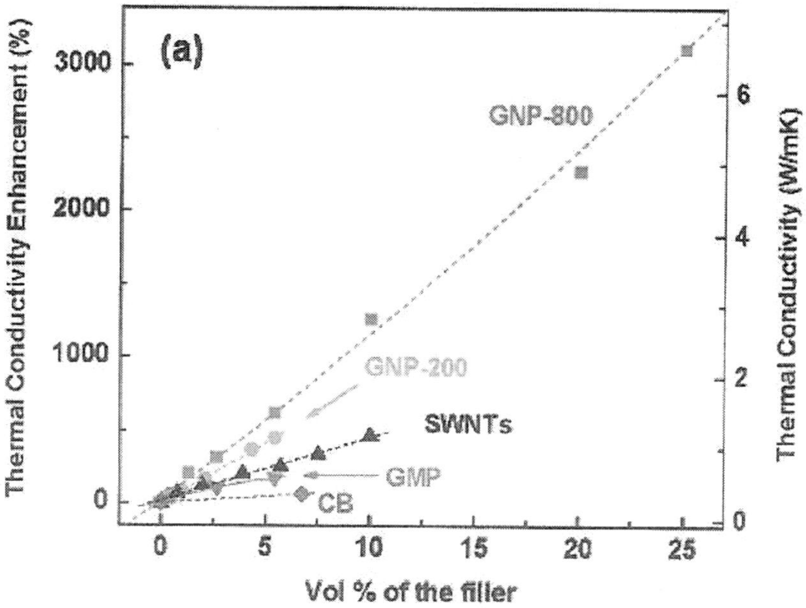

FIGURE 5 a Thermal conductivity enhancement % of epoxy based composites at 30 ^{0}C with graphitic fillers: graphite microparticles (GMT), GNPs at 200 ^{0}C (GNP-200) and GNPs at 800 ^{0}C (GNP-800), carbon black (CB) and purified single walled nanotubes (SWNTs) (taken from Aiping Yu et al, 2007)

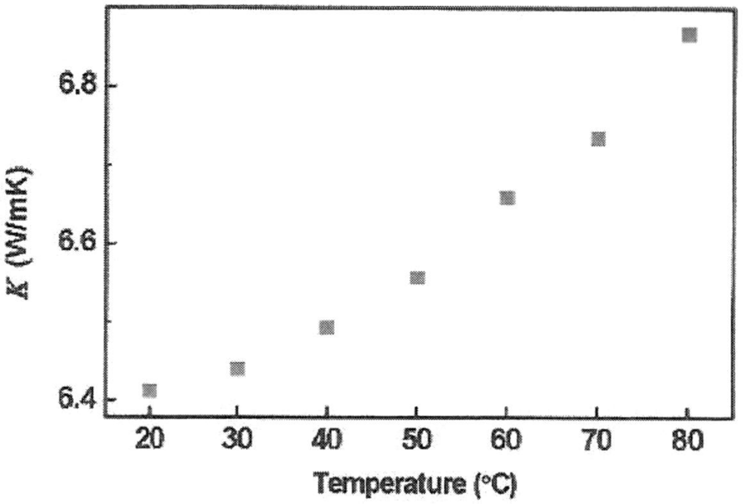

FIGURE 5b Temperature dependence of thermal conductivity (K) for the composites with 25 % loading of GNP-800 by volume (taken from Aiping Yu et al, 2007)

1.1.3 Graphite/silicone rubber composite to enhance thermal conductivity and storage modulus

Nylon 6/layered silicate nanocomposites were first prepared by intercalation polymerization. Since then, layered silicate nanocomposites have progressed and methods have been identified to homogenously disperse layered silicates such as montmorillonite in polymeric matrix in nanosheets to improve desirable properties of the matrix such as poor electrical and thermal conductivity (Alexandre M. & Dubois P., 2000; Saujanya C. and Radhakrishnan S., 2001; Weng W. P. and Pan C.Y., 2004).

Mu Q. and Feng S. (2007) studied silicone/EG nanocomposites prepared via solution intercalation and normal melt mixing with methyvinylsilicone gum (M_n 5.8 x 10^5; mole content of vinyl group, 0.15%), silica and 2,5-bis(tert-butyl peroxy) 2,5-dimethly hexane (DBPMH) and EG of average diameter 300 μm.

For melt mixing (MM) method, methylvinylsilicone gum, silica, EG and some other additives were rolled on a twin-roller at room temperature (RT) for certain time. They were then compress molded at 190 ^{0}C under a pressure of 9.8 MPa for 20 min to obtain 120 mm x 40 mm plates with thickness of 2 mm.

In solution intercalation (SI) method, the EG was dispersed in toluene to prepare a suspending dispersion with certain EG content. The methylvinylsilicone gum was dissolved in toluene by heating to a backflow temperature of toluene (~100 ^{0}C). The EG suspending dispersion was added drop by drop into the methylvinylsilicone gum solution at the backflow state. After a backflow of 2 h, the heating was stopped and a portion of toluene was extracted under vacuum during the cooling process. The resultant mixture was dried under vacuum and pressed to composites under similar condition as MM method.

Composites of different EG filler content were prepared by the two methods. The thermal conductivity of the composites prepared using these two methods (MM and SI) were measured at 50 ^{0}C by using Fourier law with the assumption that the heat flows only in the direction of measurement (i.e. one-direction). Figure 6 shows that the thermal conductivity increases as the EG content increases. However, the composites prepared by SI method showed higher thermal conductivity compared to the composites prepared by MM method with same EG content. For 9 per hundred rubber (phr), the thermal conductivity of composite prepared using SI method was calculated to be 0.32 W/m K whereas the composite at same phr prepared using MM method showed thermal conductivity of 0.24 W/m K which is the conductivity level of composed prepared by SI method at 4 phr (Mu Q. and Feng S., 2007). This was attributed to the surface-to-volume ratio which in SI prepared composites was found to be larger than MM prepared composites. Hence, in SI prepared composites, the EG having larger surface-to-volume ratio can abut or contact and then form a conducting path network at a lower EG content thus playing an important role in improving thermal conductivity of the composites (Mu Q. and Feng S., 2007).

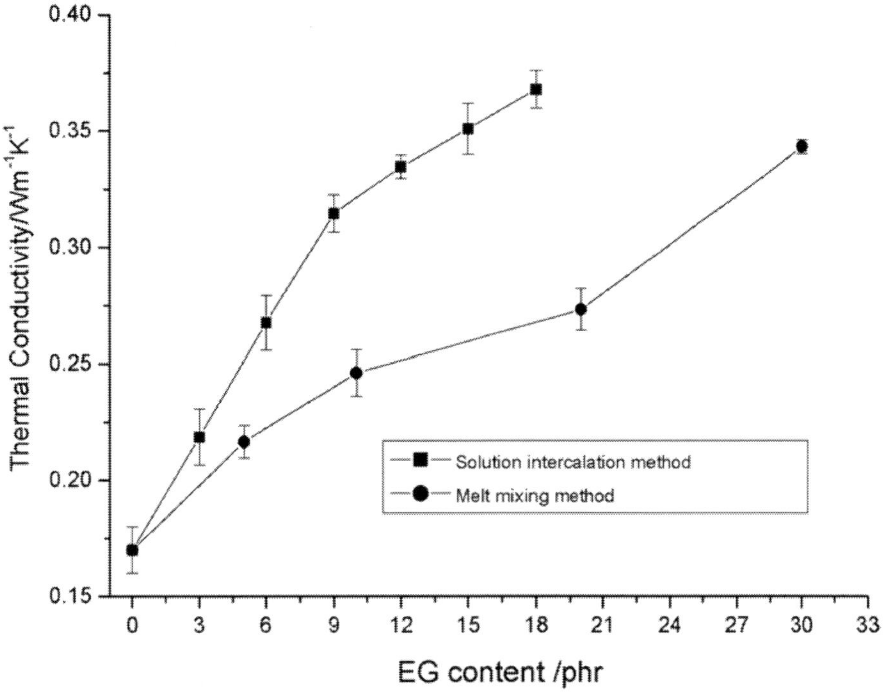

FIGURE 6 Thermal conductivity versus fraction of expanded graphite for silicone/EG composites prepared by two methods at 50 ^{0}C (taken from Mu Q. and Feng S., 2007).

Similarly, storage modulus E' for composites prepared using SI method at 10 phr were relatively higher than the E' of composites prepared using MM method at same phr (Fig. 7). It has been reported that there exists a correlation between the conductivity network formation and the stiffening effect arising from the EG dispersion in the polymer solution (Zheng W. & Wong S.C., 2003).

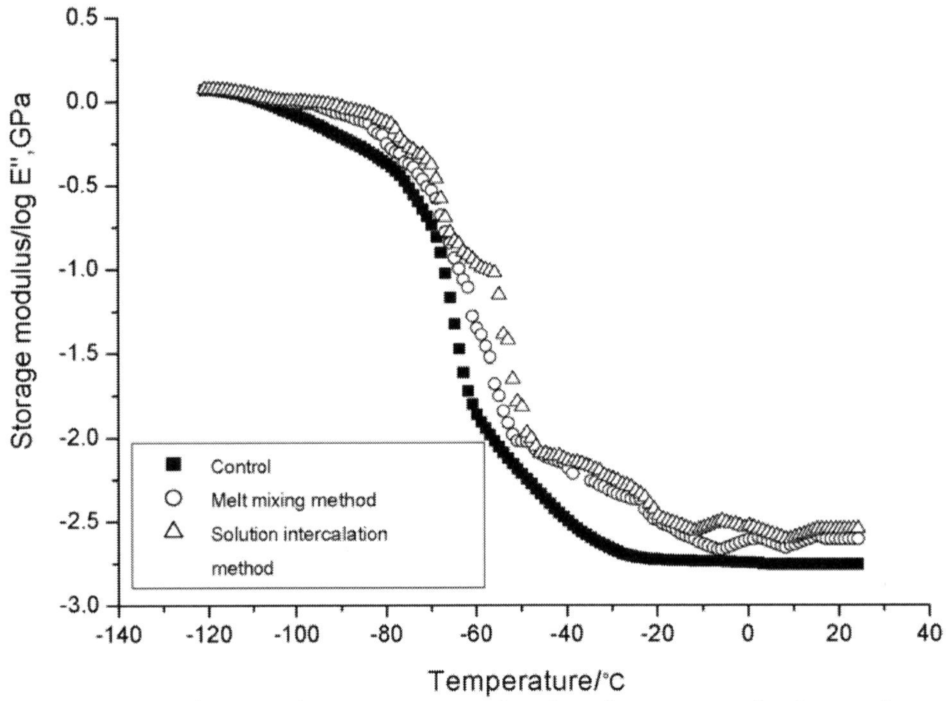

FIGURE 7 Storage modulus at 10 phr EG content as a function of temperature for the samples prepared by two methods (taken from Mu Q. and Feng S., 2007)

1.1.4 Exfoliated graphite nanoplatelet/paraffin composites to enhance thermal conductivity in PCMs

Paraffin wax is one of the most attractive materials for latent heat storage PCM as it is not only commercially available but has desirable properties of high latent heat, chemical inertness and no phase segregation. However, due to its low thermal conductivity, it decreases the overall power of the thermal storage device (Sharma A. et al, 2009; Zalba B., Marin J. M. and Cabeza LF., 2003; Farid MM. et al, 2004). Attempts to improve the thermal conductivity by adding metal foams or fins to paraffin wax increases the weight of the composite and cost to the storage systems and some such methods of enhancements are not compatible with PCMs (Sharma A. et al, 2009; Lafdi K., Mesalhy O. and Shaikh S., 2007).

Many studies have been carried out where EG is reported to be inserted into the paraffin wax to improve thermal conductivity (Karaipekli A. et al, 2007; Zang Z. and Fang X., 2006; Alawadhi E.M. and Amon C.H., 2003; Mills A. et al., 2006; Pincemin S. et al, 2008; Zang Y. et al, 2006). Kim S and Drzal L.T. (2009) prepared composites with xGnP (exfoliated graphite nanoplatelet) with varying mass fraction (1%, 2%, 3%, 5% and 7%) in paraffin wax to understand the relationship between the thermal conductivity of the composite PCM and the mass fraction of graphite nanoplatelets. The paraffin wax was melted by heating it at 75 °C and then the xGnP was mixed into the liquid paraffin. After being filtered and dried, the paraffin/xGnP composite PCM was obtained. To ensure the availability of PCMs as continuous PCMs, the samples were remelted and electrical and thermal conductivity of the samples was measured. The electrical resistivity was found to show a sharp transition from electrical insulator to an electrical conductor (Fig. 8). The resistivity of the second melted sample showed similar behaviour as first melted samples.

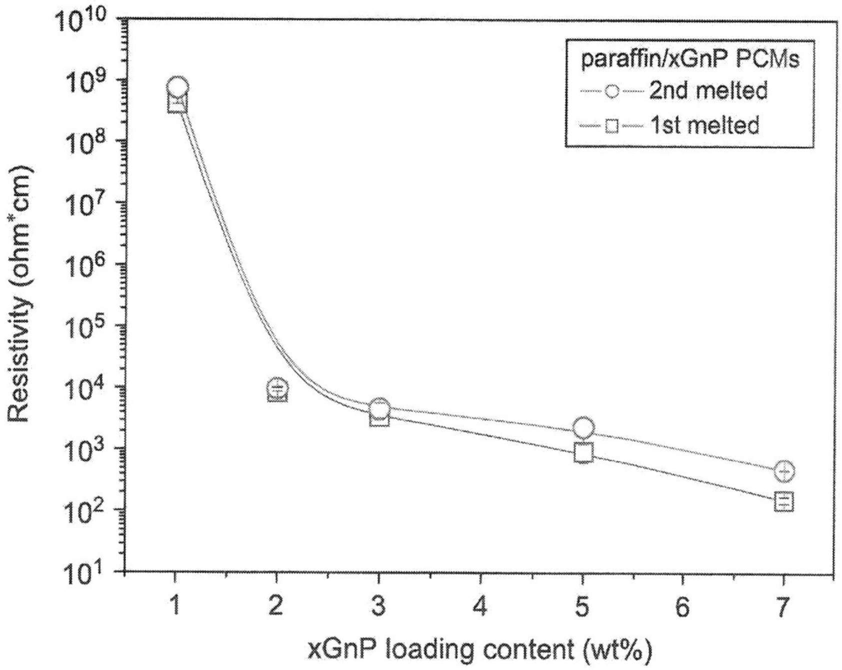

FIGURE 8 Resistivity of paraffin/xGnP composite PCMs by melting times (taken from Kim S and Drzal L.T., 2009)

The thermal conductivity of the xGnP/paraffin PCM composite is found to significantly improve compared to that of pure paraffin and increase with an increase in xGnP content. For 7% wt of xGnP, the conductivity went up to 0.8 W/m K (pure paraffin k=0.26 W/m K). Paraffin/xGnP composites PCMs as continuous PCMs for thermal conductivity were remelted and the second melted samples showed higher conductivity than the first samples (Fig. 9).

Although, previous results showed a decrease in latent heat with an increase in graphite loading content (Zang Z. and Fang X., 2006; Cai Y. et al, 2008), the latent heat of the paraffin/xGnP composites PCMs showed no significant difference from the latent heat of the pure paraffin. Hence, xGnP can be considered as an effective heat diffusion promoter to improve thermal conductivity of PCMs without reducing its latent heat storage capacity.

Zong Y. et al (2010) prepared composites with compressed expanded natural graphite (CENG) with different densities in paraffin wax of thermal conductivity 0.35 W/m K, melting point 50 °C and latent heat of 167 J/g. The expanded graphite was placed in a cubical mode and pressed to obtain CENG of varying bulk densities ranging from 0.07 to 0.26 g/cm^3. The CENG prepared were wrapped in paraffin wax and heated above the melting point of paraffin wax in vacuum oven for 3 h. When the CENG were submerged in the melted paraffin wax, the system was allowed to cool until the wax solidified under normal atmosphere. Depending on the bulk density of the CENG matrix, mass ratio of the paraffin wax in these composites varied from 74% to 92%. The dispersion of the expanded graphite in the paraffin wax was found to be uniform and the interfaces of the two phases (expanded graphite/paraffin wax) were seen to compactly combine due to the high wetting ability of paraffin wax. The pore structure and the thermal conductivity were found to be anisotropic for the CENG matrix. The thermal conductivity in both, the direction perpendicular to the pressing direction (axial direction) and parallel to the direction of pressing (radial direction) was measured by using a steady state method. As expected, the thermal conductivity of the composite was found to increase with an increase in the bulk density of the CENG. An almost linear relationship was seen between the density and the thermal conductivity and the plot represents that the correlation coefficient R is 0.98 (Zong Y. et al, 2010) (Fig. 10).

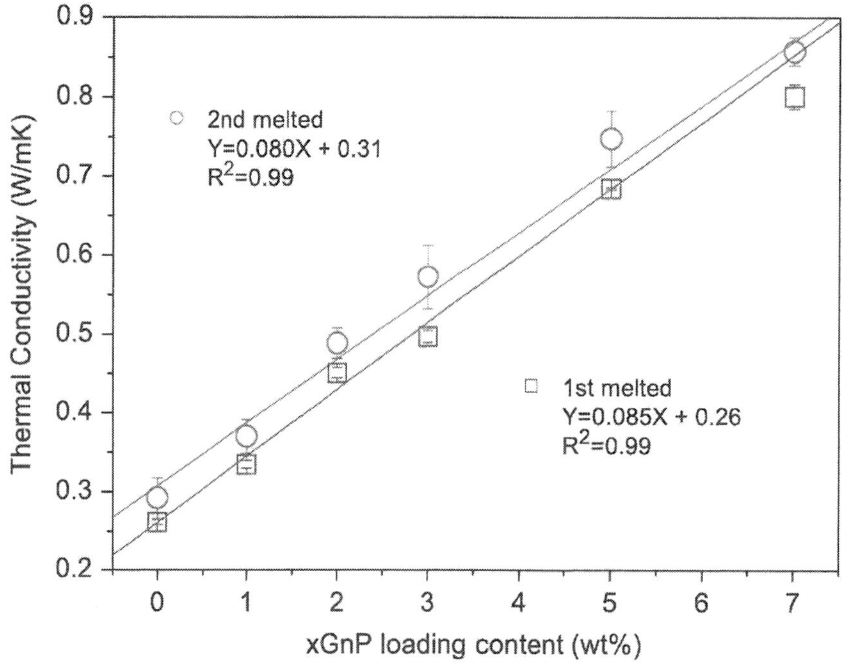

FIGURE 9 Thermal conductivity of paraffin/xGnP composite PCMs by melting times (taken from Kim S and Drzal L.T., 2009)

To measure the temperature distribution as a function of time, the samples were heated on a hot plate with a constant temperature of 70 ^{0}C (higher than the melting point of paraffin wax). A photo shot of the samples was taken every 3 s interval by thermal imager (IR Flex Cam Thermal Imager Ti45HT-20). At a given time, the conductivity of the composite in the two directions (axial and radial) was measured by comparing the thermal images of composites with heat transferred perpendicular to the compression force and along the direction of compression. When the thermal conductivity in the two directions (axial and radial) was compared, the temperature response was same in both directions for first 380 s as the temperature of the composite was below the melting point of paraffin wax. However, as the response time increased, the thermal conductivity was found to be dominated by the direction of orientation of the pores in the CENG matrix. Zong Y. et al (2010) concluded that it was due to this anisotropy of the pores and the existence of natural convection that significantly reduces the melting time of the paraffin wax and thus increases the thermal conductivity of the CENG.

1.1.5 EG/pitch composites to enhance conductivity

Pitches are known to be excellent binding materials in carbon/carbon composites due to their high carbon yield, high fluidity and excellent capacity for producing graphitisable materials (Mendez A. et al., 2005).

They are made of hundreds of polycyclic aromatic compounds and hence making them an insulator with specific resistance being ~ 10^{11} Ωm (Zander M., 1987; Grigoriev I. S., 1991). Hence, Afanasov I. M. et al (2009) prepared composites of coal tar pitch with exfoliated graphite to improve the thermal and electrical conductivity of the pitch.

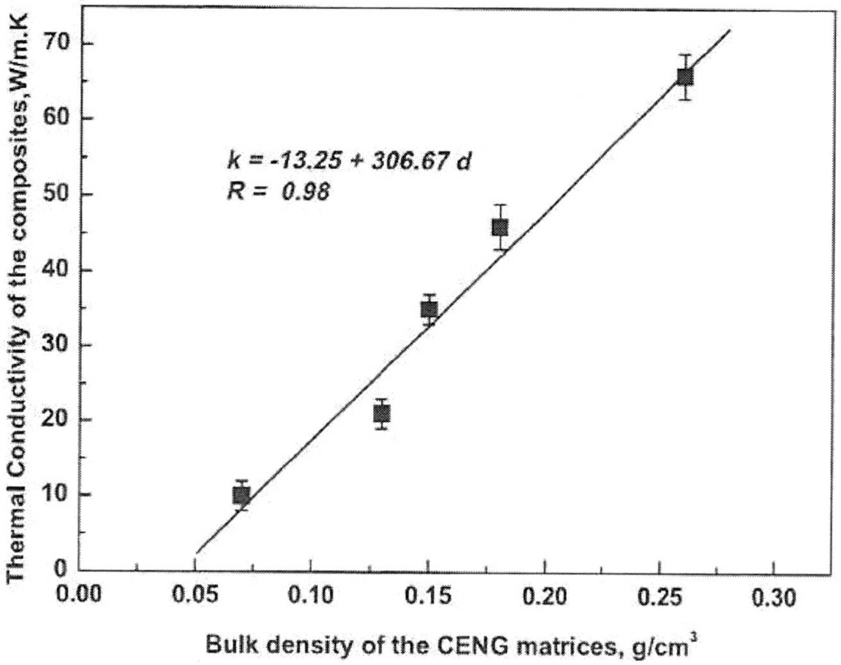

FIGURE 10 Variation in thermal conductivity of the composites with the bulk density of CENG matrices (taken from Zong Y et al, 2010)

Three types of EG with different bulk densities were prepared and referred to as EG1, EG2 and EG3.The fourth type was prepared by mechanically crushing EG1 and referred to as EG4. Two commercial binder coal tar pitches CPT1 and CPT2 were used which were milled to obtain a pitch powder with an average size of less than 500 μm. The EG and pitch were mixed by direct stirring of EG and pitch for 1 min in ambient temperature with 0.3- 10 wt % of EG. The mixture was kept at 200 °C for 3 h to ensure formation of uniform composite (Afanasov I. M. et al, 2009).

The volume resistance with different weight content of EG was studied for composites prepared with pitches CTP1 and CPT2. It can be seen from figure 11 that there is no significant difference in volume resistance based on the type of EG in the composites (CPT1 /EG2, CPT1/EG3 and CPT1/EG4). However, there was decrease of 10 magnitudes when ~1.5% weight of EG is incorporated in the pitch matrix which is similar to the variation in resistivity seen in paraffin/xGnp composites (Fig. 8). This sudden decrease was attributed to the conductive network made by EG particles. The higher aspect ratio of the exfoliated graphite plays an important role in the formation of the conductive network within the pitch matrix. With further increase in EG content, the electrical resistance decreased slowly. Since, there is a close relationship between electrical resistance and thermal conductivity, it can be concluded that these composites showed similar pattern in thermal conductivity.

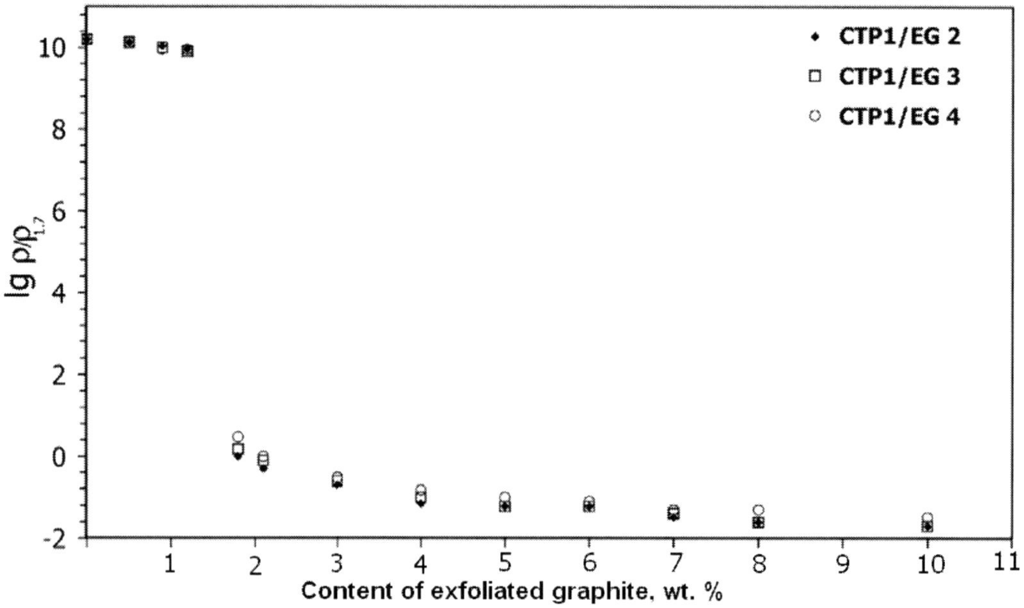

FIGURE 11 Volume resistance of the composites of CTP1 with different exfoliated graphite versus the EG content (taken from Afanasov I.M. et al., 2009.

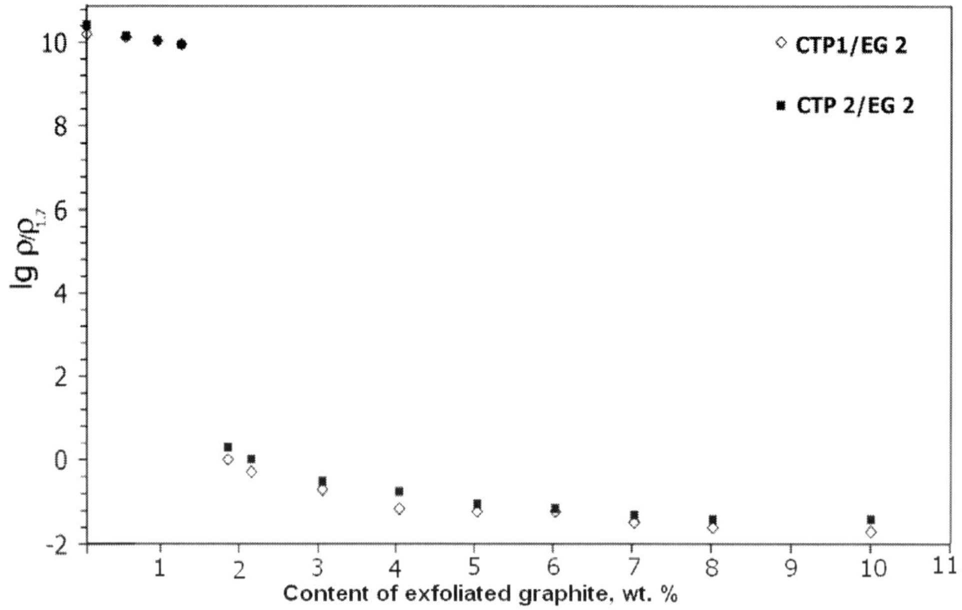

FIGURE 12 Volume resistances of the composites of two coal tar pitches with EG2 versus EG2 content by weight (taken from Afanasov I.M. et al., 2009)

When compared to composites with CPT2 Fig. 12, similar trend was seen that is, a sudden decrease in electrical resistance in composites with EG ranging between 1.2- 1.7%. Hence, it can be concluded that the type of coal tar pitch and type of EG did not have significant role in the behaviour of conductivity. It was the critical weight content of EG in the composite that showed a sudden decrease in the resistance in the composite (Afanasov I.M. et al., 2009).

The thermal gravimetry (TG) graphs of the initial pitches and composites prepared from EG2 is shown in Fig. 13. It can be seen that the composite with 1.2% EG shows onset of degradation 20 °C higher than the pure matrix and with 10 % wt of EG, the degradation is almost 50 °C higher than that of pure matrix for both pitches. Hence, addition of EG showed higher thermal stability compared to pure pitch. This was assumed to be due to the heat shielding effect achieved from EG or the polymerisation processes of the pitch components on the EG particles (Afanasov I.M. et al., 2009).

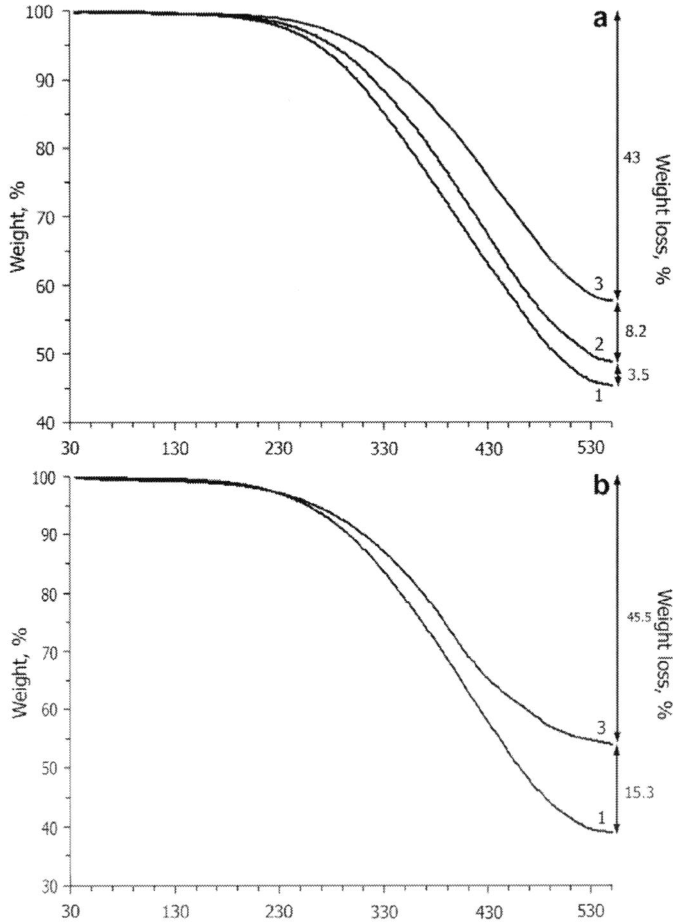

FIGURE 13 TG curves of pure coal tar pitches and CTP/EG2 composites for CTP1 (a) and CTP2 (b): pure CTP (1), CTP with 2 wt.% EG2 (2) and CTP with 10 wt.% EG2 (3) (taken from Afanasov I.M. et al., 2009)

1.1.6 Natural graphite flakes, ammonium lingo-sulfonate and mesophase based composites
Desai S. (2006) prepared and studied composites with natural graphite three different average flake sizes; small (avg. flake size: 180 μm), medium (avg. flake size: 300 μm) and large (avg. flake size: 600 μm). The mesophase pitch supplied in the form of pellets was ground into fine powder and passed through a sieve of 35 micron. The composites were made by mixing graphite flakes of known average flakes size, mesophase pitch powder and ammonium ligno-sulfonate (ALS). The mixture was degassed at 350 ^{0}C for 90 min and hot-pressed to obtain a composite.

Composites were prepared with a given flake size and same starting volume fraction of graphite (75%) and a set of composites were prepared with varying starting volume fraction of large flakes graphite using same procedure. The composites were heat treated to different temperatures (1000 ^{0}C, 1600 ^{0}C and 2400^{0}C/graphitization temperature) and characterized at room temperature.

Thermal conductivity of the composite was calculated using the relation between thermal diffusivity, bulk density and specific heat. The specific heat of the samples was estimated using Spencer's formula, the density was calculated from the weight and dimensions of the sample and the thermal diffusivity was measured using the conventional laser flash method at Tyndall Institute, Cork.

Thermal conductivity of composites with 75% starting volume fraction of graphite heat-treated to 1000 °C prepared from medium and large flakes were found to have diffusivities in the range between 14×10^{-5} m^2/s to 15×10^{-5} m^2/s and 21×10^{-5} m^2/s to 25×10^{-5} m^2/s respectively. The samples prepared from small flakes and heat-treated to 1000 °C are assumed to have poor conductivity since no signal could be detected while measuring thermal diffusivity. Thermal diffusivities and the estimated thermal conductivities of the samples heat-treated to 1600 °C are summarized in Table 1. It can be again seen that the thermal diffusivity and hence the thermal conductivity increases significantly with the flake size.

TABLE 1 Diffusivity and estimated thermal conductivity of samples HT to 1600 °C of different flake sizes

Flakes forming of the composite	Average diffusivity ($\times10^{-5}$ m^2/s)	Thermal conductivity of the sample (W/m K)
S	4.54±01	52
M	15.47±01	220
L	24.98±01	325

When the thermal conductivity of the graphitized composites was calculated, it was seen that the thermal conductivity increases as the flake size increases (Fig. 14). The graphitised samples prepared using large flakes were found to have thermal conductivity of ~655 W/m K which is more than one and half times that of copper and with density of only one fourth of that of copper.

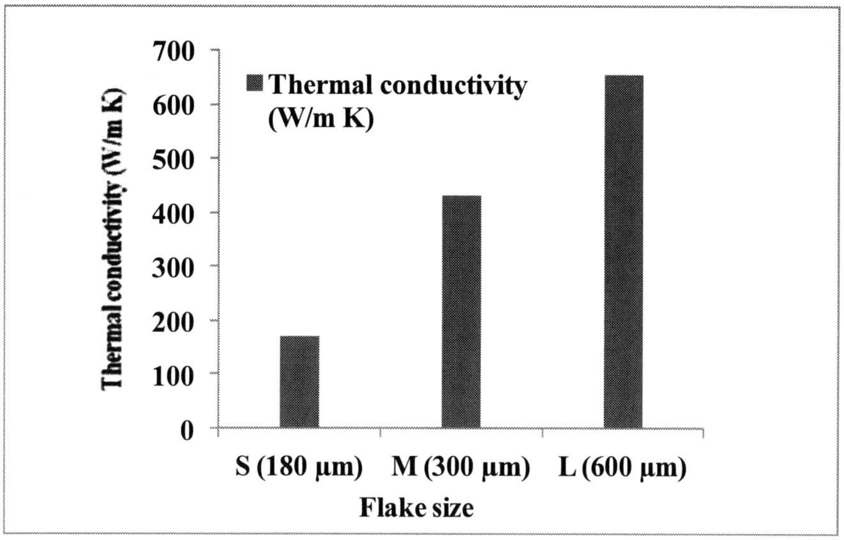

FIGURE 14 Thermal conductivity of graphitised composites (75% starting volume fraction of graphite) versus flake sizes (taken from Desai S., 2006)

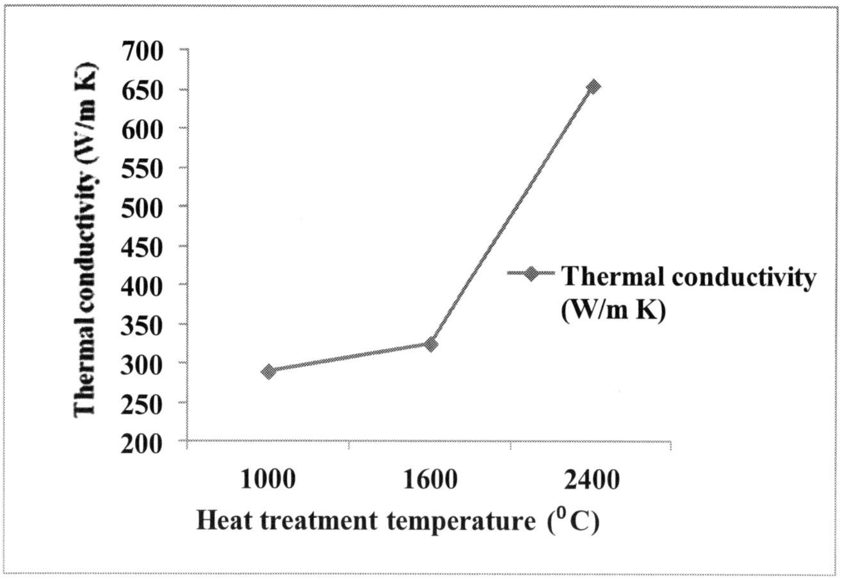

FIGURE 15 Graph of thermal conductivity versus heat-treatment temperature of samples fabricated using large flakes (taken from Desai S., 2006)

The conductivity of the composite prepared from same average flake size and heat-treated to different temperatures was studied. It was found that the conductivity of a composite for a given flake size increases as the heat-treatment temperature increases (Fig. 15). A significant increase in conductivity was observed when the heat-treatment temperature increased from 1600 °C to graphitization temperature. This large improvement in the thermal conductivity is attributed to the graphitization of the binder which provides thermally conducting links.

The graphitised composites with varying starting volume fraction of graphite flakes (large) showed that the thermal conductivity increases with an increase in starting volume fraction of graphite (Desai S., 2006).

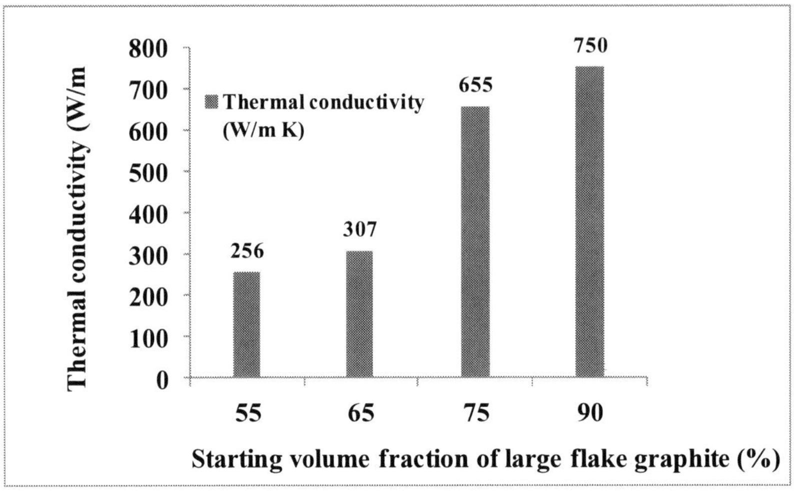

FIGURE 16 Thermal conductivity of large flake graphitised composites prepared with different starting volume fraction of flakes (taken from Desai S., 2006)

Desai S. (2006) concluded that the thermal conductivity was dominated by the flake sizes and heat-treatment temperature. As the volume fraction of graphite increases in the composite, superior properties are observed (Fig 16). The thermal conductivity of sample prepared from 90% starting volume fraction of flakes is found to have a thermal conductivity as high as ~750 W/m K which is nearly twice that of copper with density(1.78 g/cc) approximately one fourth that of copper. Hence, this material is particularly attractive for thermal management problems.

1.1.7 Graphene based composites

Since materials scientists are continuously examining materials with improved properties that will dimensionally be more suitable in the field of nano-science and technology. The discovery of graphene and graphene-based polymer nanocomposites is an important step forward. Graphene is regarded as the thinnest material in the universe with tremendous application potential due to its remarkable properties such as superior mechanical properties, high electrical transport properties and high thermal conductivity (Geim A. K. and MacDonald A. H., 2007; Si Y. and Samulski T., 2008; Wang G. et al, 2008 & 2009; Blake P. et al, 2008; Dreyer R.D. et al, 2010). It is the basic structural unit of some carbon allotropes including graphite, carbon nanotubes and fullerenes. In 2004, Geim A. K. and co- workers at Manchester University successfully identified the single layers of graphene and other 2-D crystals.

Although pristine graphene materials are unsuitable for intercalation by large species such as polymer chains (graphene has a tendency to agglomerate in a polymeric matrix), the agglomeration can be prevented by different chemical modification methods (Geng Y. et al, 2009). Initially, graphite oxide is prepared from natural graphite. After oxidation, a number of methods have been identified to obtain soluble graphene. Some of them include covalent modification by the amidation of the carboxylic groups

(Worsley K. A. Et al,2007; Niyogi S. et al, 2006), non-covalent functionalisation of reduced graphene oxide (Bai H. et al, 2009; Salavagione H. J. et al, 2009; Stankovich S. et al, 2006), nucleophilic substitution to epoxy groups (Bourlinos A. B. et al, 2003), diazonium salt coupling (Lomeda JR et al, 2008) and reduction of graphite oxide in a stabilisation medium (Park S. et al, 2008).

Ghosh S. et al. (2008) prepared a large number of graphene layers by the mechanical exfoliation of bulk highly oriented pyrolitic graphite (HOPG) using the standard technique (Geim K. and Novoselov K.S., 2007; Zhang Y.B. et al, 2005). Si/SiO_2 substrate with array of trenches of 300 nm depths and 1-5 μm width were fabricated by reactive ion etching. A long single layer with relatively constant width was placed across the trenches on the Si/SiO_2 wafer. Since conventional thermal conductivity measurement techniques could not be used in this set-up, Ghosh S. et al (2008) developed a non contact technique based on confocal micro-Raman spectroscopy. The graphene flakes in the system were found to have a thermal conductivity ranging between 3080 to 5150 W/m K and the phonon mean free path was calculated to be ~775 nm near RT. Hence, graphene can be looked upon as a thermal management material in future nanocircuits and similar applications that require high thermal transport properties.

2 Conclusion

It can be seen from the above that graphite has become popular in enhancing thermal conductivity of composites with different end applications. The discovery of graphene ' the thinnest material in the universe' with superior thermal properties has now attracted researchers to find methods to fully extract the properties of graphene in thermal management applications.

3 References

Afanasov I.M. et al (2009), *Carbon*, v **47**, p263
Aiping Yu et al (2007). *The Journal of Physical Chemistry C letters*, pg7565
Alawadhi E.M. and Amon C.H. (2003), *IEEE Trans. Components Packag. Technol.*, **v26**, pg116
Alexander M. & Dubois P. (2000), *Mater. Sci. Eng.*, **v28** pg1
Alysworth J. W. (1916), Expanded graphite and composites thereof. US Patent 1,137,373
Bai H., Xu Y., Zhao L., Li C. and Shi G.(2009), *Chem Commun*,pg1667
Blake P. et al (2009), *Nano Lett*, **v8**, pg1704
Bourlinos A. B. et al (2003), *Chem Mater*, **v19**, pg6050
Cabeza L F et al (2002), *Appl. Therm. Eng.*, **v 22**, pg1141
Cai Y. et al (2008), *Applied Energy,* **v85**, pg765
Chen G.H. et al (2001), *J. Appl. Polym. Sci.*, **v82**, pg2506
Chen G. H. et al (2003), *Carbon*, **v41**, pg619
Debelak B. and Lafdi K. (2007), *Carbon*, v45 (9), pg1727
Desai S. (2006), Fabrication and analysis of highly conducting graphite flake composites, PhD Thesis, University of Leeds, Leeds, UK
Dreyer R. D. et al (2010), *Chem. Soc Rev*, **v39**, pg228
Farid M. M. et al. (2004), *Energ Convers Manage*, **v45**, pg1597
Fukai J. et al (2000), *Proceedings of 3rd European Thermal Science Conference, Pisa*
Fukushima H. (2003), Graphite nano reinforcements in polymer nanocomposites, PhD. Thesis, Michigan State University, East Lansing, MI, USA
Ganguli S. et al (2008), *Carbon*, **v46**, pg80
Geim A.K. and K. S. Novoselov (2007), *Nat. Mater.*, **v6**, pg18
Geim A. K. and McDonald A. H. (2007), *Physics Today*, **v60** (8), pg35
Geng Y. et al (2009), *J Colloid Interface Sci*, **v336**, pg592

Ghosh S. et al. (2008), *Applied Physics Letters*, **v92**, pg151911
Giannelis E.P. (1996), *Adv Mater*., v**8**, p29
Grigoriev I.S. (1991), editor. Handbook on physical magnitudes. Moscow: Energoatomizdat, Pg551
Guohua C. et al (2003), *European Polymer Journal*, v **39**, p2329
Kalaitzidou K. (2006), Exfoliated graphite nanoplatelets as reinforecement for multifunctional polypropylene nanocomposties, PhD. Thesis, Michigan State University, East Lansing, MI, USA
Kalitzidou K., Fukushima H., Drzal L. T. (2007). *Compos Sci Technology*, v **67**, pg 2045
Karaipekli A. et al (2007), *Renew Energy,* **v32,** pg2201
Kelly B. T. (1981), *Physics of Graphite*, Applied Science Publishers, London
Kim S. and Drzal L. (2009), *Solar Energy Materials and Solar cells*, **v**93 pg136
Lafdi K., Mesalhy O. and Shaikh S., (2007), *J Appl Phys*, **v102**, pg083849
Lebaron P. C., Wang Z. and Pinnavaia T. (1999), *J Appl Clay Sci*., v**15**, p11
Li J., Kim J. and Sham M. (2005), *Scripta Materialia,* **v53**, pg 235
Lincoln V. F. and Claude Z. (1983) Inventors, organic matriz composites reinforced with intercalated graphite. US Patent 4,414,142
Lomeda JR et al (2008), J Am Chem Soc., v130, pg16201
Mariner J. T. et al (January 1999), *High thermal conductivity composite and method*, USP no. 5,863,467
Matzek MD (2004), Polymeric carbonnanocomposites: Physical preoperties and osteoblast adhesion studies. Thesis. April 2004
Mendez A. et al. (2005), *Wear,* **v258**, pg1706
Mills A. et al (2006), *Appl Therm. Eng.,* pg1652
Niyogi S. et al (2006), *J Am Chem Soc*., **v128**, pg7720
Norley J. and Chen G. (2002), Graftech Inc., *PCIM 2002 Power Electronics-Thermal management and Packaging*, May 14-16, 2002.
Norley J., *Natural Graphite based Spreader Shields* ™ *Thermal Management products*, Graftech International Ltd. Also see, Norley J., -Chen G., Cooling Zone Submit 2003, Graftech International Ltd.
Park S. et al (2008), Chem Mater., v20, pg6592
Pincemin S. et al (2008), *Sol. Energy Mater. Sol C,* **v92**, pg603;).
Prieto R. et al (2008), *Scripta Materialia*, **v 59**, pg11
Salavagione H. J., Gomez MA and Martınez G. (2009) *Macromolecules*, **v42**, pg6331
Saujanya C. & Radhakrishnan (2001), *Polymer,* **v42**, pg6723
Sharma A. et al. (2009), *Renew Sust Energy Rev*, **v13**, pg318
Si Y. and Samulski T. (2008), *Nano Lett*, **v8**, pg1679
SpreaderShield introductory Presentation, Feb. 2004, Graftech International Ltd.
Stankovich S. et al (2006), *J Mater Chem*., **v16**, pg155
Technical Bulletin 268M, eGraf 1200 Electronic Thermal management Products
Wang G. et al (2009), *J Phys Chem C*, **v112**, pg8192
Wang W. P. and Pan C. Y. (2004), *Polymer*, **v45**, pg3987
Worsley K. A. et al (2007), *Chem Phys Lett*., **v445**, pg51
Usuki A, Kojima Y., Kawassumi M., Okada A. and Fukushima Y. (1993), *J Mater Res*, v**8**, p1179
Zalba B., Marin JM., and Cabeza LF. (2003), *Appl Therm Eng*, **v23**, pg251
Zander M. (1987), *Fuel,* **v 66** (11), pg1536
Zhang Y. et al (2005), *Nature London*, **v438**, pg201
Zhang Y. et al (2006), *Sol. Energy Mater Sol C,* **v90**, pg1962
Zhang Z. and Fang X (2006), *Energy Convers Manage*, **v47**, pg303

Zheng W.G. et al (2002), *Polymer*, **v43**, pg6767
Zheng W. & Wong S.C. (2003), *Compos. Sci.Technol.*, **v63** pg2225
Zong Y. et al. (2010), *Carbon*, **v48**, pg300

IOP Conf. Series: Materials Science and Engineering **40** (2012) 012018 doi:10.1088/1757-899X/40/1/012018

Graphene-polymer composites

G Carotenuto[1], V Romeo[1], I Cannavaro[2], D Roncato[2], B Martorana[2] and M Gosso[2]

[1] Institute of Composite and Biomedical Materials. National Research Council. Piazzale Tecchio, 80 – 80125 Napoli. Italy.
[2] Centro Ricerche Fiat. Strada Torino, 50 – 10050 Orbassano (TO). Italy.

Corresponding author's e-mail address: giancaro@unina.it

Abstract. Graphene is a novel nanostructured material that can be conveniently used as filler for thermoplastic polymers or thermosetting resins, and the resulting nanocomposite material has unique mechanical and chemical/physical properties. Industrial production of graphene/polymer materials requires the availability of a chemical route to produce massive amount of graphene. Natural graphite flakes can be the best starting material for a bulk-production of graphene to be used in the polymeric nanocomposite preparation.

1. Introduction

Graphene is a rigid planar nanostructure made of a single layer of carbon atoms arranged in an hexagonal crystal lattice [1,2]. Graphene can be also considered as a sort of two-dimensional macromolecule, where benzene is the repeating structural unit. In particular, this molecular structure is made of sp2-hybridized carbon atoms involved in both σ and π C-C covalent bonds delocalized to the full molecular structure. The graphene nanostructure can be advantageously used as a filler for thermoplastic polymers and thermosetting resins since it provides polymers of a number of functional characteristics which are not accessible by using other types of fillers [3-24].

The main characteristic of this special type of filler is the unique aspect ratio (the height of a graphene plane is of only a few Angstroms while the other two dimensions are of hundred of microns). Graphene has a number of unusual characteristics, for example its molecular structure is not permeable also to very small molecules like H_2. and noble gasses [25], it is characterized by the best mechanical properties known for a solid material (better than carbon nanotubes and steel), high electrical/thermal conductivity (for the absence of electron scattering phenomena), it shields electromagnetic waves (ultraviolect, visible, infrared and microwaves), etc. Some of these very interesting characteristics of the graphene molecule are given in Table I. As a consequence, graphene-polymer composites are unique materials that can be exploited for both functional and structural applications in different technological fields.

The graphene molecules have been synthesized by chemical vapor deposition (CVD) and epitaxial growth on nickel or SiC substrates, however high-quality graphene and graphene-based materials (e.g., graphene-polymer composites, grafene intercalated by metal nanoparticles or other nanostructures like fullerene and/or CNTs) can be simply obtained starting from natural graphite.

There have been a number of efforts to make very thin films of graphite by mechanical exfoliation (starting from 1990 and continuing until after 2004) but nothing thinner than 50 to 100 layers was produced during these years. A key advance in the science of graphene came when Andre Geim and

Kostya Novoselov at Manchester University managed to extract single-atom-thick crystallites (graphene) from bulk graphite in 2004. The Manchester researchers pulled out graphene layers from graphite and transferred them onto thin SiO_2 on a silicon wafer in a process sometimes called micromechanical cleavage or, simply, the Scotch tape technique. The SiO_2 electrically isolated the graphene, and was weakly interacting with the graphene, providing nearly charge-neutral graphene layers. However, graphite micromechanical exfoliation cannot be used for bulk-scale graphene production and approaches based on the production of colloidal suspensions must be used.

2. Graphene bulk-production

Weak physical interactions (Van der Waals forces) are involved in the natural graphite, and material cohesion is possible only because of the very large number of interactions simultaneously acting among neighbor carbon sheets. Therefore, the graphite structure is made of perfectly piled flat carbon sheets touching each other to allow induction of instantaneous dipoles among the neighbor carbon planes. Such a well-ordered structure is the basic condition for cohesion in graphite and the disordering of this piled structure significantly decreases the material cohesion.

Owing to the high thermodynamically stable chemical structure graphite is an inert substance, and only a few chemical reactions are possible using graphite as a reactant. Graphite flakes can be chemically modified by treatment with very strong oxidizing species leading to a chemical compound known as graphite oxide [26-36]. A mixture of sulfuric acid (H_2SO_4), fuming nitric acid (HNO_3), and potassium permanganate ($KMnO_4$), or a mixture of sulfuric acid, fuming nitric acid and a perchlorate salt (e.g., $KClO_3$), or simply fuming nitric acid and a perchlorate salt system are just some of the oxidizing conditions that can be used to transform mineral graphite flakes into graphite oxide (other oxidizing species used for the graphite oxide production are: CrO_3, $K_2Cr_2O_7$, $(NH_4)2S_2O_8$, H_2O_2, O_3, etc.). These oxidizing mixtures contains a very low amount of water (less than 1% by weight) and therefore are extremely active and able to produce graphite oxide species with a carbon/oxygen ratio of 2-4, depending on the applied reaction conditions (treatment time and temperature). When graphite is treated by so strong oxidizing chemical species, oxygen atoms in form of hydroxyls groups (OH), ether bridges (-O-), carbonyls (-CHO), and carboxylic groups (-COOH) appear everywhere in the carbonous structure. The formation of such functional groups involves the full graphite structure. Sulfuric acid is used as a solvent in these oxidative mixture, since it easily penetrates (intercalate) in the graphene structure, thus facilitating the oxidation process. Typically -OH, -CHO, epoxy (ether bridges), and -COOH groups are bonded to the partially unsaturated graphite network both at edges and the inner regions of the plane. However, the treatment of mineral graphite flakes with some less aggressive liquid oxidizing mixture (for example, an aqueous solution of sulfuric and nitric acid, H_2SO4/HNO_3(sol. 65%) in the ration of 9:1 by volume) allows to achieve a special type of graphitic material known as expandable graphite, which is a graphite oxide with a C:O atomic ratio higher than the ordinary graphite oxide. Expandable graphite is a black material quite similar to the original graphite powder, while graphite oxide is a colored (from yellow to dark-green, depending on the oxidation level) water-soluble material. This type of graphite oxide can be conveniently used to prepare graphene with good thermal and electrical conductivity.

The presence of these oxygenated groups reduces the extension of the conjugation in the expandable graphite, therefore making this type of graphite a chemical species less thermodynamically stable than the starting graphite. When expandable graphite is heated at a convenient temperature (higher than 200°C) the organic structure eliminates the oxygenated groups in form of small molecules (H_2O, and CO_2), thus restoring the more thermodynamically stable (aromatic) structure. When water and carbon dioxide are released, the high ordered structure of graphite is completely destroyed and a foam-like material, named expanded graphite, is obtained. In this new type of graphite structure, cohesion is very low and the carbon sheets can very easily move away from each other by application of some thermal or mechanical energy. Therefore, during the expansion process the oxidized carbon planes are thermally reduced to graphite sheets and only a limited percentage of residual ether-bridges leaves in the final compound (expanded graphite). Such residual epoxy groups can be eliminated by chemical reduction (for example with hydrazine, dimethyl-hydrazine, hydroquinone, $NaBH_4$, etc.) or by treatment with strong alkaline solution (KOH, NaOH, etc.). The extension of graphite foaming, and therefore the lowering of the material cohesion, depends on the rate of the H_2O and CO_2 molecule release, therefore it is of a fundamental importance to apply a significant thermal shock to the material. Faster is the graphite oxide heating process lower is the residual cohesion in the thermally-reduced product. The combination of oxidation and thermal reduction treatments allows to deeply disorder the graphite structure, thus decreasing the physical interactions among the carbon sheets. Carbon sheets in the expanded graphite can be separated each other by simply applying some thermal or mechanical energy to this material. In particular, expanded graphite needs to be wet by an adequate organic medium that allows to uniformly apply mechanical energy to the full mass and the obtained liquid suspension can be heated under stirring or some mechanical energy can be applied by a piezoelectric generator of ultrasounds. The elastic wave propagate to the foam structure (expanded graphite) and the resulting oscillations of the carbon planes quickly disaggregates the low coherent solid structure. When heat is applied, the thermal vibrations of carbon atoms are involved in the separation of carbon planes in the low coherent solid structure, however such mechanism is less effective than the sonication treatment and therefore a combination of heat and sonication is the most effective process. Heat has also the important role to prevent recombination of graphen planes after the exfoliation process.

Also the nature of the sonication medium has an important role in strongly solvating the carbon sheets and avoid their recombination. In particular, thermally reduced graphite has an anphiphilic nature (both non-polar regions and polar functional groups are present in the molecular structure), consequently the best sonication medium should be made of molecules containing both a non-polar part able to interact with carbon planes by Van der Waals interactions and a polar group that may interact with residual oxygenated groups present in the carbon structure. Alcohols, glycols and polyols represent the best choice for the presence of hydroxyl groups and aliphatic hydrocarbon segments in their molecular structure, but also other liquids phase can be conveniently used (e.g., N-methylpyrrolidone, dimethyl-formammide, etc.). For a complete exfoliation of the expanded graphite, the sonication should be applied to a very dilute system in order to reduce the importance of the graphene sheets recombination process (dilution increase the average inter-plane distance during the sonication treatment, thus making difficult the recombination of carbon planes).

When sonication is removed the graphene sheets may slowly recombine, settling at container bottom, however these planes are solvated by molecules of the liquid dispersing medium and therefore they can be fully redispersed by applying again ultrasounds. To avoid recombination of the carbon sheets it is required to intercalate them by some intercalantes. Polymers soluble in alcohols can be dissolved in the colloidal suspension and the system allowed to concentrate by solvent evaporation. When the solvent (alcohol, N-methylpyrrolydone, etc.) has been completely removed by evaporation, the carbon planes are separated by the enveloping polymer molecules and their recombination to form graphite cannot take place anymore. In the case of non-polar polymers the liquid sonication medium must be selected among the compounds acting as solvents for the polymer. Frequently chloroform and ethers (e.g., dioxane) are used. Such new type of discontinuous composite materials allows the fabrication of several devices for future carbon-based nanoelectronics.

3. Graphene in automotive application

Graphene has been investigated with great interest for use as electrode material in electrochemical supercapacitors in electrical/hybrid automobiles because of their unique combination of properties such as high surface area, lightweight, good electrical conductivity, compatibility with other materials, and controlled pore size distribution [37-44]. This material has recently been used in supercapacitor devices to replace conventional carbon electrodes and have shown very good performance[45-49]. Typically, graphitic carbon-based materials are randomly oriented with respect to the current collectors in a conventional stacked geometry in supercapacitors leading to a lower utilization of the electrochemical surface area of graphene layers and consequently limits the extent of the

EDL(electrical double layer) formed at the interface. Most recently, Yoo et al.[50] demonstrate the utilization of pristine and multilayer graphene as electrodes in an "in-plane" device geometry for use in supercapacitors in order to offers new opportunities for the electrolyte ions toenhance interaction with all the carbon layers with a full utilization of the high surface area offered by graphene layers.

4. Conclusion

The mild oxidation treatment of natural graphite flakes by a H_2SO_4/HNO_3 aqueous solution combined with thermal expansion at very high temperature and exfoliation by applying sonication in liquid dispersing medium represents a very convenient approach for a massive graphene production. In addition, polymer intercalation in liquid phase is also an adequate way to avoid recombination of the single graphene sheets produced during the exfoliation stage. The resulting isoorientation of grafene sheets in the polymer matrix is an important characteristic and most physical properties of this new type of material originates from such special topological characteristic.

References

[1] A. Geim, "Graphene: Status and Prospects", *Science* **324**(2009)1530.
[2] D.D.L. Chung, "Review Graphite", *J. Mater. Sci.* **37**(2002)1475-1489
[3] P. Mukhopadhyay, R.K. Gupta, "Trends and frontiers in graphene-based polymer nanocomposites", Plastic Engineering (2011)32-42.
[4] D.R. Paul, L.M. Robeson, "Polymer nanotechnology: Nanocomposites", *Polymer* **49**(2008)3187-3204.
[5] G. Chen, D. Wu, W. Weng, C. Wu, "Exfoliation of graphite flake and its nanocomposites", *Carbon* **41**(2003)579-625.
[6] H. Kim, A.A. Abdala, C.W. Macosko, "Graphene/polymer nanocomposites", *Macromolecules* **43**(2010)6515-6530.
[7] K. Kalaitzidou, H. Fukushima, L.T. Drzal, "A new compounding method for exfoliated graphite polypropylene nanocomposites with enhances flexural properties and lower percolation threshold", *Composite Science and Technology* **67**(2007)2045-2051.
[8] S. Stankovich, D.A. Dikin, G.H.B. Dommett, K.M. Kohlaas, E.J. Zimney, E.A. Stach, R.D. Piner, S.T. Nguyen, R.S. Ruoff, "Grephene-based composite materials", *Nature* **442**(2006)282-286.
[9] G. Zheng, J. Wu, W. Wang, C. Pan, "Characterization of expanded graphite/polymer composites prepared by in situ polymerization", *Carbon* **42**(2004)2839-2847.
[10] H. Kim, Y. Miura, C.W. Macosko, "Graphene/polyurethane nanocomposites for improved gas barrier and electrical conductivity", *Chem. Mater.* **22**(2010)3441-3450.
[11] P. Liu, K. Gong, P. Xiao, M. Xiao, "Preparation and characterization of poly(vinyl acetate)-intercalated graphite oxide nanocomposite", *J. Mater. Chem.* **10**(2000)933-935.
[12] H. Kim, C.W. Macosko, "Processing-property relationships of polycarbonate/graphene composites", *Polymer* **50**(2009)3797-3809.
[13] H. Kim, C.W. Macosko, "Morphology and properties of polyester/exfoliated graphite nanocomposites", *Macromolecules* **41**(2008)3317-3327.
[14] K. Kalaitzidou, H. Fukushima, L.T. Drzal, "Multifunctional polypropylene composites produced by incorporation of exfoliated graphite nanoplatellets", *Carbon* **45**(2007)1446-1452.
[15] S. Kim, I. Do, L.T. Drzal, "Multifunctional xGnP/LLDPE nanocomposites prepared by solution compounding using various screw rotating systems", *Macromol. Mater. Eng.* **294**(2009)196-205.
[16] Y. Xu, W. Hong, H. Bai, C. Li, G. Shi, "Strong and ductile poly(vinyl alcohol)/graphene oxide composite films with layered structure", *Carbon* **47**(2009)3538-3543.
[17] M.A. Osman, V. Mittal, M. Morbidelli, U.W. Suter, "Polyurethane adhesive nanocomposites as gas permeation barrier", *Macromolecules* **36**(2003)9851-9858.

[18] N.K. Lape, E.E. Nuxoll, E.L. Cussler, "Polydisperse flakes in barrier films", *Journal of Membrane Science* **236**(2004)29-37.
[19] H.M. Ju, S.H. Choi, S.H. Huh, "X-ray diffraction patterns of thermally-reduced graphenes", *Journal of the Korean Physical Society* **57(6)**(2010)1649-1652.
[20] X. Jiang, L.T. Drzal, "Multifunctional high density polyethylene nanocomposites produced by incorporation of exfoliated graphite nanoplatelets 1: morphology and mechanical properties", Polymer Composites (2010)1091-1098.
[21] T. Kuilla, S. Bhadra, D. Yao, N.H. Kim. S. Bose, J.H. Lee, "Recent advances in graphene based polymer composites", *Progress in Polymer Science* **35**(2010)1350-1375.
[22] A. Yasmin, J.J. Luo, I.M. Daniel, "Processing of prepared graphite reinforced polymer nanocomposites", *Composites Science and Technology* **66**(2006)1179-1186.
[23] P. Xiao, M. Xiao, K. Gong, "Preparation of exfoliated graphite/polystyrene composite by polymerization-filling technique", *Polymer* **42**(2001)4813-4816.
[24] G. Chen, W. Weng, D. Wu, C. Wu, J. Lu, "Preparation and characterization of graphite nanosheets from ultrasonic powdering technique", *Carbon* **42**(2004)753-759.
[25] J. Scott Bunch, S.S. Verbridge, J.S. Alden, A.M. van der Zande, J.M. Parpia, H.G. Craighead, P.L. McEuen, "Impermeable atomic membranes from graphene sheets", *Nano Letters* **8(8)**(2008)2458-2463.
[26] B. Brodie, *Phil. Trans.* 149 (1869) 249.
[27] U. Hofman, *Z. anorg. allg. Chem.* 234 (4), 311 (1937).
[28] H. Hamdi, Kolloid Briheffc, **64**(1943)654.
[29] W.S:Hummers, "Preparation of graphitic acid", U.S. Patent No. 2,798,878 (1957).
[30] A.V. Yakovlev, A.I. Finaenov, S.L. Zabud'kov, E.V. Yakovleva, "Thermally expanded graphite: synthesis, properties, and prospects for use", *Russian Journal for Applied Chemistry* **79(11)**(2006)1741-1751.
[31] S. Park, R.S. Ruoff, "Chemical methods for the production of graphenes", *Nature Nanotechnology* **58**(2009)1-8.
[32] A. Celzard, J.F. Mareche, G. Furdin, "Modelling of exfoliated graphite", *Progress in Materials Science* **50**(2005)93-179.
[33] K.E. Carr, "Intercalation and oxidation effects on graphite of a mixture of sulphuric and nitric acids", *Carbon* **8**(1970)155-166.
[34] D.R. Dreyer, S. Park, C.W. Bielawski, R.S. Ruoff, "The chemistry of graphene oxide", *Chem. Soc. Rev.* **39**(2010)228-240.
[35] S. Stankovich, D.A. Dikin, R.D. Piner, K.A. Kohlhaas, A. Kleinhammes, Y. Jia, Y. Wu, "Synthesis of graphene-based nanosheets via chemical reduction of exfoliated graphite oxide", *Carbon* **45**(2007)1558-1565.
[36] M. Hirata, T. Gotou, S. Horiuchi, M. Fujiwara, M. Ohba, "Thin-film particles of graphite oxide 1: High-yield synthesis and flexibility of the particles", *Carbon* **42**(2004)2929-2937.
[37] Conway, B. E. Electrochemical Capacitors - Scientific Fundamentals and Technological Applications; Kluwer: New York, 1999.
[38] Pandolfo, A. G.; Hollenkamp, *A. F. J. Power Sources* **157** (2006)11–27.
[39] E. Frackowiak, *Phys. Chem. Chem. Phys.* **9**(2007)1774–1785.
[40] P. Simon, Y. Gogotsi, *Nat. Mater.* **7**(2008)845–854.
[41] L. L Zhang, X. S. Zhao, *Chem. Soc. Rev.* **38**(2009)2520–2531.
[42] R. Kotz, M. Carlen, *Electrochim. Acta* 2000, 45, 2483–2498.
[43] M. D. Levi, G. Salitra, N. Levy; D.Aurbach, J. Maier, *Nat. Mater.* **8**(2009)872–875.
[44] A. C. Dillon, *Chem. Rev.* **110**(2010)6856–6872.
[45] S. R. C. Vivekchand, C. S. Rout, K. S. Subrahmanyam, A.Govindaraj, C. N. R. Rao, *J. Chem. Sci.* **120**(2008) 9–13.
[46] M. D.Stoller, S.Park, Y.Zhu, J. An, R. S. Ruoff, *Nano Lett.* **8**(2008)3498–3502.
[47] Y.Wang, Z. Shi, Y.Huang, Y.Ma, C.Wang, M.Chen, Y.Chen, *J. Phys. Chem. C*

113(2009)13103–13107.
[48] D.W.Wang, F.Li, Z.S. Wu, W. Ren, H.M. Cheng, *Electrochem. Commun.* **11**(2009)1729–1732.
[49] C.Liu, Z.Yu, D.Neff, A.Zhamu, B. Z. Jang, *Nano Lett.* **10**(2010)4863–4868.
[50] J.J.Yoo, K. Balakrishnan, J. Huang, V. Meunier, B.G. Sumpter, A. Srivastava, M.Conway, A.L.M. Reddy, J. Yu, J. Vajtai, and P.M. Ajayan *Nano Lett.* **11**(2011)1423–1427

IOP Conf. Series: Materials Science and Engineering **40** (2012) 012019 doi:10.1088/1757-899X/40/1/012019

Evaluating the electrical discharge machining (EDM) parameters with using carbon nanotubes

M M Sari[1,2], M Y Noordin[2] and E Brusa[1]

[1] Department of Mechanical and Aerospace Engineering, Politecnico di Torino Corso Duca degli Abruzzi, 24 – 10129 Torino, ITALY
[2] Faculty of Mechanical Engineering, Universiti Teknologi Malaysia, 81310 UTM Skudai, Johor, MALAYSIA

E-Mail (corresponding author): mehdi.mohammadzadeh@polito.it

Abstract. Electrical discharge machining (EDM) is one of the most accurate non traditional manufacturing processes available for creating tiny apertures, complex or simple shapes and geometries within parts and assemblies. Performance of the EDM process is usually evaluated in terms of surface roughness, existence of cracks, voids and recast layer on the surface of product, after machining. Unfortunately, the high heat generated on the electrically discharged material during the EDM process decreases the quality of products. Carbon nanotubes display unexpected strength and unique electrical and thermal properties. Multi-wall carbon nanotubes are therefore on purpose added to the dielectric used in the EDM process to improve its performance when machining the AISI H13 tool steel, by means of copper electrodes. Some EDM parameters such as material removal rate, electrode wear rate, surface roughness and recast layer are here first evaluated, then compared to the outcome of EDM performed without using nanotubes mixed to the dielectric. Independent variables investigated are pulse on time, peak current and interval time. Experimental evidences show that EDM process operated by mixing multi-wall carbon nanotubes within the dielectric looks more efficient, particularly if machining parameters are set at low pulse of energy.

1. Introduction

Electrical Discharge Machining (in the following simply EDM) is one of the most accurate manufacturing processes available for creating complex geometries within parts and assemblies [1,2]. EDM erodes material in the path of electrical discharges generating an arc between an electrode tool and the work piece. EDM manufacturing is quite affordable and suitable when high accuracy is required. Nevertheless, EDM process has some problems such as generation of high heating on the electrically discharged material and formation of white layer in work piece, decreasing the quality of surface roughness and the product life. Carbon nanotubes (CNTs) display unexpected strength and unique electrical properties and are effective thermal conductors. They can absorb the heat from the electrical discharged material and decrease the amount of white layer. CNTs came only recently and lately to the manufacturing industries to improve the machining performance. Among the literature dealing with CNTs some specialized contributions were recently focused on the EDM improvement. Guu [3] proposed the EDM process for machining the surface morphology and roughness of the AISI D2 steel; Prabhu et al. [4,5] analyzed the surface characteristics of tool steel material by using multi-wall CNTs to improve the surface finish of material to the nanometer size. Guu [6] based his analysis on the effect of EDM on surface characteristics and damage of the AISI D2 tool steel, while Prabhu [5] proposed an approach to generate a nanosurface of Inconel 600 material by grinding process, by using single-wall CNTs. Pecas [7] proposed a EDM process with powder mixed dielectric and compared it with the normal condition of EDM to analyse the surface quality. A deeper investigation about the improvement of the EDM parameters in case of multi-wall CNTs application was not yet proposed. Therefore, this paper evaluates through some experiments the benefits introduced by CNTs in the EDM process.

2. Selected materials

A preliminary investigation of the results obtained through the traditional EDM process was performed. AISI H13 steel was selected as a tool material. It is a chromium-molybdenum-vanadium-alloyed steel. EDM is extensively used as non-conventional material removal process and consists of machining electrically conductive materials by using some precisely controlled sparks occurring between an electrode and a work piece, in presence of a dielectric fluid. It is a thermal process, therefore material is removed by heat, generated by the spark. EDM is only applicable to electrically conductive materials, like hardened steel, titanium, hastelloy, kovar, inconel and carbide. The AISI H13 steel is hot work tool steel with overall good ductility, toughness, wear resistance, and harden ability and machinability. It is evenly known as DIN 1.2344 and EN X40CrMoV5. It assures a good resistance to abrasion at both low and high temperatures, high level of toughness and ductility, uniform and high level of machinability and polish ability. Moreover it shows a good high-temperature strength and resistance to thermal fatigue, excellent through-hardening properties and a very limited distortion during hardening. Selection of this material was motivated by the number of applications as tools for extrusion, dies, backers, die holders, liners, dummy blocks, stems, plastic molding applications, injection molding of thermoplastics, long production runs and molding of parts in thermosetting plastic. Main properties are summarized in Table 1.

Table 1. Properties of the AISI H13 tool steel.

Property	Description	
Composition [%]	C 0.39; Si 1.0; Mn 0.4; Cr 5.3; Mo 1.3; V 0.9	
Delivery condition	Soft annealed (approx. 185 HB)	
Colour code	Orange/Violet	
Hardness	52 HRc	45 HRc
Tensile strength	1820 MPa	1420 MPa
Yield strength	1520 MPa	1280 MPa

Table 2. Properties of MWCNTs used in this study.

Property	Description
Outer diameter	< 8 nm
Length	10 to 30 μm
Purity	> 95 wt%
Ash	< 1.5 wt%
Specific Surface Area	> 500 m^2/g
Electrical conductivity	> 10^{-2} S/cm

CNTs were used for this investigation. They are allotropes of carbon with a cylindrical nanostructure [8]. These cylindrical carbon molecules have novel properties that make them potentially useful in many applications as electronics, optics and other fields of materials science.

They exhibit extraordinary strength and unique electrical properties and are efficient thermal conductors. Among several configurations, MWCNTs consist of multiple rolled layers of graphite. Interlayer distance is close to the distance between graphene layers in graphite, i.e. approximately equal to 3.4 Å. Relevant properties of the CNTs used in this experimental activity are described in Table 2.

3. Experimental setup

To perform the EDM and measuring its relevant parameters, some facilities were used at the Universiti Teknologi Malaysia. The arithmetic surface roughness was measured on the machined surface by using the Mitutoyo-Formtracer CS 5000 shown in Fig.1. This equipment measures the surface roughness on plane, groove and cylinder. A Scanning Electron Microscope (SEM) was used to analyze and measure the thickness of the altered surface layer such as recast layer, heat affected zone and microcracks (Fig.2). Each specimen was made of the AISI H13 tool steel. The work pieces were prepared to size of 10x10x10 mm. Hardness of specimen was measured as 53 HRC. The electrode material is copper, having circular section of diameter d=4.9 mm. The dielectric fluid used in the EDM process was kerosene. Up to 80 litres of liquid were used to fill the tank connected to the EDM equipment. Specimens were sparked on the die-sinking EDM machine Charmilles Roboform 100 (Fig.4). Copper electrode was used as a negative polarity during the EDM process, while the specimen was the positive one.

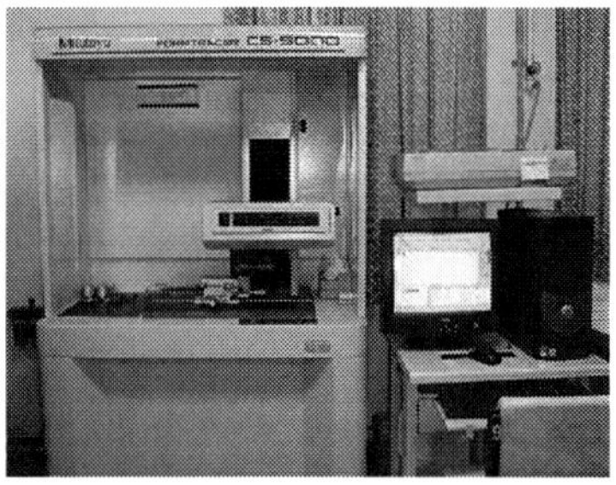

Figure 1. Formtracer CS–5000.

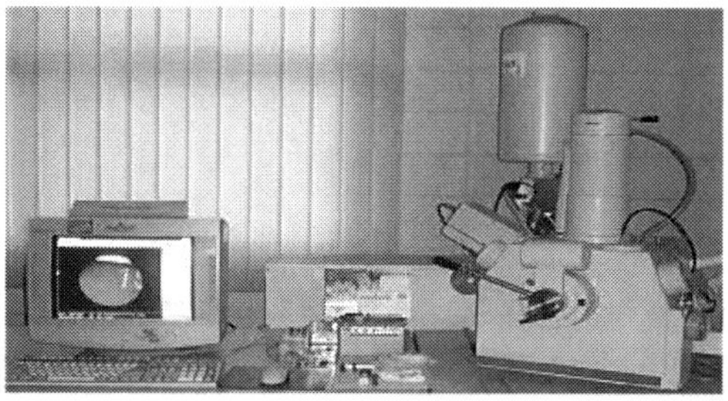

Figure 2. Scanning electron microscope (SEM)

Figure 3. EDM Charmilles Roboform 100.

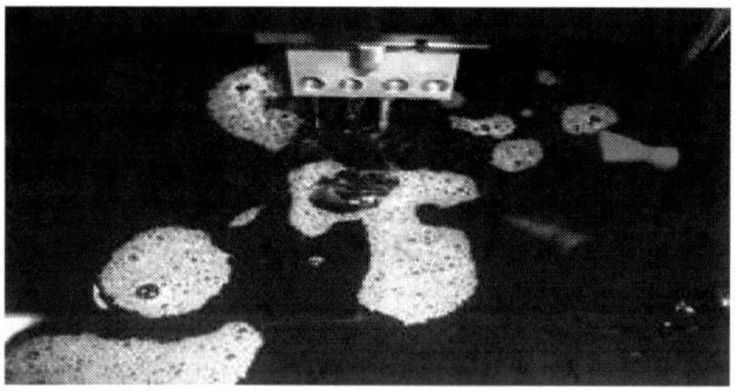

Figure 4. Mix of kerosene and MWCNTs in the dielectric used for the experiments.

4. Experimental investigation

EDM process is characterized when some relevant parameters are measured. The peak current, the pulse on time and the interval time completely define the operating conditions of the EDM machine. Response in terms of machining effectiveness is detected if the material removal rate, the electrode wear rate, the surface roughness and the recast layer are known. Design of Experiments was based on the factors and levels described in Table 3. Experiments were carried out according to the ANOVA analysis. Three factors were selected for each step of the investigation, leading to perform up to eight experiments. Table 4 summarizes the procedure followed in each run.

Table 3. Factors and levels for the EDM process analysis.

Factor	Name	Units	Type	Low level (-1)	High level (+1)
P	Peak current	Ampère	Numeric	8	24
A	Pulse on time	μs	Numeric	12.8	50
B	Interval time	μs	Numeric	12.8	50

Table 4. List of the EDM runs performed and analysed.

Std	Run	P (Peak current) [A]	A (Pulse on time) [μs]	B (Interval time) [μs]
1	6	8	12.8	12.8
2	2	8	12.8	50
3	8	8	50	12.8
4	4	8	50	50
5	7	24	12.8	12.8
6	1	24	12.8	50
7	5	24	50	12.8
8	3	24	50	50

In the first step of the investigation, experiments were performed by implementing the conventional method, without using carbon nanotubes. A precision balance was used to measure the weight of work piece and electrode, before and after each machining process. Electrode wear and material removal rates for each run of the performed experiments were measured. Machined surface roughness of each specimen was detected by the surface roughness tester. Thickness of recast layer on the surface of specimens was measured by the Scanning Electron Microscope (SEM). In the second step of this study, the multi-wall carbon nanotubes were mixed in the dielectric fluid in proportion of 1 g of MWCNTs for each liter of kerosene (Fig.4). A separate tank was used for machining with using MWCNTs. Even in this case relevant parameters were measured as previously described. Results were finally compared to determine the effect of using multi-wall carbon nanotubes during the EDM process.

5. Results and discussion

Maximum value of Material Removal Rate (MRR) is a significant indicator of the efficiency and cost of the EDM process. It can be calculated by dividing the difference between the weight of work piece before (w_b) and after (w_a) machining, respectively, found by weighing each specimen, against the machining time (t_n). CNTs in the dielectric flow increase the MRR. Benefit is appreciable, as Table 5 shows.

Table 5. Material removal rates.

Run	1	2	3	4	5	6	7	8
A) Without MWCNTs [g/min]	0.1850	0.0039	0.1192	0.0149	0.1540	0.0039	0.0726	0.0194
B) With MWCNTs [g/min]	0.1915	0.0040	0.1579	0.0188	0.2519	0.0099	0.0890	0.0233
Difference (B-A) [g/min]	0.0065	0.0001	0.0387	0.0039	0.0979	0.006	0.0164	0.0039

Electrode wear depends on the dielectric flow in the machining zone. If flow is turbulent, an increasing in electrode wear is appreciated. Electrode Wear Ratio (EWR) can be defined as the ratio of the weight of material used of the electrode (EWW) and of work piece (WRW), respectively, being expressed as a percentage. As Table 6 shows, CNTs mixed in dielectric decrease the electrode wear rate, since heating is lower. It was demonstrated that during the EDM, CNTs can absorb the heat leading to have a better flowing of the electric current.

Table 6. Electrode wear rate ratios.

Run	1	2	3	4	5	6	7	8
A) Without MWCNTs	0.2433	0.2182	0.2946	0.1283	0.2812	0.2765	0.3972	0.1351
B) With MWCNTs	0.2911	0.1983	0.2822	0.1084	0.2180	0.1978	0.3572	0.1199
Difference (B-A)	0.0478	-0.0199	-0.0124	-0.0199	-0.0632	-0.0787	-0.0400	-0.0152

On the surface of the work piece a large number of craters is formed by the discharge energy produced by the EDM process. Therefore surface quality depends upon the energy per spark, peak current, pulse on and off time. Performance of the EDM with CNTs was evaluated in this case by resorting to the computation of the arithmetic surface roughness value. Measurements were performed on the surface of the holes by means of the Mitutoyo-Formtracer CS 5000.

Table 7. Surface roughness.

Run	1	2	3	4	5	6	7	8
A) Without MWCNTs [μm]	3.704	2.284	5.094	2.768	3.055	2.405	2.689	2.675
B) With MWCNTs [μm]	3.013	2.090	3.354	2.537	2.760	2.348	2.285	2.390
Difference (B-A) [μm]	-0.691	-0.194	-1.74	-0.231	-0.295	-0.057	-0.404	-0.285

Table 7 demonstrates that thermal properties of MWCNs allow a good absorption of heat, during the EDM, thus leading to reduce the thickness of recast layer and increase the quality of surface roughness. When a larger amount of heat is transferred into the work piece, dielectric looks increasingly unable to clear away the molten material. Surface roughness is consequently decreased.

Table 8. Machining parameters of different runs during the EDM.

Run	P: Peak Current [A]	A: Pulse on time [μs]	B: interval time [μs]
1	24	12.8	50
2	8	12.8	50
3	24	50	50
6	8	12.8	12.8
7	24	12.8	12.8

Recast layer produced on the surface of work piece induces the aging of material and affects the surface roughness. It was measured directly by the SEM to compare the results obtained with and without using CNTs. Before the inspection by SEM, each work piece was machined, mounted, grinded and polished. Etching was evenly performed, with a solution of 100 ml of Ethanol, 5 ml of HCL and one gram of Picric Acid. Effects upon the recast layer were analysed on specimens n.1,2,3,6 and 7. Thickness of recast layer was compared as it looks at SEM, like in Fig.5. Each run was characterized as is described in Table 8. Moreover, results shown in Table 9 make clear that using CNTs mixed in the dielectric flow improves and minimizes the thickness of recast layer. As it was observed in case of the surface roughness, the heat absorption operated by MWCNTs, during the EDM, leads to thinner recast layer. Table 9 summarizes the average thickness of white layer.

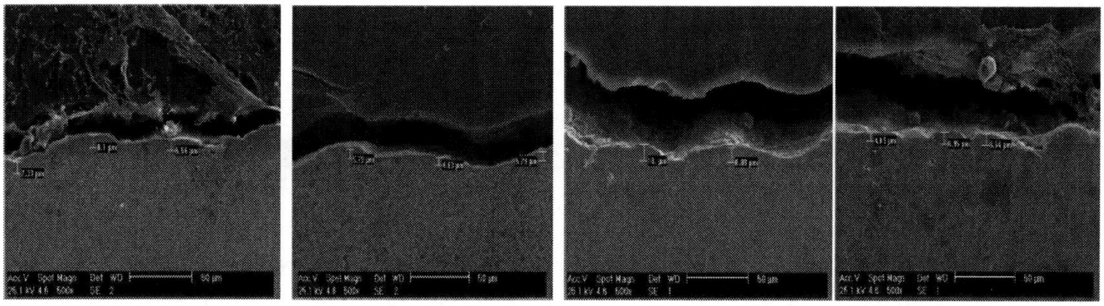

Figure 5: Sequence of recast layer measurements performed by SEM in: run 1 without CNTs, with using CNTs, in run 2 without CNTs and with using CNTs.

Table 9. Comparison of white layer thickness without and with using CNTs in EDM.

Thickness of white layer cast [μm]	Without using Carbon nanotubes	With using Carbon nanotubes
Sample 1	9.4	6.05
Sample 2	7.33	5.07
Sample 3	11	6.95
Sample 6	5.20	3.95
Sample 7	7.59	5.53

6. Conclusion

A technological improvement of the Electrical Discharge Machining with copper electrode was proposed by including Multi wall Carbon Nanotubes into the dielectric liquid. Several machining parameters were considered and measured. Experimental evidences show that carbon nanotubes give better surface compared to the traditional EDM process. Thickness of recast layer is smaller when CNTs are used. Electrode Wear Rate and Material Removal Rate are better and heat can be effectively absorbed by the CNTs, if machining parameters are set to have a low pulse of energy.

7. References

[1] Jamson E C 2000 *Electrical Discharge Machining* (Dearborn, Michigan, USA: Society of Manufacturing Engineers)

[2] Ho K H, Newman S T 2003 State of the Art Electrical Discharge Machining. *Int. J. Machine tools and Manufacture* Vol. **43**(13) pp 1287–1300

[3] Guu Y H 2005 AFM Surface imaging of AISI D2 tool steel machined by the EDM process. *Applied Surface Science* Vol.**242** pp 245–250

[4] Prabhu S, Vinayagam B K 2010 Analysis of surface characteristics of AISI D2 tool steel material using Electrical Discharge Machining process with single wall Carbon nanotubes. *IACSIT Int.J.Eng.Tech.* Vol. **2**(1) pp 35–41

[5] Prabhu S, Vinayagam B K 2009 Nano surface generation of Inconel 600 material by grinding process using single wall Carbon nano tubes. *Int.J.Nanotech.App* Vol. **3**(1) pp 7–16

[6] Guu Y H, Hocheng H, Chou C Y, Deng C S 2003. Effect of Electrical Discharge Machining on surface characteristics and machining damage of AISI D2 tool steel. *Mat.Sc.Eng.* Vol.**358**(1-2)

[7] Pecas P, Henriques E 2003. Influence of silicon powder mixed dielectric on conventional Electrical Discharge Machining. *Int. J. Machine Tools & Manufacture* vol.**43** pp 1465–1471.

[8] Henrich F, Chan C, Moore V, Rolandi M, O'Connell M 2006 *Carbon nanotubes, Properties and Applications* (New York, USA: Taylor & Francis).

Twin-screw extrusion of multi walled carbon nanotubes reinforced polycarbonate composites: Investigation of electrical and mechanical properties

C Mack[1,2], S Sathyanarayana[1,3,*], P Weiss[1], I Mikonsaari[1], C Hübner[1], F Henning[1,3] and P Elsner[1,2]

[1] Polymer Engineering Division, Fraunhofer Institute for Chemical Technology (ICT), Joseph-von-Fraunhofer Strasse 7, 76327, Pfinztal, Germany
[2] Institute for Applied Materials (IAM-WK) - Faculty of Mechanical Engineering, Karlsruhe Institute of technology (KIT), Engelbert-Arnold-Straße 4, 76128, Germany.
[3] Institute for Vehicle System Technology (FAST) - Faculty of Mechanical Engineering, Karlsruhe Institute of technology (KIT), Germany.

E-mail: *shyam.sathyanarayana@ict.fraunhofer.de

Abstract. 1, 3 and 5 wt.% multi walled carbon nanotubes (MWCNT) reinforced polycarbonate (PC) composites were processed in a twin-screw extruder (L/D=52) with two different screw speeds, throughputs and screw configurations. Extruded strands were characterized for dispersion and measurement of electrical resistivities while the pelletized extrudates were injection molded to produce samples for mechanical and further electrical property measurements. The absolute resistance of the melt was recorded with an online melt resistance setup developed by our group. The volume resistivity of pure PC (10^{17} Ω.m) was lowered to 10^4 - 10^5 Ω.m on an injection molded PC-1 wt. % MWCNT composite. 3 wt.% MWCNT incorporated composites showed volume resistivity less than 1 Ω.m independent of process conditions. At lower filler contents the volume resistivity of injection molded samples were higher than those observed on the extruded strands and this effect diminished with increasing MWCNT loadings; owing to the loss of CNT network contacts due to shear induced filler orientation and core-skin effects. The quality of dispersion was exceptional for all filler concentrations at any process condition owing to the affinity of MWCNT towards PC due to the lower interfacial energy difference between the reactants and high polarity of PC. The modulus and strength of the composites increased with filler addition, however at 5 wt.% filler loading the strength of the composites processed with lower SMEs was less than that observed on the 1 wt.% MWCNT reinforced PC composite. The elongation of the composites at maximum tensile strength were comparable to that of neat PC except for composites with 5 wt.% MWCNT loading processed with lower SMEs. Composites with identical filler loadings which were processed with higher SMEs showed higher notched impact strength values principally because of the ability of very well dispersed filler fractions to inhibit crack propagation. The significance of the results obtained in this work stems from the fact that we were able to produce composites with substantial improvements in electrical properties with improvements/without significant loss in mechanical properties. This is one of the few exceptions to results achieved with commercially available MWCNT reinforced polymer composites processed on a large scale.

1. Introduction

The inherent mechanical, electrical and thermal properties of carbon nanotubes (CNT) [1] at a very high surface to volume ratio causing reinforcements at very low filler loadings place them among the elite filler categories for polymers. These characteristics make them attractive towards development of high strength, light weight and conductive multi-functional composites for automotive, aerospace and electronics applications. Commercially available MWCNT come as severely entangled agglomerates and hence their inherent strength can be translated into those of the composites' only if these agglomerates are dispersed efficiently.

Many processing methodologies have been adopted for the preparation of CNT based thermoplastic composites [2-7]. Of all those existing methods, twin-screw extrusion is the most realistic from the perspective of production scale up and commercialization of these composites. Understanding the influence of extrusion process variables for every matrix-CNT composite is important for tailoring dispersion qualities which affect the morphology and consequently the macroscopic behavior of the composites. Compression molding and injection molding, one of which would follow extrusion as the secondary process also mandate understanding of a range of process variables to fabricate the final part with desired functionality. Compression molding is ideally suited for producing parts with good conductivity owing to its creation of a favorable environment for the re-agglomeration of previously dispersed tubes to form a conductive pathway; however would result in poor mechanical behavior. Injection molding on the other hand is a complex process, understanding and tailoring of which could lead to good multi-functional behavior and facilitate mass production.

Though there are many issues associated with the compatibility of CNT with polymers, the intrinsic characteristic of CNT like the geometrical characteristics [8], strength of the agglomerates [9], the nature of surface chemistry etc. [10-14] have a drastic influence on the macroscopic properties of the composite. It is imperative to achieve optimum dispersion qualities of CNT in the matrix for good macroscopic behavior, but the trade-off of reduction in their aspect ratios is a major concern due to high magnitudes of shear generated within the twin-screw extruder. Good dispersion qualities created at high shear rates need not certainly mean the best, but the reduction in the length of CNT during processing might tilt the percolative filler concentration to higher MWCNT loadings. In addition to good dispersion quality and minimal damage to CNT during processing, good wetting of the CNT by the matrix and a good interfacial adhesion between them are principal requirements for mechanical property enhancements. Chemical functionalisation has shown to be effective in improving the mechanical properties of the composites, but it also results in the loss of structural integrity of the CNT leading to increased electrical percolation thresholds [10-12]. Also, to produce large quantities of chemically functionalized tubes as required for processing with a twin-screw extruder is expensive and demanding.

While a good distribution of individualized CNT provide the most efficient load transfer, a CNT network in the composite morphology is necessary for good electrical properties. The conductive network of CNT could be formed by many single tubes and/or a combination of agglomerates. The electrical percolation threshold is dependent on the type of CNT, their synthesis method and aspect ratio, nature of surface functionalisation, type of polymer, melt viscosity, dispersion method, composite processing methodology, etc. Many literatures have been published reporting electrical properties of CNT reinforced thermoplastics and thermosets and close to 150 of them have been compiled in the review of electrical properties of CNT filled polymers of Bauhofer and Kovacs [15].

Considerable amount of research has gone into MWCNT filled PC composites, but most of them have been principally aimed at achieving good electrical properties and have been processed with lab scale equipments [16,17]. In this work we attempt to investigate the potential of MWCNT reinforced PC as a multi-functional composite processed on a twin-screw extruder. Different MWCNT loadings were incorporated into PC by twin-screw extrusion with varying process parameters. All the extruded composite granulates were injection molded in identical conditions and were tested for electrical and mechanical properties. The resistance of the melt that was measured in the extruder and the volume resistivity of the extruded strands are correlated with those of the injection molded composite samples in an attempt to understand how different processes affect electrical behavior of MWCNT filled composites.

2. Experimental

2.1. Materials

Wonderlite PC-122 purchased from CHIMEI-ASAHI Corporation, Taiwan was used as the PC matrix. It has a MFI of 22 g/10 min (ASTM D 1238). Nanocyl NC7000 was used as MWCNT and was purchased from Nanocyl S.A., Belgium. It had a carbon purity of 90%, average length of 1.5 µm, average diameter of 9.5 nm and surface area of 250-300 m^2/g according to the producers. PC was dried for 4 hours at 120 °C before processing and the MWCNT was used as received.

2.2. Processing

2.2.1. *Compounding.* 1, 3, and 5 wt.% MWCNT reinforced PC composites were processed on a Leistritz ZSE 27HP-52D twin-screw extruder with an L/D ratio of 52. Two different screw designs namely "L" and "H" were employed for processing at a bulk extruder barrel temperature of 260 °C. The "L" design was made up of kneading elements and transportation elements and the "H" design had some kneading elements of the "L" design replaced by a combination of back flow and special mixing elements. Figure 1 shows the extrusion parameters with which each composite was processed.

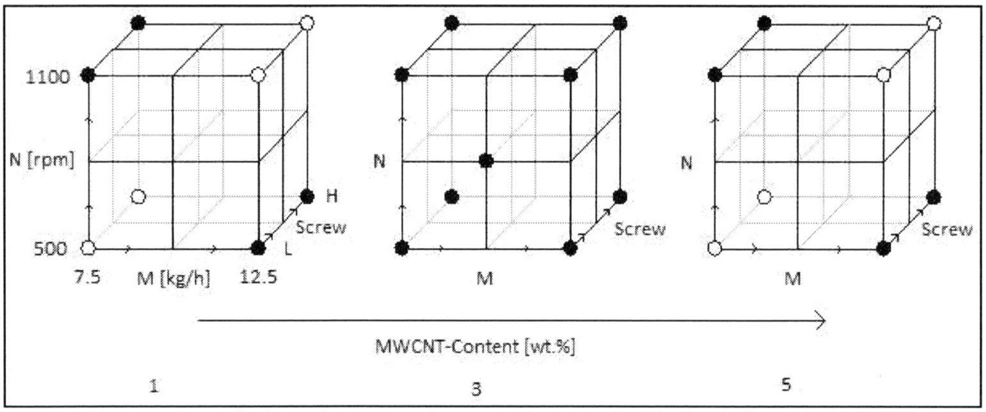

Figure 1. Depiction of processed composite samples (only darkened circles).

2.2.2. *Injection Molding.* Extruded granules were injection molded into specimens for tensile and impact testing and plates for electrical measurements on an Engel ES 200/60 HL ST injection molding machine. The back pressure was maintained at 18 bars, while the holding pressure was 70 bars and 100 bars for the mechanical test specimens and plates respectively. The temperature at the nozzle was 285 °C and 320 °C respectively for the geometries. Corresponding injection speeds were set as 40 mm/s and 10 mm/s respectively. The parameters were tailored to achieve a good surface finish.

2.3. Characterization

2.3.1. *Composite Morphology.* Optical microscopy was carried out on extruded strands. 5 µm thick slices were cut on extruded strands perpendicular to the extrusion direction on at least 3 different strands using a diamond knife and an area of 3 mm^2 was investigated. Image analysis was performed using macros developed in house with Image J software. Undispersed MWCNT area fraction (A_f) was quantified and agglomerates having an area of less than 20 μm^2 were ignored in the analysis according to ISO 18553 standards. Field Emission Scanning Electron Microscopy (FESEM) was performed on broken charpy specimens (sputter coated with Gold for 120 s at 0.05 mbar and 40 mA) using Zeiss SUPRA 55 VP microscope, Carl Zeiss GmbH, Germany.

2.3.2. *Electrical measurements.* Inline melt resistance measurements in the extruder, volume resistivity of the extruded strands and injection molded plates (60*60*2 mm) were measured with the help of a FLUKE 8846A Digital Multimeter. Measurements are effective in the region between 10 and 10^9 ohms with this setup. Higher resistances were measured using a HAMEG HM8118 200 kHz LCR measurement bridge at a frequency of 500 Hz and 0.5V and can be approximated to be the DC resistance.

2.3.3. *Mechanical testing.* The injection molded composite samples were used for mechanical testing. The notched impact strength of the composites was evaluated using ISO 179-1/1eA norm on a CEAST Resil Impactor while tensile properties were evaluated with ISO 527-1 standard on a Universal testing machine Inspekt Table 50KN from Hegewald & Peschke.

3. Results and Discussions

3.1. Process and morphology

Figure 2 presents the specific mechanical energy (SME) inputs during the processing of various composites. SME is a critical parameter while working with an extruder as it takes into account the effect of all parameters which affect composite processing. Although higher screw speeds and lower throughputs contribute to higher SME values, only a small difference in the SME between 3 wt.% and 5 wt.% MWCNT filled composite was observed. Increasing MWCNT contents tend to increase the viscosity of the melt, but the loss in the quality of dispersion (a good dispersive morphology getting modified to an agglomerative morphology) will certainly have a strong influence on the viscosity of the melt which could be attributive to the aforementioned. The SME values were not significantly influenced by employing different screw configurations. The reproducibility of the process was demonstrated by identical values of SME achieved on 3 wt.% MWCNT reinforced MWCNT composite samples processed at 800 rpm and 10 kg/h on two different days.

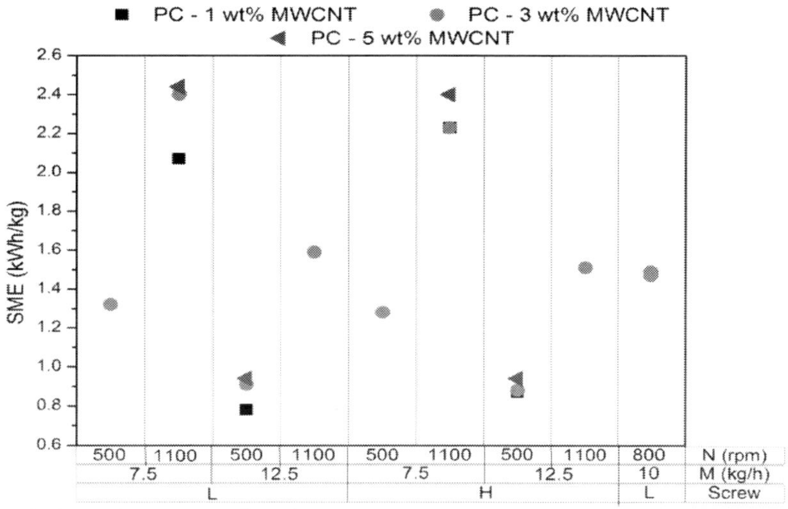

Figure 2. Variation of SME as a function of process parameters for different MWCNT loadings.

The principal objective to employ varying process parameters was to identify parameters that contributed to optimum dispersion qualities for good macroscopic behavior. Representative optical micrographs of a few composite samples are presented in figure 2. The undispersed MWCNT area fraction (A_f) which is quantified as the fraction area of MWCNT agglomerates visible on the micrographs did not change significantly as a function of the process parameters for a defined MWCNT loading in the composite. For all the prepared composites this was less than 0.8% of the investigated area. A notable observation was that the A_f of the 5 wt.% MWCNT filled PC composite was not very different than those observed with a 1 wt.% MWCNT filled PC composite. However, the number of visible agglomerates per mm^2 varied significantly depending upon the process parameters. Composites with similar MWCNT loadings processed at higher SME inputs showed low counts but large sized agglomerates while the lower SME processing led to increased fractions of smaller agglomerates (Refer to images (c) & (d) in figure 3). The former points to excessive shear induced re-agglomeration of dispersed tubes and the latter reflects the maximum potential of the agglomerates to get dispersed due to their intrinsic agglomerate strength. The good quality of dispersion and much lower A_f in the optical micrograph of the different composites visualized in this work indicate a better compatibility of the CNT with the PC principally because of latter's polarity and lower interfacial energy difference with the filler. This however was not the case for composites processed with identical conditions in our earlier works with MWCNT reinforced PS [18] and PP [19].

As no conclusive differences were observed in the optical micrographs, FESEM was used for understanding the dispersion on a nanoscale. Figure 4 shows 1 wt.% and 5 wt.% MWCNT reinforced PC composites. On a closer observation of different areas of the sample, no noticeable differences in the dispersion quality was observed. The variations in the MWCNT concentration were easily observable and the biggest agglomerate observed had a size < 30 μm (assuming a circular geometry in 2-D) irrespective of MWCNT concentration. It was also very interesting to note that employing different detectors for capturing the emitted electrons in the SEM could give a completely different picture of the dispersion qualities and hence it is necessary to exert caution while interpreting the images. Inline detectors are quite useful for extracting topographical information, while the SE detectors could be used for understanding the morphology deep into the sample where image quality and the obtained level of information is a direct function of the applied voltage.

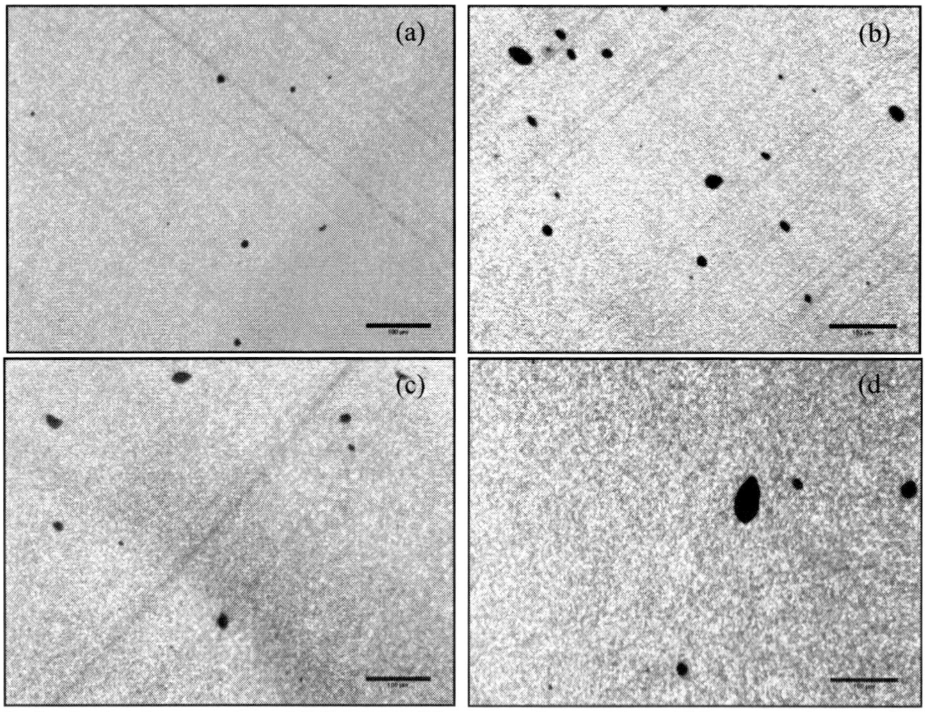

Figure 3. Optical micrographs of PC-MWCNT composites with a 100 μm scale bar: PC-1 wt.% MWCNT (a) & PC-5 wt.% MWCNT (b) processed at 500 rpm and 12.5 kg/h; PC-3 wt.% MWCNT processed at 12. 5 kg/h and 500 rpm (c), 1100 rpm (d)

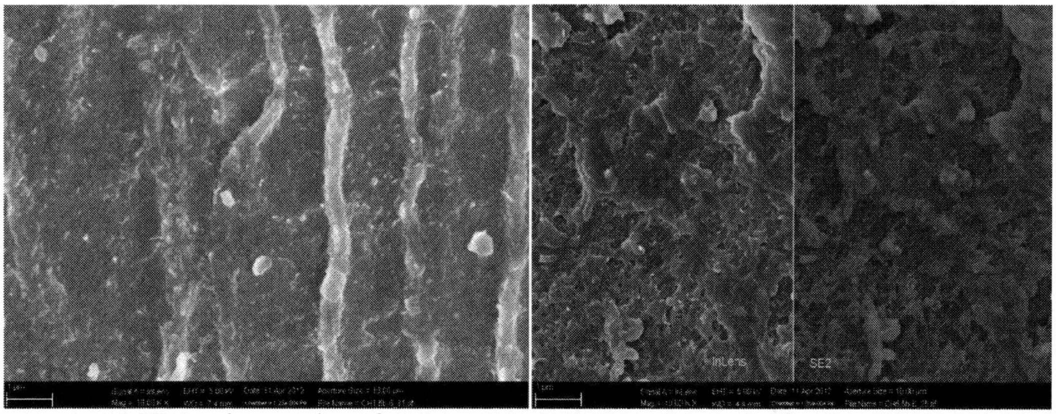

Figure 4. FESEM micrographs of PC-1 wt.% MWCNT (L) and PC- 5 wt.% MWCNT (R) composites. The latter shows the difference in contrast for observing dispersion by using two different detectors

3.2. Measurement of electrical properties

The resistance of the melt is plotted alongside the volume resistivity of the extruded strands for the different MWCNT loadings in PC in figure 5. Measurement of melt resistance is used as a tool for understanding the influence of process parameters on composite conductivities and as a measure of the quality of the produced composites. It has also been used in our earlier works with MWCNT filled PP [19] and PS composites [20]. The order of magnitude of volume resistivity of the extruded strands was always less than the resistance of the melt at 3 wt.% and 5 wt.% MWCNT loadings. A good correlation of these observations with the processing parameters show that the inline measurements serve as a valuable guide in understanding offline resistivity characteristics without actually having to measure them. However, the observations on a PC-1 wt.% MWCNT composite were different. The inline melt resistance values are local measurements and are continuous with time while those observed on the strands are representative of the bulk specimen. The orientation of the CNT as the melt passes through the nozzle and the cooling behaviour when the melt passes through the water bath (after which extruded strands are collected) are factors to be considered while interpreting these results. Measurement conditions are very sensitive to the conductive network created by CNT especially at low concentrations. At higher filler loadings (above the percolation threshold) the online and the offline values are not expected to vary by a higher order of magnitude because of the existence of multiple conductive pathways and high probabilities for pathways to be formed by a network of agglomerates.

Figure 6 shows the volume resistivity plots of the extruded strands plotted alongside the injection molded plates. The volume resistivity of injection molded pure PC is measured to be 10^{17} Ω.m. The injection molded 1 wt.% MWCNT filled PC composites (processed under identical molding parameters) shows volume resistivities in the order of 10^{4}- 10^{5} Ω.m. The 3 wt.% and 5 wt.% MWCNT reinforced PC composites show a good correlative pattern between the volume resistivity of the extruded strands and injection molded plates. The electrical percolation of compression molded MWCNT reinforced PC composite occurs between 0.875 - 1 wt.% of MWCNT loading [21]. The percolation threshold achieved on an injection process is expected to be higher than those achieved on compression molding principally owing to the ability of the latter to present a favourable environment for CNT network formation by facilitating re-agglomeration of previously dispersed tubes. The 1 wt.% MWCNT in our work could be considered to be the percolation regime on the injection molded composites as we observe a 12 order drop in magnitude of resistivity compared to pure PC. The volume resistivity of these samples is strongly dependant on the process parameters for a certain degree of dispersion which could make or break the CNT contacts and hence a deviation from the correlative behaviour with the extruded strands could be expected. We could however not go less than

1 wt.% MWCNT content to study lower concentrations in the composite by feeding as-received MWCNT as dry powder addition due to limitations imposed by the feeding equipment. At 3 wt.% and 5 wt.% MWCNT loadings, the volume resistivities of the injection molded samples did not vary significantly at a defined loading, but the pattern of variation as observed in the extruded strands is replicated here indicating the need to optimize the extrusion parameters. Overall, the volume resistivities of the injection molded samples were higher than those observed on extruded strands. The loss of CNT network contacts due orientation of the filler under the action of shear and the rapid cooling of the melt on contact with the cold surface of the mould during injection results in a skin-core structure [22] contributing in an overall reduction in bulk conductivity.

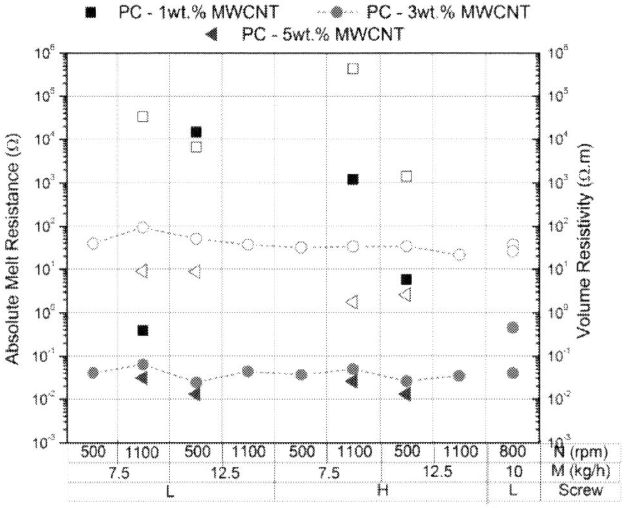

Figure 5. Absolute melt resistance (open symbols) and volume resistivity of the extruded composite strands (filled symbols) as a function of process parameters for different MWCNT loadings. The dotted lines are for ease of comparison. Standard deviations were too low to be plotted.

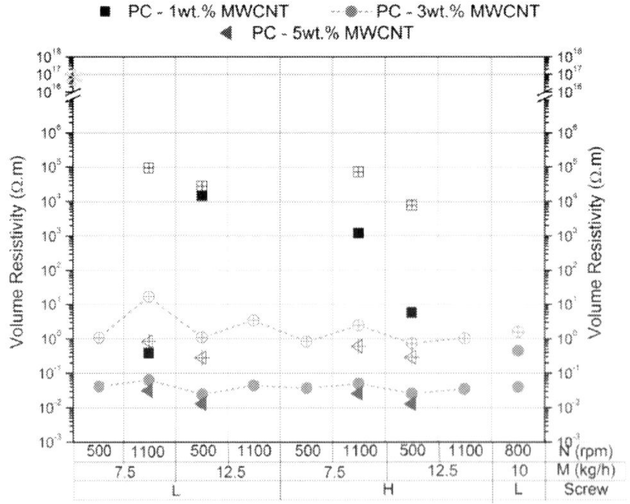

Figure 6. Volume resistivity of the extruded composite strands (filled symbols) and injection molded plates (crossed symbols) as a function of process parameters for different MWCNT loadings. The dotted lines are for ease of comparison. Standard deviations were too low to be plotted. Neat PC is plotted on the ordinate.

3.3. Mechanical properties of the composite

The elastic modulus of PC (2320 MPa) increased to a maximum of 2433 MPa when it was reinforced with 1 wt.% MWCNT, 2681 MPa with 3 wt.% MWCNT and 2908 MPa with 5 wt.% MWCNT incorporations (figure 7). As the aspect ratio of CNT and their orientation are bound to suffer under shear intensive mixing like the compounding process and an additional shear induced orientation during injection molding, achieving theoretical capabilities of these composites is almost impossible. Orientation effects have been stated to be negligible with nanocomposites due to the requirement of bulk properties [23], but the aspect ratio plays a key role in determining the composite properties. Hence, there is a huge compromise between the rate of shear applied for dispersing the agglomerates in the extruder to get an optimum dispersion quality and the destruction of tube lengths consequently. With increase in the volume fraction of the filler, the modulus is expected to increase, and we see an increasing trend of modulus with increased MWCNT loading in figure 7. It is very difficult to increase the modulus of an already stiff material like PC and the result achieved here is quite attractive.

The enhanced surface area presented by good dispersion and a good chemical compatibility arising from a polymer-filler interaction are key factors determining the strength of a composite. It is seen from figure 7 that there is a slight increase in tensile strength of PC (62.75 MPa) with increasing MWCNT loadings. At lower loadings the observed tensile strength is almost independent of the process parameters, while the composites with 5 wt.% MWCNT loadings show a distinctive pattern very dependant on the process parameters. The samples processed at 1100 rpm and 7.5 kg/h at both the screw configurations perform better compared to those processed at 500 rpm and 12.5 kg/h at these loadings. The SMEs of the former were about 2.8 times higher than the latter (Refer to figure 2). With increased loadings of MWCNT in the composite, higher shear forces are required to decrease the extent of agglomeration. Lack of afformentioned resulted in a highly agglomerated network of CNT, which potentially has created increased stress concentration zones in the composites. The quality of dispersion that has been achieved in this work is quite good at least as observed from optical micrographs and electron microscopy images. However, the chances of a large agglomerate existence cannot be ruled out as there is a big complexity in investigating domains that lie in between the resolutions of a optical and a electron microscope. Nanotube pullout from the matrix has been observed in the SEM pictures, which indicates weak interfacial adhesion. This is expected as we did not employ a coupling agent or functionalized CNT for this work. Also the presence of agglomerates result in a higher probability of intra-filler interaction and present low available CNT surface area to the polymer for interaction which could have prevented efficient load transfer from the polymer to the filler. Abbasi et al. [24] report lowering of tensile strength of neat PC on MWCNT incorporations..

The strain at maximum tensile strength of PC (6.15 %) and its composites are reported in figure 8. Addition of 1 wt.% MWCNT shows a strain of 6 % irrespective of the processing conditions. The PC - 5 wt.% MWCNT composites processed at 500 rpm and 12.5 kg/h with both screw configurations show a very low strain at maximum tensile strength (also happens to be the strain at break) around 3.1-3.5 % compared to those processed at 1100 rpm and 7.5 kg/h. The agglomeration effect as explained earlier could be attributed to this effect. The stress-strain plots of pure PC and 1 wt.% MWCNT incorporated PC (not presented here) show necking before failure, while the composites with 3 wt.% and 5 wt.% MWCNT loadings exhibit increased brittleness and fracture immediately on the load on the composite reaching the maximum tensile strength. The increased contents of MWCNT lowered the strain at maximum tensile strength compared to pure PC, but the rate of decrease is not significant (with a couple of exceptions stated earlier) which may principally be due to the achievement of good dispersion qualities.

Pure PC showed impact strength of 12.5 kJ/m^2 (figure 8). A significant variation in the impact strength of the composites is observed depending on how they were processed. On a general note it can be summarized that composites with identical MWCNT loadings processed with higher SMEs tend to show better tendencies to resist fracture propagation from the notch. Even though high shear processing would result in tube breakage and facilitate re-agglomeration of previously dispersed tubes, the prospect of them creating better dispersion qualities is higher. This could have critically led to improvements in the toughness. With increasing MWCNT concentrations, the density of agglomerates increases and act as supplementary stress creators in addition to the applied impact force.

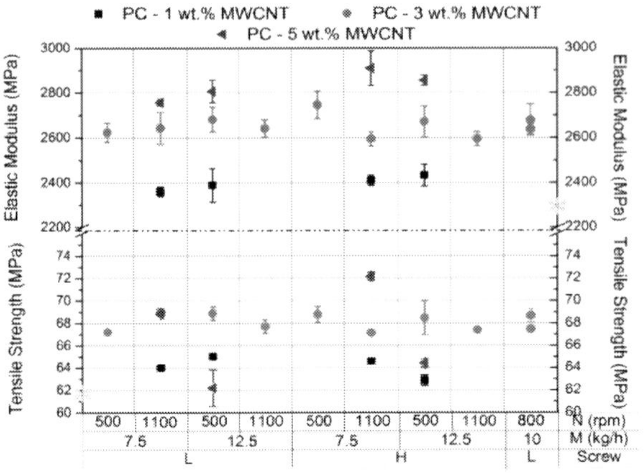

Figure 7. Elastic modulus and tensile strength of the composites as a function of process parameters for different MWCNT loadings. Neat PC plotted on the ordinate.

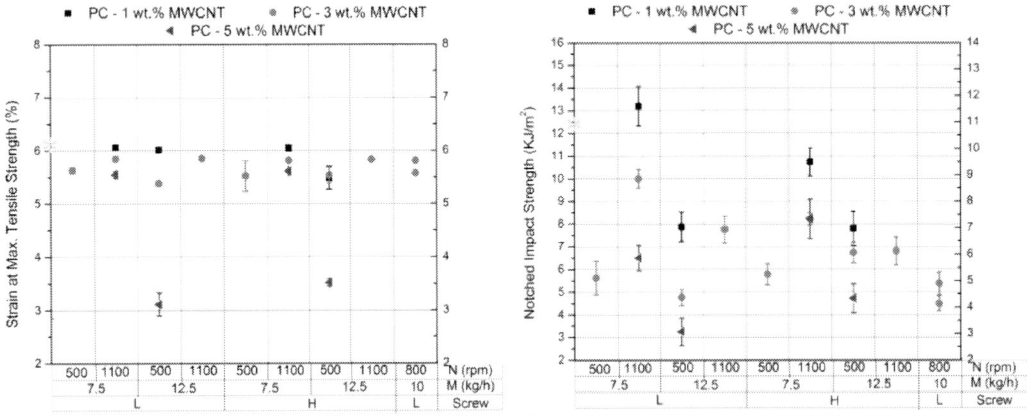

Figure 8. Strain at maximum tensile strength (L) and notched impact strength (R) of the composites as a function of process parameters for different MWCNT loadings. Neat PC plotted on the ordinate.

4. Summary

MWCNT reinforced PC composites produced by twin-screw extrusion and injection molding were tested for electrical and mechanical properties. Electrical percolation threshold was observed with 1 wt.% MWCNT reinforcement into PC. The volume resistivities of injection molded sample were higher than those measured on extruded strands due to loss of CNT contacts arising out of shear induced filler orientation and skin-core effects during injection. A good correlative pattern is observed with the volume resistivities of the injection molded bars and extruded strands at MWCNT loadings of 3 wt.% and 5 wt.%, the magnitude of injection molded bars being higher. However, a slight deviation is observed with 1 wt.% MCWNT reinforced PC composite as this sample lies in the percolation regime and the influence of process parameters are very sensitive on volume resistivities. While the elastic modulus and tensile strength of the composites increased with higher filler loadings, 5 wt.% MWCNT reinforced PC processed with lower SMEs during extrusion showed poor tensile behaviour and lower values of strain at maximum tensile strength. Composites processed with higher SME inputs during extrusion showed much better resistance to crack propagation during impact testing. From this work we understand that it is very important to optimize the extrusion parameters to get an optimal injection molded specimen, although the configuration of these parameters doesn't have to be necessarily the same to get the best electrical and mechanical properties. We have also demonstrated an insight into developing multi-functional composites with MWCNT reinforced thermoplasts. Future work would involve optimizing the injection molding process to understand its influence on composite properties.

5. Acknowledgements

The research leading to these results has received funding from the European Community's Seventh Framework Programme (FP7-NMP-2010-1.2-1) under grant agreement number 263406. Shyam Sathyanarayana would like to thank the European Community's Seventh Framework Programme (FP7-PEOPLE-ITN-2008) for his funding through the Marie Curie Fellowship under grant agreement number 238363.

References

[1] Dresselhaus M S, Dresselhaus G, Charlier J C and Hernández E 2004 *Phil. Trans. R. Soc. Lond. A* **362** 2065-98
[2] Safadi B, Andrews R and Grulke E A 2002 *J. Appl. Polym. Sci.* **84** 2660-9
[3] Andrews R, Jacques D, Minot M and Rantell T 2002 *Macromol. Mater. Eng.* **287** 395-403
[4] Lin T S, Cheng L Y, Hsiao C C and Yang A C M 2005 *Mat. Chem. Phys.* **94** 438-43
[5] Vigolo B, Pénicaud A, Coulon C, Sauder C, Pailler R, Journet C, Bernier P and Poulin P 2000 *Science* **290** 1331-4
[6] Xia H, Wang Q, Li K and Wu G H 2004 *J. Appl. Polym. Sci.* **93** 378-86
[7] Sen R, Zhao B, Perea D E, Itkis M E, Hu H, Love J, Bekyarova E and Haddon R C 2004 *Nano Lett.* **4** 459-64
[8] Martone A, Faiella G, Antonucci V, Giordano M and Zarrelli M 2011 *Compos. Sci. Technol.* **71** 1117-23
[9] Krause B, Mende M, Pötschke P and Petzold G 2010 *Carbon* **48** 2746-54
[10] Qiu J, Zhang C, Wang B and Liang R 2007 *Nanotechnology* **18** 275708-19
[11] Saeed K and Park S Y 2007 *J. Appl. Polym. Sci.* **104** 1957-63
[12] Koval'chuk A A, Shevchenko V G, Shchegolikhin A N, Nedorezova P M, Klyamkina A N and Aladyshev AM 2008 *Macromolecules* **41** 7536-42
[13] Li X, Wong S Y, Tjiu W C, Lyons B P, Oh S A and He C B 2008 *Carbon* **46** 829–31
[14] Ramasubramaniam R, Chen J and Liu H 2003 *Appl. Phys. Lett.* **83** 2928-30
[15] Bauhofer W and Kovacs J Z 2009 *Compos. Sci. Tech.* **69** 1486–98
[16] Kasaliwal G, Göldel A, Pötchke P and Heinrich G 2011 *Polymer* **52** 1027-36
[17] Krause B, Villmow T, Boldt R, Mende M, Petzold G and Pötschke P 2011 *Compos. Sci. Technol.* **71** 1145-53
[18] Sathyanarayana S, Olowojoba G, Weiss P, Caglar B, Pataki B, Mikonsaari I, Hübner C and Henning F 2012 Accepted by *Macromol. Mat. Eng.* DOI = 10.1002/mame.201200018
[19] Zamora D B, Sathyanarayana S, Weiss P, Hübner C, Diemert J, Henning F and Elsner P To appear in the *proceedings of SPE ANTEC 2012* (Orlando, USA, 2-4 April 2012)
[20] Sathyanarayana S, Olowojoba G, Weiss P, Hübner C and Henning F To appear in the *proceedings of SAMPE 2012* (Baltimore, USA, 21-24 May 2012)
[21] Kasaliwal G, Göldel A and Pötschke P 2009 *J. Appl. Polym. Sci.* **112** 3494-3509
[22] Villmow T, Pegel S, Pötschke P and Wagenknecht U 2008 *Compos. Sci. Technol.* **68** 777-89
[23] Schaefer D W and Justice R S 2007 *Macromolecules* **40** 8501-17
[24] Abbasi S, Derdouri A and Carreau P J 2011 *Polym. Eng. Sci.* **51** 992-1003

IOP Conf. Series: Materials Science and Engineering **40** (2012) 012021 doi:10.1088/1757-899X/40/1/012021

High-performance Supercapacitor cells with Activated Carbon/MWNT nanocomposite electrodes

F. Markoulidis[1], C. Lei[1], C. Lekakou[1], E.Figgemeier[2], D. Duff[2], S. Khalil[2], B.Martorana[3], and I.Cannavaro[3]

[1] Mechanical, Medical and Aerospace Engineering, Faculty of Engineering and Physical Sciences, University of Surrey, Guildford GU2 7XH, UK,
[2] Bayer Technology Services GmbH, 51368 Leverkusen, Germany,
[3] Centro Ricerche Fiat S.C.p.A., Strada Torino 50, 10043 Orbassano (TO), Italy

Email: foivos.markoulidis@surrey.ac.uk, c.lei@surrey.ac.uk, c.lekakou@surrey.ac.uk, daniel.duff@bayer.com, sofie.khalil@bayer.com, brunetto.martorana@crf.it, irene.cannavaro@tirocinanti.crf.it.

Abstract. The purpose of this work was to investigate and improve the performance of supercapacitor cells with carbon-based nanocomposite electrodes. The electrode structure comprised activated carbon (AC), four types of multi-wall nanotubes (MWNTs) and two alternative polymer binders, Polyvinyl alcohol (PVA) or Polyvinylidene fluoride (PVDF). Electrode fabrication involved various stages of mixing and dispersion of the AC powder and carbon nanotubes, rolling and coating of the AC/MWNT/binder paste on an aluminium substrate which also served as current collector. The organic electrolyte utilised was 1M tetraethylammonium tetrafluoroborate ($TEABF_4$) fully dissolved in propylene carbonate (PC). All devices were of the electrochemical double layer capacitor (EDLC) type, incorporating four layers of tissue paper as separator material. The surface topography of the so fabricated electrodes was investigated with scanning electrode microscopy (SEM). Overall cell performance was evaluated with a multi-channel potentiostat/galvanostat/impedance analyser. Each supercapacitor cell was subjected to Cyclic Voltammetry (CV) at various scan rates from 0.01 V/s to 1 V/s, Charge-Discharge at a fixed current steps (2 mA) and Electrochemical Impedance Spectroscopy (EIS) with frequency range from 10 mHz to 1 MHz. It was established that an AC-based supercapacitor with 0.15%w/w MWNT content and 30 μm roll-coated, nanocomposite electrodes provided superior energy and power and energy densities while the cells was immersed in the electrolyte; well above those generated by the AC-based EDLC cells.

1. Introduction

Electrochemical double layer capacitors (EDLCs) are designed to optimize their supercapacitor behaviour by using nanomaterials and nanocomposite coatings as electrodes of high surface area and an appropriate pore size distribution as well as a separator to halve the distance between the electrodes and, hence, double the capacitance. Activated carbon (AC) powder of large surface area has been the main electrode material in this study processed into a coating using the typical PVDF(polyvinylidene fluoride) binder. However, the large surface area of AC is associated with large porosity which reduces the conductivity of the activated carbon electrode. Hence, conductive additives are required such as carbon black, multiwalled carbon nanotubes (MWNTs), and conductive polymer binders.

Carbon black is usually added at 5-10 wt% [1,2] to decrease the resistance, although its small particle size creates a compact particle network filling the AC pores and possibly decreasing the overall specific capacitance. PEDOT:PSS used as a binder [1] increases the specific capacitance by contributing pseudocapacitance to the cell but it is not as conductive as the carbon black or MWNTs. As a result, the aim of this study is to investigate the use of MWNTs as the conductive additive, given the high conductivity and open nature of their networks. Dispersion is a key requirement for carbon nanotubes (CNTs) whereas the high aspect ratio of these nanomaterials means that percolation network is formed at very low loadings. As a result, in contrast to previous studies(15 wt% MWNT in AC in [3]] only 0.15 wt% MWNT in AC/PVDF nanocomposite coating has been used in this study. Polyvinyl alcohol (PVA) has also been investigated as a binder.

2. Materials and Experimental Procedures

The activated carbon (AC) powder (Sigma Aldrich) has a surface area of 1000 m^2/g and a particle size distribution of 10-15 % greater than 74 tm and 70-75 % greater than 10 tm. The carbon nanotubes were Baytubes® (Bayer Technology Services) of 4 nm inner diameter and 13 nm outer diameter, and length greater than 1 tm. Different types of these Baytubes® have been tried, the main difference being their powder density: C70P of 45-95 kg/m^3 and purity >95 wt%, C150P of 130-150 kg/m^3 and purity >95 wt%, C150PW same as C150W but washed, C150HP of 140-230 kg/m^3 and purity >99 wt%.

Fig.1: MWNT slurry in the homogeniser and nanocomposite coating on Al foil

The MWNTs were dispersed in 1-methyl-2-pyrrolidinone (NMP) by first sonicating for half hour and then vigorously stirring for 1 hour using a Wiggenhauser homogenizer (Fig.1). The MWNT solution was then added into the AC/binder-NMP solution. The ingredients were mixed to form a slurry, which was coated onto aluminium foil using a Film Applicator. The coating was then dried in an oven at 120°C for 4 hours (Fig.1).

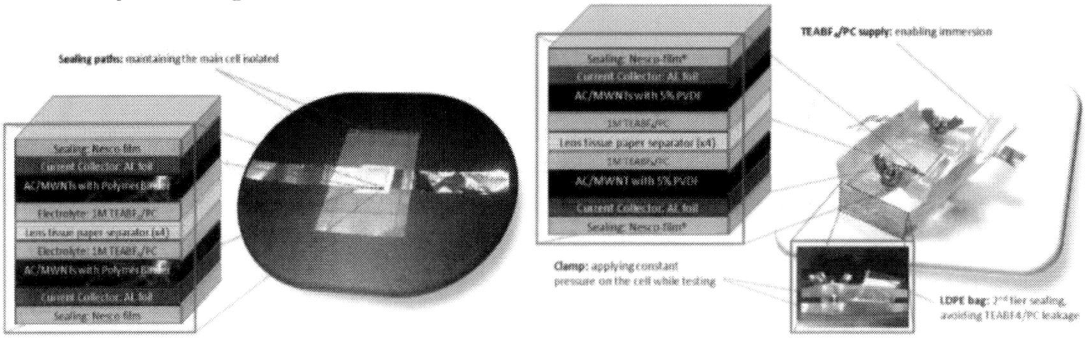

Fig.2. Capacitor cell assembly: cell tested as just sealed and cell immersed in the electrolyte

Capacitor cells were prepared using 1 M $TEABF_4/PC$ electrolyte, as the organic electrolyte has relatively large voltage window to Vmax = 3.3V (used up to 3 V in this study). Lens tissue was used as separator. The cells were tested using a VersaSTAT MC Potentiostat and were subjected to impedance spectroscopy, cyclic voltammetry and galvanostatic charge-discharge. After the cell fabrication and sealing, cells were tested as just sealed or in some cases the cell was immersed in the electrolyte contained in the bag with a feed-tube to replenish the electrolyte supply if needed.

IOP Conf. Series: Materials Science and Engineering **40** (2012) 012021 doi:10.1088/1757-899X/40/1/012021

3. Results and Discussion

Fig.3 presents the Nyquist plots of the capacitor cells with the different types of MWNTs in their carbonaceous nanocomposite electrodes, where all of them have an electrode coating thickness of 120 tm whereas the standard AC electrode has a thickness of 30 tm. The smaller thickness of the AC electrode reduces its resistance substantially, hence, in a way this is an unfair comparison. Nevertheless, MWNTs added at 0.15 wt% to the electrode reduce the resistance in almost all cases, even if their corresponding electrodes are four times thicker than the AC-based reference electrode. The low density MWNT powder C70P leads to the highest resistance cell whereas the high density MWNT powder C150HP seems to lead to the lowest resistance cell. On the other hand, the AC-based cell has the highest capacitance.

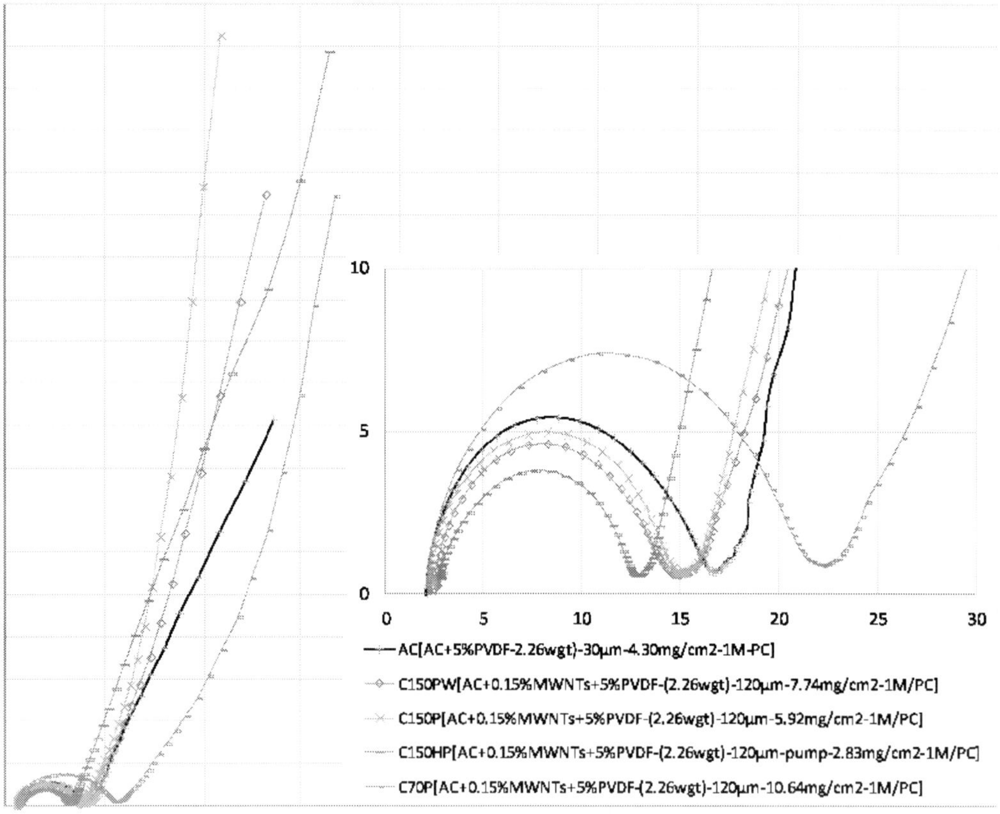

Fig.3: Impedance plots of capacitor cells with AC/MWNT/PVDF nanocomposite electrodes for different types of MWNTs (Baytubes®); frequency 1Mz-10mHz

5

Fig.4 presents the Nyquist plots for various electrode thicknesses (30, 60, 120 tm) where it is clear that the smallest electrode thickness of 30 tm in the cell immersed in the electrolyte has the highest performance, including the smallest resistance, highest specific capacitance and smallest loss (most vertical line).

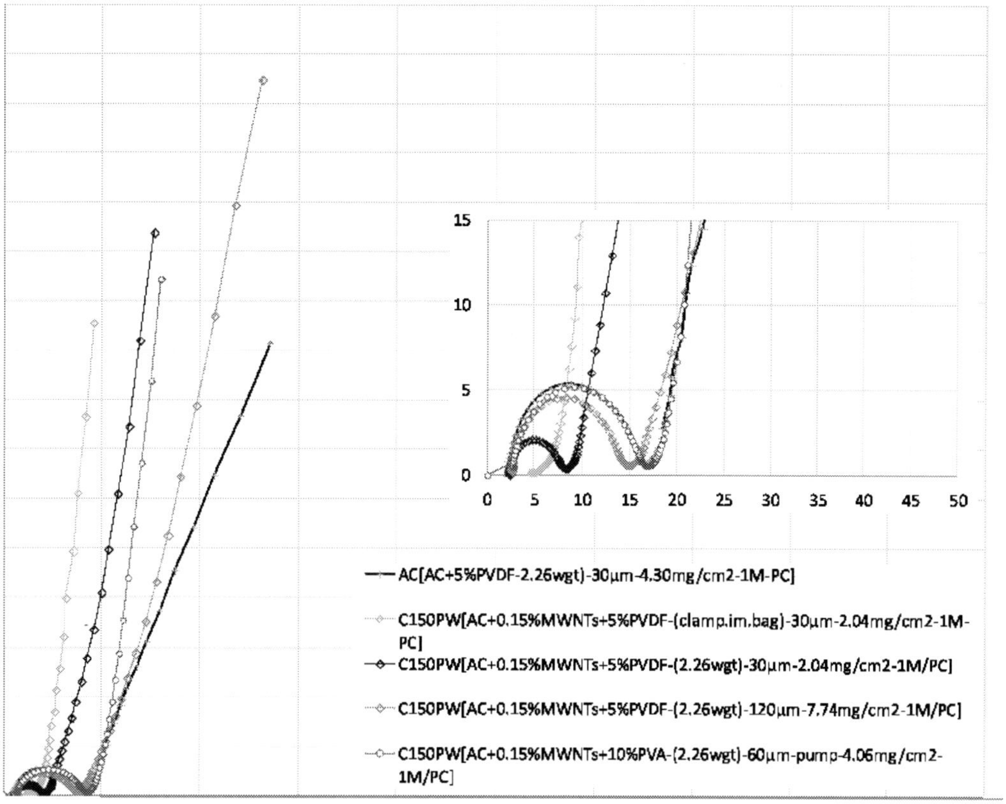

Fig.4: Impedance plots of capacitor cells with AC/MWNT/binder nanocomposite electrodes of different thickness, with PVDF or PVA binder; cells just sealed or immersed in the electrolyte; frequency 1Mz-10mHz

The excellent performance of the AC/MWNT 0.15 wt%/PVDF-based cell is confirmed in the CV tests (Figure.5) where the specific capacitance of the electrode of 30 tm thickness, immersed in the electrolyte, doubles and more than triples that of the AC-based cell at the rates of 0.01 and 0.1 V/s, respectively. The charge-discharge data in Fig.6 display a large voltage drop for the AC-based cell, due to its high equivalent in series resistance (ESR), which is really reduced when 0.15 wt% MWNTs is added while the capacitance of the AC/0.15 wt% MWNT/PVDF cell is significantly higher than that of the AC-based cell at 20 and 50 A.

The SEM micrographs in Fig.7 show a large amount of macropores in the AC carbon coating.When 0.15 wt% MWNTs is added the carbon particles are laced with a MWNT network, which, apartfrom increasing conductivity, reduces the macropore size and seems to increase the specific surface area which results in a significant increse of capacitance.

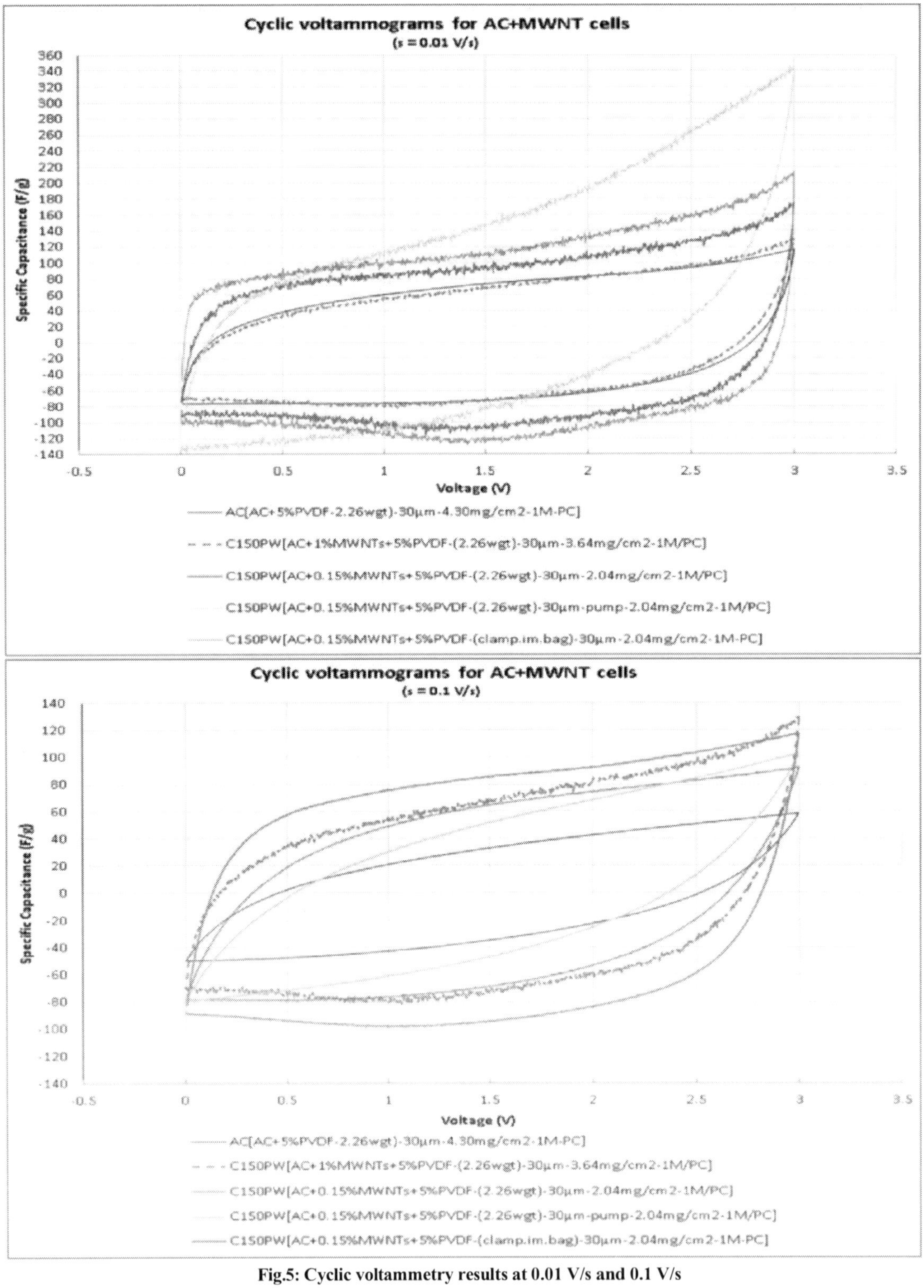

Fig.5: Cyclic voltammetry results at 0.01 V/s and 0.1 V/s

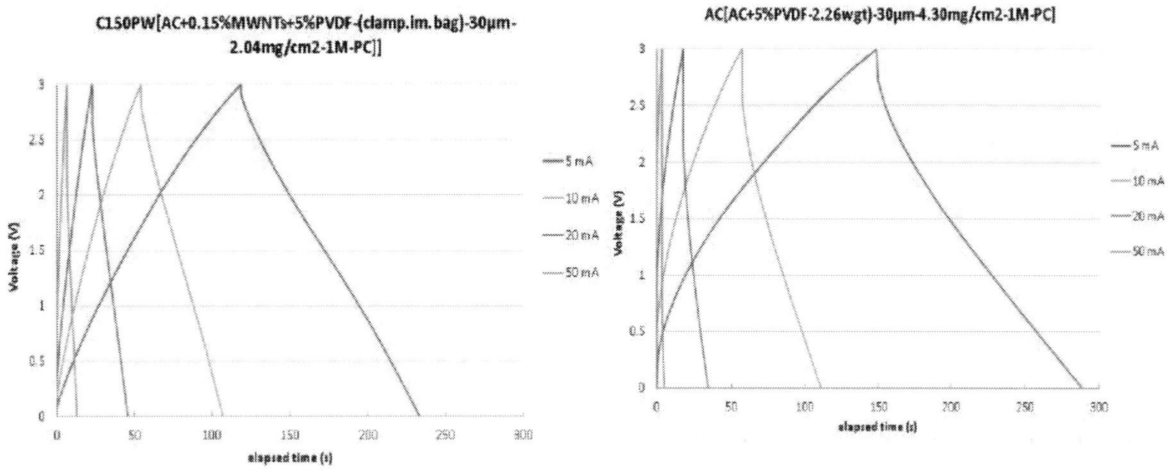

Fig.6: Charge-discharge results for different currents

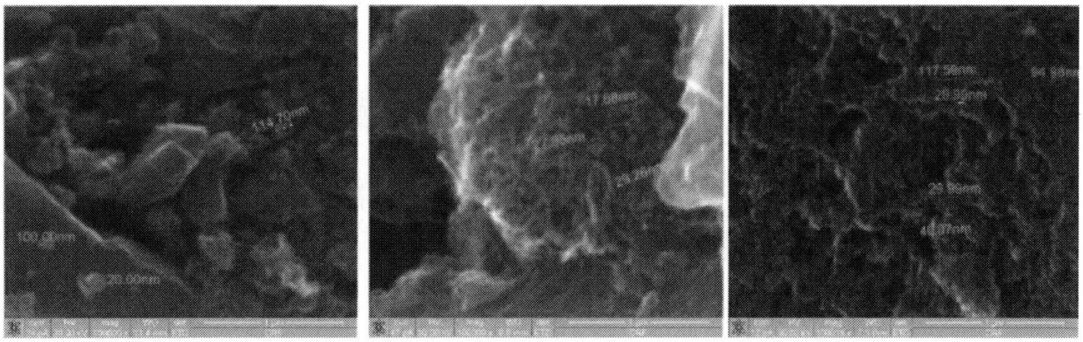

(a) (b) (c)

Fig.7: SEM micrographs of (a) AC+5wt%PVDF; (b) AC+0.15wt%C150HP+5wt%PVDF); (c) AC+0.15wt%C150PW+5wt%PVA

4. Conclusions

Novel EDLC supercapacitors have been presented based on AC/MWNT/polymer binder nanocomposite electrodes where the addition of just 0.15 wt% MWNTs increases not only conductivity but also specific capacitance two- to more than three-fold in comparison to AC/polymer binder electrodes-based cells. Such dramatic increase of capacitance (as well as conductivity, as expected) was attributed to the MWNT network lacing the AC particle surface and creating more pores. It must be noted that the MWNTs used in this study are of the long and thin type, where their high aspect ratio and length contribute greatly to the high conductivity of the MWNT network while they also create a well entangled network that greatly contributes to the increase of the specific capacitance. However, the dispersion process of such MWNTs is critical for the improved performance of the supercapacitor device.

Acknowledgements

The research is part of the FP7 project AUTOSUPERCAP, funded by the European Community under the Green Car Programme.

References

[1] C.Lei, P.Wilson and C.Lekakou, J. Power Sources, 196(18), 2011, 7823-27
[2] P.Kossyrev, J. Power Sources, 201, 2012, 347-52
[3] P-L.Taberna, G.Chevallier, P.Simon, D.Plee and T.Aubert, Mat.Res.Bull., 41(3), 2006, 478-484

IOP Conf. Series: Materials Science and Engineering **40** (2012) 012022 doi:10.1088/1757-899X/40/1/012022

Reinforcement of poly ether sulphones (PES) with exfoliated graphene oxide for aerospace applications

K Balasubramanian

Head: Materials Engineering, DIAT(DU), Girinagar, Pune- 411025, Maharashtra, India

E-mail: balask@diat.ac.in, meetkbs@gmail.com

Abstract. Composite materials have been used for aerospace for some time now and have gained virtually 100% acceptance as the materials of choice. Speciality polymers like poly ether sulphones (PES), poly ether ether ketones(PEEK) , poly ether imides (PEI) are highly preferred materials as plastic matrix due to their superior temperature performance, excellent wear & friction resistance, excellent dimensional accuracy, high tensile strength, high modulus, precise machinability and chemical resistance. In recent years nanoadditives like single and multiwall carbon nanotubes, graphenes and graphene oxides(GO) are finding huge market potential in aerospace and automobile industries. But manufacture related factors such as particle/ matrix interphases, surface activation, mixing process, particle agglomeration, particle size and shape may lead to different property effects. In this research GO/PES composites were prepared by high shear melt blending technique. GO monolayers were exfoliated from natural graphite flake and dispersed homogeneously in PES matrix for the GO content ranging between 0.5 to 2.0 volume percentage with a high shear twin screw batch mixer. These melt blended nanocomposites were injection moulded for mechanical property validation of tensile strength, flexural modulus and impact resistance. Addition of 0.5 volume percentage of GO enhanced the tensile strength and flexural modulus by 40% and 90% respectively. The results show that addition of GO to PES increase mechanical properties due to the formation of continuous network, good dispersion and strong interfacial interactions. The strong interfacial interactions were accounted for the increase in glass transition temperature. Also there was a significant improvement in the impact resistance of the PES/ GO nanocomposite. The injection moulded samples were tested for stealth performance by measuring the electromagnetic shielding property.

1. Introduction

The excellent mechanical, thermal and electrical property, combining with its ultrahigh surface area and economical sources (graphite), makes graphene a promising nanoscale filler for the fabrication of polymer composites. However, since graphene sheets are intrinsically stacked in graphite due to the high van der Waals forces of attraction between the adjacent layers, the exfoliation and incorporation of it into polymer matrix to synthesise single layer graphene reinforced polymer, which has been considered as the best utilization of the excellent in plane properties of graphene is rather difficult [1]. In the past several years, various methods for producing graphene have been developed, such as micromechanical exfoliation, chemical vapour deposition, epitaxial growth and chemical synthesis [2]. Recently there are several reports related to graphene/ polymer composites [3-8]. When incorporated, graphene improved significantly the electrical, thermal and mechanical properties of the host polymer at a low concentration. Presence of functional groups like hydroxyl, carbonyl and carboxyl groups are found to make graphene oxide sheets strongly hydrophilic, which allows graphene oxide to readily swell and disperse in water.

In addition, the oxygen functionality allows the improvement in interactions with the polar polymer matrices. But surface modification is required to disperse graphene oxide in nonpolar polymer [9, 10]. Carbon nanotube reinforced polymer composite have been widely investigated in recent years while application of graphene or graphene oxide as a reinforcing phase has been mainly focussed on thermoplastics or thermoset polymers in which the strength and modulus increased but flexibility decreases. In the previous work, it was found that addition of multiwall carbon nanotubes enhanced the conducting properties of EPDM [11]. Extensive literature is not available on the polymer nanocomposites of graphene and graphene oxides of speciality and engineering polymers like poly ether ether ketones, liquid crystal polymers and polysulphones. In this work a modified Hummers method was used to produce chemical exfoliated graphene oxide nanosheets. These nanosheets were melt blended with a high shear twin screw compounder and injection moulded to validate the mechanical and electromagnetic properties.

2.Experimental is

2.1. Materials

Flake graphite with average particle size of 20μm was purchased from Research Lab Mumbai. Powder grade PES was procured from Solvay speciality Chemical. Sodium nitrate and potassium permanganate was procured from Ranchem, India. Concentrated hydrochloric acid and concentrated nitric acid were procured from Merck, India. Hydrogen peroxide was purchased from Fischer Scientific.

2.2.Preparation of Graphene Oxide (GO)

3g of flake graphite, 3 g of sodium nitrate and 150ml of concentrated sulphuric acid were taken in a beaker and ultrasonicated in an ice water bath for a period of 1 hour. 25g of potassium permanganate was added slowly to the mixture over two hours. 500ml of aqueous sulphuric acid solution was added and ultrasonicated for 2 hours. To the ultrasonicated mixture 20 ml of H_20_2 was added and stirred at temperature of 70^0C. The product was washed distilled water for several times. This solution was then freeze dried.

2.3. Preparation of graphene oxide/PES nanocomposite

PES was dried in an dehumidifying chamber for a period of 8 hours at 160^0C. The surface treated graphene oxides were tumble blended with the resins. Compounding was done using a twin screw mixer (screw speed 40 rpm) with zonal temperatures ranging between 240^0C to 280^0C.

The compounded samples were injection moulded for testing the mechanical and shielding properties.

3. Testing and Characterisation

3.1. Mechanical Testing

Tensile properties were determined for samples according to BS 903: Part A: 1971 using Lloyd Tensile testing machine, type 200R fitted with a 500N load cell. The grip separation speed was 10mm/min at room temperature. The average of five results was reported. Flexural testing was carried out, according to ISO 178 with the test spped of 10mm/min using Lloyd tensile test machine. The Dapmat computer programme software calculated all flexural properties automatically. The average of five samples per formulation was reported.

3.2. Shielding Efficiency

The sample geometry as shown in figure 1 washers were electrically connected between the inner and outer conductor of a specially designed coaxial fixture, which is placed between the signal source and the detector. The ratio of the power of the signal received with the sample present to the power received with no sample is the shielding effectiveness of the sample. The sample must make good electrical contact at both the inner and outer surfaces, or the measured shielding effectiveness will be lower than the sample is capable of providing. The microwave transmission losses of the filled plastic materials within the range of 0.1-20 GHz, were measured at room temperature. A major drawback of this test is that it would not identify any property anisotropy which is common in filled polymer nanocomposite.

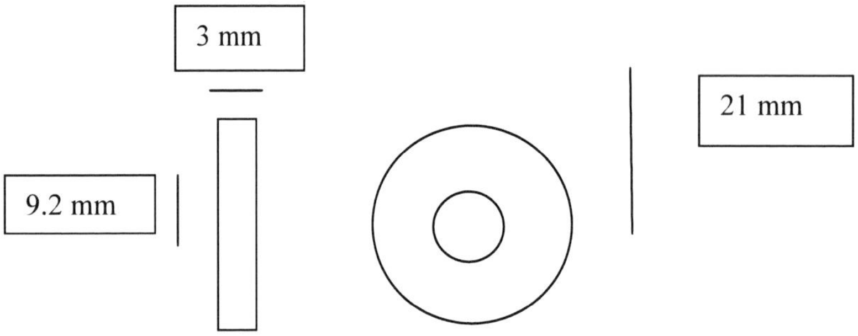

Figure 1. Sample Dimensions for EMI test.

4. Results and Discussion

4.1. Properties

Table 1 presents the FTIR spectroscopy of graphene oxide. Peaks corresponding to the presence of hydroxyl and carboxyl were observed at 3458 and 1727 cm^{-1}respectively. These peaks indicate the oxidation of graphite by concentrated acid. The functional groups allow the graphene oxide to disperse in the polymer system during high shear mixing. X ray diffraction peak as in figure 2 in the result of surface treated graphene. This clearly indicates that the interlayer spacing of graphite is expanded after the chemical modification.

Table 1. FTIR peaks of Graphene oxide.

Wave number cm^{-1}	**Group**
3458	Hydroxyl
1727	Carboxyl

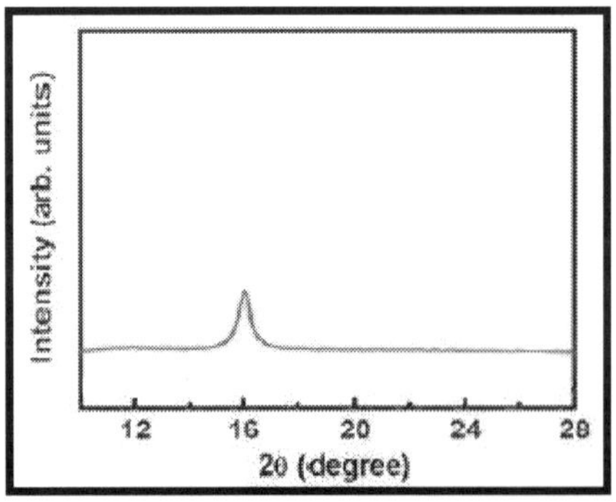

Figure 2. XRD peek of Graphene oxide.

Table 2. Comparison of Mechanical Properties of GO/PES nanocomposite.

Composition	**Tensile Strength (MPa)**	**Flexural Modulus (MPa)**
PES	102	5918
PES+ 0.5 Vol% GO	138	8216
PES+1.0 Vol% GO	142	8354
PES+1.5 Vol% GO	147	8693
PES+2.0 Vol% GO	148	8742

Table 3. Comparison of Shielding effectiveness of GO/PES nanocomposite.

Composition	**Shielding Effectiveness(dB)**
PES	-
PES+ 0.5 Vol% GO	9.1
PES+1.0 Vol% GO	26
PES+1.5 Vol% GO	38
PES+2.0 Vol% GO	42

Figure 3. SEM of GO in PES.

Tensile and flexural data for the composite produced are shown in Table 2. Mechanical properties increases which proved that high shear mixing and surface functionalisation of graphenes help in molecular level adhesion. The test samples are assumed to be isotropic and homogeneous. The shielding effectiveness of 0.5 vol % was found to be 9 dB for the frequency range of 750-2000MHz. The value was found to be consistent for all the five samples. The shielding effectiveness of 1 vol% and 1.5vol%of graphene was found to be 26dB and 38dB respectively for the frequency range of 750-2000MHz. With 2 vol% the values were found to be 42dB between the frequency range of 750-2000MHz.

Figure 3 shows the scanning electron microscopy of graphene oxide dispersed in PPS. From the SEM no obvious graphene oxide aggregates were observed from the cross section of the composite.

4. Conclusions

Dispersion of graphene oxide was possible with different volume percentage loading in PES. Graphene oxide sheets containing abundant carboxylic and hydroxylic groups can form strong interactions with PES matrix. Use of graphene oxide help in improving the mechanical and electromagnetic shielding properties. From scanning electron microscopy it is clear that graphene oxides have good adhesion with PES. It can also be inferred that at low concentration of graphene oxide percolation threshold can be easily achieved. High shear mixing of graphene oxide with PES increased the interfacial interactions between the functional groups of graphene oxide and poly ether sulphones. Such high performance composite can be explored for automotive and aerospace applications since they are light weight, corrosion resistance and such systems can be used to develop self repairing components.

5. Acknowledgments

Author wish to acknowledge encouragement from Dr Prahlada, Vice Chancellor, DIAT (DU), Pune, India.

References

[1] Yan W, Zixing S, Jianhua F, Hongjie X, JYin, Carbon 49 (2011), 1199-1207.
[2] Young S Y, Yo Han B, Do Hyeong K, Jae Yun L, In Joo C, Hyoung Joon J, Carbon (2011), doi: 10.1016/j.carbon.2011.04.055.
[3] Krpa I, Chodak I, Eur. POlym. J. 2001; 37(11):2159-68.
[4] Yun Y S, Kwon HI, Bak H, Lee EJ, Yoon JS, Jin HJ, Macromol Res 2010; 18(9): 828-33.
[5] Lee HS, Yun CH, Kim HM, Lee CJ, J Phys Chem C 2007; 111(50): 1882-7.
[6] Lee HS, Yun CH, Kim SK, Choi JH, Lee CJ. Jin HJ, Appl. Phys Lett 2009; 95(13): 134103-4.
[7] Song PC, Liu CH, Fan SS, Appl Phys Lett 2006; 88(15): 153111-3.
[8] Dikin DA, Stankovich S, Zimney EJ, Piner RD, Dommett GHB, Nature 2007; 448(7152): 457-60.
[9] Li D, Muller MB, Gilje S, Kaner RB, Wallace GG, Nature Nanotech 2008; 3(2): 101-5.
[10] Stankvich S, Dikin DA, Dommett GHB, Kohlaas KM, Zimney EJ, Stach EA, Nature 2006; 442(7099): 101-5.
[11] Balasubramanian K, Erik B, Didier G, Technical Programme for Nanomaterials and Nanostructures Symposium, 2006, 322-326.

IOP Conf. Series: Materials Science and Engineering **40** (2012) 012023 doi:10.1088/1757-899X/40/1/012023

Nylon 6,6 electrospun fibres reinforced by amino functionalised 1D and 2D carbon

F Navarro-Pardo[1,2], G Martínez-Barrera[1], A L Martínez-Hernández[2,3], V M Castaño[2], J L Rivera-Armenta[4], F Medellín-Rodríguez[5], C Velasco-Santos[2,3,6]

[1]Universidad Autónoma del Estado de México, Facultad de Química, Laboratorio de Investigación y Desarrollo de Materiales Avanzados, Km. 12 de la carretera Toluca-Atlacomulco, C.P. 50200, San Cayetano, Toluca, Estado de México, México.
[2]Universidad Nacional Autónoma de México, Centro de Física Aplicada y Tecnología Avanzada, Boulevard Juriquilla No. 3001, C.P. 76230, Santiago de Querétaro, Querétaro, México.
[3]Instituto Tecnológico de Querétaro, División de Estudios de Posgrado e Investigación, Av. Tecnológico s/n, esq. Gral. Mariano Escobedo, Col. Centro Histórico, C.P. 76000, Santiago de Querétaro, Querétaro, México.
[4]Instituto Tecnológico de Ciudad Madero, División de Estudios de Posgrado e Investigación, Juventino Rosas y Jesús Ureta, Col. Los Mangos, C.P. 89440, Ciudad Madero, Tamaulipas, México.
[5]Universidad Autónoma del San Luis Potosí, Facultad de Ciencias Químicas, Av. Manuel Nava 6, Zona Universitaria, C.P. 78210, San Luis Potosí, San Luis Potosí, México.

[6]E-mail address: cylaura@gmail.com

Abstract. Nylon 6,6 electrospun nanocomposites were prepared and reinforced with 0.1, 0.5 and 1wt.% of 1D and 2D carbon. Both carbon nanotubes and graphene were functionalised with amino groups (f-CNT and f-Ge respectively). The morphology and graphitization changes of carbon nanomaterials were evaluated by transmission electron microscopy (TEM) and Raman spectroscopy; functional groups of modified nanomaterials was analysed by infrared spectroscopy. The mechanical response and the crystallinity of the fibres were measured by dynamical mechanical analysis, differential scanning calorimetry and wide angle x-ray diffraction. The morphology and dispersion of the nanomaterials in the nanofibres was studied by scanning electron microscopy and TEM. The storage modulus was improved by 118% for f-CNT and 116% for f-Ge. The mechanical response of the nanocomposites exhibited different behaviour upon loading of 1D and 2D carbon. This trend is consistent with the crystallinity of the nanofibres. This study showed f-CNT resulted in better mechanical properties at the lowest loading. On the other hand f-Ge showed improved reinforcing effect by increasing the filler loading. The two-dimensional structure of graphene was an important factor for the higher crystallinity in the electrospun nanofibres.

1. Introduction

The recent discoveries of 1D and 2D carbon structures have attracted the attention to carbon based nanocomposites. The use of these nanomaterials with improved physicochemical properties has generated a wide variety of research concerning the fabrication and properties of the resulting nanocomposites [1-3]. One-dimensional carbon nanotubes (CNTs) are considered as a graphene sheet rolled into a cylindrical shape. Two-dimensional graphene is built from a hexagonal array of sp^2 carbon atoms.

Pristine nanometric carbon is extremely difficult to disperse and align in a polymer matrix [1]. Functionalisation provides efficient stress transfer from the polymer matrix to the nanometric carbon by preventing aggregation of them, providing a better dispersion of the nanomaterials in the polymer matrix [4-7]. It also increases the polymer–nanomaterial physical contact and can form chemical bonds between the nanomaterial and the polymer matrix [4].

Nylon 6,6 is a semicrystalline polymer suitable for electrospinning processing. This approach is a simple and versatile method to produce micro to nanometric fibres [8]. Electrospinning offers the possibility to incorporate nanomaterials to fibres [9].

This work presents the reinforcement effect provided by 1D and 2D functionalised carbon to electrospun nanofibres in order to find the features this two different structures and functionalisation provide to the nanocomposites obtained. A relationship between the crystallinity properties found in the nanocomposites is also presented.

2. Experimental work

2.1. Amino functionalisation

Chemical vapour deposition multiwalled carbon nanotubes (Sunnano Company), with 10-30 nm in outer diameter, 1-10 μm length and purity above 80%, were refluxed for three hours at 80ºC in a 3:1 molar solution of nitric acid (HNO_3,70%, Sigma-Aldrich) and sulphuric acid (H_2SO_4, 98%, J.T. Baker). The solution was filtered and washed with distilled water until neutral pH. The oxidised CNTs were dried overnight at 80°C.

Graphene oxide (GO) was prepared by oxidation of graphite using the modified Hummers method as described elsewhere [10, 11]. The graphite oxide was dried overnight at 60°C. The obtained powder was re-dispersed into water and sonicated for three hours in an ultrasound bath in order to obtain graphene oxide. The GO was filtered and dried overnight at 60°C.

An acid–base titration with sodium hydroxide (NaOH, 98%, Sigma-Aldrich) was used to determine the concentration of acidic surface groups [12]. The oxidised carbon materials were added into a 25 ml of a 0.04 N NaOH solution and stirred for 48 h to allow the solid material to equilibrate with the basic solution. The mixture was titrated with a 0.04 N HCl solution to determine the excess NaOH in the solution and the concentration of the carboxylates on CNTs and GO. The concentrations of acidic surface sites were 6.5 mmol/g in the oxidised CNT and 5.4 mmol/g in the GO. The nanomaterials were functionalised as follows: a solution of 1:1 molar ratio of carbon material and 1-ethyl-3-[3-dimethylaminopropyl]carbodiimide hydrochloride) (EDAC, 99%, Sigma Aldrich) was kept under stirring at 80°C in dymethilamine excess for 6 h. The material obtained was washed until neutral pH and dried overnight at 60°C. The functionalised carbon nanotubes and graphene were labelled as f-CNT and f-Ge respectively.

2.2. Polymer solution preparation

The polymer solution was obtained by dissolving 5.5 g of nylon 6,6 (Ultramid® A3K, BASF) in 20 g of formic acid for two hours at 70°C. Afterwards the filler was added and stirred for two hours more. Three filler loadings were chosen: 0.1, 0.5 and 1 wt. %.

2.3. Electrospinning

The polymer solution was fed into a 5 ml syringe (21 gauge 1" needle). The flow rate of the solution was controlled using a syringe pump and kept constant at 5 ml/hr. A voltage of 20 kV was applied directly to the needle; the tip-collector distance was 15 cm. A grounded copper collecting screen was used. The nanocomposites were labelled by the weight percent of the filler followed by the nomenclature of the carbon nanomaterials.

2.4. Characterisation

The carbon nanomaterials were analysed by Fourier transform infrared (FTIR) spectroscopy on a Bruker Optics Vector 33 spectrometer. Raman measurements were obtained using a Dylor LabRaman II with an excitation line of 632.8 nm. Transmission electron microscopy (TEM) micrographs of the carbon nanomaterials and nanocomposites were taken on a JEOL TEM 1010 microscope operating at 80 kV.

The thermo-mechanical response of the nanocomposites was measured using a dynamic mechanical analyser (DMA), TA Instruments DMA 2980. The analyses were performed on samples of 30x5x0.05 mm^3 under tension film mode in a temperature range of room temperature to 200 °C at a frequency of 1 Hz and a heating rate of 5°/min.

The crystallinity properties were obtained in a differential scanning calorimeter (DSC), Perkin Elmer DSC-7. The samples were heated up to 280°C for five minutes and then they were cooled at 10°C/min. Wide angle X-ray diffraction (WAXD) patterns were obtained in PANalytical X'Pert Pro X-ray diffraction equipment with Cu Kα radiation (k =0.154 nm). The scanning rate was 0.05°/s.

3. Results and discussion

Figure 1a shows the amino functionalised carbon materials. The peaks between 750 cm^{-1} and 980 cm^{-1} in the spectra are due to the wagging of the N-H bond. The C-N stretching is seen ~1220 cm^{-1} [4]. The amide group (C=O) N-H features ~1700 cm^{-1} [5]. This is also related to the in-plane amide N-H deformation mode at ~1570 cm^{-1} [6]. The N-H bend of Amide I is seen at ~1641 cm^{-1} and the N-H stretching at ~1740 cm^{-1} [5]. The peaks at ~1400 cm^{-1} and ~1450 cm^{-1} are due to the attachment of additional methyl groups, attributed by the asymmetric bending of CH_3 groups and the peaks at ~2850 and ~2920 cm^{-1} corresponding to the C-H stretching vibration [7].

The usual characteristics of carbon materials in Raman spectra are the G band (~1580 cm^{-1}), which is usually attributed to the E_{2g} phonon of C sp^2 atoms, and the D band (~1320 cm^{-1}) as a breathing mode of κ-point phonons of A_{1g} symmetry, which is attributed to local defects and disorders, particularly located at the edges of graphene [13]. The Raman spectra of f-CNT and f-Ge are shown in figure 1b. Both materials display the G and D bands. Moreover a major fingerprint of graphene is the D' peak at ~2700 cm^{-1}.

The morphology of the carbon materials is shown in the figure 2. In f-CNT is possible to observe a rough surface on the walls of the tubes. In the f-Ge, a mixture of graphene layers is observed; the sheets are rippled and entangled with each other like a silk weave form. Figure 2 also shows the micrographs of the nanofibres at the maximum filler loading. Electrospinning technique is known for the alignment of the nanomaterials along the fibre [9]. Polymer nanocomposites require homogeneous dispersion and strong interfacial interaction between carbon nanomaterials and the polymer matrix for the enhancement of the mechanical and thermal properties [1-7]. Functionalisation is expected to reduce agglomeration of the nanotubes. However at high loading the f-CNT agglomerated showing poor dispersion of the nanotubes. In contrast, at the same loading a graphene sheet is seen embedded and aligned along the fibre axis showing the good affinity between the polymer matrix and the f-Ge. The agglomeration at high loading of f-CNT in the nanocomposites could be explained by the one-dimensional structure of the nanotubes that allows them to easily bend and interact with the amino groups on their surface.

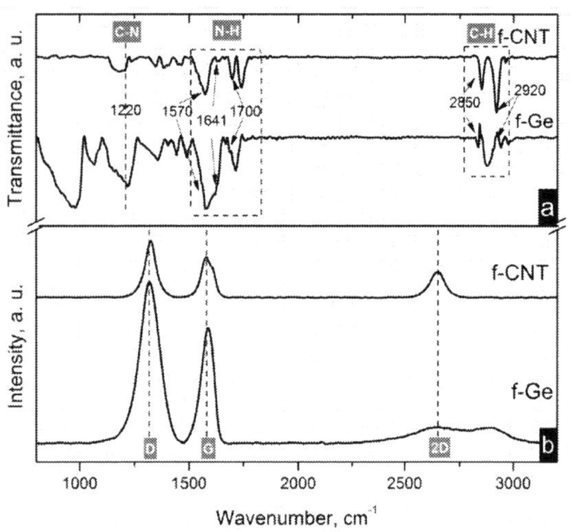

Figure 1. a) IR and b) Raman spectra, of the carbon nanomaterials.

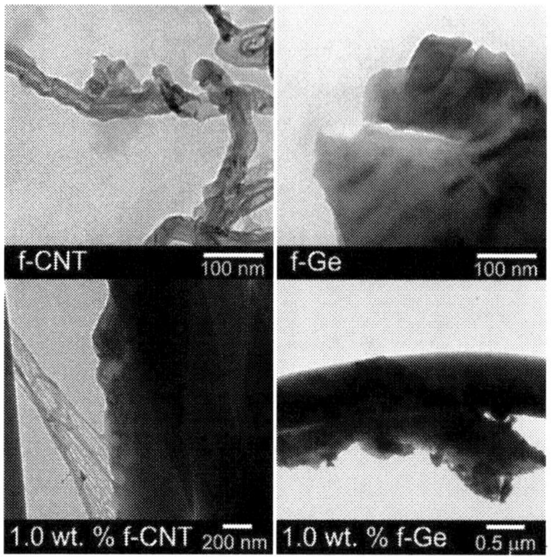

Figure 2. TEM micrographs of the carbon nanomaterials and nanocomposites.

Figure 3 shows the SEM images of the electrospun nanofibres oriented randomly in a non-women mat. Uniform fibres with no beads were obtained. The nanometric carbon embedded within the fibres is not visible. However the deposited carbon materials on the surface of the fibres are observed in the bright areas of the images.

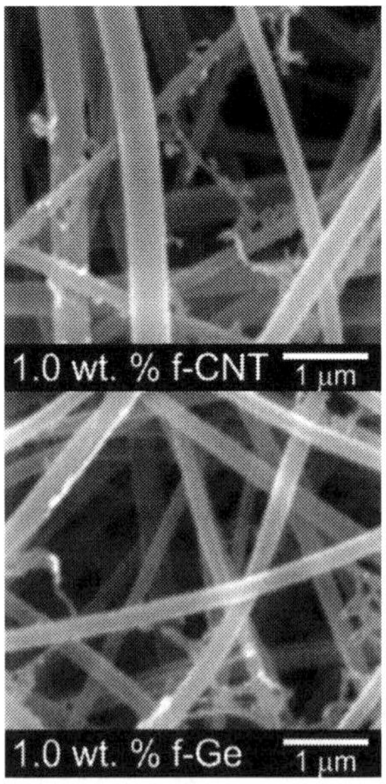

Figure 3. SEM photographs of nanocomposites.

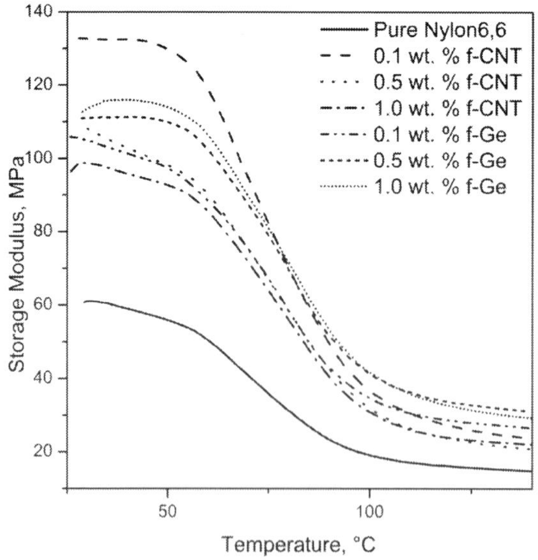

Figure 4. Storage modulus of the pure nylon 6,6 nanofibres and nanocomposites.

The storage modulus of pure nylon 6,6 and the nanocomposites is presented in figure 4. The improved mechanical response of the nanofibres is due to the strengthening effect of the carbon nanomaterials [1]. At the lowest loading of f-CNT the modulus reaches a 118% increase compared to pure nylon 6,6 nanofibres. However when the filler loading is increased this property is decreased. This behaviour can be attributed to agglomeration of the nanotubes in the nanofibre, as seen in Figure 2. On the other hand the storage modulus in the f-Ge reinforced composites showed a tendency to increase as the loading of the filler increased. The modulus was improved from a 70% for a 0.1 wt. % to a 116% for the maximum loading of f-Ge. The values of the glass transition temperature in table 1 were obtained from tan δ peak maximum of the DMA. This property was enhanced up to 7°C for the f-CNT nanocomposites and up to 5°C for the f-Ge nanocomposites at the highest loading.

Crystallinity and molecular orientation provide unique physical features to the nanocomposites and play an important role in mechanical properties of the fibres [14]. The crystallinity properties of the nanofibres are also displayed in table 1. The degree of crystallinity (X_C) was calculated from the heating thermograms of DSC [15] and the crystallinity index (CI) from the WAXD patterns. The crystal sizes L_1 and L_2 were obtained applying the Scherrer equation to the two diffraction signals of the nylon 6,6 that are a project value of the interchain distance within the hydrogen-bonded sheet (L_1) and the the intersheet distance (L_2) [16].This results are associated with the effect imparted by the carbon nanomaterials on the crystal structure and hence with the mechanical response of the nanocomposites. The crystallinity in the f-CNT reinforced nanofibres decreases as the filler content increases .The opposite behaviour is observed in the f-Ge based nanofibres. This is in agreement with Wong et al. who found that a higher degree of crystallinity and molecular orientation of fibres resulted in improved mechanical strength and stiffness [8]. Moreover, the crystal sizes of the nanocomposites are in accordance with these results. This means the carbon nanomaterials with a high nucleation density provide more defects in the crystal indicating lower crystal perfection [14]. The better dispersion in f-Ge resulted in smaller crystal size of the nanocomposites.

Table 1. Crystallinity, crystal size and glass temperature of the nanocomposites.

Nanofibres	Xc, %	CI	L_1, nm	L_2, nm	Tg, °C
Nylon 6,6	39.38	0.383	6.451	3.410	79
0.1 wt. % f-CNT	44.31	0.435	5.350	3.105	83
0.5 wt. % f-CNT	38.90	0.390	5.930	3.233	85
1.0 wt. % f-CNT	41.13	0.388	5.982	3.880	86
0.1 wt. % f-Ge	41.12	0.403	6.291	2.974	83
0.5 wt. % f-Ge	41.73	0.417	5.667	3.187	84
1.0 wt. % f-Ge	43.09	0.431	5.763	2.848	84

4. Conclusion

The electrospun nanocomposites obtained in this work showed the different behaviour of functionalised 1D and 2D carbon at different loadings. One of the findings of the study between the two structures of the fillers was the one-dimensionality of the f-CNT favours the aggregation of the nanotubes at high loadings. In contrast the two-dimensional f-Ge structure enhanced the mechanical response when using higher loading of this filler. This explanation is confirmed by the nanocomposites crystal properties. The nanocomposites reinforced with graphene showed the highest crystallinity and lowest crystal perfection due to the nucleation provided by the better dispersion of this nanomaterial in the polymer matrix.

Acknowledgments
The authors are grateful to Dr G. Hernández Padron for her assistance in IR spectroscopy, to Dr S. J. Jiménez Sandoval for Raman facilities at CINVESTAV and F. Rodríguez Melgarejo for his technical assistance. We also appreciate the support of M. A. Palma Tirado for acquiring TEM images, A. del Real López for her assistance in SEM, O. Davalos Montoya for her help in DSC analyses and A. Torres Martínez for her assistance in WAXD measurements. PhD student F. Navarro-Pardo is thankful to CONACYT for grant No. 213733 and to Universidad Autonóma del Estado de México for the use of facilities.

References

[1] Martínez-Hernández, Velasco-Santos C and Castaño V M 2010 Carbon nanotubes composites: Processing, grafting and mechanical and thermal properties *Curr Nanosc* **6** (1) 12-39.
[2] Velasco-Santos C, Martínez-Hernández A L, Fisher F, Ruoff R and Castaño V M 2003 Dynamical–mechanical and thermal analysis of carbon nanotube–methyl-ethyl methacrylate nanocomposites J. *Phys. D: Appl. Phys.* **36** 1423-1428.
[3] Velasco-Santos C, Martínez-Hernández A L, Fisher F T, Ruoff R and Castaño V M 2003 Improvement of thermal and mechanical properties of carbon nanotube composites through chemical functionalization *Chem Mater* **15** (23) 4470-4475.
[4] Shen J,. Shi M, Yan B, Ma H, Li N, Hu Y et al 2010 Covalent attaching protein to graphene oxide via diimide-activated amidation *Colloids Surf., B* **81** 434–438.
[5] Shen J, Huanga W, Wu L, Hu Y and Ye M 2007 Study on amino-functionalized multiwalled carbon nanotubes *Mater. Sci. Eng.* **464** 151-156.
[6] Shen J, Yan B, Shi M, Ma H, N Li, Hu Y and Ye M 2011 Synthesis of graphene oxide-based biocomposites through diimide-activated amidation *J.Colloid Interface Sci.* **356** 543–549.
[7] Chen X, Chen X, Lin M, Zhong W, Chen X and Chen Z 2007 Functionalized multi-walled carbon nanotubes prepared by in situ polycondensation of polyurethane. *Macromol. Chem. Phys.* **208** 964–972.
[8] Wong S C, Baji A, Leng S 2008 Effect of fiber diameter on tensile properties of electrospun poly(ε-caprolactone) *Polymer* **49** (21) 4713-4722.
[9] Baji A, Mai Y, Wong S C, Abtahi M and Chen P 2010 Electrospinning of polymer nanofibers: effects on oriented morphology, structures and tensile properties *Compos. Sci. Technol.* **70** (5) 703-718.
[10] Park S, An J, Piner R D, Jung I, Yang D, Velamakanni A, et al 2008 Aqueous suspension and characterization of chemically modified graphene sheets *Chem Mater* **20** (21) 6592-6594.
[11] Shen X, Jiang L, Ji Z, Wu J, Zhou H, Zhu G 2011Stable aqueous dispersions of graphene prepared with hexamethylenetetramine as a reductant. *J. Colloid Interface Sc*i. **354** (2) 493-497.
[12] Datsyuk V, Kalyva M, Papagelis K, Parthenios J, Tasis D, Siokou A, et al 2008 Chemical oxidation of multiwalled carbon nanotubes *Carbon*, **46** (6) 833-840.
[13] Ferrari A C and Robertson J 2000 Interpretation of raman spectra of disordered and amorphous carbon *Phys Rev B* **61** (20) 14095-14107.
[14] Jog J P 2006 Crystallisation in polymer nanocomposites *Mater Sci Technol* **22** (7) 797-806.
[15] Chavarria F and Paul D R 2004 Comparison of nanocomposites based on nylon 6 and nylon 66 *Polymer* **45** (25) 8501-8515.
[16] Lu Y, Zhang Y, Zhang G, Yang M, Yan S, Shen D 2004 Influence of thermal processing on the perfection of crystals in polyamide 66 and polyamide 66/clay nanocomposites *Polymer* **45** (26) 8999-9009.

Efficient optical resolution of amino acid by alanine racemaze chiral analogue supported on mesoporous carbon

D Jang, K Kim, D Park and G Kim

Department of Chemical Engineering, Inha University, 253 Younghyun-dong, Nam-gu, Incheon 402-751, Republic of Korea.

E-mail: kimgj@inha.ac.kr

Abstract. Optically pure D-amino acids are industrially important chiral building blocks for the synthesis of pharmaceuticals, food ingredients, and drug intermediates. Chemoenzymatic dynamic kinetic-resolution processes have recently been developed for deracemization of amino acids. S-ARCA would be a good candidate for the selective adsorption of D amino acid through the imine formation reaction. The organic phase containing S-ARCA adsorbent, TPPC or Ionic Liquid (as a phase transfer catalyst) in MC were coated on the surfaces of mesoporous carbon C-SBA-15(CMK). The aqueous solution of racemic D/L-amino acid and NaOH were added to the carbon support coated with ARCA. The D/L ratios on ARCA and in solution were determined with increasing reaction time. S-ARCA has a unique property for the selective adsorption of D- amino acid (up to 90% selcetivity) in the racemic mixture. The fixed bed reactor containing ARCA/carbon support was also adopted successfully for the selective separation of amino acid.

1. Introduction

Amino Acid is a molecule containing unique structure coexisted acidic COOH and basic -NH_2. Except glycine exsisting in nature, 19 kinds of Amino acids are divided as L-form and D-form by optical activity. L-form of amino acid is digested and synthesized in the living body, but D-form having an opposite optical activity can not be. Also these abilities and effects often depend on the optical avtivity. Therefore proper asymmetric synthesis method would provide economically and scientifically valuable molecules with high optical purity. D-amino acid(D-AA) in the racemic D/L amino acid mixture can be selectively attached to (S)- ARCA through imine formation [1]. (S)-ARCA (Alanine Racemase Chiral Analogue) adsorbs amino acid of D-AA selectively, and (R)-ARCA does L-type amino acid. (S)-ARCA [2] is an imitated organic compound injected chiral properties on Alanine Racemase enzyme existed in the wall of bacteria cell[3]. That chiral compound may be widely applied on the production of variety amino acid[4], and it is possible to convert from L-(D-)AA to D/L-AA and in addition to seperate as one optical form of amino acid form the racemate, using ARCA as an efficient chiral catalyst or adsobate. CMK (Ordered Mesoporous Carbon) has interesting properties because of its unique pore structure, and they are used widly in many applications[5,6]. In this work, we have designed and prepared various types of S-ARCA heterogenized on carbon supports for direct separation of D-AA from racemic D/L-AA. In the batch type system, the chiral ARCA supported on the carbons could easily seperate the amino acid in a optically high purity. The fixed bed reactor containing granular carbon support was also adopted successfully for the selective separation of amino acid in this study.

2.Experimental

2.1 Sysnthesis of Ordered Mesoporous Carbon(CMK)
A high quality SBA-15 was prepared using the triblockcopolymer Pluronic P123 (Aldrich) as a surfactant and tetraethylorthosilicate(TEOS, 98%, Aldrich) as a silica source, modifying the synthetic procedure reported by Zhao et al.[7]. C-SBA-15 (CMK) was synthesized with silica templates SBA-15 and sucrose as carbonprecursor following the procedure in the literature [5,6]. After carbonization in furnace at 80 °C under vacuum, silica template was removed by dissolving with HF solution (50 wt%; Duksan PureChemicals Co. Ltd.). The final mesoporous carbon products were dried at 100°C. The mesopore structure of carbon support was investigated by TEM. They have very high surface area and pore volumes with uniform hexagonal sutictre as shown in Figure 1.

2.2 Separation of D-AA from racemic D/L-AA by adsorption

2.2.1 Batch type adsorption. Organic compounds containing 1.0eq ARCA((AMINOLOGICS, 87%, 3g), 1.4eq of Ionic Liquid or 1.4eq of TPPC (Tetraphenyl phosphonium chloride), based on (S)-ARCA, in 10ml MC were mixed and stirred until the full dissolution of ARCA. The structures of those compounds are listed in Figure 2. The organic solution was coated on the mesoporous carbon CMK(5g) and the MC solvent was evaporated partly. For 1st adsorptive separation reaction, the aqueous basic solution of 2.0 eq. D/L-phenylalanine (D/L-PA; AMINOLOGICS, 93%), based on ARCA amount, was prepared by dissolving D/L-phenylalanine, 2.0eq of NaOH and 1.8eq of Diisopropyl amine (DEAJUNG, 99%, DIPA) in 20ml of distilled water. That aqueous solution was mixed and stirred with solid ARCA coated carbon support. The concentration of amino acid remained in the aqueous solution was determined by HPLC with prolonged reaction time. The adsorpted amount of AA on the ARCA/support was also analysed after treatment of adsorbate with 3M HCl. The adsorpted AA on ARCA through imine formation reaction between aldehyde of ARCA and NH_2 in AA was hydolyzed to release D-AA into the acidic aqueous solution. Organic layer coated on support was washed with HCl solution and used again for the 2nd adsorptive separation reaction. In the repeated use of ARCA/support for the separation of chiral AA, the fresh basic solution of D/L-AA, NaOH, DIPA in water was prepared, and then it was mixed and reacted with ARCA/carbon. Purification of D-AA from solid adsorbate was followed the same procedure mentioned above. The schematic diagram for batch type reactor using CMK powder support is shown in Figure 3. D-AA in acidic solution could be isolated as a crystallized powder in methanol solvent with 3h stirring.

2.2.2 Fixed Bed type Adsorption. Organic solution of 10g of (S)-ARCA, 1.4eq of Ionic Liquid, 20ml of MC was mixed until ARCA was fully dissolved. This solution was mixed with CMK powder and then it was coated on the macroporous silica matrix(Figure3). In comparison, that solution was directly impregnated on the granular carbon support. After partial evaporation of solvent, they were packed it in the tube type reactor. The aqueous solution of 2.0eq D/L-phenylanine, 2.0eq NaOH and 1.8eq DIPA dissolved in 200ml of distilled water was pumped and circulated through the reactor packed with (S)-ARCA/carbon support for adrsorption of D-AA. At the begining, the aqueous layer was milky white, but it became trasparent during the continueous circulation. The adsorption was performed for one hour, and after changing the circulating solution as 3M HCl, the hydrolysis of adsopted AA was done for additional 1 h with circulation. The typical diagram for fixed bed type reactor adopted in this work is shown in Figure 7.

2.3 Characterization
The C-SBA-15 morphology and microstructures of as-prepared samples were characterized by field

emission transmission electron microscopy (FE-TEM, S-4200), and field emission scanning electron microscopy (FESEM, JEM-2100F). The concentration of amino acid was analyzed by Alliance HPLC (Waters 2695 system) using chiral column (CH SCA(-)-51002546; 250 x 6mm), in the elluent of MeOH 60%, Water 40%, $HClO_4$ 5mmol with , 1.0ml/min.

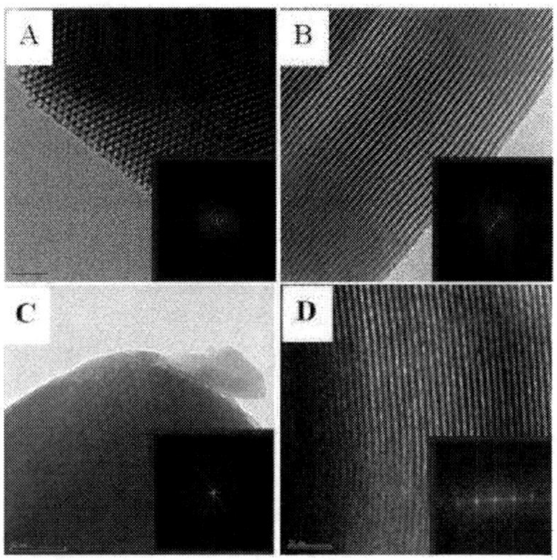

Figure 1. TEM images of SBA-15(A and B) and ARCACMK(C and D).

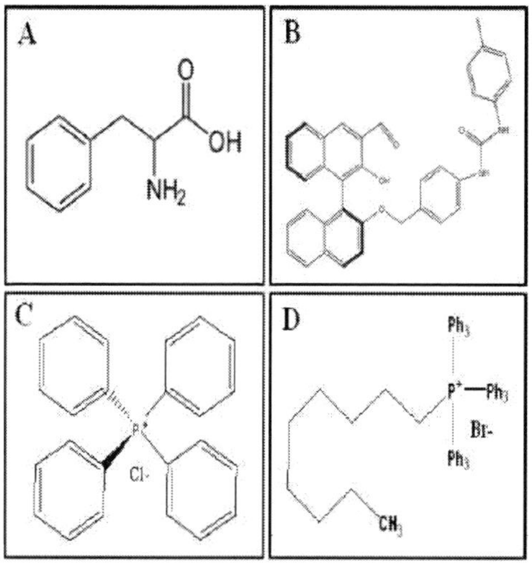

Figure 2. (A)D/L-phenylalanine (B) (S)-(C) TPPC (D) Ionic Liquid

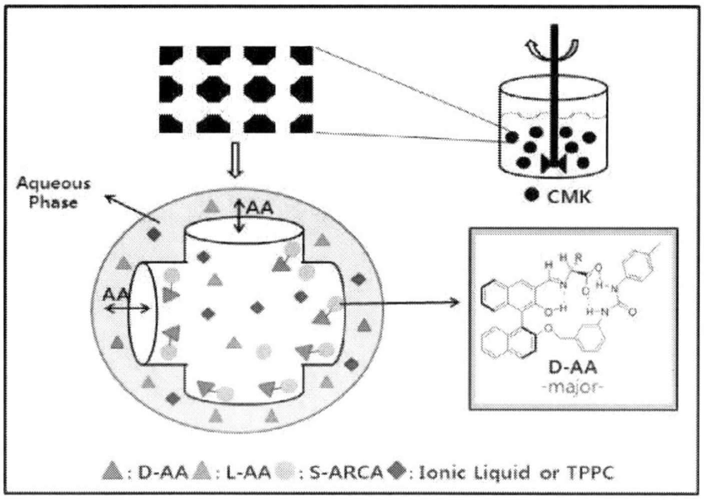

Figure 3. Schematic diagram for Batch type reactor using ARCA/CMK powder adsobate.

3. Results and Discussion

3.1 Batch type adsoption of D-AA on (S)-ARCA

The racemic D/L-AA is consisting of 50% D-AA and 50% L-AA. (S)-ARCA, as shown in Figure 4, forms an imine group with D-AA in major. As a result, only 1.0 eq D-AA reacts with 1.0 eq D-AA theoretically from 2eq D/L-AA, leaving 1.0 eq of L-AA in the aqueous layer. Methylene chloride (MC) was usually used as a solvent in liquid-liquid system for ease separation from aqueous phase. The aqueous layer contains D/L-AA, and ARCA is dissolved in the organic solvent with PTC. Since MC has a low boiling point, it should be resupplied during the repeated recycle because of severe loss in amount. However, in the view point of solvent effect, the use of ethylene dichloride (EDC) was efficient as well as MC. A small amount of MC or EDC may be supplied additionally into the reaction system to prevent the full dryness of ARCA on the carbon support. At the high loading of new ionic liquid, a full dissolution and homogeneous coating of ARCA were performed. In the heterogenized ARCA system, phase separation into organic layer and aqueous layer is unnecessary. To increase the adsorption rates, (S)-ARCA and PTC were coated on the carbon support as thin layers, maintaining the separated particles without forming a big aggregated lump.

Figure 4. Before (left) and after(right) coating of ARCA/CMK mixture on macroporous silica substrate(A), and granule type active carbon coated with ARCA and IL (B).

Table 1. Effect of additives concentration for optimum reaction conditions to obtain higher adsorption capacity.

Entry	NaOH (eq.)	DIPA(eq.) [a]	IL(eq.) [b]	TPPC(eq.)	1st Reaction; Adsorpted amount of D-PA(%)	2nd Reaction; Adsorpted amount of D-PA(%)
1	2.0	-	-	1.2	80	71
2	2.0	-	-	1.4	75	70
3	2.0	1.8	-	1.2	82	74
4	2.0	-	1.4	-	78	88
5	2.0	1.8	1.4	-	82	93
6	5.0	1.8	1.4	-	62	62

Basis; 2.0 eq of D/L-PA was used as a reactant, based on ARCA a.diisopropylamine, [b] Ionic Liquid

The adsorption ability of ARCA/carbon was compared in the presence of PTC such as IL and TPPC with different mole ratios. Additionally the effect of base additive on the activity of ARCA in the adsorptive separation of PA was investigated, and the results are listed in Table 1. The high loading of NaOH was not efficient since pH of aqueous solution became increased too high. Then the addition of co-base such as DIPA in the presence of NaOH was effective for increasing the amount of D-PA adsorpted on (S)-ARCA, as shown in Table 1. In the liquid-liquid system without using the support, TPPC is typically used as a phase transfer catalyst. It has good property as PTC, but in regenerated system, a large amount of TPPC was lost during the repeated use in two liquid phases because of its diffusion into the aqueous phase. This phenomenon can be found also in the batch type separation of D-PA form D/L-PA using ARCA/carbon system. As can be seen in Figure 4, during the hydrolysis of D-PA adsorpted on ARCA by HCl, TPPC was diffused into the aqueous phase, and it was detected on HPLC spectra as in Figure 5(B). However the Ionic Liquid (IL) synthesized newly in this work has brought the better result than TPPC in separation, and it acted as a good phase transfer catalyst (PTC), showing no extraction during the adsorption/desorption process (Figure 6(A)). IL was so stable that the unique TPC property was maintained during the repeated use in the adsorptive separation of D-PA.

The ability of IL and TPPC as PTC was compared in the reused system, and the adsorpted amount of D-AA on (S)-ARCA/carbon was determined with time. The obtained result is summarized in Figure5. The optical purity of L-phenylalanine determined in aqueous layer was 80% for both case of IL and TPPC, meaning that 80% of D-phenylalanine was adsorpted on ARCA/carbon. The adsorpted amount of D-AA was also determined after hydrolysis treatment of ARCA/carbon with HCl. However at the 2nd regenerated adsorption, the adsorpted amount of D-AA on (S)-ARCA was increased to 90% in the use of IL. But TPPC showed a reducing D-AA adsorption property at the repeated 2nd use. That is contributed by reduction of TPPC in concentration during the hydrolysis with HCl. However, IL had better separation property than TPPC, providing 93% adsorption of D-form in the 2nd regeneration reaction. Ionic Liquid continuously worked as an efficient PTC transfer catalyst in the repeated regeneration for 5 times.

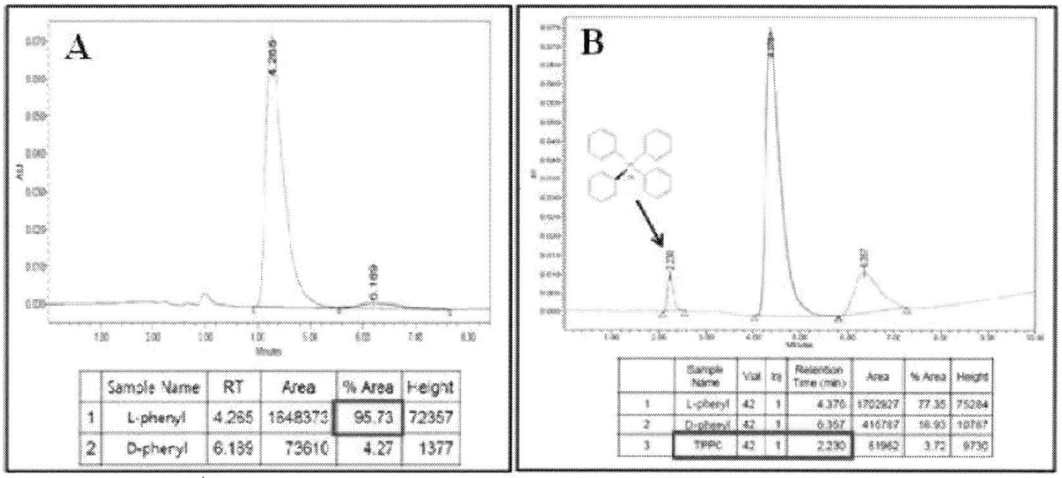

Figure 5. HPLC analytical result for batch type separation of D-PA using ionic liquid (A) and TPPC (B) as a phase transfer catalyst (determination of aqueous phase after HCl hydrolysis).

The ability of IL and TPPC as PTC was compared in the reused system, and the adsorpted amount of D-AA on (S)-ARCA/carbon was determined with time. The obtained result is summarized in Figure5. The optical purity of L-phenylalanine determined in aqueous layer was 80% for both case of IL and TPPC, meaning that 80% of D-phenylalanine was adsorpted on ARCA/carbon. The adsorpted amount of D-AA was also determined after hydrolysis treatment of ARCA/carbon with HCl. However at the 2nd regenerated adsorption, the adsorpted amount of D-AA on (S)-ARCA was increased to 90% in the use of IL. But TPPC showed a reducing D-AA adsorption property at the repeated 2nd use. That is contributed by reduction of TPPC in concentration during the hydrolysis with HCl. However, IL had better separation property than TPPC, providing 93% adsorption of D-form in the 2nd regeneration reaction. Ionic Liquid continuously worked as an efficient PTC transfer catalyst in the repeated regeneration for 5 times.

3.2 Fixed Bed type adsoption of D-AA on (S)-ARCA

For the fixed bed adsorption system (Figure 7), (S)-ARCA/CMK coated macroporous silica ball or (S)-ARCA-IL coated active carbon granule(Figure2 A& B) was packed in the tube. The optimum condition chosenfor the batch type system was adopted in the circulating fixed bed. First of all, the aqueous solution of 2.0eq D/L-phenylanine, 2.0eq NaOH and 1.8eq DIPA dissolved in 200ml of distilled water was pumped and circulated with prolonged time for adrsorption of D-AA. Under this reactant concentration, the adsorpted amount of D-PA on to (S)-ARCA/CMK coated macroporous silica ball was 75% for 1h circulation, and 82% for 2h and 3h. The similar result was obtained using (S)-ARCA-IL coated granular active carbon. The loading amount of ARCA was controlled as same for both cases. However, with increasing concatration of D/L-PA in the reactant solution, the adsorpted amount of D-PA has increased at the same pumping speed. When the mole ratio of (D/L-PA)/ARCA was fixed as 15 instead of 2, the adsorpted amount of D-PA on to (S)-ARCA/CMK coated macroporous silica ball was 92% for 2h circulation. To investigate the dependance of initial D/L ratio of PA on the separation property, D-PA rich reactant, such as 7:3 D-PA/L-PA, was injected into the fixed bed system. In this case, 93% D-PA was adsorpted on ARCA. Fixed bed system has provided the easy and simple process for adsorption and desorption of chiral PA. After saturated adsorption, HCl solution was injected into the bed for hydrolysis of imine group to isolate the adsorpted D-PA. The repeated separation of D-PA from racemic D/L-PA could be performed in the fixed bed in convinience.

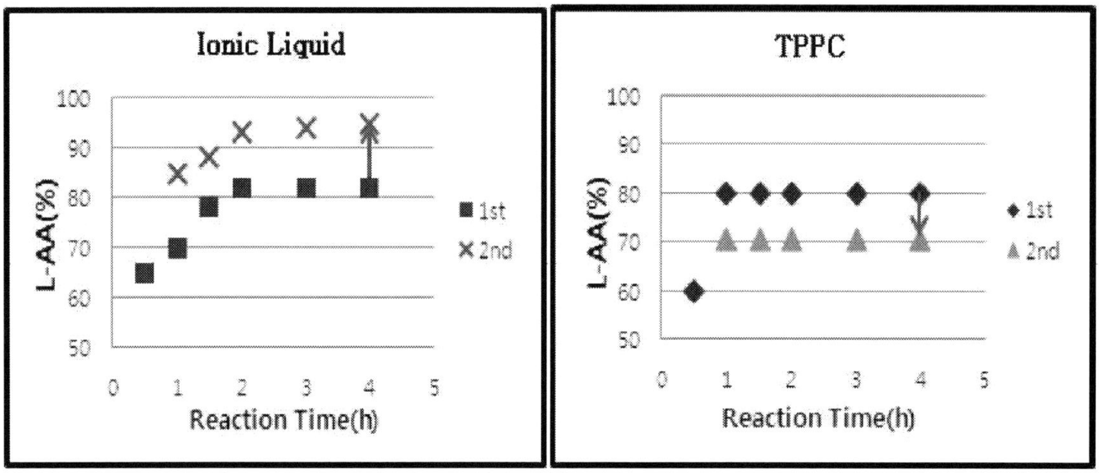

Figure 6. The performance comparison of D-AA adsorptive separation using Ionic Liquid and TPPC as PTC.

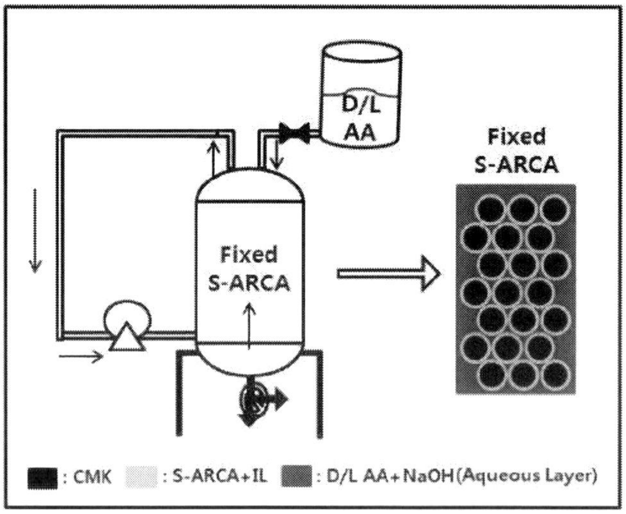

Figure 7. Typical diagram of circulating Fixed bed adsoption system for D-AA separation.

4. Conclusion

Important factors in adsorption of D-PA was to maintain pH between 10 to 12. The mesoporous support provided excellent property for homogeneous loading of chiral ARCA on the surfaces in thin layer.

The newly synthesized IL could be applied successfully as PTC in the repeated adsorption and separation of D-PA in the batch and fixed bed system. It exhibited superior ability to conventional TPPC, showing no loss during the extration process of adsorpted D-PA. In the repeated adsorption, maximum adsorption of 93% D-amino acid was achieved in the batch and contineous circulation system. The high purity purity of D-PA up to 99% would be achieved after simple purification.

References

[1] Kreil, G. *Science* **1994**, *266*, 996-997.(b) Corrigan, J. J. *Science* **1969**, *164*, 142-149.
[2] Kim, K. M.; Park, H.; Kim, H.-J.; Chin, J.; Nam, W. *Org. Lett.* **2005**, *7*, 3525-3527.
[3] Stryer, L. *Biochemistry*, 4th ed.; W. H. Freeman and Co.: New York, 2000; pp 201-202.
[4] Mendel Friedman, *J. Agric. Food Chem.* **1999,** *47,* 3457-3479
[5] Ryoo, R. Joo, S.H. Jun, S. *J. Phys. Chem. B* **1999**, 103, 7743.
[6] Lee, J. Yoon, S. Hyeon, T. Oh, S.M. Kim, K.B. *Chem. Commun.* **1999**, 2177.
[7] Zhao, D. Huo, Q. Feng, J. Chmelka, B.F. Stucky, G.D. *J. Am. Chem. Soc.* **1998**, 120, 6042.

IOP Conf. Series: Materials Science and Engineering **40** (2012) 012025 doi:10.1088/1757-899X/40/1/012025

Modeling of short fiber reinforced injection moulded composite

A Kulkarni, N Aswini, C R Dandekar and S Makhe

Eaton Technologies Pvt Ltd,EIEC, Kharadi, Pune, India

E-mail: AbhijitKulkarni@Eaton.com , AswiniN@Eaton.com , ChinmayaDandekar@Eaton.com , SunilMakhe@Eaton.com

Abstract. A micromechanics based finite element model (FEM) is developed to facilitate the design of a new production quality fiber reinforced plastic injection molded part. The composite part under study is composed of a polyetheretherketone (PEEK) matrix reinforced with 30% by volume fraction of short carbon fibers. The constitutive material models are obtained by using micromechanics based homogenization theories. The analysis is carried out by successfully coupling two commercial codes, Moldflow and ANSYS. Moldflow software is used to predict the fiber orientation by considering the flow kinetics and molding parameters. Material models are inputted into the commercial software ANSYS as per the predicted fiber orientation and the structural analysis is carried out. Thus in the present approach a coupling between two commercial codes namely Moldflow and ANSYS has been established to enable the analysis of the short fiber reinforced injection moulded composite parts. The load-deflection curve is obtained based on three constitutive material model namely an isotropy, transversely isotropy and orthotropy. Average values of the predicted quantities are compared to experimental results, obtaining a good correlation. In this manner, the coupled Moldflow-ANSYS model successfully predicts the load deflection curve of a composite injection molded part.

1. Introduction

Thermoplastics have many advantages over metal parts including weight, ease of fabrication and economy. Although the advantages have allowed them to proliferate in various industries, the lack of structural load carrying capacity has hindered there use in automotive and aerospace industries. To mitigate the disadvantage, short fibers such as glass or carbon fiber etc., are added to these polymers to improve the elastic modulus, strength to weight ratio, creep resistance, and dimensional stability. This increase in the load carrying capacity is what allows these composites to be widely explored in the Automobile & Aerospace component industry. However, the application of fiber-filled thermoplastic materials has been limited in many cases due to the inability to accurately predict performance and durability as the behavior of the polymer composites depends primarily on the fiber length [1] and the fiber orientation distribution [2]. Prediction of the fiber orientation is a complex problem and there is a lack of models to do the same. Injection molding is a popular method of manufacturing of these composites. During injection molding, the fiber alignment (or orientation) distributions show a layered nature. This alignment is affected by numerous parameters, chief amongst them are the mold filling speed, processing conditions, material viscosity, fiber aspect ratio and the volume fraction or concentration of the fibers. Hence without a proper understanding of the fiber behavior, there is a tendency to significantly overestimate the orientation levels which will result in significant variation in mechanical properties of the molded product.

Fiber orientation is one of the major factors that determine the mechanical (elastic) strength as well as the stiffness of a molded part. Moldflow, a commercial modeling software predicts the fiber orientation based on a Folgar-Tucker model [6]. This model allows for significantly improved orientation prediction accuracy over a range of materials and fiber contents and is therefore used in this study to predict the fiber orientation.

Many methods have been developed to predict the influence of the reinforcement (fiber) and matrix on the behavior of the composite material as a whole. Some methods are reliant on the assumptions regarding the specific details of the geometric arrangement of the reinforcement phase in the matrix while other models are not sensitive to the reinforcement arrangement [9]. Two methods exist in calculating the effective elastic properties of the composite: exact methods and energy methods. Exact methods deal with specific geometries of the arrangement of the reinforcement and then the elastic properties are solved analytically or numerically [9]. Energy methods on the other hand involve variational calculus and are used to obtain upper and lower bounds for a particular composite. An exact solution is determined only if the upper and lower bounds coincide, which usually does not happen except at the 0% and 100% volume fraction of reinforcement. Models range from simple slab models (rule of mixtures) to more complex self consistent models. One of the most popular homogenization theories is by Halpin and Tsai [4, 5] who developed a simple and generalized equations to approximate the results of more exact micromechanical techniques so as to readily describe the composite material. Another theory for predicting the linear elastic behavior of the composite is by Tandon-Weng [3]. Many micromechanics based theories are available in literature which can be applied to fiber reinforced composites and it is beyond the scope of the current study to list them. For the present case of interest, Halpin-Tsai is used to predict longitudinal and transverse elastic modulus as well as the shear modulus (longitudinal-transverse) & Poisson's ratio. The composites being anisotropic in nature, for the present case a linear orthotropic model, is chosen. The linearity assumption is based on the consideration that, 'with addition of fibers in the matrix, failure observed is mostly linear elastic'.

Moldflow fiber orientation (with mechanical properties) results are given as input to carry out the structural linear static analysis on a 30 % volume fraction carbon fiber reinforced PEEK plate in ANSYS. Simulated results are then compared to experiments. A comparison is also made between linear elastic, isotropic, transversally isotropic and orthotropic material models. The results indicate that the orthotropic material model shows good correlation with experiments.

The modeling methodology can be broken down into the three major steps: (i) Prediction of mechanical properties of carbon fiber reinforced PEEK (ii) fiber orientation prediction using Moldflow and (iii) Exporting the properties as per fiber orientation to ANSYS to execute the structural analysis. These three steps are explored in the below subsections.

1.1. Mechanical Property Prediction of Carbon Fiber Reinforced PolyEtherEtherKetone (PEEK)

The linear elastic properties of the composites are predicted based on the individual properties of the fiber and the matrix, summarized in Table 1 and the fiber volume fraction of, using Halpin-Tsai micromechanics [5] based equations as listed in the following equations.

$$E_L = E_m \frac{1 + 2(\frac{l_f}{d_f})\eta_L V_f}{1 - \eta_L V_f} \tag{2.1}$$

$$E_T = E_m \frac{1+2\eta_T V_f}{1-\eta_T V_f} \tag{2.2}$$

$$G_{LT} = G_m \frac{1+2\eta_G V_f}{1-\eta_G V_f} \tag{2.3}$$

$$\nu_{LT} = V_f \nu_f + V_m \nu_m \tag{2.4}$$

where

$$\eta_L = \frac{E_f/E_m - 1}{E_f/E_m + 2(lf/df)} \qquad \eta_T = \frac{E_f/E_m - 1}{E_f/E_m + 2}$$

$$\eta_G = \frac{G_f/G_m - 1}{G_f/G_m + 1}$$

E_L = Longitudinal Elastic Modulus	E_f = Elastic Modulus of Fiber
E_T = Transverse Elastic Modulus	E_m = Elastic Modulus of Matrix
G_{LT} = Shear Modulus	l_f & d_f = Length & diameter of fiber
ν_{LT} = Poisson's Ratio	η_L, η_T, η_G = Constants derived by Halpin-Tsai
ν_f = Poisson's ratio of fiber	ν_m = Poisson's ration of matrix
V_f = Fiber Volume Fraction	V_m = Matrix Volume Fraction

1.2. Flow Simulation- Fiber Orientation Prediction Using Moldflow

The processing of the thermoplastics is simulated using Moldflow (MPI) software [Refer Moldflow manual]. Moldflow has the capability to predict the fiber orientations based on the processing related inputs such as gating location, time for filling etc. and a digital model is created. For illustration the digital model developed for a plate whose injection molding process parameters are known are shown in Figure1 and the corresponding fiber orientation plot are shown in Figure 2.

1.3. Moldflow-ANSYS Interface Establishment

The interfacing of, Moldflow & ANSYS is illustrated in Figure 3. In this, first the FE model is set up in ANSYS, and with help of AMSA (Autodesk Moldflow Structural Alliance) model is exported to Moldflow as geometry format (.stl). Flow analysis is performed in Moldflow and fiber orientation is predicted. The results (orientation and material properties per orientation) are written as .lsp format which are imported in ANSYS using AMSA and further structural analysis is performed.

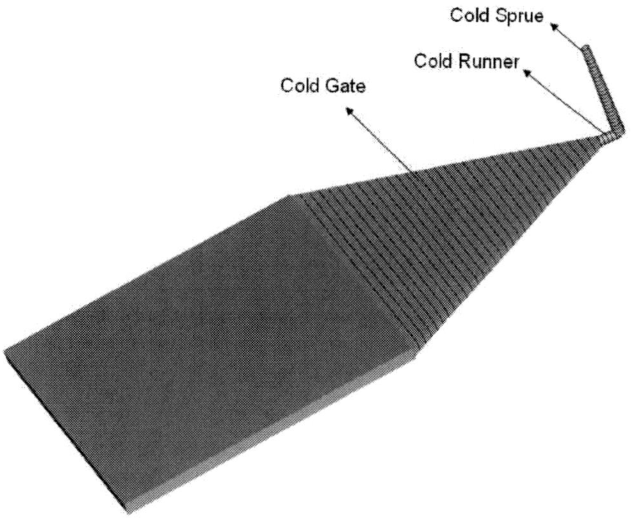

Figure 1. Injection Moulding Process - Set Up.

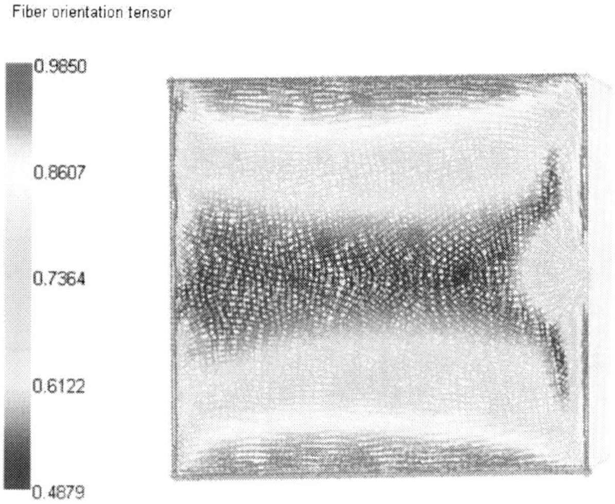

Figure 2. Fiber Orientation Plot in the I principal direction.

1.4. Structural Analysis

For illustration consider a carbon fiber reinforced square plate composed of composites, which is held at one edge and displacement applied at the opposite end. The required fiber and matrix properties are available in [7] & [8]. The composite properties and fiber orientation are estimated as described in the subsections 1.1 – 1.3 and the model is solved using ANSYS. The resulting model has an orthotropic configuration. Also two other cases are analyzed for the same boundary conditions, (i) Assuming complete isotropic (E_f = 231 GPa , v_f = 0.2) configuration (ii) assuming transversely isotropic configuration with properties indicated in Table 1. For the later case alone the flow orientation information is coupled. The deflection at the loaded edge is computed in each case and the load – deflection curve obtained for the orthotropic configuration and for the other two simplified cases are plotted in Figure 6 along with the experimental results.

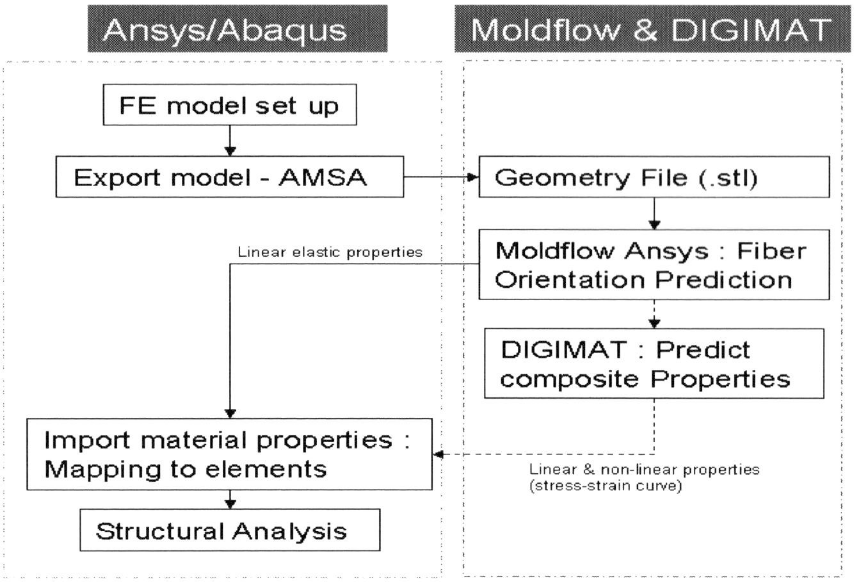

Figure 3. Moldflow-ANSYS Interface.

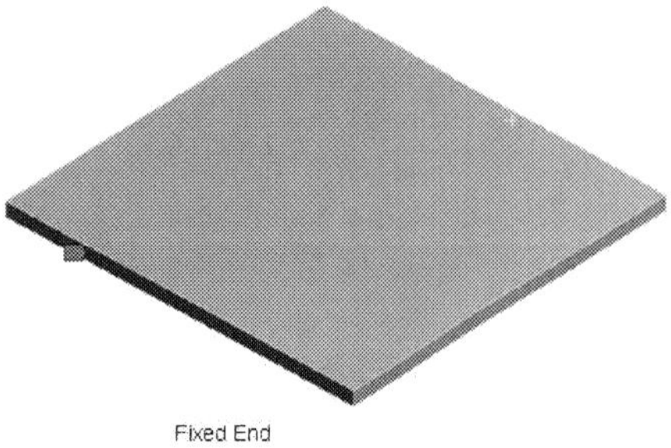

Figure 4. Fixed End.

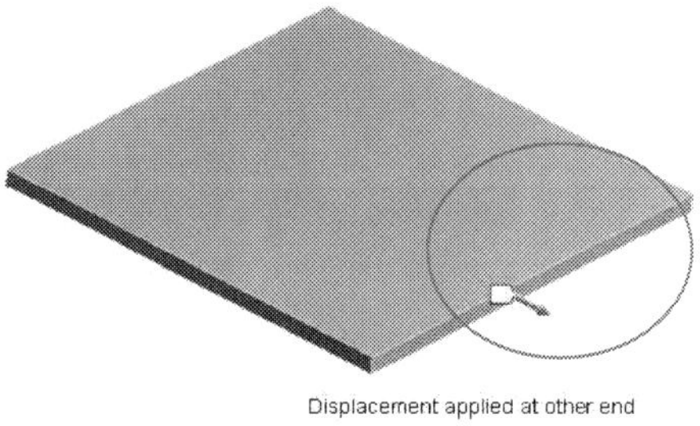

Figure 5. Displacement End.

Table 1: Carbon – Transversely isotropic properties

	Carbon fiber (Thronel 300)
E_{11} (Gpa)	231
E_{22} (Gpa)	13.8
G_{23} (Gpa)	5.52
G_{12} (Gpa)	12.4
ν_{12}	0.2
ν_{23}	0.25
ρ (g/cm^3)	1.75

2. Results and Discussion

It can be seen in Figure 6 that the isotropic results are 70% off as compared to experimental results . This is due to the non realistic assumption that the stiffness is equal in all directions. It also confirms fiber orientation has substantial influence on the stiffness of the short fiber composite, which can be seen for results of the transversely isotropic & orthotropic material model. The transversely isotropic model is far better than the isotropic case, but still the deviation is large. Relatively the orthotropic model exhibits a better correlation with more than 92% accuracy. This is because of the closeness to the actual scenario that the stiffness is different in all 3 directions.

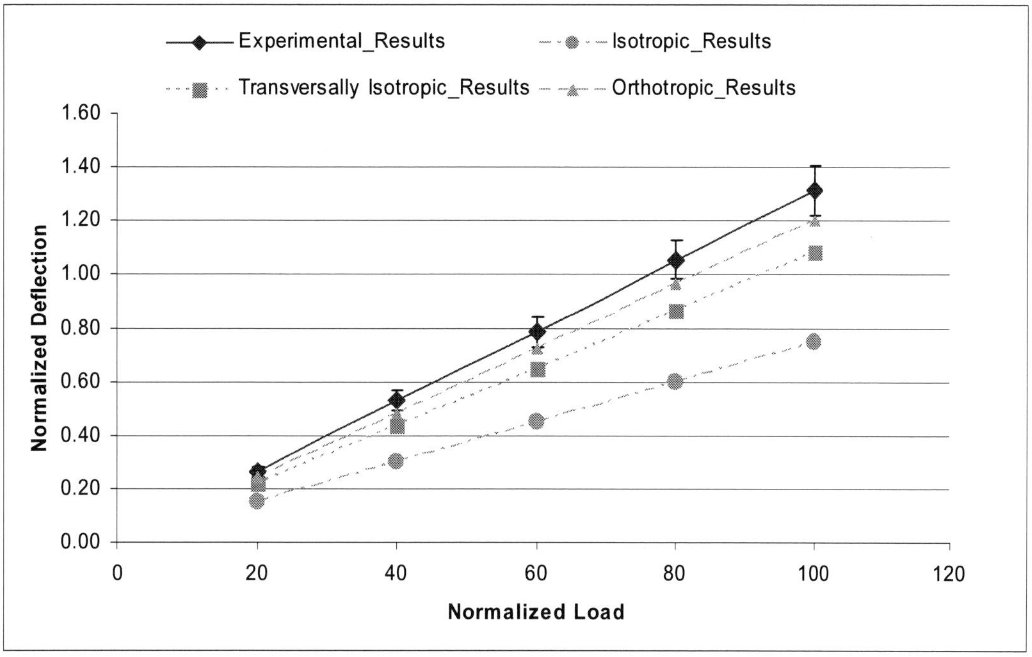

Figure 6. Comparison- Load Vs Deflection.

3. Conclusion

This work demonstrates the influence of the material configuration and fiber orientation over the structural analysis of the composite components. Unlike conventional materials, where simple isotropic models can be employed, for the current case transversely isotropic and orthotropic configurations are to be employed to gain a better agreement with the real time experimental results. The key requirement is that there must be a precise modeling of the experimental setup to predict the results accurately. The current approach involving ANSYS-Moldflow interface yield a good accuracy of more than 92%. This implies that the methodology can be extended for other cases also, thus replacing the need for time consuming and expensive physical testing involved in arriving at the optimum material and design configuration.

References

[1] Rodney Andrews and Eric GNlke, *"MECHANICAL PROPERTIES OF CARBON FIBER COMPOSITES FOR ENVIRONMENTAL APPLICATIONS"*

[2] Tsu-Wei Chou, Seiichi Nomura,*" Fiber orientation effects on the thermoelastic properties of short-fiber composites",* Fiber Science and Technology, Volume 14, Issue 4, June 1981, Pages 279-291

[3] [3] Tandon GP, Weng GJ, *The effect of aspect ratio of inclusions on the elastic properties of unidirectionally aligned composites*. Polym Compos 1984; 5:327–33.

[4] Jones, R. F, "Guide to Short Fiber Reinforced Plastics", Hanser Publishers: Munich, 19
[5] Halpin, J. C. and Kardos, J. L., "*The Halpin-Tsai Equations: A Review",* Polymer Engineering and Science, 16(5), p.344 (1976)
[6] AMI Fiber Orientation Analysis, Autodesk Moldflow Insight 2012
[7] http://k-mac-plastics.net/data%20sheets/victrex_peek_data_sheet_450g.htm
[8] http://www.shtnc.com/news/UploadFile/Picture/201132815164571394.pdf
[9] Hyer, M.W., and Waas, A.M., 2000, "Micromechanics of Linear Elastic Continuous Fiber Composites," Comprehensive Composite Materials, Volume 1, pp. 345 – 375

IOP Conf. Series: Materials Science and Engineering **40** (2012) 012026 doi:10.1088/1757-899X/40/1/012026

Preparation, characterization and FE-simulation of the reinforcement of polycaprolactone with PEGylated silica nanoparticles

N Moussail[1], I Viejo[1], J M Bielsa[1], C Crespo[1], S Irusta[2], C Yagüe[2] and J G Meier[1]

[1] Instituto Tecnológico de Aragón (ITA), María de Luna 8, 50018 Zaragoza (Spain)
[2] Instituto de Nanociencia de Aragón, Mariano Esquillor s/n, University of Zaragoza, 50018 Zaragoza, (Spain)

Email: jmeier@ita.es

Abstract. We recently published the preparation and characterization of polycaprolactone (PCL) nanocomposites with a 45% increased modulus reinforced with only 4 wt% PEGylated silica (polyethylene-glycol/SiO_2) nanoparticles obtained by melt-extrusion [1]. The achieved reinforcement is related to an excellent dispersion of the nanoparticles due to the polyethylene-glycol graft of the nanoparticles which was obtained by a simple one-pot synthesis. X-ray photoelectron spectroscopy (XPS) and infrared spectroscopy (FTIR) analyses identified the location of the PEG at the PCL/silica interface. However, the extension of the interface could not be resolved. In an attempt to describe the effect of the interface on the reinforcement we applied several analytical micromechanical models. Models considering core-shell systems fitted the experimental data well and gave estimations of the modulus and extension of the interphase. However, different sets of parameters gave equally good representations. In an alternative approach, 3D representative volume elements (RVE) of the composite with spherical nanoparticles including the shell were built-up from the morphological data to carry out computational micromechanics based on finite elements (FE). The interphase was modeled in the RVE. Both approaches demonstrated the need of an interphase extension of roughly twice the radius of the particle. The FEM approach estimates the interface-modulus much higher than the analytical models.

1. Introduction

Polycaprolactone (PCL) is generally considered as a green polymer on account of its biocompatibility and biodegradability by microorganisms. Its commercial availability makes it an attractive substitute for commodity non-biodegradable polymers such as in packaging or in the agricultural sector. Additionally, due to its biocompatibility, PCL can play a significant role in medical applications and controlled drug release systems. However, a wider use of PCL is hampered by its low melting temperature (Tm~65°C), low elastic modulus and low abrasion. Additionally, its poor barrier properties to water and gases is a further impediment for its use for instance as biodegradable packaging material. Consequently, there is a significant interest in improving the pure PCL by chemical modifications and by compounding with other polymers and by reinforcement of PCL with various types of fillers.

With respect to mechanical and barrier properties nanoparticles, especially silica nanoparticles and organically modified layered clays are the most promising filler systems. Low additions of these fillers, typically less than 5 wt%, often yield remarkable mechanical reinforcement and improvements of the barrier properties in polymers and already at a competitive price-level. Attaining comparable improvements with conventional fillers would require loadings of 30-50 wt%. Thus, compounding with silica nanoparticles is a viable way of improving the properties of PCL.

The extraordinary effects are related to the small size of the particles and the resulting high number density of particles and the ultra-large interfacial area density between the nanoinclusion and the matrix. Already at small filler volume fractions the interparticle distance in the composite reaches the nanometre scale assuming ideal dispersion. Additionally, the polymer in contact with the filler surface has different dynamic properties or different crystallization behaviour than the bulk polymer leading to substantial changes of the matrix morphology. This interfacial layer can extent up to several radii of gyration of the polymer coil which is in the order of several nanometres and at sufficient filler concentration also in the order of the interparticle distance. Consequently, the properties of polymer nanocomposites are dominated by their interface properties.

We recently reported the preparation of PCL nanocomposites reinforced with surface modified silica-nanoparticles [1]. The surface modifier of the silica particles was poly-ethylene-glycol (PEG) – a biodegradable hydrophilic polymer that is compatible with PCL. The PEGylated silica nanoparticles were prepared in a facile, one-pot synthesis. The nanocomposites were prepared by simple melt-extrusion, i.e. adding the dry surface-modified nanoparticle powder to the melt of PCL; systematically varying the filler concentrations in the range from 1-4 wt%. In the tested range the modulus increased systematically with increasing filler concentration reaching its highest value of 45% at the highest tested concentration of 4 wt% PEGylated silica. As a reference we prepared by the same processing method a composite of PCL with 3 wt% of non-modified silica particles of same size. The measured modulus only increased marginally. The achieved reinforcement is related to an excellent dispersion of the nanoparticles due to the polyethylene-glycol graft of the nanoparticles confirmed by TEM and SEM studies. X-ray photoelectron spectroscopy (XPS) and infrared spectroscopy (FTIR) analyses identified the location of the PEG at the PCL/silica interface. However, the extension of the interface could not be resolved.

In an attempt to describe the effect of the interface on the reinforcement we applied various published analytical micromechanical models considering spherical inclusions covered with an interphase layer embedded in a polymer matrix. The models are all based on effective medium theories following either the Mori-Tanaka [2-4] approach or the generalised self consistent scheme by Christensen and Lo [5, 6]. Fits of the modulus increment vs. filler volume fraction yield in all cases apparently well representations of the experimental data. However, different sets of parameters gave equally good representations for each of the models and we found significant variations for the parameters of the interphase between the two classes' homogenization methods.

In an alternative approach, 3D representative volume elements (RVE) of the spherical nanoparticles including the shell were built-up from the morphological data to carry out computational micromechanics based on finite elements (FE). In order to obtain an optimised mesh and an accurate prediction, a convergence study of RVEs and a sensitive analysis of boundary conditions and RVE-size were carried out. The effective properties obtained with the FE approach were compared with the mechanical properties of the composites.

2. Experimental

PEGylated silica particles were synthesized according the procedure described by Xu et al.[29]. In brief, 2 g of poly(ethylene glycol) with a molecular weight of 3000 g/mol was dissolved in 6 ml ammonium hydroxide solution (30 wt% NH_3) and 24 ml methanol in a first step. 0.2 ml of tetramethyl orthosilicate (TMOS) was added dropwise and after stirring for 4 h, the samples were filtered and washed with ethanol. The pellet was re-dispersed in distilled water and freeze-dried using lyophilization to stabilize it for storage and later use. All chemicals were purchased from Sigma-Aldrich and used as-received.

Polycaprolactone (PCL) with a molecular weight around 120000 g/mol was used to prepare nanocomposites with, 2, 3 and 4 wt. % of filler (dry PEGylated SiO_2 nanoparticles) by loading into a twin-screw mini-extruder (DSM-Xplore 15 Micro compounder, Model 2005). An intimate mixture was achieved at 100 °C for 10 min, with a rotation speed of 100 rpm and under nitrogen atmosphere. In the remainder of this work, samples are named PCL/x SiO_2, where x is the wt. % of PEGylated particles added to the polymeric matrix. For comparison, a PCL composite filled with 3 wt. % of non-modified silica (named PCL/SiO_2) has also been prepared using the same experimental conditions except for the presence of PEG.

BET surface areas and pore-size distributions of the silica nanoparticles, based on N_2-adsorption/desorption isotherms, were obtained with a Micromeritics ASAP 2020 V1 device at 77 K. For the determination of the particle-size distribution the nanoparticles were dispersed in water and subjected to photon correlation spectroscopy (PCS) using a Malvern Zetasizer 3000 HS.

The morphology and size distribution of PEGylated nanoparticles were examined by scanning electron microscopy, SEM (JEOL JSM 6460 LV and Hitachi S2300). Statistical size-distribution histograms for the resulting nanoparticles were produced from SEM images using Image J software (sample size = 30). Assessment of the filler dispersion in the composites with transmission electron microscopy (TEM) was done on ultra-thin sections of the nanocomposites with thicknesses of approximately 60–80 nm. Specimens were cut under cryogenic conditions with a Leica Ultramicrotome equipped with a diamond knife. The TEM was a JEOL 2000 FXII using an acceleration voltage of 200 kV. No staining to enhance contrast was needed due to the high electron density difference between the silica and the polymer.

Dynamic mechanical properties of the base PCL and of the PCL nanocomposites with different loadings were measured with a DMA+450 Metravib, using a bar specimen of 25 mm length, 7 mm width and 2 mm thickness. Excitation mode was in uniaxial tension/compression at an oscillation frequency of 1 Hz at temperatures ranging from 30 to 60 °C and with a heating rate of 2 °C /min.

3. Results and Discussion

3.1. Particle characterisation and nanocomposite morphology

An aqueous suspension of the PEGylated silica nanoparticles at a neutral pH showed a particle-size distribution centred at 147.4 nm. On the other hand, in aqueous suspension the non-modified SiO_2 particles could not be measured by PCS indicating the existence of agglomerates of a size larger than 2 μm – the limit of the PCS equipment. SEM images of the synthesized PEGylated particles show uniform spherical shaped particles with a size distribution centred at 140 nm, coincident with the PCS measurements (c.f. Figure 1 in [5, 6]). It indicates that the particles remain dispersed in water without agglomeration. The PEGylated nanoparticles showed a modest BET surface area of 136.3 ± 1.5 m^2/g and a type II isotherm characteristic of non-porous materials with a small hysteresis between the adsorption and desorption branches, indicating a limited mesoporosity. The sample did not show any microporosity according to the t-plot results [5, 6].

Transmission electron microscopy (TEM) was used to examine the morphology of the nanocomposites based on PCL and PEGylated / non modified (non PEGylated) silica nanoparticles. Figure 1 shows TEM micrographs of PCL containing 3 wt. % of PEGylated silica and non modified silica, respectively. The silica particles appear as dark domains. A nanoscale dispersion of individual PEGylated silica particles, with uniform size, is observed (Figure 1a), while the non-modified silica nanoparticles appear highly agglomerated in the matrix (Figure 1b). These observations point to the fact that the PEGylated silica particles are easily dispersed due to the PEG-graft introduced during the particle synthesis. Additionally, we may postulate that the PEG-graft acts as an interfacial compatibilizer, capable of reducing the interfacial tension between the silica particles and the PCL preventing particle re-agglomeration and giving rise to a homogeneous distribution of the nanofillers.

In contrast, the non-modified silica particles show a high number of large and dense particle clusters in PCL. The observation that in an aqueous dispersion of the non-modified silica particles a particle size and distribution could not be determined by PCS, although water should be able to wet the silica particle surface and due to its high mutual affinity should result in a complete break-up of agglomerates, point towards the formation of stable, partially fused aggregates during synthesis. Therefore, the partially fused nature of the particle aggregates prevents effective filler dispersion in the case of non-modified silica particles.

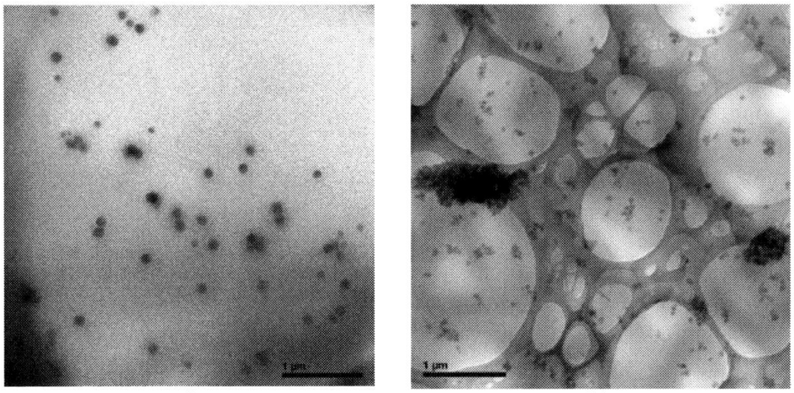

a) b)

Figure 1. TEM micrographs of PCL with 3 wt% of a) PEGylated silica nanoparticles and b) without coating.

3.2. Dynamic Mechanical Analysis (DMA) of the composites and micromechanical modelling

In Figure 2a, the real part of the extensional dynamic modulus varying with temperature for the different nanocomposites is presented. The PCL-composite containing 3 wt. % of non-modified silica behaves almost the same as the neat PCL. The PEGylated silica nanocomposites exhibit a clear increase of E′ with increasing weight percentage of the fillers. Figure 2b shows a representation of E' as function of the filler concentration at different temperatures together with the data for the elongation at break of the composites at T=32°C. The storage modulus (E′) of the nanocomposite containing 4 wt. % of PEGylated silica are approximately 45 to 50 % higher compared to the PCL matrix at a given temperature (32°C-47°C). On the other hand the elongations at break (measured at 32°C) of all of these nanocomposites are similar to that of the neat PCL and remain independent of the silica loading in the studied filler concentration range. PCL nanocomposites with PEGylated SiO_2 show a high stiffness at least up to a silica loading of 4 wt. % while maintaining a good ductility in the studied loading range.

Several material models relate the modulus of the composite to the moduli of its constituents and the morphology. They are typically valid for non-aggregated isolated hard particles suspended in a soft matrix in the limit of small filler volume fractions. The microscopic analysis of the SiO_2-PEG composites shows that these composites fulfill the conditions very well.

Probably the most famous one is Einstein's theory of viscosity. It gives the first analytical account on the reinforcement of a soft medium with hard spherical particles (Eq. (1)). Smallwood [1] adapted Einstein's formula to elastic materials in modulus-representation. They read as follows:

$$\eta^* = \eta_o\left(1+5/2\phi\right); \quad E = E_m\left(1+5/2\phi\right) \tag{1}$$

predicting that the enhancement of viscosity or modulus is independent of the size of the filler particles and is linearly related to loading in the limit of low loading, where η^*, E and η_0, E_m are the viscosities / moduli of the suspension and the pure matrix, respectively, and ϕ is the volume fraction of the "hard" inclusion. Guth and Gold [1] developed an extension to somewhat higher filler volume fractions taking into consideration the arising interactions between filler particles, obtaining Eq. (2). In this rigid filler limit, the elastic energy is stored in the distorted strain field around the particles.

$$E = E_m(1 + 2.5\phi_f + 14.1\phi_f^{\,2}) \tag{2}$$

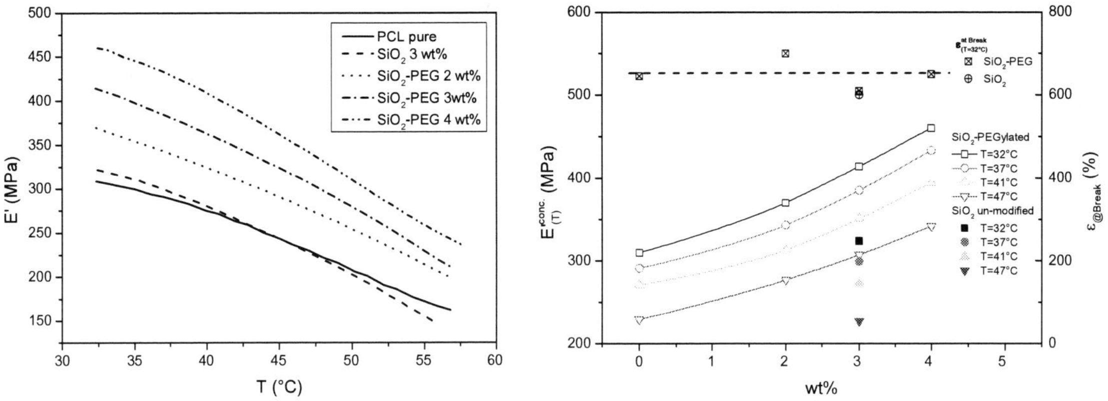

a b

Figure 2. a) Real part of the tensile modulus of the pure polymer matrix and the nanocomposites as function of temperature; b) Representation of the data as function of nanoparticle concentration together with the measured elongation at break of the composites.

Figure 3 shows the representation of the modulus increment varying with filler volume fraction. The moduli – measured at different temperatures – coincide, demonstrating that the modulus increment is independent of the temperature in the measured range. From the TEMs of the SiO_2-PEG composites the conditions of the validity of the Einstein-Smallwood equation – non-aggregated isolated spheres, low filler volume fraction – seems to be well fulfilled. However, in the graph of Figure 3 *only* the SiO_2-composite fits Eq.(1) or (2). The composites with the surface-modified filler show clearly higher reinforcement as predicted by Einstein-Smallwood or Guth-Gold.

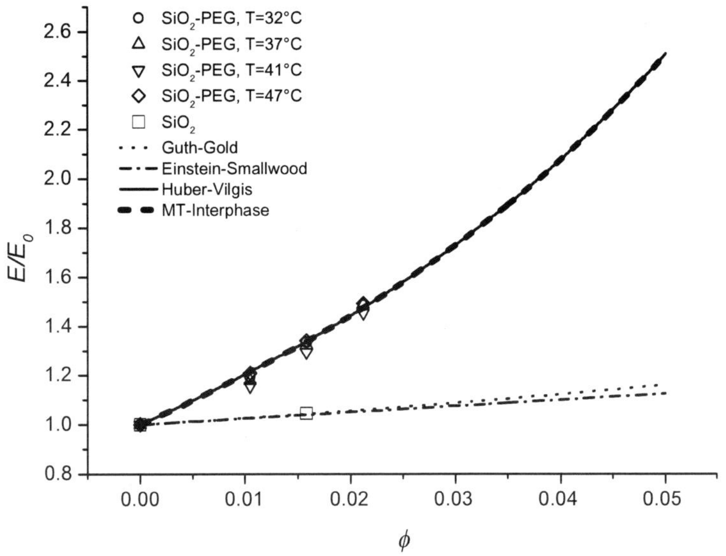

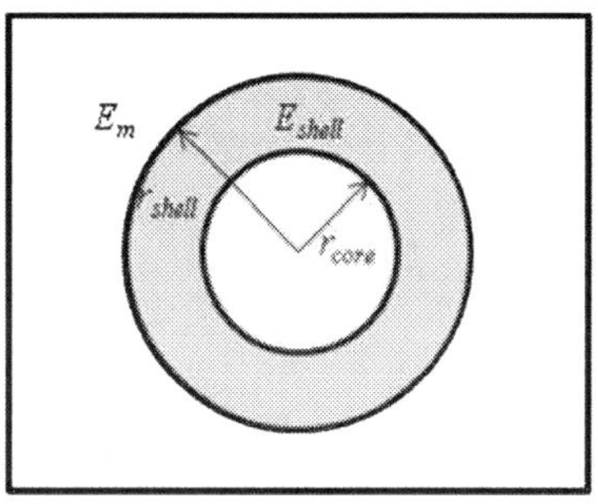

Figure 3. Modulus increment of the nano-composites in dependence of the filler volume fraction. The lines are fits according to the different analytical models

Figure 4. Scheme of the definitions of the core, shell and matrix radii and moduli

This indicates the possible existence of a third phase known as 'interphase' located between matrix and inclusions. This hypothesis of an interphase considered as a third phase around the particles is reasonable for the nanocomposites with functionalised nanoparticles, such as this case. The interphase phenomenon is complex and is affected by a great number of parameters as nanoinclusion type, nanoinclusion-polymer interactions, polymer chemistry, etc. Generally, the polymer layer around the inclusion has different dynamic properties than the bulk-polymer and hence, different mechanical properties. In case of attractive interactions between filler and particle the interphase dynamics slow down, giving rise to an increase of the glass transition temperature and hence resulting in a stiffening of the interfacial polymer layer compared to the bulk polymer [7]. However, the direct experimental measurement of the interphase' mechanical properties and extension is extremely difficult.

All of the above discussed models only consider two-phase materials, which are composed by a matrix and a reinforced-phase. In nano-reinforced materials a third-phase with different properties – the interphase – has a relevant influence and should be considered. The interphase-problem can be addressed by introducing a further generalization of the Smallwood-Einstein theory, which is obtained by relaxing the "hard sphere" condition to fillers that have elasticity. The general theory for such systems has been derived by Felderhof and Iske [8, 9] giving a general result for the effective modulus:

$$E/E_m = 1 + \left(\mu\phi / 1 - 0.4\mu\phi \right) \tag{3}$$

where μ denotes the intrinsic modulus and ϕ is the filler volume fraction. The theory is based on a mean field approximation analogous to the Lorentz local field method in the theory of dielectrics, where it leads to the famous Clausius-Mosotti equation for the effective dielectric constant. The result clearly goes beyond the limits of Einstein- Smallwood, since two-body interaction (excluded volume) is included, leading to the strong increase of the modulus at high volume fraction. The only variable to determine is the intrinsic modulus which depends on the elasticity of the spherical inclusion. For rigid and spherical filler particles at low volume fraction, the Einstein-Smallwood formula is recovered because of $\mu = 5/2$ (the intrinsic modulus μ follows from the solution of a single-particle problem). To model an interphase one needs to consider that the interphase has an additional reinforcing effect of the matrix but with a finite modulus, effectively changing the volume fraction and effective modulus of the hard inclusions(c.f. Figure 4). Consequently, we can try to extract some values from models that consider such an interphase.

For the special cases of uniform soft sphere [10-13], configurations of soft-core – hard-shell[14] and hard-core – soft-shell[15] analytical solutions have been obtained. We reproduce here the algebraic expression for the intrinsic modulus of the hard-core – soft-shell model as published by Huber and Vilgis in [6],

$$\mu(\tilde{\mu},\tilde{r}) = \frac{5}{2} - 25\tilde{\mu}\frac{\left\{(1-\tilde{r}^3)\left(8\tilde{\mu}+\frac{19}{2}\right)+(\tilde{\mu}-1)\left(-42\tilde{r}^3+84\tilde{r}^5-50\tilde{r}^7+8\tilde{r}^{10}\right)\right\}}{\left\{(3\tilde{\mu}+2)(16\tilde{\mu}+19)+(\tilde{\mu}-1)\left[-300\left(\tilde{\mu}+\frac{3}{4}\right)\tilde{r}^3+168(3\tilde{\mu}+2)\tilde{r}^5-100(3\tilde{\mu}+2)\tilde{r}^7+48(\tilde{\mu}-1)\tilde{r}^{10}\right]\right\}} \tag{4}$$

with $\tilde{\mu} = E_m/E_{shell}$ and $\tilde{r} = r_{shell}/r_{core}$ and the definitions of the particle and matrices as depicted in Figure 4

Employing the generalization of Einstein-Smallwood, Eq. (3), an excellent fit can be obtained yielding an intrinsic modulus of $\mu \approx 18.8$. When using Eq. (4) for μ we obtain various combinations of $\tilde{r}$ and $\tilde{\mu}$ depending on the start values and the way of adjustment, i.e. fixing first some reasonable but arbitrary $\tilde{\mu} \approx 0.1$ and adjusting first $\tilde{r}$ followed by a second run fitting both $\tilde{\mu}$ and $\tilde{r}$ gives some reasonable values of $\tilde{\mu}_{Fit} = 0.099$ and $\tilde{r}_{Fit} = 4.62$. Thus, while Eq. (3) gives an apparently reasonable fit of the experimental data, the estimations of the ratios of matrix to shell modulus and particle shell to core radius according to Eq. (4) is quite unstable and no concise information on the interphase can be obtained.

A forth model is based on the Mori-Tanaka homogenization method to first calculate an effective shear modulus of the particle surrounded by its shell [2-4]. In the limit of an isotropic, incompressible material with spherical inclusions the formula to calculate the shear modulus of the composite reads:

$$\mu = \mu_m + \frac{c}{\frac{1}{\mu_e - \mu_m} + \frac{2}{5}(1-c)}, \text{ with } c = \phi \frac{r_{shell}^3}{r_{particle}^3} \quad (5)$$

where, μ_e is the effective shear modulus of the particle with its surrounding shell and μ_m is the shear modulus of the matrix, ϕ is the volume fraction of the particles relative to all phases. Introducing as fixed values $\mu_m = 1.033 \times 10^8$ Pa and $r_{Particle} = 70$ nm, the fit yields $\mu_e = 9.769 \times 10^8$ Pa and $r_{shell} = 149$ nm; i.e. corresponding to a $\tilde{\mu} = 0.106$ and $\tilde{r} = 2.129$. We note that fitting with the Mori-Tanaka Interphase model appears to be numerical more stable.

3.3. FEM Modelling

Another way of obtaining more detailed information on the local nanosctucture and the interphase is offered by finite element modeling (FEM). In contrast to the most available micromechanical models, that make rigorous assumptions, restricting its utility to certain heterogeneous composite structures, FEM is very flexible provided that moduli of the constituent phases and the geometries can be known or estimated with sufficient accuracy. To assess the average properties of the composite, however, the limits of continuum mechanics have to be reached. That means that the number of individual inclusions and the extension of the "simulation box" has to be big enough and sufficiently densely meshed to obtain a statistical representative ensemble and a trade-off versus computing time has to be found. Thus, the general procedure for modeling such complex heterogeneous composites relies on generating representative volume elements (RVE) and an appropriate mesh generation.

We used the combination of the commercial software: Digimat and Abaqus. The 3D-geometry of the composite is built in Digimat using the morphological characterization. Subsequently, the FE model is defined in Abaqus code with the mechanical properties of the constituents, including the boundary conditions that are representative of the desired macroscopic strain state. The macro behaviour in terms of stresses and strains is obtained by a volume averages procedure. This average is executed over the local stress and strain fields of the complete RVE. In this study, the material model was considered analogue to the analytical models as linear elastic, incompressible and isotropic for both, the matrix and the particles. The bonding between constituent phases was supposed perfect. The inclusions were considered as spheroids, which mean ellipsoids of aspect ratio equal to the unity. Representative volume elements (RVE) of the nanocomposite models were designed for each fraction by weight (2%, 3% and 4%).

Prior to any further analysis we tested our model with respect to size of the "simulation box", influence of the mesh typology and boundary condition. In short: The analyses were carried out on the composite with 3 wt% SiO_2-PEG. RVEs of different sizes, ranging from 1.95 μm^3 to 19.68 μm^3, which corresponds to 32 to 270 inclusions, were submitted to three different boundary conditions (BC): Dirichlet-BC, Mixed-BC and Periodic-BC. The prediction of the elastic properties for each model type and size was obtained showing that Mixed- and Periodic-BC gave similar predictions for all modelled RVE-sizes, so the convergence was achieved even for the smallest analysed RVE. Nevertheless, the chosen parameters for further simulations were the Periodic-BC and a volume size of 3.375 μm^3 of the RVE having about 60 inclusions and being slightly larger than the smallest RVE tested.

So far we only considered a two-phase system; that is the polymer matrix and the silica nanoparticles. In the next step we constructed 3D-models including the interphase surrounding each silica particle for the different filler volume fractions. Each of the RVE had about sixty inclusions, varying in function of the filler volume fraction in size from 4.913 μm^3, 3.375 μm^3 and 2.197 μm^3. We took the estimated values from the analytical models for the interphase modulus and extension and fed them into the 3D-FEM model. In Figure 5, the predictions obtained with FEA are compared with the analytical MT-Interphase model values and the experimental results. The predictions with FEA and the analytical model differ less than 2% from the experimental values for all concentrations and the numerical values of the obtained interphase modulus and thickness for the three different models is summarized in Table 1.

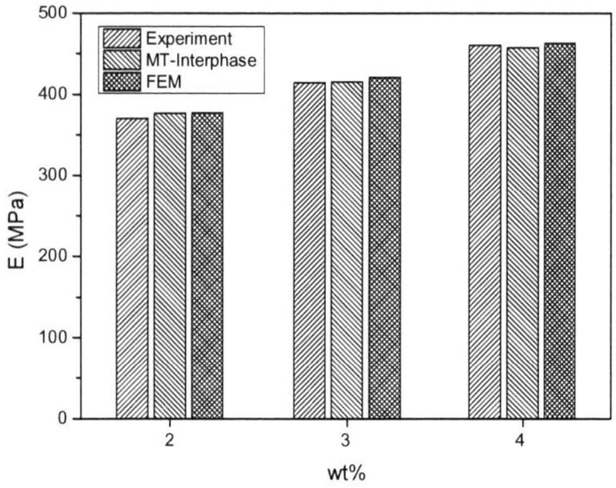

Figure 5. Comparison of the module obtained by MT-Interphase FEM-model

Table 1. E-modulus and thickness of the interphase obtained from the different model approaches.

Model	E-modul interphase	Thickness interphase
FEM	42.342×10^9 Pa	68 nm
Huber-Vilgis	3.131×10^9 Pa	253 nm
MT-Interphase	2.931×10^9 Pa	79 nm

Although the predictions obtained by the finite element simulations and the analytical models are very similar, the FEA methodology admits detailed analysis in terms of local shear stresses near the nanoparticles or the effects of strain shielding in multi-inclusion cases. Additionally, the FEA studied in this paper can be adapted to consider other mechanical phenomena like plasticity, debonding, crack propagation and damage accumulation or the prediction of electrical and thermal properties.

4. Conclusions

PCL/PEG/SiO_2 nanocomposite materials were successfully prepared by melt-extrusion, using PEGylated silica nanoparticles prepared in a facile, one-pot synthesis. A homogeneous distribution of the individual silica particles was obtained when the PEGylated silica was used, while the same silica nanoparticles aggregated severely when they were not functionalized. From the dynamic mechanical measurements, a strong increase of the elastic moduli of the nanocomposites with increasing PEGylated silica content was observed while maintaining the elongation at break of these nanocomposites at room temperature remained similar to that of the neat PCL over all the loading range.

Different micromechanical models were tested to represent the experimental data of which an Interphase model based on the Mori-Tanaka homogenization method performed best and yield reasonable modulus values and extensions of the interphase. Furthermore, a FEA methodology for the development of RVE and the prediction of mechanical properties of nano-reinforced material has been proposed. The RVE were built from the material morphology (reinforcement/inclusion geometry, volume fraction, constituent mechanical properties, etc) obtained by means of SEM, TEM and experimental test. The boundary conditions and minimum RVE size was chosen by means of an influence study, using in the final models Periodic BCs and 60 inclusions. The prediction with FEA including the interphase differed less than 2% than experimental characterization for all concentrations.

In spite of fact that the prediction obtained by means of the FEA and the analytical models are analogous, the FEA methodology admits detailed analysis in terms of local values. On the other hand, the present methodology was applied in the prediction of lineal elastic properties but it is extendable to analyse no-linear effects as plasticity, crack, debonding, damage, crack propagation, etc.; obtaining the macro stress-strain behaviour.

5. Acknowledgements

The authors are grateful to the "Ministerio de Ciencia & Innovación" (XDEX-60900-2008-17) and the European Regional Development Fund (ERDF) for the financial support for this research. One of the authors, N.M. is indebted to the "Fundación Agencia Aragonesa para la Investigación y el Desarrollo (ARAID)" for supporting this research.

References

[1] Moussaif N, Irusta S, Yagüe C, Arruebo M, Meier J G, Crespo C, Jimenez M A and Santamaría J 2010 Mechanically reinforced biodegradable nanocomposites. A facile synthesis based on PEGylated silica nanoparticles *Polymer* **51** 6132-9

[2] Mori T and Tanaka K 1973 Averge stress in matrix and averge elastic energy of materials with misfitting inclusions *Acta Metallurgica* **21** 571-4

[3] Lombardo N and Ding Y 2005 Effect of inhomogeneous interphase on the bulk modulus of a composite containing spherical inclusions. . In: *Proceedings of the 4th Austalasian Congress on Applied Mechanics,* pp 523–9

[4] Lombardo N 2007 Properties of Composites Containing Spherical Inclusions Surrounded by an Inhomogeneous Interphase Region. In: *School of Mathematical and Geospatial Sciences,* (Melbourne: RMIT University) p 204

[5] Christensen R M and Lo K H 1979 Solutions for the effective shear properties in 3 phase sphere and cylinder models *Journal of the Mechanics and Physics of Solids* **27** 315-30

[6] Huber G and Vilgis T A 2002 On the Mechanism of Hydrodynamic Reinforcement in Elastic Composites *Macromolecules* **35** 9204-10

[7] Smallwood H M 1944 Limiting law of the reinforcement of rubber *J. Appl. Phys.* **15** 758-66

[8] Guth E and Gold O 1938 On the Hydrodynamical Theory of the Viscosity of Susyensions. *Physical Review* **53** 322

[9] Guth E 1945 Theory of Filler Reinforcement *J. Appl. Phys.* **16** 20-5

[10] Tsagaropoulos G and Eisenberg A 1995 Dynamic-Mechanical Study of the Factors Affecting the 2 Glass-Transition Behavior of Filled Polymers - Similarities and Differences with Random Ionomers *Macromolecules* **28** 6067-77

[11] Tsagaropoulos G and Eisenberg A 1995 Direct Observation of 2 Glass Transitions in Silica-Filled Polymers - Implications for the Morphology of Random Ionomers *Macromolecules* **28** 396-8

[12] Grohens Y, Hamon L, Reiter G, Soldera A and Holl Y 2002 Some relevant parameters affecting the glass transition of supported ultra-thin polymer films *Eur. Phys. J. E* **8** 217-24

[13] Fragiadakis D, Pissis P and Bokobza L 2005 Glass transition and molecular dynamics in poly (dimethylsiloxane)/silica nanocomposites *Polymer* **46** 6001-8

[14] Felderhof B U and Iske P L 1992 Mean-field approximation to the effective elastic moduli of a solid suspension of spheres *Phys. Rev. A* **45** 611-7

[15] Jones R B and Schmitz R 1983 Isotropic elastic medium containing a spherical particle: I. Incompressible media *Physica A: Statistical Mechanics and its Applications* **122** 105-13

Primitive path network, structure and dynamics of SWCNT / polymer nanocomposites

A. Karatrantos[1], R.J. Composto[2], K.I Winey[2] and N.Clarke[1]

[1]Department of Physics and Astronomy, University of Sheffield, Sheffield S3 7RH, UK

[2]Department of Materials Science and Engineering, University of Pennsylvania, Philadelphia, Pennsylvania, 19104, USA

E-mail: a.karatrantos@sheffield.ac.uk, n.clarke@sheffield.ac.uk

Abstract. We investigate the entanglements, structure and dynamics of monodisperse polymer melts in the presence of a single wall carbon nanotube (SWCNT) in comparison to inclusion-free polymer melts by molecular dynamics simulations. The SWCNT has an infinite aspect ratio and radius smaller than the polymer radius of gyration. In the presence of SWCNT with or without attractive interactions, the contour length of the primitive path increases indicating more entanglements. We also find that the overall configuration, as characterized by the radius of gyration, is not perturbed by either the interaction energy between the polymer and the SWCNT or by variations in the SWCNT radius. We also find that there is a large heterogeneity in the polymer dynamics of the polymer melts with a SWCNT due to the polymers in contact with the SWCNT.

1. Introduction

Polymer nanocomposites, that is nanoparticles (spheres, cylinders, plates) dispersed in an entangled matrix, have attracted substantial academic and industrial interest [1, 2]. Whereas polymer-nanoclay composites appear to be reaching maturity, interest in systems with cylindrical nanoparticles, such as single wall carbon nanotubes (SWCNT) has been rapidly accelerated since the widespread availability of carbon nanotubes in the late 1990s. Understanding the polymer structure and dynamics of such systems in the presence of nanoparticles is very important to design materials with improved mechanical, rheological, tribological properties. The dynamics of long polymers is limited by "entanglements" which are topological constraints imposed by the other chains. These can dramatically change the polymer viscosity.

In this paper we study a dilute SWCNT nanocomposite system when the polymer radius of gyration is larger than the radius of the nanotube [3]. We focus on simple but unresolved questions in the area of polymer/SWCNT nanocomposites: (i) How the primitive path (shortest path connecting the two ends of the polymer chain subject to the entanglements) is affected by the SWCNT? (ii) What is the effect of SWCNT on polymer chain conformation? (iii) What is the effect of SWCNT on polymer diffusion?

2. Simulation method

We performed molecular dynamics simulations of monodisperse polymer melts and SWCNT nanocomposites [4]. The polymer melt is simulated by using a coarse grained polymer model [5].

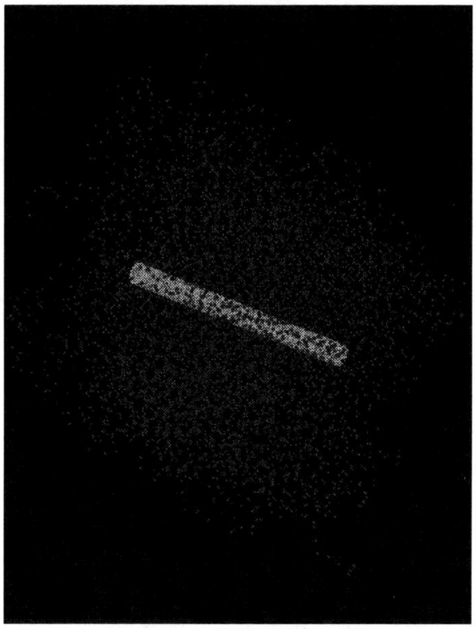

Figure 1. Model of the nanocomposite system (SWCNT in polymer melt). For clarity, the connectivity of the beads is not shown.

Polymers are composed of bead–spring chains of Lennard–Jones (LJ) monomers m of diameter $\sigma_m = 1$ and mass $m_m = 1$. The melt density $\rho^* = (N_m/V)\sigma_m^3 = 0.85$ (where N_m is the total number of monomers in the system and V is the volume of the polymer melt).The radius of a SWCNT is of the order of the Kuhn length of polystyrene (PS). Consequently in these simulations an atomistic model was used for the SWCNT. Three different SWCNTs ((12,0), (17,0), (22,0) of radius $r_{SWCNT} = 0.46, 0.66, 0.85\ \sigma_m$ respectively) are considered and span the simulation cell with its atoms held fixed in a centered position in the simulation cell, and is oriented along the z axis. A molecular model of the nanocomposite is shown in Fig. 1.

Non-bonded monomer-monomer as well as monomer (m)-carbon (c) interact via a truncated and shifted Lennard–Jones (LJ) potential with a cut-off radius $r_c = 2.3\sigma_m$, given by:

$$U_{LJ}(r_{ij}) = \begin{cases} 4\epsilon \left[\left(\frac{\sigma_{ij}}{r_{ij}}\right)^{12} - \left(\frac{\sigma_{ij}}{r_{ij}}\right)^{6} - C(r) \right] & \text{for } r < r_c \\ 0 & \text{for } r_c \geq r \end{cases} \tag{1}$$

where $C(r)$ [6] is a 3rd degree polynomial so that the potential vanishes continuously at $r = r_c$, ϵ is a characteristic interaction energy between particles and r_{ij} is the distance between particles i and j. The interaction energy between monomers is $\epsilon_{mm} = k_B T = 1$ (where k_B is Boltzmann's constant and T is temperature). The interaction energy between monomer and carbon, $\epsilon_{mc} = k_B T$. For the simulations where the polymers have no attractive interaction with the nanotube surface, only the repulsive term of Eq. 1 is retained and in order to have a very slight repulsion, $\epsilon_{mc} = 5{\times}10^{-4} k_B T$. More simulation details can be found in [4].

3. Primitive path network

Understanding if the SWCNT influences the entanglement length, N_e, is important because N_e relates changes in chain structure to rheological properties [7]. The entanglement molecular length N_e can be calculated by the estimator of Everaers *et al.* [7] which is determined by statistical properties of the primitive path as a whole ('coil') and evaluated for a given N as follows:

$$\mathcal{N}_e(N) = (N-1)\frac{\langle R_{ee}^2\rangle}{\langle L_{pp}\rangle^2} \tag{2}$$

where R_{ee} is the end-to-end vector distance of a polymer and L_{pp} is the contour length of its primitive path. It is known that using the estimator Eq. (2) following [7], one underestimates $N_e \equiv \lim_{N\to\infty}\mathcal{N}_e(N)$, that is $\mathcal{N}_e(N)$ monotonically increases with N. To obtain a more accurate N–independent value, we use an ideal N_e estimator ("M-coil") [8]:

$$\left(\frac{C(x)}{x}\right)_{x=\mathcal{N}_e(N)} = \frac{d}{dN}\left(\frac{\langle L_{pp}\rangle^2}{R_{RW}^2(N)}\right) \tag{3}$$

where $C(x) \equiv \langle R_{ee}^2\rangle / R_{RW}^2(x)$ is the characteristic ratio [9] for a chain with x monomers, and $R_{RW}^2(x) = (x-1)r_0^2$ is the reference square end–to–end distance of a random walk. The derivative of Eq. 3 means that the $\langle L_{pp}\rangle$ needs to be measured (using the geometrical Z1 algorithm [10] in this study) as function of N. Figure 2 shows the "M-coil" estimator for all the polymer melts and SWCNT nanocomposites. For the polymer melt used in this study $N_e \approx 98$ from "M-coil". Figure 2 shows that N_e decreases in the polymer melt systems with a SWCNT compared to its melt value. The decrease of N_e in the polymer melt with SWCNT is due to the larger contour length of the primitive path, L_{pp}, that is created by the presence of the carbon nanotube. In addition, when there is an attractive interaction between the polymers and the carbon nanotube surface, the N_e decreases even further ($N_e = 58.8$) due to the higher monomer density in the vicinity of the carbon nanotube.

4. Dimensions of polymers in contact with the SWCNT

A static property that gives information about the conformation of the polymers is the radius of gyration $< R_g >$ [9], the average squared distance between monomers in given conformation and the polymer's center of mass. In Fig. 3 we show the ratio of the radii of gyration of polymers which are in contact with the SWCNT to the value in the corresponding pure melt at different interaction strengths between the polymers and SWCNT. We observe that R_g of the contact polymers retain the melt values for both unentangled and entangled polymers. Similarly, in the inset of Fig. 3 we show the same ratio as a function of SWCNT radius. Remarkably, the inset of Fig. 3 seems in agreement with Fig. 3, where the R_g of the contact polymers retain (within 6% deviation) its melt value for both unentangled and entangled polymers at any radius of SWCNT smaller than the R_g. We do not observe any stretching of the polymers that are in contact with the nanotube because the polymer radius of gyration is larger than r_{SWCNT}, in contrary to the previous studies [11, 12] where the nanoparticle radius of curvature is large compared to the R_g, and has an influence on the structure of the polymers in the vicinity of the nanoparticle.

Thus, we can conclude that in polymer/SWCNT nanocomposites with radius of SWCNT smaller than the R_g of polymers, the polymers are unperturbed by the presence of the SWCNT for an isolated SWCNT. These results are in agreement with a recently studied experimental system of PS with spherical well dispersed [13] or aggregated nanoparticles [14].

5. Polymer dynamics

In this section, we will discuss the dynamic behaviour of polymers in the presence of a SWCNT for different monomer-SWCNT interactions and in the vicinity of SWCNT. The dynamic properties of polymers are estimated through the self diffusion coefficient D_0 which is calculated through the mean square displacement [15]:

$$D_0 = \frac{1}{6}\lim_{t\to\infty}\frac{d}{dt}\left\langle |\mathbf{r}_i(t) - \mathbf{r}_i(0)|^2\right\rangle \tag{4}$$

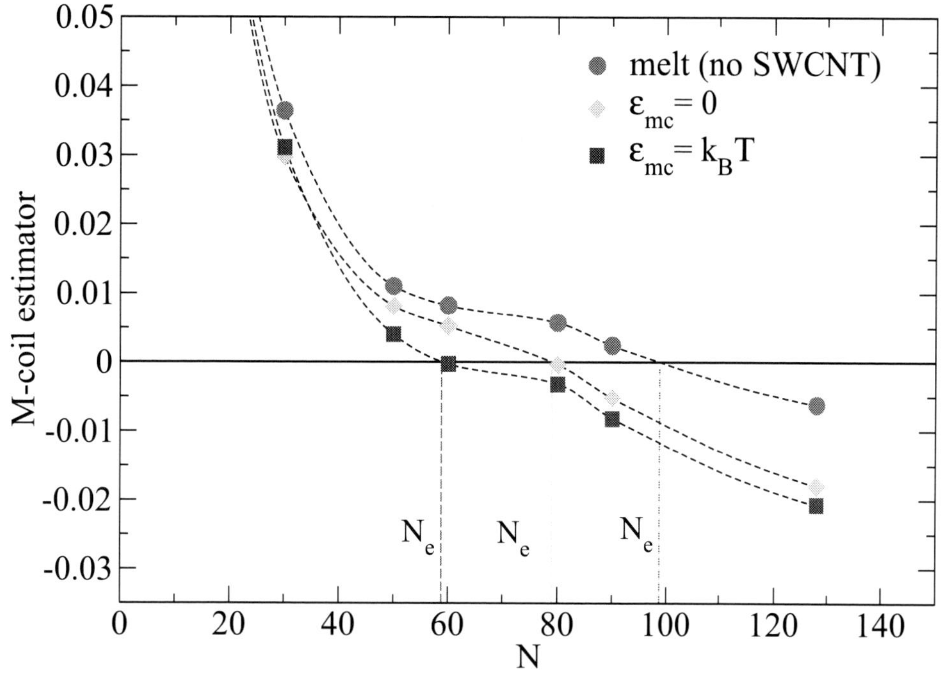

Figure 2. N_e estimated from $\mathcal{N}_e(N)$ using Eq. 3 for polymer melts with a SWCNT ($r_{SWCNT} = 0.66$) and melts. (i) polymer melts(filled circles), (ii) polymer melts with a SWCNT in the absence of attractive interaction between polymers and SWCNT surface (filled diamonds) (iii) polymer melts with a SWCNT and attractive (k_BT) interaction between polymers and SWCNT surface (filled squares). Dashed lines interpolating between data points have been added to guide the eye.

where $\langle|\mathbf{r}_i(t) - \mathbf{r}_i(0)|^2\rangle$ is the time dependent displacement of a given atom averaged over time and atoms of the ensemble ($\mathbf{r}_i$ is the position of a monomer i). In this case of no attractive interaction between polymers and SWCNT the self diffusion coefficient of the polymers with a SWCNT, D scaled with its melt value D_0 is depicted for different number of monomers per chain in Fig. 4. The polymer diffusivity in the presence of a SWCNT is similar to its melt value. Consequently, there is no alteration of the polymer diffusivity due to geometric effects when the polymer radius of gyration is greater than the SWCNT radius.

When there is an attractive interaction, modelled by the LJ potential, Eq. 1, between polymers and SWCNT. In Fig. 4 we show the self diffusion coefficient of the polymers in the polymer melt with a SWCNT system scaled with its melt value, $\frac{D}{D_0}$, for different N. As can be seen, as N increases there is a decrease in the polymer chain diffusivity compared to the melt in the case of entangled polymer melts with a SWCNT. In the case of unentangled polymer ($N = 10$) with a SWCNT, the polymer chain diffusivity retains its melt value. In order to understand the diffusion behaviour of the polymers in the presence of the SWCNT we show in the inset of Fig. 4 the mean square displacement of the polymers that are always in contact with the SWCNT surface (chains that diffuse around or along the SWCNT) and compare it with the rest of the polymer chains (not in contact) and also its melt value (without SWCNT). As can be seen, there is a large heterogeneity in the polymer chain dynamics in the presence of the SWCNT. Additionally, the mean square displacement of the polymers ($N = 60$) that are

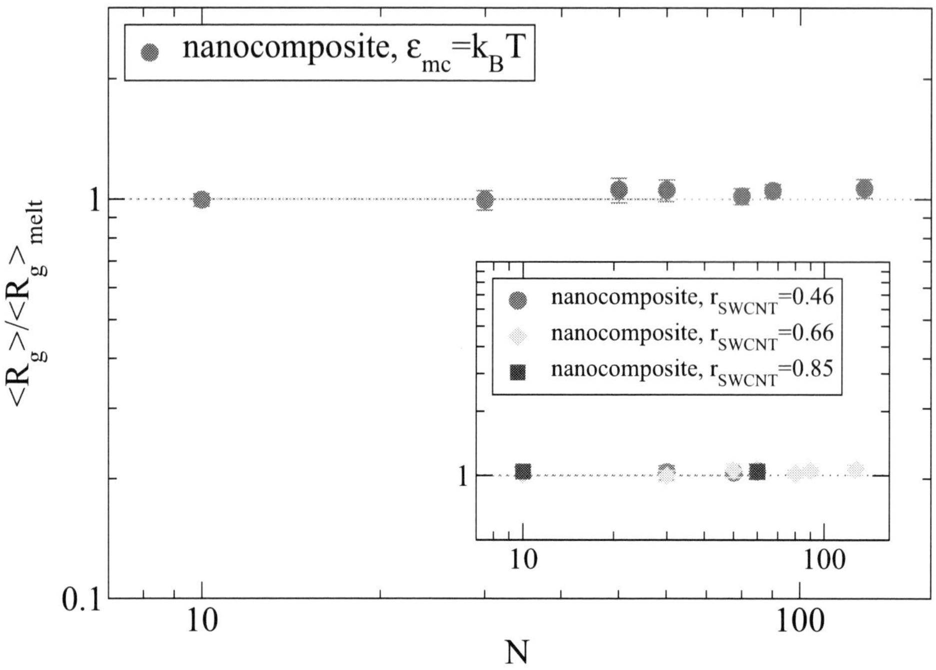

Figure 3. Radius of gyration of polymers in contact with the SWCNT of different molecular weight normalized with its bulk value at different interaction strengths between polymers and SWCNT ($r_{SWCNT} = 0.66$) surface: (i) $\epsilon_{mc} = k_BT$ (red circles). Inset: Radius of gyration of polymers in contact with the SWCNT of different molecular weight normalized with its bulk value for difference radius of SWCNT: (i) $r_{SWCNT} = 0.46$ (red circles), (ii) $r_{SWCNT} = 0.66$ (green diamonds), and (iii) $r_{SWCNT} = 0.85$ (blue squares).

not always in contact with the SWCNT is smaller than its melt value, showing that polymers which are adsorbed and desorbed from the SWCNT are influenced by the interaction with the nanotube surface.

6. Conclusions

The entanglements, structure and dynamics of polymers of SWCNT / polymer nanocomposites at low volume fractions of SWCNT (below the percolation threshold) were investigated by means of molecular dynamics simulations for both unentangled and entangled systems. An atomistic model was used for the SWCNT with an infinite aspect ratio and radius smaller than the polymers radius of gyration. We found that the entanglement length decreases both with and without attractive interactions between the polymers with the SWCNT surface. Also, we found that the large scale structure, as quantified by the radius of gyration remains unaffected in the vicinity of the SWCNT. Furthermore, the overall polymer diffusion is affected by the interaction between monomers and SWCNT surface and there is a large heterogeneity in the polymer dynamics due to the polymers in contact with the SWCNT.

Acknowledgments

We thank Dr. Martin Kröger for providing the Z1 algorithm and for useful discussions. This research was funded by the EPSRC/NSF Materials Network program EP/5065373/1 (EPSRC: NC, AK) and DMR-0908449 (NSF: KIW, RJC)

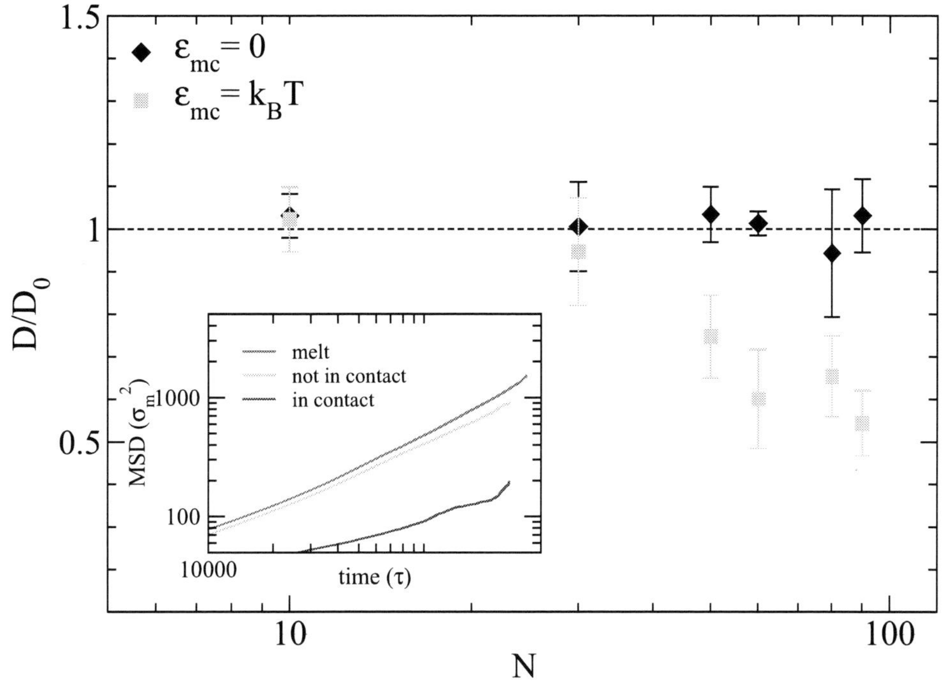

Figure 4. Self-diffusion coefficient of polymers as a function of monomers per chain for a polymer melt with a SWCNT ($r_{SWCNT} = 0.66$) normalized by its melt value: (i) Polymer diffusion in a melt with a SWCNT without attractive interaction between polymers and SWCNT (diamonds) and (ii) polymer diffusion in a melt with a SWCNT with attractive interaction ($\epsilon_{mc} = k_BT$) between polymer and (squares). Dashed line has been added to guide the eye. Inset: Mean square displacement of polymers for $N = 60$, interacting with attractive k_BT energy with the SWCNT surface: all polymers in the melt without SWCNT (red line), polymers detached from the SWCNT (green line), and polymers in contact with the SWCNT (blue line).

References

[1] Usuki A, Kawasumi M, Kojima Y, Okada A, Kurauchi T and Kamigaito O 1993 *J. Mater. Res.* **8** 1174
[2] Kojima Y, Usuki A, Kawasumi M, Okada A, Fukushimay Y, Kurauchi T and Kamigaito O 1993 *J. Mat. Res.* **8** 1185
[3] Mu M, Clarke N, Composto R J and Winey K I 2009 *Macromolecules* **42** 7091
[4] Karatrantos A, Composto R J, Winey K I and Clarke N 2011 *Macromolecules* **44** 9830
[5] Peter S, Meyer H and Baschnagel J 2006 *J. Polym. Sc.:Part B: Polymer Physics* **44** 2951
[6] van der Spoel D, Lindahl E, Hess B, Groenhof G, Mark A and Berendsen H 2005 *J. Comp. Chem.* **26** 1701
[7] Everaers R, Sukumaran S K, Grest G S, Svaneborg C, Sivasubramanian A and Kremer K 2004 *Science* **303** 823
[8] Hoy R, Foteinopoulou K and Krøger M 2009 *Phys. Rev. E.* **80** 031803
[9] Flory P 1969 *Statistical Mechanics of Chain Molecules* (New York: Cornell University Press)
[10] Shanbhag S and Krøger M 2007 *Macromolecules* **40** 2897
[11] Starr F W, Schrøder T B and Glotzer S C 2002 *Macromolecules* **35** 4481
[12] Ndoro T V M, Voyiatzis E, Ghanbari A, Theodorou D N, Bohm M C and Muller-Plathe F 2011 *Macromolecules* **44** 2316
[13] Mackay M E, Tuteja A, Duxbury P M, Hawker C J, Van Horn B, Guan Z, Chen G H and Krishnan R S 2006 *Science* **311** 1740
[14] Sen S, Xie Y, Kumar S K, Yang H, Bansal A, Ho D L, Hall L, Hooper J B and Schweizer K S 2007 *Phys. Rev. Lett.* **98** 128302
[15] Allen M and Tildesley D 1987 *Computer Simulation of Liquids* (Oxford: Clarendon Press)

IOP Conf. Series: Materials Science and Engineering **40** (2012) 012028 doi:10.1088/1757-899X/40/1/012028

Prediction of energy absorption characteristics of aligned carbon nanotube/epoxy nanocomposites

D Weidt[1,2,3], Ł Figiel[1,3] and M Buggy[2,3]

[1]Department of Mechanical, Aeronautical and Biomedical Engineering, University of Limerick, Ireland
[2]Department of Civil Engineering and Materials Science, University of Limerick, Ireland
[3]Materials & Surface Science Institute, University of Limerick, Ireland

E-mail: Lukasz.Figiel@ul.ie

Abstract. This research aims ultimately at improving the impact performance of laminates by applying a coating of epoxy containing carbon nanotubes (CNTs). Here, 2D and 3D computational modelling was carried out to predict energy absorption characteristics of aligned CNT/epoxy nanocomposites subjected to macroscopic compression under different strain rates (quasi-static and impact rates). The influence of the rate-dependent matrix behaviour, CNT aspect ratio and CNT volume fraction on the energy absorption characteristics of the nanocomposites was evaluated. A strong correlation between those parameters was found, which provides an insight into a rate-dependent behaviour of the nanocomposites, and can help to tune their energy absorption characteristics.

1. Introduction

Carbon fibre reinforced plastic (CFRP) laminates are commonly used in the aircraft industry because of their high specific stiffness, high specific strength, and good fatigue and corrosion resistance (when compared to metals). Unfortunately, they frequently behave as brittle materials when exposed to various types of impact loading as they cannot absorb/dissipate enough energy. As a result, matrix cracks can develop and in combination with a weak interlaminar strength, can lead to delaminations and catastrophic failure of CFRP-based structures.

Several attempts have been undertaken to enhance the damage and failure resistance of CFRP laminates by using carbon nanotubes (CNTs). This is due to CNT's excellent mechanical properties (e.g. stiffness, resilience) [1-2], their large specific surface area [3], and their ability to reinforce polymers at very low weight loadings. Hence, some researchers have used CNTs to enhance the matrix material's mechanical performance [4], others have used them to strengthen the interface between the CFRP pre-preg lamina [5], while others have grown them on the surfaces of fibres to produce a stitching effect [6].

We believe that the incorporation of a nanocomposite coating onto the laminate surface can help to enhance overall energy absorption characteristics of the laminates. This is to result from: (1) the large interfacial area of CNTs, which becomes potentially available for fracture propagation, CNT pull-out and crack bridging [7-9]; and (2) an enhanced nonlinear deformation of the nanocomposite matrix due to the presence of nanoparticles.

Therefore, the ultimate objective of this research project is to predict an impact response of CFRP laminates coated with CNT/epoxy nanocomposites using multiscale modelling validated by experiments [10]. This study is a part of our project, and it is focused on the prediction of mechanical behaviour of aligned CNT/epoxy nanocomposites across different strain rates. In particular, the emphasis is on the influence of aligned CNTs on the compressive macroscopic stress-strain response of the CNT/epoxy nanocomposite at strain rates ranging from quasi-static to impact rates. They are then used to evaluate energy absorption characteristics of the nanocomposites. For this, a nonlinear and rate-dependent constitutive model for epoxy is implemented into the finite element (FE) framework (ABAQUS), and combined with the representative volume element (RVE) and numerical homogenisation. FE predictions of the nanocomposite response at different strain rates are compared with unfilled epoxy, to demonstrate the effect of CNTs (aspect ratio, volume fraction) on energy absorption characteristics of CNT/epoxy nanocomposites.

2. Modelling methodology

It was assumed that the macroscopic response of the nanocomposite can be predicted using the RVE concept and numerical homogenization. For that purpose it was assumed that the continuum theory holds down to the scale at which the nanocomposite morphology can be captured with an RVE. This offers a compromise between accuracy and computational time.

It was further assumed that the CNT morphology is globally periodic and that the macroscopic deformation is uniform. This enabled the application of RVEs with imposed periodic boundary conditions. Two types of RVEs were considered here (see Figure 1): (1) simple axisymmetric RVE (A) and (2) three-dimensional cuboidal (3D) RVE (B). The axisymmetric RVE was used because of its simplicity, and ability to capture regular, well dispersed morphologies of CNTs. On the other hand, the 3D RVE was chosen to account for an irregular morphology of randomly distributed and aligned CNTs. A cuboidal RVE was chosen over a cubic RVE in this work because of the assumed length of CNTs, CNT volume fraction and a small number of CNTs (10). 3D RVEs were generated using an in-house developed Python script.

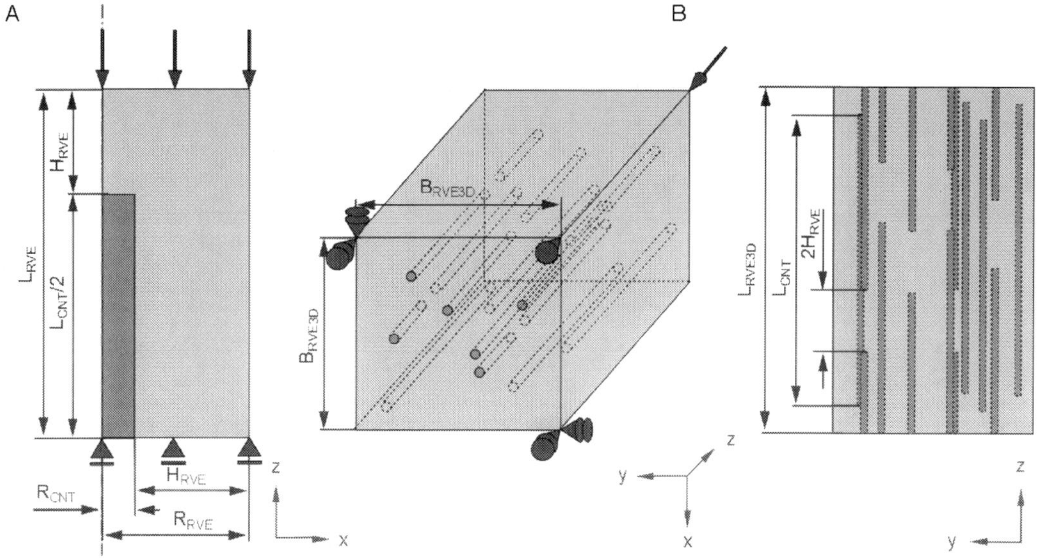

Figure 1. Axisymmetric (A) and three-dimensional RVE (B) of CNT/epoxy nanocomposites; Epoxy matrix (light grey), homogenised CNT (medium grey), boundary conditions (dark grey).

Homogenised CNTs with effective elastic properties were used in this work as computationally more efficient than their dicrete tubular continuum counterparts. In order to calculate the effective elastic properties of CNTs, the following formula was used:

$$E_{CNT}^{eff} = E_{CNT}\frac{2R_{CNT}T_{CNT} - T_{CNT}^2}{R_{CNT}^2} \tag{1}$$

where E_{CNT}, R_{CNT} and T_{CNT} are the Young modulus, outer radius and thickness of a discrete CNT, respectively. In this work, E_{CNT}=3.36TPa was used as obtained from the elastic shell theory for the given wall thickness T_{CNT}=0.1nm and Poisson's ratio of ν=0.2 [12]; the outer radius of the CNTs was assumed to be equal to R_{CNT} =5nm. The effective Young modulus of homogenised CNTs was calculated from Eq. (1) to be 133.056GPa in both axial and transverse directions. However, it must be mentioned that Eq. (1) was derived based on the axial CNT direction. Hence, this is only an approximation to the transverse response of CNTs. In our former study [11], we found negligible differences between the stress-strain response of RVEs with homogenised CNTs and their discrete counterparts. However, this result was obtained from simulations on RVEs with perfectly aligned CNTs under compressive loading in CNT axis direction [11], and it will be verified in the future for randomly oriented CNTs.

Perfect bonding between CNTs and the matrix was assumed in this work. Hence, a possible debonding and pull-out of CNTs from the matrix was excluded as possible energy dissipation/absorption mechanisms. As a result, plastic deformation of the epoxy matrix was expected to be the major cause of energy absorption.

The non-linear and rate-dependent matrix behaviour was represented using a physically-based constitutive model proposed by Buckley et al. [13] for thermosetting resins. The model accounts for the strain softening and adiabatic heating effect (active upon applied impact loads), and enabled capturing basic phenomena of epoxy matrices subjected to varying strain rates. The mechanical constitutive behaviour was implemented as a user subroutine UMAT (ABAQUS/Standard), with the model parameters obtained by Buckley et al [13,14] for the standard bisphenol A resin.

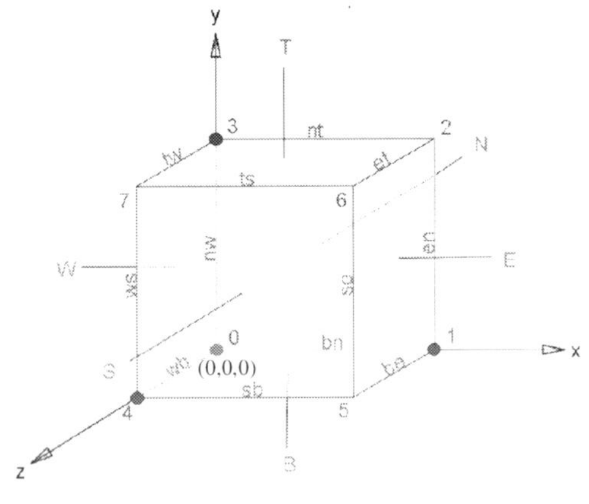

Figure 2. Schematic of a 3D RVE with notations of faces (N, W, S...), edges (nt, tw, ts...) and corner nodes (0,1,2...); periodic boundary conditions were applied according to equation (2); in addition, displacements were applied/constrained at corner nodes 0, 1, 3 and 4 (as shown in Figure 1), e.g. corner node 0 at (x,y,z)=(0,0,0) is constrained in all directions to exclude rigid body motion.

Uniform displacement boundary conditions were applied to the axisymmetric RVE in the y-direction – uniform deformation of the right edge in the x-direction (see Figure 1A) was enforced by setting the corresponding nodal displacements (right edge) as equal to the top right corner node of the RVE (using the ABAQUS '*Equation' command).

The complete set of periodic boundary conditions (PBCs) was used for 3D RVEs as follows:

$$\boldsymbol{u}_F^+ - \boldsymbol{u}_F^- = \boldsymbol{u}_{(I)}^+ - \boldsymbol{u}_{(I)}^-, \text{ and } \boldsymbol{u}_E^+ - \boldsymbol{u}_E^- = \boldsymbol{u}_{(I)}^+ - \boldsymbol{u}_{(I)}^- \text{ and } \boldsymbol{u}_{(I)}^+ - \boldsymbol{u}_{(I)}^- = \boldsymbol{u}_{(J)}^+ - \boldsymbol{u}_{(J)}^- \tag{2}$$

where $\boldsymbol{u}_F$ and $\boldsymbol{u}_E$ are the displacement vectors of face and edge nodes, while $\boldsymbol{u}$ with the subscript I or J refers to a corner node; symbols '+' and '-' represent two opposite faces, edges or nodes of the RVE. In addition, external nodal displacements representing macroscopic deformations (macroscopic compressive strains) were applied to relevant corner nodes (1, 3 and 4 – see Figure 2). Those boundary conditions result in tractions $\boldsymbol{t}$ (due to PBCs) and nodal forces $\boldsymbol{f}$ (due to external displacements) acting on the RVE. The macroscopic Cauchy stresses $\boldsymbol{\sigma}_M$ can then be calculated as follows [15]:

$$\boldsymbol{\sigma}_M = \frac{1}{V_{RVE}} \int_{V_{RVE}} \boldsymbol{\sigma}_{RVE}(\boldsymbol{x})\, dV = \frac{1}{V_{RVE}} \sum_I \boldsymbol{f}_{(I)} \otimes \boldsymbol{x}_{(I)}, \tag{3}$$

where V_{RVE} is the current volume of an RVE, and $\boldsymbol{f}_{(I)}$ and $\boldsymbol{x}_{(I)}$ are current nodal forces and nodal positions of a corner node *I*. For the macroscopic compressive strains applied uniaxially in the z-direction (see Figure 2), the resulting stresses in z-direction, σ_{zz}, can be calculated from Eq. (3) as $\sigma_{zz} = V_{RVE}^{-1} f_{z(4)} x_{z(4)}$, where $f_{z(4)}$ and $x_{z(4)}$ denote force and position of the node 4 (see Figure 2).

The axisymmetric RVEs (matrix and CNTs) were discretized using 8-node quadratic axisymmetric quadrilateral elements CAX8. 3D RVEs were discretized using 10-node quadratic tetrahedron elements C3D10 (matrix region) and 20-node quadratic brick elements C3D20 (CNT region). CNT and matrix regions were tied together via surface-based tie constraints.

3. Results and discussion

Predicted compressive stress-strain curves and energy absorption characteristics of CNT/epoxy nanocomposites subjected to different strain rates are shown and discussed in this section.

Effects of CNT volume fraction (VF) and varying strain rates on the stress-strain response are shown in Figure 3. As expected, the incorporation of CNTs into epoxy produced a considerable stiffening effect in the linear portion of the curve – the Young modulus of the nanocomposite increased by around 15%, 30%, 46% and 77% (CNT VF of 0.5%, 1%, 1.5% and 2.5%, respectively) compared to pure epoxy. Hence, CNT loading exhibits almost a linear effect on the Young modulus of the nanocomposite. As expected, no change in the nanocomposite Young's modulus was predicted with increasing strain rate. This is because the constitutive model for epoxy used in this work does not account for the rate-dependent stiffness. However, experimental data across a wide range of compressive strain rates for an RTM-6 epoxy system reported in Gerlach et al. (2008) [16] shows that there is no significant change in Young's modulus up to true strain rates ~ 10^3 s^{-1}. The latter is greater than the maximum magnitude of nominal strain rates considered in this work, as the true strain rate $\dot{\varepsilon}_M$ is related to the applied (nominal) strain $\dot{\varepsilon}_{M(app)}$ through $\dot{\varepsilon}_M = \dot{\varepsilon}_{M(app)} / \left(1 + \varepsilon_{M(app)}\right)$.

The stiffening effect caused by CNT loading on the stress response continues beyond the elastic limit - for the given magnitude of applied strain, a gradual increase in the macroscopic true stress with CNT loading was predicted. Moreover, the start of nonlinear deformation is shifted to a slightly lower value of applied strain with increasing CNT volume fraction – in all cases a higher yield stress is predicted when compared with the pure epoxy.

The post-yield strain softening, which was predicted (and validated experimentally [13]) for the pure epoxy, is no longer present for VF equal to or greater than 1%. Furthermore, an increase in the CNT VF resulted in an increased post-yield strain-hardening effect. Hence, assuming that the energy absorbed by the material system can be represented by the area under the stress-strain curve, it is postulated that the improved post-yield behaviour of the nanocomposite can largely contribute to the improved energy absorption of the CNT/epoxy nanocomposite in compression. This increase in energy absorption characteristics results from an enhanced nonlinear deformation (pronounced by strain stiffening) of the matrix, caused by the presence of CNTs.

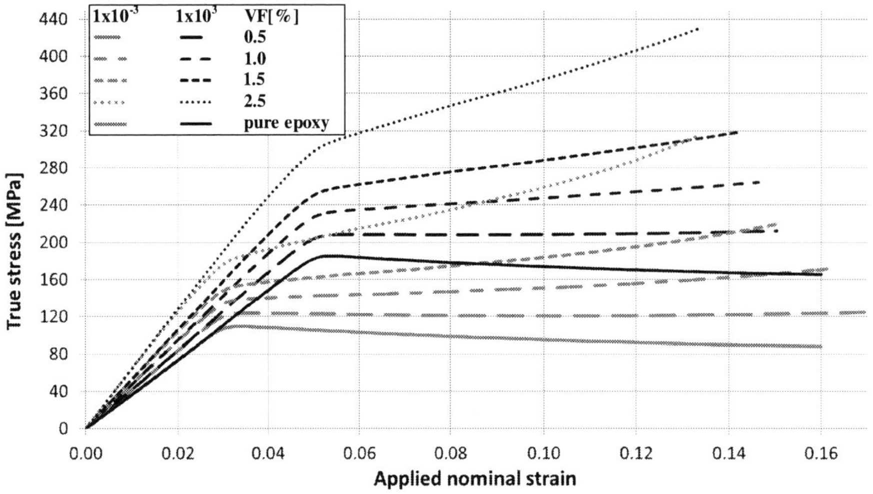

Figure 3. Predicted stress-strain curves for pure epoxy and nanocomposites for different CNT volume fractions (VF) and applied strain-rates; results obtained from axisymmetric RVEs; CNT aspect ratio AR=50.

Effects of CNT aspect ratio (AR) and varying strain rates on the stress-strain were also investigated. In general, qualitatively similar trends were found as for the case of increasing CNT loading. However, quantitatively, only a slight increase of the nanocomposite modulus was predicted with increasing CNT AR within the investigated range from 50 to 250. As indicated by different curves, the effect of CNT AR on the stress-strain response tends to saturate with increasing aspect ratio - less significant improvement of the overall stress-strain response was obtained for aspect ratios greater than 100. However, in general there is an enhanced post-yield behaviour (increasing strain stiffening) with increasing CNT AR, which suggests that energy absorption characteristic are enhanced only at large deformations. Again, the major contribution to that comes from the enhanced nonlinear deformation (strain stiffening) of the matrix caused by the CNT presence.

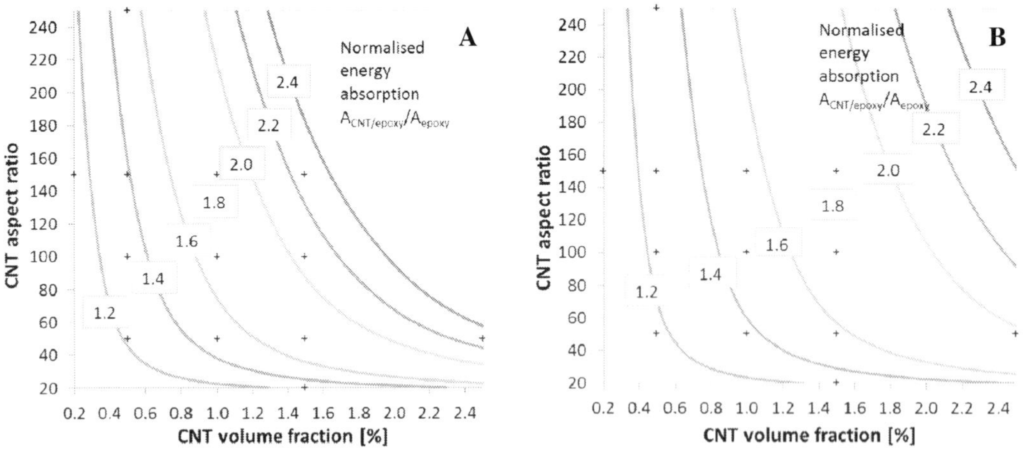

Figure 4. Predicted normalised energy absorption characteristics as a function of CNT aspect ratio (AR) and CNT volume fraction (VF) for quasi-static $1x10^{-3}$ s^{-1} (A) and high strain rates $1x10^{3}$ s^{-1} (B). The normalised energy absorption characteristics at the data points (crosses) were used to determine the coefficients of the nonlinear regression functions via the least squares method.

Normalised energy absorption characteristics were then evaluated based on the compressive stress-strain curves. Areas under those curves were evaluated up to strains ~13%, both for the nanocomposites and matrix, at two strain rates ($1x10^{-3}s^{-1}$ and $1x10^{3}s^{-1}$), and with different volume fractions and aspect ratios. Our predictions are shown in Figure 4, where, in general, the normalised energy absorption is enhanced due to the presence of CNTs at both strain rates. A significant correlation between CNT aspect ratio and CNT volume fraction is predicted - an increase in CNT volume fraction contributes to a considerable increase in normalised energy absorption under quasi-static loading ($1x10^{-3}s^{-1}$) for CNT aspect ratios AR=150. In comparison, predictions for aspect ratios AR=50 do not show such a significant increase in the normalised energy absorption with increasing volume fraction. Hence, our results suggest what CNT loading must be used for a given average CNT aspect ratio to improve the energy absorption of CNT/epoxy nanocomposites. This is under the assumption that the nonlinear matrix deformation is the primary energy absorption mechanism. Future studies will include other mechanisms such as debonding [10] and pull-out of CNTs from the matrix.

4. Conclusions

Computational modelling carried out here for aligned CNT/epoxy nanocomposites subjected to different compressive strain rates demonstrated: (1) increase of the nanocomposite stiffness; (2) strain stiffening in the post-yield regime; and (3) the improvement of energy absorption characteristics, with increasing CNT volume fraction and CNT aspect ratio.

Acknowledgements

The project is supported by the Irish Research Council for Science, Engineering and Technology. Computational facilities are provided by the Irish Centre for High-End Computing.

References

[1] Treacy M M J, Ebbesen T W and Gibson J M 1996 *Nature* **381** 678-680
[2] Cooper C A, Young R J and Halsall M 2001 *Compos. Part A-Appl. S.* **32** 401-411
[3] Peigney A, Laurent Ch, Flahaut E, Bacsa R R and Rousset A 2001 *Carbon* **39** 507-514
[4] Yokozeki T, Iwahori Y, Ishiwata S and Enomoto K 2007 *Compos. Part A-Appl. S.* **38** 2121-2130
[5] Hu N, Li Y, Nakamura T, Katsumata T, Koshikawa T and Arai M 2012 *Compos. Part B-Eng.* **43** 3-9
[6] Wicks S S, de Villoria R G and Wardle B L 2010 *Compos. Sci. Technol.* **70** 20-28
[7] Gojny F H, Wichmann M H G, Fiedler B and Schulte K 2005 *Compos Sci Technol* **65** 2300-2313
[8] Zhang W, Picu R C and Koratkar N 2007 *Appl. Phys. Lett.* **91** 193109
[9] Wernik J M and Meguid S A 2010 Appl. Mech. Rev. **63** 050801 **70** 20-28
[10] Weidt D, Figiel Ł and Buggy M 2012 *Mater. Sci. Forum* **714** 3-11
[11] Weidt D, Figiel Ł and Buggy M 2012 *Proc. 15th Int. Conf. of Deformation, Yield and Fracture of Polymers (Kerkrade, the Netherlands, 1-5 April 2012)* pp 278-281
[12] Wang C Y and Zhang L C 2008 *Nanotechnology* **19** 195704
[13] Buckley C P, Dooling P J, Harding J and Ruiz C 2004 *J. Mech. Phys. Solids* **52** 2355-2377
[14] Buckley C P, Harding J, Hou J P, Ruiz C and Trojanowski A 2001 *J. Mech. Phys. Solids* **49** 1517-1538
[15] Figiel Ł, Dunne F P E, Buckley C P 2010 *Modelling Simul. Mater. Sci. Eng.* **18** 015001
[16] Gerlach R, Siviour C R, Petrinic N, Wiegand J 2008 *Polymer* **49** 2728-2737

Multi-scale modeling of the progressive damage in cross-ply laminates under thermal and mechanical loading

D Yang[1, 3], Y Sheng[1], J Ye[2] and Y Tan[3]

[1] School of Civil Engineering, University of Leeds, Leeds, LS2 9JT, UK
[2] Department of Engineering, Lancaster University, Lancaster LA1 4YW, UK
[3] School of Mechanical Engineering, Xiangtan University, Hunan, 411105, China

E mail: D.Yang@leeds.ac.uk

Abstract. The progressive damage in cross-ply laminates was modeled by discrete element method (DEM). A particle radius expansion method was used to account for thermal loading applied to cross-ply laminates in which nominal fibers were introduced in the 0° plies so as to achieve the anisotropic thermal expansion behaviors. A series of convergence and validation tests of both mechanical and thermal properties of the 0° plies with nominal fibers have been carried out in order to validate the method. The DEM results of interfacial stress distribution of cross-ply laminates under pure thermal loading and under coupled thermal/mechanical loading were compared with other theoretical predictions. Microstructure of 90° plies was also studied by the DEM model. Transverse cracking which was formed by the coalescence of micro cracks in matrix and at fiber/matrix interface has been observed in the modeling results together with the ply-ply delaminations. It was found that the DEM model can predict not only the stress distribution but also the progressive damage initialized from the constituent failure due to its multi scale nature.

1. Introduction

Due to the complexity of the physical microstructure of fiber reinforced composite materials, the onset of damage does not cause instantaneous failure of the entire structure, and there exists a progressive process from the damage initiation to final structural collapse. Transverse cracking and delamination are typical damages taking place before the final catastrophic failure when the cross-ply laminates subject to transverse tensile loading. Transverse cracking can be initiated from the matrix cracking and/or fiber/matrix interfacial debonding caused by the high tensile stress in 90° ply. Delamination is normally intrigued by the shear stress concentrating between the two neighboring plies due to the dissimilar material properties of the two adjacent plies, under the action of transverse loadings or free-edge stresses. Therefore, the analysis of stresses distribution induced by thermal and/or mechanical loadings is primary and also critical for the predication of damage initiation before the application of proper failure criteria. Through the literature, theoretical solutions of stress distribution have been achieved by using different analytical models, such as classical laminate theory (CLT) [1] and full layer-wise theory [2]. Finite element method (FEM), as one of the typical numerical methods for stress analysis, has also been used to compute the interlaminar stress distribution [3]. However, FEM is sometimes time consuming and even unreliable for free-edge of multi-ply laminates [4]. Ye et al. [5] developed a state space finite element method, which is a semi-analytical method, to solve the stress singularities in the vicinity of free-edge or localized traction free surface by combining the traditional finite element approximation and the recursive formulation of state space equation. Even though many methods can properly predict the stress distribution in composite laminates under various loading conditions, few of them can be applied to describe the progressive damage behaviors, e.g., initiation and propagation of transverse cracking and/or delamination. The generation of transverse cracking is normally predicted by the damage analysis models of homogeneous using either strength based theory [6] or critical fracture energy release rate [7] as failure criterion. Delamination is normally predicted by applying a cohesive zone model (CZM) in FEM codes [8]. The interaction of transverse cracking and delamination is essential to be taken into account by a numerical model intended to simulate progressive damage process of cross-ply laminates. Berthelot and Corre [9] presented a statistical analysis model, by which the initiation of transverse cracking and delamination is evaluated according to local stress values. Okabe et al. [10] used a embedded process zone (EPZ) model in FEM to simulate the transverse cracking and interlaminar delaminations by assuming that the damage propagates along the pre-defined embedded process zone.

Actually, the transverse cracking as well as delamination are collectively formed by smaller damage events, such as matrix cracking and fiber/matrix debonding, which somehow take place randomly cross the whole material. Therefore, the evaluation of transverse cracking and delamination could be more realistic if they are statistically characterized by smaller collective damage events. On the other hand, the consideration of the material microstructure, at least in the 90° plies is also essential to model the progressive damage from matrix cracking and fiber/matrix debonding at micro scale to transverse cracking and delamination at meso scale.

The purpose of this paper is to establish a thermal/mechanical DEM model which can predict the interfacial stress distribution as well as material progressive damage. First, thermal expansion in homogeneous DEM model was introduced. Calibration of mechanical behavior of a heterogeneous DEM model was also carried out and was then combined with the previous homogeneous DEM thermal model to eventually construct a heterogeneous DEM thermal/mechanical model. Interfacial stress distribution of cross-ply laminate under thermal and/or mechanical loading was analyzed and compared with other theoretical or numerical results. Transverse cracking and delamination in the DEM models of cross-ply laminates with/without microstructure under thermal-mechanical loading were both simulated with results discussed.

2. Thermal expansion in DEM

In our previous work, a discrete element method (DEM) was proposed to analyze the interfacial stress distribution and to predict the crack density in cross-ply laminates under pure mechanical loading [11]. As also reported in other literature, thermal residual stress produced during the laminate fabrication process may affect the final stress distribution and material damage under later applied mechanical loading [12, 13]. Thermal stress has also to be considered to analyze thermal induced cracking in some applications where the environment temperature is significant changed [14-16]. Unlike continuum mechanics based models in which the material thermal behavior can be introduced directly by adding a thermal item in the stress-strain equation, the thermal expansion in DEM model is normally represented by correspondingly changing the geometrical size of each DEM element, namely the radius of each particle. To account for the thermal expansion of fiber reinforced composite which is a typical heterogeneous material with two constituents of significantly different thermal properties, the material's microstructure has to be considered based on our previous homogenous DEM model [11].

In this paper, the particle radius expansion method will be used. Thermal strains can be accounted by the thermal expansion of particles or bonds in DEM. The particle radius, R, is changed when a temperature change, ΔT, is given, such that:

$$\Delta R = \alpha R \Delta T \tag{1}$$

Where α is the coefficient of linear thermal expansion (CTE) of the particle. If the particles were bonded by a parallel bond, the expansion of the bond is accounted by assuming that only normal force carried by the parallel bond will be affected by the temperature change, and the bond length $\bar{L}$ is effectively changed [17]. The change of the parallel bond normal force is modeled as:

$$\Delta \bar{F}^n = -\bar{k}^n A \Delta U^n = -\bar{k}^n A(\bar{\alpha}\bar{L}\Delta T) \quad (2)$$

Where $\bar{k}^n$ is the normal stiffness of parallel bond, A is the cross section area of parallel bond, $\bar{\alpha}$ is the CTE of the parallel bond which is equal to the average value of the CTEs of the two contacting particles, $\bar{L}$ is the bond length and ΔT is the temperature change.

2. DEM modeling of thermal expansion in orthotropic laminae

2.1 Mechanical convergence

The construction of a DEM micromechanical model for laminae is straightforward by adding them together and introducing the fiber/matrix interface properties. Since the stress and strain are averaged from those of particles at the surface of the DEM model, this method may be sensitive with the particle size if the model size is constant. So it is essential to carry out a convergence study of this method to guarantee that this size effect should be eliminated when the DEM model of laminae is constructed by adding the DEM models of fiber and matrix together.

A matrix material of LY-556 epoxy, which elastic properties are E=3.35 GPa and υ=0.33, is taken as an example to study the convergence of DEM model in both X and Y directions in terms of elastic modulus and Poisson's ratio. When the loading is applied in X (or Y) direction, the number of particles in X (or Y) direction is fixed, and particle's number in Y (or X) direction is increased gradually. The convergence results from these tests will be applied in the later laminae model to guarantee the particle's number for fiber, matrix and laminae are all within the convergence area. 300 particles are used in X direction when loading is applied in X direction. Figure 1 shows the results from which we find that the Poisson's ratio of the model is converged when 15 particles is used in the thickness (Y direction). The elastic modulus seems not sensitive to the model size in Y direction, though the value is slightly increased gradually from 2.9126 GPa to 2.9325 GPa which just has a relative error of 0.68%. The elastic modulus and Poisson's ratio are both found to be sensitive to the model size when loading is applied in Y direction where the number of particles in X direction is fixed as 30, and they are both converged when the particle number in Y direction was increased to 300, as shown in Figure 2.

The DEM model of laminae is built up by adding the models of fiber and matrix together. The contacts between the particles of fiber (in blue color) and the particles of matrix (in yellow color) are treated as fiber/matrix interface, as shown in Figure 3. Longitudinal and transverse loading tests are then carried out to validate the interface properties by comparing the DEM results with the theoretical calculations from rule of mixtures for laminae as following [18]:

$$E_L = E_f V_f + E_m V_m \quad (3)$$

$$E_T = \frac{E_f E_m}{E_f V_m + E_m V_f} \quad (4)$$

$$\upsilon_{LT} = \upsilon_f V_f + \upsilon_m V_m \quad (5)$$

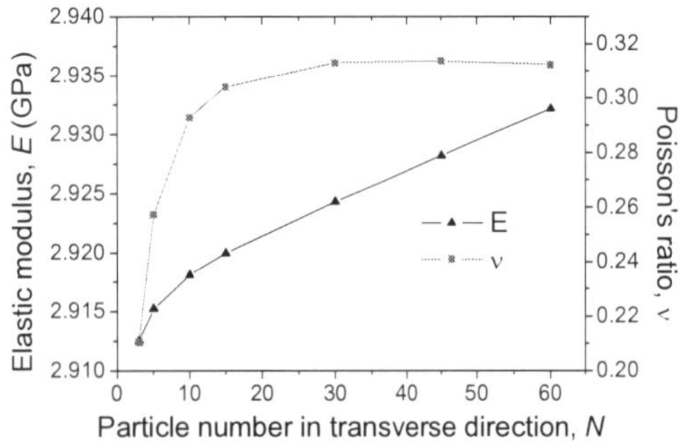

Figure 1. Convergence in Y direction (300 particles in X direction)

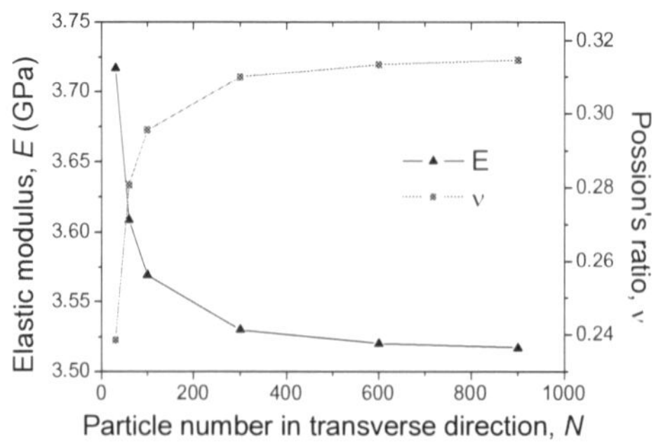

Figure 2. Convergence in X direction (30 particles in Y direction)

The number of particles used for fiber and matrix in transverse (indicated as '*T*') and longitudinal (indicated as '*L*') directions is based on the previous convergence study, as listed in Table 1. The laminae material is Glass/ LY-556 and the fiber volume fraction is 0.6. The matrix is assumed to be aligned equally at both sides of the fiber with a same volume fraction of 0.2. Each of the constituents has been tested by tension tests in both longitudinal and transverse directions to get its mechanical properties. In the longitudinal tension test, two types of interfacial stiffness are assessed. It is found when same stiffness as of matrix was used, smaller error was given, as shown in Table 2. Although one may achieve a more accurate interfacial stiffness by fitting the theoretical values more closed, here we intend to get a general solution which could be used in other kind of composite laminae. Similar test was carried out in transverse direction by using matrix stiffness for interface and only a very small error was produced, as shown in Table 3 and Table 4.

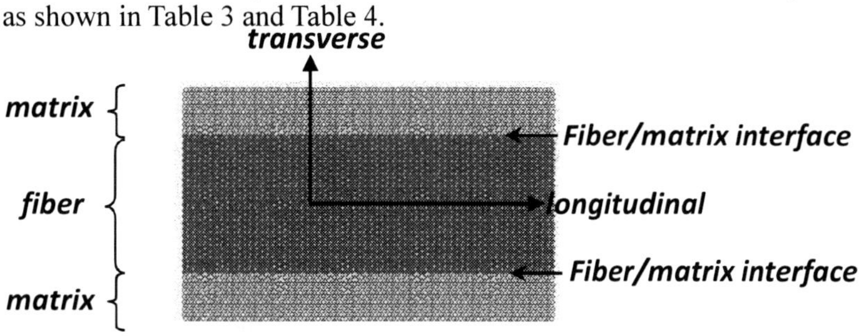

Figure 3. A representative element of laminae with inclusion of microstructure.

Table 1. Material constituents in laminae longitudinal tension test.

Materials	Dimension ($L \times T$)	E (GPa)	υ	Volume Fraction
Matrix	300×15	2.91997	0.30401	0.2
Fiber	300×45	64.3119	0.18934	0.6
Matrix	300×15	2.91997	0.30401	0.2

Table 2. Laminae elastic properties measured in longitudinal tension test.

Laminae	K^n_{int}	K^s_{int}	E_L (GPa)	Error (%)	υ_{LT}	Error (%)
Laminae (Theoretical)			39.76484		0.24115	
Laminae-1	K^n_m	K^s_m	39.75513	**-0.024**	0.23521	**-2.46**
Laminae-2	K^n_f	K^s_f	39.86546	**+0.253**	0.21828	**+9.48**

Table 3. Material constituents in laminae transverse tension test.

Materials	Dimension ($L \times T$)	E_T(GPa)	Volume Fraction
Matrix	30×300	3.52976	0.2
Fiber	30×900	78.0004	0.6
Matrix	30×300	3.52976	0.2

Table 4. Laminae elastic properties measured in transverse tension test.

Laminae	K^n_{int}	K^s_{int}	E_T(GPa)	Error (%)
Laminae(Theoretical)			8.26348	
Laminae-3	K^n_m	K^s_m	8.23864	**-0.03**

2.2 Validation of thermal expansion

The DEM model of laminae which micromechanical properties were calibrated is then further used to study its thermal expansion behavior. The radius expansion method used in the previous single material has been applied to each individual constituent, and results are compared with the theoretical prediction by Schapery [18]:

$$\alpha_L = \frac{E_f\alpha_f V_f + E_m\alpha_m V_m}{E_f V_f + E_m V_m} \quad (6)$$

$$\alpha_T = (1 + \nu_f)\alpha_f V_f + (1 + \nu_m)\alpha_m V_m - \alpha_L(V_f\nu_f + V_m\nu_m) \quad (7)$$

Same material as in the previous mechanical models is used, and the CTEs of fiber and matrix are 4.9e-6 /°C and 58e-6 /°C, respectively. In order to obtain the anisotropic CTEs of the whole laminae, constraint thermal expansion test (under plane stress) on both longitudinal and transverse directions are simulated and the corresponding stress of particles on the constraint boundaries were recorded which are later then used to calculate the CTEs by:

$$\Delta\sigma_x = -\alpha_L E \Delta T \quad (8)$$

$$\Delta\sigma_y = -\alpha_T E \Delta T \quad (9)$$

It is shown in Table 5 that good agreement of longitudinal CTE can be seen from the comparison between DEM numerical result and theoretical prediction, with only a small deviation of 1.7%. A relatively large difference of transverse CTE was also found, and the DEM result is almost 20% smaller. As reported in the literature [19], the theoretical prediction of transverse CTE is diverse, and the Schapery's theory seems to predict a larger value (nearly 20% as well) compared to other theories. This might be the reason for the large error of CTE in transverse direction.

Table 5. Laminae CTE measured in DEM tests.

Laminae	α_L (×10^{-6}/°C)	α_T(×10^{-6}/°C)	Error (%)
Laminae(Theoretical)	6.46005	3.22302	
Laminae-1	6.35006		**-1.7**
Laminae-3		2.60740	**-19.1**

3. DEM analysis of interlaminar stress in cross-ply laminates

To validate the DEM model of thermal expansion in cross-ply laminates, two numerical tests were carried out to analyze the interlaminar stress distribution in $[90/0]_s$ and $[0/90]_s$ under pure thermal loading and thermal-mechanical loading, respectively, with results compared to those from other theoretical and numerical methods.

3.1 Pure thermal loading

A $[90°/0°]_s$ cross-ply composite laminate under pure thermal loading was simulated by DEM to analyze the interlaminar stress distribution. Material properties of the cross-ply laminate are as follows:

E_L=137.9 GPa E_T=14.48 GPa υ_{TT}=0.21 υ_{LT}=0.21 G_{LT}=5.86 GPa

α_L=0.36×10^{-6}/°C α_T=28.8×10^{-6}/°C

The laminate has a length of L and a thickness of H, with L/H=4. Each ply has an equal thickness of $h=H/4$. In order to represent the anisotropic thermal expansion behavior, nominal fibers were used in the 0° plies to account for the anisotropic properties, as shown in Figure 3. Since the mechanical properties given above is in laminae level, Eqs.3~5 are used to derive the mechanical properties for fiber and matrix, assuming the fiber volume fraction is 0.6. The Schapery's equations for CTEs were also used to get the CTEs for fiber and matrix, respectively. It is noticed that transverse CTE, α_T, was scaled by a factor of 0.8 before it was put into the Schapery's equation since Schapery's equations over-predict the α_T.

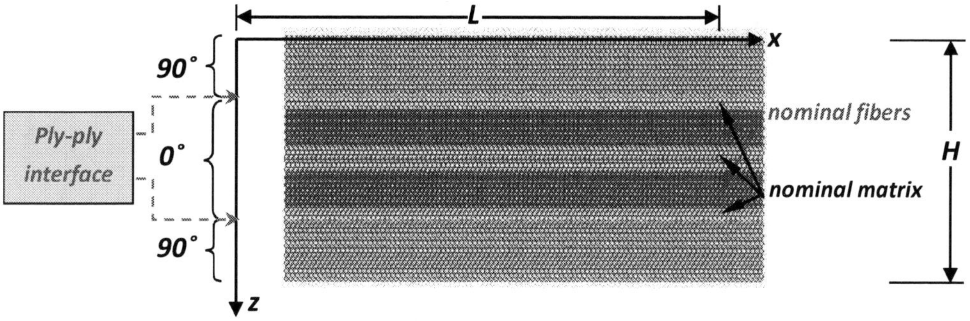

Figure 3. A representative volume of DEM model of [90°/0°]$_s$ cross-ply laminate.

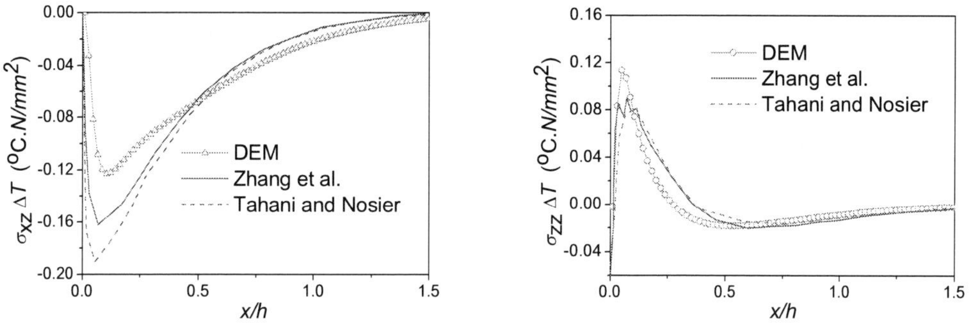

(a) shear stress (b) normal stress

Figure 4. Interlaminar stress distribution at 90°/0° interface under thermal loading.

The norminal fibers have the same mechanical and thermal properties as the real fiber except the diameter dimension. Same values of interfacial stiffness were used for fiber/matrix interface as well ply/ply interface. Infinite strengths were set for both interface and plies to avoid any damage, and a unit temperature change ΔT was applied. After solving the DEM model into an equilibrium state, the normal and shear stress along the 0°/90° interface are obtained and are plotted in Figure 4. It can be seen that the DEM results agree well with other theoretical predictions [12, 20].

3.2 Thermal-mechanical loading

A representative element of [0°/90°]$_s$ cross-laminate with the consideration of ply cracking under thermal-mechanical loading was simulated by DEM in which the material was assumed to be, as shown in Figure 5. The model was under temperature change of ΔT = -120 °C and a mechanical loading stress of σ_0 = 0.2 GPa. The material properties are as follow [21]:

E_L=144.78 GPa E_T=9.58 GPa υ_{TT}=0.31 υ_{LT}=0.31 G_{LT}=4.785 GPa

α_T==-0.72×10^{-6}/°C α_L=27×10^{-6}/°C

The trough-thickness Poisson's ratio of 90° was set as υ_{TT}=0.31 instead of 0.55 which was used in the references [12, 22], considering the maximum Poisson's ratio for DEM model (plane stress) is 0.33.

The DEM model was first under a pure thermal loading and reached an equilibrium condition before the start of mechanical loading. A constant loading velocity v=0.05 m/s, which is found to be a quasi-static loading velocity [23], was then applied on the 0° plies and an average loading stress was measured to reach the final loading stress. The interfacial shear and normal stresses were plotted and compared with other theoretical predictions [12, 22] in Figure 6. The general trends of the results were found in good agreement, though the DEM seems to slightly under-predict the peak values.

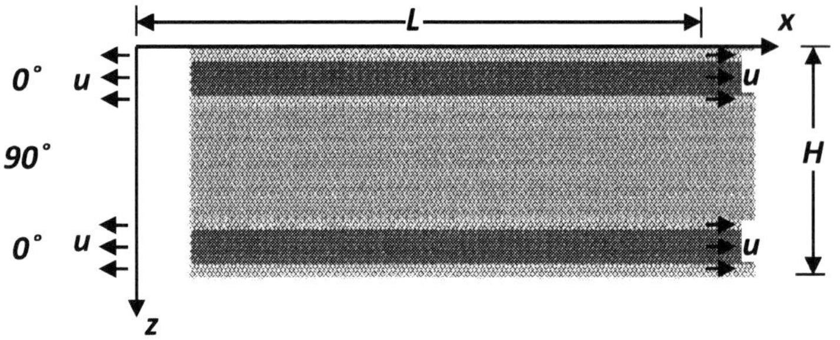

Figure 5. A representative volume of DEM model of [0°/90°]$_s$ cross-ply laminate under thermal-mechanical loading.

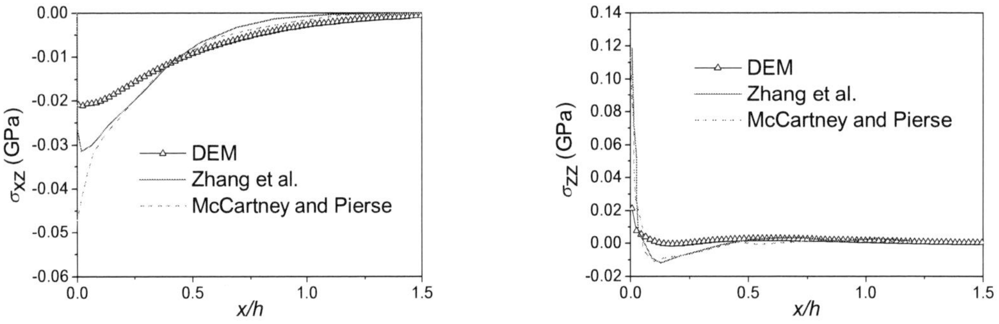

(a) shear stress (b) normal stress

Figure 6. Interlaminar stress distribution at 0°/90° interface under thermal-mechanical loading.

4. Progressive damage in cross-ply laminates

The microstructure in 90° plies of cross-ply laminates can be included by randomly generating *real* fibers according a corresponding fiber volume fraction. Therefore the fiber/matrix interface behavior is concerned and fiber/matrix debonding could take place during the formation of transverse cracking under thermal and/or mechanical loadings.

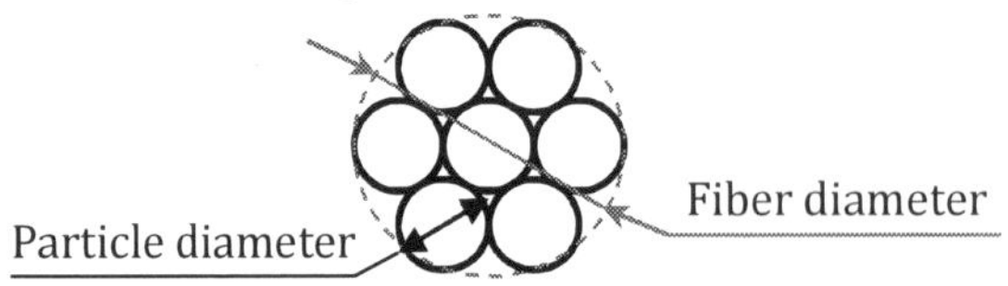

Figure 7. Schematic of a fiber with seven particles.

E-Glass/ LY3506 epoxy [0°/90°]$_s$ laminate is taken as an example, detailed material properties of which is listed in Table 6 [24]. The fiber volume fraction is 0.4 in this study, which is relatively small, for the purpose of successfully generating random fibers in 90° plies and remaining a large amount of particles for matrix so that the matrix cracking could onsite and propagate. The strength of the materials in 0° plies and fibers in 90° plies is assumed to be infinite in this model to avoid any breakage. Each fiber in 90° plies is formed by seven particle elements, as shown in Figure 7. The matrix and fibers are simplified as brittle elastic materials and are modeled by using parallel bonds. The interlaminar interface as well as the fiber/matrix interface are both assumed to the same with values in [25]. A deviation of 10% is assigned to both the interface and matrix strength in 90° plies according to Gauss distribution law so that damage could occur at random positions within the model. The model has a length of 2 mm and thickness of 0.5 mm with 82,221 particles.

Table 6. Constituents' material properties in E-Glass/ LY3506 epoxy cross-ply laminate.

Fiber volume fraction	0.40
Fiber diameter (μm)	11
Fiber modulus (GPa)	74
Fiber Poisson's ratio	0.2
Fiber CTE (10^{-6}/°C)	4.9
Matrix modulus (GPa)	3.35
Matrix Poisson's ratio	0.33
Matrix tensile strength (MPa)	80
Matrix shear strength (MPa)	54
Matrix CTE (10^{-6}/°C)	58
Ply thickness (mm)	0.125
Ply-ply interface properties	G_I=257 N/m, G_{II}=856N/m
Fiber/matrix interface properties	G_I=257 N/m, G_{II}=856N/m
Temperature change	ΔT=-120 °C

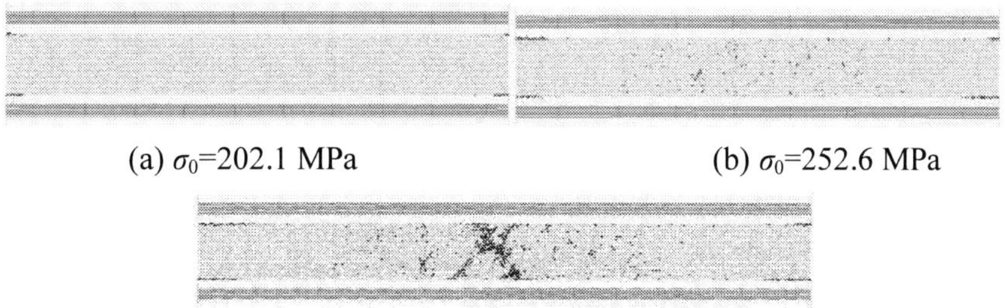

(a) σ_0=202.1 MPa (b) σ_0=252.6 MPa

(c) σ_0=281.2 MPa

Figure 8. Progressive damage in $[0/90]_s$ cross-ply laminate under thermal and axial mechanical loading (Red dots indicate matrix cracking; Black dots indicate interfacial yielding/failure).

Figure 8 shows the progressive damage when the mechanical loading, σ_0, was increasing after thermal loading. Delamination at the free-edges which was formed by both matrix cracking and ply-ply interfacial yielding of the material occurred at around 200 MPa loading (Figure 8 (a)). Lots of fiber/matrix interfacial debonding, compounding with few matrix cracking, were concentrated in the middle part of the model with the increase of loading (Figure 8 (b)). Transverse cracking was finally generated by the connection and bridge of matrix cracking and fiber/matrix debonding in the middle of the model, and reached the ply-ply interfaces causing a large amount of matrix cracking as well as interlaminar delamination along the ply-ply interfaces. A magnified view of the transverse cracking, as shown in Figure 9, shows that matrix cracking takes more occupations than fiber/matrix debonding. This may due to the low volume fraction of fiber. Apart from the transverse cracking, the fiber/matrix debonding mostly occurred at either left or right side of the fibers where large tensile stress exists.

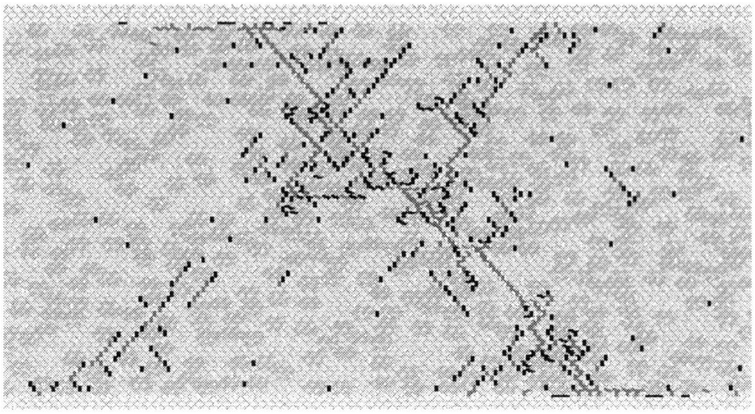

Figure 9. A magnified view of the transverse cracking in 90° plies.

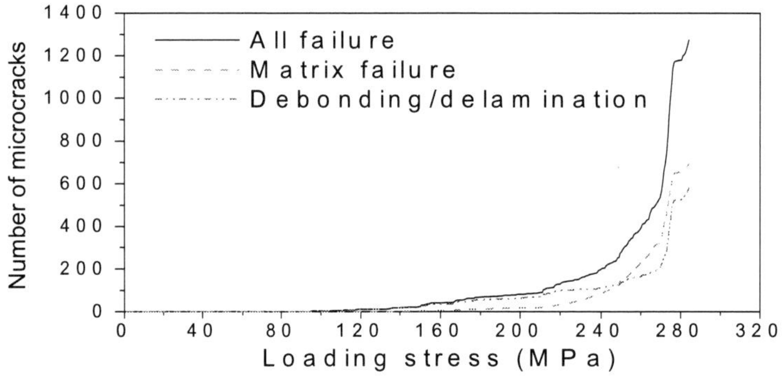

Figure 10. Microcrack number versus loading stress under thermal-mechanical loading.

A record of the interfacial failure (including both fiber/matrix debonding and ply-ply delamination) and matrix cracking was plotted in Figure 10, to give a quantitative assessment of the microcracking events. The number of microcracking for ply-ply delamination is much smaller the one for fiber/matrix interfacial debonding, since the delamination is mainly formed by matrix cracking, as shown in Figure 8 (c). The matrix failure seems less than the fiber/matrix debonding in the middle of the model before the generation of transverse cracking, which also can be seen in Figure 8 (b). Both the amounts of matrix cracking and fiber/matrix debonding were sharply increasing at a loading value of 270 MPa to form a transverse cracking. The amount of matrix cracking began to be larger than the one of fiber/matrix debonding at about 280 MPa loading when the transverse cracking reached the ply-ply interfaces and propagated along them with the failures of matrix material at the vicinity of ply-ply interfaces (also seen Figure 8 (c)).

5. Concluding remarks

The interlaminar stress distributions as well as progressive damage propagation in cross-ply laminates under thermal and/or mechanical loading have been studied by DEM. The material microstructure was considered in the DEM model which has given more detailed descriptions of the microcracking events including matrix cracking and fiber/matrix debonding. In the future work, an elasto-plastic constitutive model should be further developed to better represent the plastic behavior of matrix. As a preliminary result, though only a representative element of the cross-ply laminates was simulated, it has given us confidence that this model can be extended to simulate the progressive damage within laminates in a larger scale, e.g., 20 mm long with about one million particles which is within the capacity of one single advanced computer, to give more assessments of the material behavior in terms of crack density and stiffness reduction, etc. Once the model becomes large enough, the fiber breakage as well as fiber/matrix debonding in 0° plies can also be modeled with the capability of predicting the final failure strength of laminates from crack initiation to final structural collapse.

Reference

[1] D.J. Dawe, S.S.E. Lam, Z.G. Azizian, International Journal for Numerical Methods in Engineering, 35 (1992) 1087-1110.
[2] J.N. Reddy, Communications in Applied Numerical Methods, 3 (1987) 173-180.
[3] V.-T. Nguyen, J.-F. Caron, Composites Science and Technology, 69 (2009) 40-49.
[4] J.Q. Ye, H.Y. Sheng, International Journal of Mechanical Sciences, 45 (2003) 1309-1326.
[5] J.Q. Ye, H.Y. Sheng, Q.H. Qin, Computers & Structures, 82 (2004) 1131-1141.
[6] K.W. Garrett, J.E. Bailey, Journal of Materials Science, 12 (1977) 157-168.
[7] J.A. Nairn, Journal of Composite Materials, 23 (1989) 1106-1129.
[8] N. Hu, Y. Zemba, T. Okabe, C. Yan, H. Fukunaga, A.M. Elmarakbi, Mechanics of Materials, 40 (2008) 920-935.
[9] J.M. Berthelot, J.F. Le Corre, Composites Science and Technology, 60 (2000) 2659-2669.
[10] T. Okabe, M. Nishikawa, N. Takeda, Composites Science and Technology, 68 (2008) 2282-2289.
[11] D. Yang, Y. Sheng, J. Ye, Y. Tan, Composites Science and Technology, 71 (2011) 1410-1418.
[12] D. Zhang, J. Ye, H.Y. Sheng, Composite Structures, 76 (2006) 314-325.
[13] P.P. Parlevliet, H.E.N. Bersee, A. Beukers, Composites Part A: Applied Science and Manufacturing, 38 (2007) 1581-1596.
[14] C.H. Park, H.L. McManus, Composites Science and Technology, 56 (1996) 1209-1219.
[15] D.S. Adams, D.E. Bowles, C.T. Herakovich, Journal of Reinforced Plastics and Composites, 5 (1986) 152-169.
[16] S. Nickerson, J.S. Mayes, J.S. Welsh, Engineering Fracture Mechanics, 72 (2005) 1993-2008.
[17] Itasca Consulting Group Inc, Minneapolis, Minnesota, (2004).
[18] R.A. Schapery, Journal of Composite Materials, 2 (1968) 380-404.
[19] Z.H. Karadeniz, D. Kumlutas, Composite Structures, 78 (2007) 1-10.
[20] M. Tahani, A. Nosier, Composite Structures, 60 (2003) 91-103.
[21] S. Groves, C. Harris, A. Highsmith, D. Allen, R. Norvell, Experimental Mechanics, 27 (1987) 73-79.
[22] L. McCartney, C. Pierse, In: Proceeding of the 11th International Conference on Composite Materials, Gold Cost, Australia, July 14-18 (1997) 662-671.
[23] D. Yang, Y. Sheng, J. Ye, Y. Tan, Computational Materials Science, 49 (2010) 253-259.
[24] P.D. Soden, M.J. Hinton, A.S. Kaddour, Composites Science and Technology, 58 (1998) 1011-1022.
[25] J. Chen, M. Crisfield, A.J. Kinloch, E.P. Busso, F.L. Matthews, Y. Qiu, Mechanics of Composite Materials and Structures, 6 (1999) 301 - 317.

Tannin-based flax fibre reinforced composites for structural applications in vehicles

J Zhu, H Abhyankar, E Nassiopoulos and J Njuguna*

School of Applied Science, Cranfield University, Cranfield, MK43 0AL, UK

Email: j.njuguna@cranfield.ac.uk (*Corresponded Author)

Abstract. Innovation is often driven by changes in government policies regulating the industries, especially true in case of the automotive. Except weight savings, the strict EU regulation of 95% recyclable material-made vehicles drives the manufactures and scientists to seek new 'green materials' for structural applications. With handing at two major drawbacks (production cost and safety), ECHOSHELL is supported by EU to develop and optimise structural solutions for superlight electric vehicles by using bio-composites made of high-performance natural fibres and resins, providing enhanced strength and bio-degradability characteristics. Flax reinforced tannin-based composite is selected as one of the candidates and were firstly investigated with different fabric lay-up angles (non-woven flax mat, UD, $[0, 90°]_4$ and $[0, +45°, 90°, -45°]_2$) through authors' work. Some of the obtained results, such as tensile properties and SEM micrographs were shown in this conference paper. The UD flax reinforced composite exhibits the best tensile performance, with tensile strength and modulus of 150 MPa and 9.6 MPa, respectively. It was observed that during tension the oriented-fabric composites showed some delamination process, which are expected to be eliminated through surface treatment (alkali treatment etc.) and nanotechnology, such as the use of nano-fibrils. Failure mechanism of the tested samples were identified through SEM results, indicating that the combination of fibre pull-out, fibre breakage and brittle resins failure mainly contribute to the fracture failure of composites.

1. Introduction

To meet the current EU recyclability regulation in vehicle industries, one of the important achievements is the investigation and usage of natural composites, also called bio-composites which provide potential good mechanical performances and outstanding environmental advantages. The ECOSHELL project concentrates on the use of sustainable technologies and lays emphasis on the use of flax/tannin composites for structural applications in superlight vehicles. There is a gap found in literature in this area and more research at fundamental level needs to be done.

To date, tannin resins extracted from plants are widely used and reported as adhesives, however, there are still a small number of papers about tannin matrix composites. Coir fibre-reinforced tannin composites were firstly studied by Vilmar et al. [1] to show a potential for automotive applications as internal parts. Ndazi and his co-workers [2] also manufactured composite panel boards from rice husks and mimosa tannin resins. The interest of studying tannin-based composites has been increasing very fast due to some advantages: (i) non-toxic; (ii) wide availability; (iii) fast curing rate.

Many studies [3] concentrated on flax composites and provided valuable information, some of which is exhibited in **Table 1**. From Van de Velde and Kiekens [3] investigations, polypropylene (PP)

is the most suitable thermoplastic matrix for a flax-reinforced composite due to its extraordinary advantages, such as low density, low thermal expansion, good resistance to water and recyclability. Compared to the thermoplastic-based bio-composites, thermoset composites coming from vegetable oils (e.g. soybean oil and linseed oil) are preferred in the structure applications [4]. Akesson et al. [5] produced and studied thermoset composites from Lactic Acid-based (LA) resins and flax fibres. LA resin, a polyester resin containing carbon-carbon bonds, was cured at elevated temperature, together with impregnated flax fibres during compression moulding; the mechanical properties of final composites were increased by varying the fibre content from 40 wt% to 70 wt%, but drastically reduced with a fibre ratio of 75 wt%. Hughes et al. [6] also produced unsaturated polyester resin composites reinforced by unidirectional flax fibres and directed their investigation towards the tensile deformation behaviour. As an example of characterization of physical properties, the storage modulus of LA composites reinforced with 70 wt% flax fibres is 9.32 GPa and 3.29 GPa at 20 °C and 140 °C, respectively [5]. SEM is also an important characterisation method for composite technology (e.g. failure mechanism) and thus widely reported in the literature [7].

Table 1. Collected mechanical properties of flax reinforced composites.

Fibre/ Matrix	Processing	Tensile strength	Tensile modulus	Flexural Strength	Reference
Flax/bio-thermoset (MSO[a])	Compression moulding	50-120 MPa	6-15 GPa	180 MPa (max)	[8]
Flax/bio-thermoset (MMSO[b])	Compression moulding	50-120 MPa	7-15 GPa	201 MPa (max)	[8]
Arctic Flax/Epoxy (50:50)	Resin transfer moulding	280 MPa	40 GPa		[9]
Plain-woven flax/ thermoset	Compression Moulding	280 MPa	32 GPa	250 MPa	[10]
Flax yarn/ SPC[c] resins	Pultrusion	298 MPa	4.3 GPa	117 MPa	[11]
Flax/Lactic acid resins(70:30)	Compression moulding	62 MPa	9 GPa	96 MPa	[5]
Flax/ PLA (polyactic acid)	Injection moulding	40-55 MPa	3-6 GPa		[12]

*a-Methacrylated soybean oil, b-methacrylic anhydride-modified soybean oil, c-Soy Protein Concentrate Resin.

Concerning with the incompatibility between polymer matrix (hydrophobic) and flax fibres (hydrophilic), surface treatment and nanotechnology are used to handle at the issue and hence improve the whole composite performance. Alkali treatment, esterification and silane treatment are common surface treatment on flax fibres to improve the flax/polymer surface adhesion [13; 14]. Compared to the traditional produced flax fibres containing amorphous materials, nano-flax fibres provide high crystallinity by removing hierarchical structure and show a high crystal stiffness [15]. The significant enhancements of nano-flax fibre-reinforced composites are dispersion, thermal and mechanical properties due to high aspect ratio (length/diameter), resulting in the formation of rigid filler network [16-18]. Seydibeyoglu and his coworkers [17] demonstrated that the 16.5% cellulose nanofibrils reinforced polyurethane composites not only show highest strength increase of 500% but also have a stable storage modulus even at low temperature of -31°C. Nano-fibrillate flax fibres from high pressure disintegration have been reported to show good homogeneous dispersion and enhanced mechanical properties in polyvinyl alcohol matrix [18] .

Due to the inherent environmental benefits of using natural resin (tannin) and natural fibres (flax), it was decided to put this new composite technology to use. From the literature research, this new composite is only reported by Pizzi et al. [19] who manufactured the mimosa tannin-based composites reinforced by non-woven mat of flax fibres and also studied their mechanical properties. We clearly find a big knowledge gap in the mechanical/physical properties of flax/tannin composite, together with the environmental aging problems (temperature, humidity etc.) A methodology was developed to systematically analyse and characterise the properties of the composite in various parameters like fibre fabric arrangements.

2. Experimental

2.1. Materials

The flax mat/tannin composites were provided by ENSTIB. Other samples with various fabric orientations were supplied by MaHyTec. The details of the materials are shown in **Table 2**. The sample size for different characterizations was discussed in following methodology chapter.

Table 2. Flax/tannin composite systems supplied by manufacturers.

Type	Fabric form	Supplier	Manufacturing
A	Flax mat	ENSTIB	Compression moulding
B	UD (8 plies)	MaHyTec	Compression moulding
C	$[0, 90°]_4$	MaHyTec	Compression moulding
D	$[0, +45°, 90°, -45°]_2$	MaHyTec	Compression moulding

2.2. Tensile test of composite

The flat coupon tensile test was carried out on the Instron 50/100 kN machine according to ASTM D3039 at the cross head speed of 2mm/min. The test specimens were cut by the suppliers and received with a size of 250×25mm2. The average thickness was 2.5 mm. To obtain micro-scale elongation information during test, digital image correlation (DIC) method, together with a placed extensometer, was used on each specimen (see **Figure 1**). Three specimens were tested at room temperature for every composite type, and the average modulus and strength were obtained.

Figure 1. DIC technique applied to provide micro-scale information.

2.3. Scanning electron microscopy

Cross-section surface topographies of the composites after tensile testing were examined using a XL30 SFEG analytical high resolution scanning electron microscopy (SEM), supplied by FEI. Prior to SEM investigation, the test specimens were vacuum –coated with thin layers of gold powder to avoid the electrical charging during analysis. The whole examination was carried out at room temperature. Micrographs of surfaces were produced at various magnifications.

3. Results and Discussion

3.1. Tensile properties of composites

Both macro and micro-scale information of tensile properties were obtained through the experiments. Tensile strength, maximum strain, tensile modulus were easily read from the micro-scale stress-strain curve, while the macro-scale load-elongation curves to some extent explain the macro-mechanism of failure, which was also roughly shown through the SEM micrographs.

Using DIC to reduce the factor of dimension change under stress and offer advantages of non-contact, **Figure 2**(a) shows the relationship between stress and strain at micro-scale level obtained through the DIC method. Theoretically, the stress-strain curves measured by DIC were non-linearand thus the quantitative values were plotted as scatter diagrams. UD flax fabric/tannin samples have the best tensile strength up to 150 MPa, which is very competitive to some results in the literature [5, 8], due to the unidirectional fabric form significantly improving the tensile properties in their longitudinal direction. The [0, +45°, 90°,-45°]$_2$ composite were found to have less tensile strength but larger maximum strain than the [0, 90°]$_4$ type. This indicates that samples with [0, 90°]$_4$ fabric orientation are more brittle. The tensile strength and maximum strain of flax mat/mimosa tannin composites (type A) reaches the lowest values of 55 MPa and 0.7%, respectively, which is attributed to the random distribution of flax fibres.

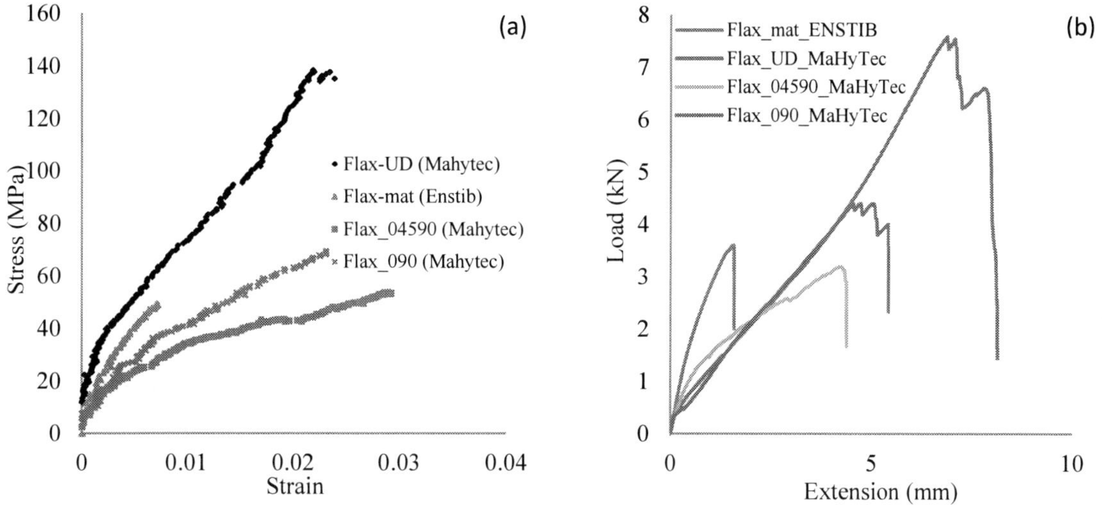

Figure 2. (a) micro-scale stress-strain curve; (b) macro-scale load-elongation curve.

Figure 3 represents the tensile modulus measured as the tangent modulus in the initial linear portion in the stress-strain curves. The portion close to zero point is not acceptable as a result of machine calibration. The young's modulus of specimen follows the trend B>C>D. Similar to tensile strength, the unidirectional fibres in type B apparently give rise to the maximum composite young's modulus of approximately 9.6 GPa. The [0, 90°]$_4$ layer arrangement (type C) produces higher Young's modulus of 4.8 GPa than the [0, 45°, 90°,-45°]$_2$ composite (type D), in accordance with tensile strength. One thing noticed is that the flax mat reinforced tannin composites (type B) shows a second highest value of Young's modulus (6.4 GPa), which is not compatible with the values of other types due to the different manufacturing details.

The failure mechanism can be indicated from **Figure 2**(b), illustrating the macro-scale load-extension curves for different tannin composites. After the maximum force, the loads applied on the flax mat composite (type A) dropped down immediately, whereas other three composite types behave in a different way (z-curves in **Figure 2**(b)). The first peak of composite B may be due to the combination of failure of tannin resins and the first occurring delamination. However, it is very likely that the progress of specimen delamination is not in a continuous and smooth fashion but rather than in an abrupt and irregular way. The load went to high levels before the sudden debond, at which the force subsequently dropped down and increased again until the next delamination; at the last peak, the specimen was flexible enough not to carry out any load. This delamination behavior indicates the poor wettability probably caused by the lack of tannin resins, which did not impregnate completely within the eight fibre plies. Another reason is the poor flax/tannin adhesion caused by the incompatibility between flax (hydrophilic) and tannin resins (hydrophobic).

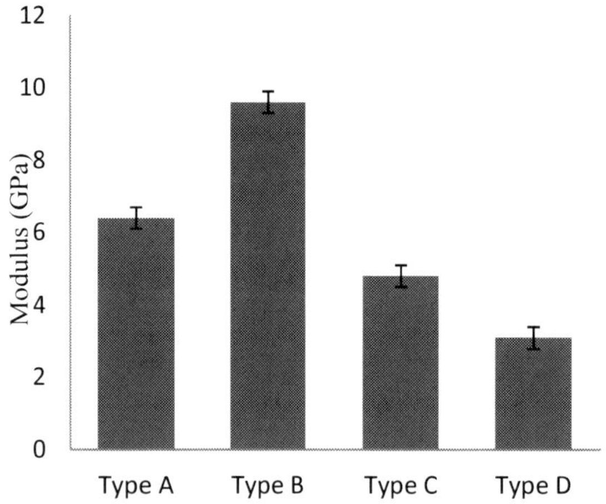

Figure 3. Tensile modulus of various flax/tannin composites.

3.2. Scanning electron microscopy

SEM micrographs for tensile surfaces of flax/tannin composites with various lay-up angles are shown in **Figure 4** and **Figure 5**. As discussed previously, composites B, C and D showed delamination behaviour without fracture during tensile testing and one of the delaminated samples were then cut into specimen and investigated by SEM. **Figure 4** (a) clearly shows the separation of fibre layers upon testing, indicating the poor surface adhesion between these fabric layers. Some noticeable gaps between matrix and fibres are indicated by 'A' which represents the non-contact fibre/matrix surface attributable to the fibre debonding and weak performances. It is advised to utilize chemical treatment like alkali treatment to enhance the fibre/matrix interfacial adhesion by purifying the fibre surface and improving the possibility for chemical bonding [20]. Also, an advanced approach for the problems is nano-scale fibres leading to remarkable fibre dispersion, hence good intermolecular interactions.

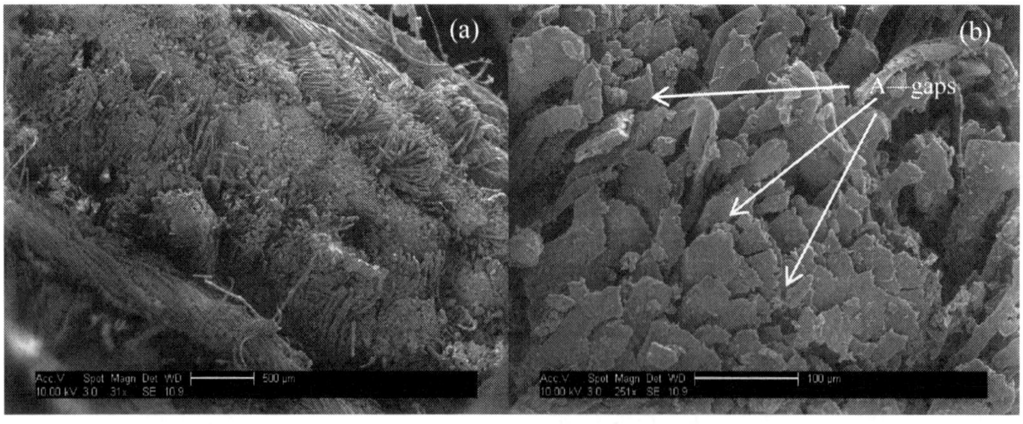

Figure 4. SEM micrographs of a cut non-fractured sample.

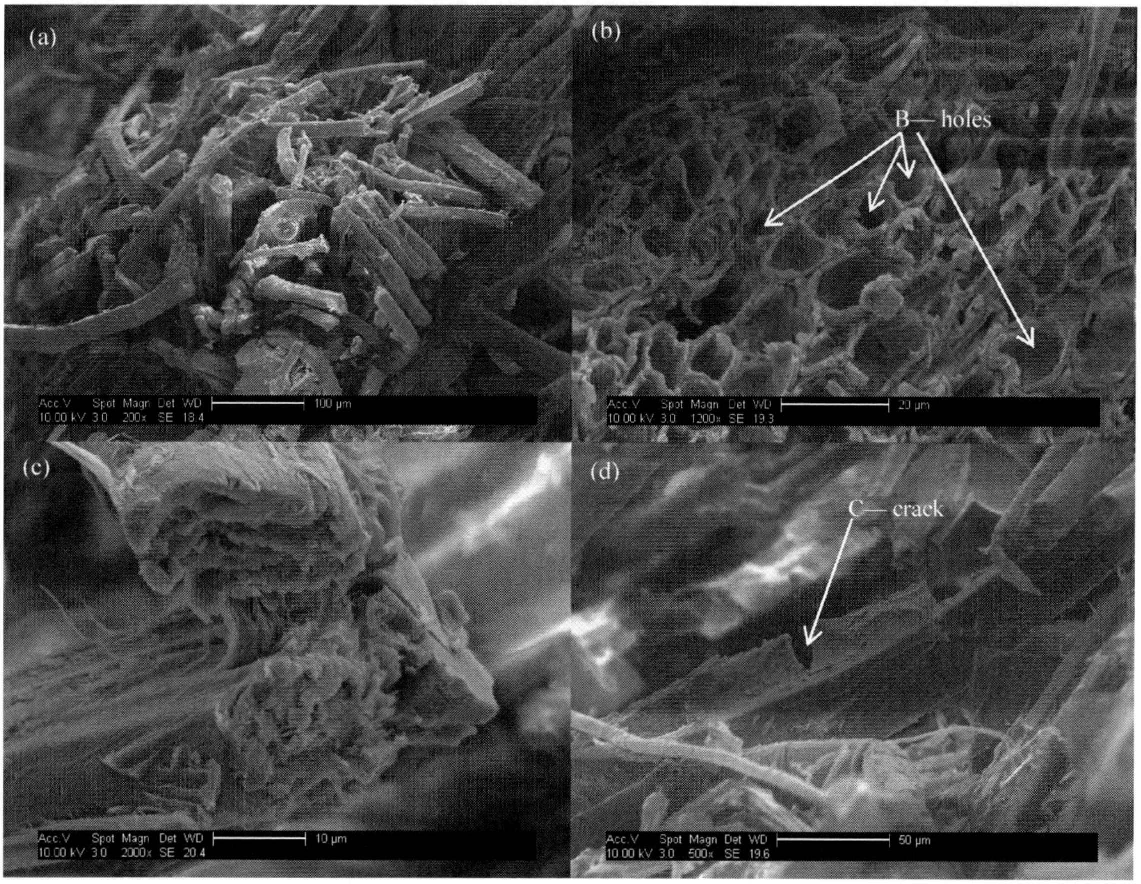

Figure 5. SEM micrographs of tensile fractured flax mat reinforced composites (Type A).

Figure 5 is the cross-sectional micrographs of tensile fracture surface. The failure mechanism shown in the figure is complicated. According to **Figure 5**(a), the bonding quality between fibre layers is fairly better than that of non-fracture specimen. **Figure 5**(b) shows the remaining holes (indicated by 'B') after fibre pull-out. The fibre breakage shown in **Error! Reference source not found.**(c) is caused by the applied stress. Symbol 'C' in **Figure 5**(d) points to a small crack within the tannin resins adjacent to fibres. This is brittle failure of tannin resins under tension force. Based on the investigations, the failure of flax fibre reinforced tannin matrix composites (ENSTIB) is governed by the combination of fibre pull-out, fibre fracture and brittle fracture of tannin resins.

4. Conclusions

Supported by ECHOSHELL project, flax/tannin composites with various fibre fabric orientations were supplied by the MaHyTec and ENSTIB. These composites were prepared by sheet compression moulding (SMC), and received by Cranfield University for further characterisation and modification. Tensile tests using DIC method and subsequent SEM characterisation on the composites were shown in the conference paper. It is apparently that the UD flax-reinforced tannin-based composites show the highest tensile strength of up to 150 MPa and elastic modulus of 9.6 MPa, as a result of longitudinal reinforcement. The trend in tensile strength partially depends on the numbers of UD oriented flax mat in the composites. The 'z' curves after failure in the load-extension figure suggest the discontinuous delamination progress probably originated from the in-sufficient wettability between flax fibres and tannin resins. SEM characterisation revealed the interface adhesion in the delaminated samples and also showed failure mechanisms (fibre pull-out, fibre breakage and brittle tannin failure) of the tensile fractured samples. These results are very competitive to the literature findings and show a good mechanical performance potential with further surface treatment (e.g. alkali treatment) or use of nano-scale flax fibres from adequate techniques (e.g. high pressure disintegration).

5. Acknowledgements

The authors are grateful to the FP 7 and ECHOSELL for the financial support for this research. We thank all the collaborative companies in ECOSHELL project and the staff/technicians in the Cranfield University for helping us on the experiments.

References

[1] Barbosa Jr., V., Ramires, E. C., Razera, I. A. T. and Frollini, E. (2010), "Biobased composites from tannin–phenolic polymers reinforced with coir fibers", *Industrial Crops and Products,* vol. 32, no. 3, pp. 305-312.

[2] Ndazi, B., Tesha, J. V., Karlsson, S. and Bisanda, E. T. N. (2006), "Production of rice husks composites with Acacia mimosa tannin-based resin", *Journal of Materials Science,* vol. 41, no. 21, pp. 6978-6983.

[3] Van de Velde, K. and Kiekens, P. (2001), "Thermoplastic polymers: overview of several properties and their consequences in flax fibre reinforced composites", *Polymer Testing,* vol. 20, no. 8, pp. 885-893.

[4] Lincoln, J. D., Shapiro, A. A., Earthman, J. C., Saphores, J. -. M. and Ogunseitan, O. A. (2008), "Design and evaluation of bioepoxy-flax composites for printed circuit boards", *IEEE Transactions on Electronics Packaging Manufacturing,* vol. 31, no. 3, pp. 211-220.

[5] Åkesson, D., Skrifvars, M., Seppälä, J. and Turunen, M. (2011), "Thermoset lactic acid-based resin as a matrix for flax fibers", *Journal of Applied Polymer Science,* vol. 119, no. 5, pp. 3004-3009.

[6] Hughes, M., Carpenter, J. and Hill, C. (2007), "Deformation and fracture behaviour of flax fibre reinforced thermosetting polymer matrix composites", *Journal of Materials Science,* vol. 42, no. 7, pp. 2499-2511.

[7] Baley, C., Perrot, Y., Busnel, F., Guezenoc, H. and Davies, P. (2006), "Transverse tensile behaviour of unidirectional plies reinforced with flax fibres", *Materials Letters,* vol. 60, no. 24, pp. 2984-2987.

[8] Adekunle, K., Cho, S. -., Ketzscher, R. and Skrifvars, M. (2012), "Mechanical properties of natural fiber hybrid composites based on renewable thermoset resins derived from soybean oil, for use in technical applications", *Journal of Applied Polymer Science,* vol. 124, no. 6, pp. 4530-4541.

[9] Oksman, K. (2001), "High quality flax fibre composites manufactured by the resin transfer moulding process", *Journal of Reinforced Plastics and Composites,* vol. 20, no. 7, pp. 621-627.

[10] Adekunle, K., Cho, S. -., Patzelt, C., Blomfeldt, T. and Skrifvars, M. (2011), "Impact and flexural properties of flax fabrics and Lyocell fiber-reinforced bio-based thermoset", *Journal of Reinforced Plastics and Composites,* vol. 30, no. 8, pp. 685-697.

[11] Huang, X. and Netravali, A. (2007), "Characterization of flax fiber reinforced soy protein resin based green composites modified with nano-clay particles", *Composites Science and Technology,* vol. 67, no. 10, pp. 2005-2014.

[12] Bax, B. and Müssig, J. (2008), "Impact and tensile properties of PLA/Cordenka and PLA/flax composites", *Composites Science and Technology,* vol. 68, no. 7–8, pp. 1601-1607.

[13] Cantero, G., Arbelaiz, A., Llano-Ponte, R. and Mondragon, I. (2003), "Effects of fibre treatment on wettability and mechanical behaviour of flax/polypropylene composites", *Composites Science and Technology,* vol. 63, no. 9, pp. 1247-1254.

[14] Kaith, B. S. and Kalia, S. (2007), "Grafting of flax fiber (Linum usitatissimum) with vinyl monomers for enhancement of properties of flax-phenolic composites", *Polymer Journal,* vol. 39, no. 12, pp. 1319-1327.

[15] Eichhorn, S. J., Dufresne, A., Aranguren, M., Marcovich, N. E., Capadona, J. R., Rowan, S. J., Weder, C., Thielemans, W., Roman, M., Renneckar, S., Gindl, W., Veigel, S., Keckes, J., Yano, H., Abe, K., Nogi, M., Nakagaito, A. N., Mangalam, A., Simonsen, J., Benight, A. S., Bismarck, A., Berglund, L. A. and Peijs, T. (2010), "Review: Current international research into cellulose nanofibres and nanocomposites", *Journal of Materials Science,* vol. 45, no. 1, pp. 1-33.

[16] Qua, E. H., Hornsby, P. R., Sharma, H. S. S., Lyons, G. and Mccall, R. D. (2009), "Preparation and characterization of Poly(vinyl alcohol) nanocomposites made from cellulose nanofibers", *Journal of Applied Polymer Science,* vol. 113, no. 4, pp. 2238-2247.

[17] Özgür Seydibeyoğlu, M. and Oksman, K. (2008), "Novel nanocomposites based on polyurethane and micro fibrillated cellulose", *Composites Science and Technology,* vol. 68, no. 3-4, pp. 908-914.

[18] Hrabalova, M., Schwanninger, M., Wimmer, R., Gregorova, A., Zimmermann, T. and Mundigler, N. (2011), "Fibrillation of flax and wheat straw cellulose: Effects on thermal, morphological, and viscoelastic properties of poly(vinylalcohol)/fibre composites", *BioResources,* vol. 6, no. 2, pp. 1631-1647.

[19] Pizzi, A., Kueny, R., Lecoanet, F., Massetau, B., Carpentier, D., Krebs, A., Loiseau, F., Molina, S. and Ragoubi, M. (2009), "High resin content natural matrix-natural fibre biocomposites", *Industrial Crops and Products,* vol. 30, no. 2, pp. 235-240.

[20] Yan, L., Chouw, N. and Yuan, X. (2012), "Improving the mechanical properties of natural fibre fabric reinforced epoxy composites by alkali treatment", *Journal of Reinforced Plastics and Composites,* vol. 31, no. 6, pp. 425-437.

A new-developed nanostructured Mg/HAp nanocomposite by high frequency induction heat sintering process

K A Khalil

Currently: Mechanical Engineering Department, College of Engineering, King Saud University P.O. Box 800, Riyadh 11421
Permanently: Materials and Mechanical Design Department, Faculty of Energy Engineering, South Valley University, Aswan, Egypt

Tel.: +966-5092120039; fax: +966-1-4676652,
E-mail address: kabdelmawgoud@ksu.edu.sa

Abstract. The objective of the present study was to investigate the effect nano-hydroxyapatite contents on the mechanical and microstructural properties of magnesium nanocomposites in order to develop a new biodegradable hard tissue substituent. Mg/HAp nanocomposite with various HAp contents (0-10 wt %) were prepared using pure magnesium and HAp nanopowder as raw materials. The starting material was super fast densified by high frequency induction heat sintering (HFIHS). The results indicated that, a uniform distribution of HAp particles was observed along the boundary between matrix particles. Nano-crystalline grains with a crystal size range of 37-50 nm were obtained. The relative densities and microhardness of the composites initially increased with increase the amount of HAp addition. Despite the short dwelling time when the current was applied, the relative density and microhardness of the sintered samples reached as high as 99.7 % and 70 HV respectively, in the composite containing 1 to 3 wt% HAp. Addition of 1 to 3 wt% of HAp improved compression strength of Mg by 16%. Addition of HAp decreases the crystal size of the nanocomposites. The mechanical properties, i.e. hardness and compressive strength are evidently increased with increasing HAp content up to 2 wt%. However, when the HAp content was larger than 2 wt%, the compressive strength decreased due to the agglomeration of HAp particles. The hardness, compressive and ultimate stress this composite

1. Introduction

Recently, magnesium (Mg) and its alloys have attracted much attention as potential biodegradable bone implant materials due to their biodegradability in the bioenvironment [1–3]. It is characterized by non-toxicity and good biocompatibility. They are considered a novel promising biodegradable biomaterial. Mg has the lowest density (1.74 g/cm3) of all the engineering metals with a density only slightly higher than that of plastics. Furthermore, Mg has excellent machineability, good recyclability and surface finish [4]. However, pure Mg has low elastic modulus, limited strength and creep resistance at elevated temperatures and also poor ductility due to its hexagonal crystal structure. Due to its insufficient mechanical strength, magnesium is not used in clinics up to now.

Thus, Mg is usually alloyed with other metals or used as composites [5]. It was found that Mg alloy with 3 wt% Al provides the optimal ductility and alloying with 6 wt% Al produces the optimum combination of strength and ductility [6]. Enhanced mechanical properties have been reported in Mg composite reinforced with nanoscaled ceramic particles such as SiC [7], Al_2O_3 [8] and AlN [9]. The demand for greater performance of the Mg alloys or composites leads to the development of Mg-based metal matrix composite (MMC) with nanocrystalline microstructure reinforced by in-situ ceramics [10]. Composites not only have the combined properties of their constituents, it can also be tailored to offer improved properties to meet different engineering requirements. Other elements alloying has been studied for developing biodegradable magnesium alloys with good mechanical and corrosion properties. Mn and Zn were selected as the alloying elements to develop Mg–Mn–Zn alloys due to the good biocompatibility of Mn and Zn [28, 29]. The addition of Mn and Zn improves both the mechanical properties and the corrosion resistance of magnesium alloys.

It is will known that, the composition of hydroxyapatite is similar to that of the mineral in human bone. It is excellent biocompatible, biodegradable, osteoconductive and non-toxic. But HAp is unsuitable for load-bearing implants due to its low fatigue resistance, brittleness, insufficient flexural strength and impact resistance. Hydroxyapatite is used as a coating for metal surgical implants (most often made of titanium and its alloys, or stainless steels), and recent studies have examined the possibility of its use in composite form, in materials that combine polymers with ceramic or metal/ceramic combinations. Considerable research has been performed on methods of coating application and in-situ synthesis of appetites, and the implications for ceramic properties and microstructure. Dense aggregates of HAp are typically produced by high temperature sintering or hot-isostatic pressing of sized powders (e.g., Halouani et al., [11]; van Landuyt et al. [12]; Layrolle et al.0 [13]). These studies have focused on optimizing the mechanical properties of the aggregates. However, to our knowledge, Mg/HAp nanocomposites have not been fabricated before in the literature.

On the other hand, an important reported aspect of the HFIHS process is the role of the rapid heat transfer to the product via electromagnetic waves. In previous papers [14- 21], we reported on an investigation on the consolidation and mechanical properties of different nanostructured bioceramic materials by high-frequency induction heat sintering. Unfortunately, the grain size of the sintered parts was higher than 100 nanometer. Furthermore, an exact interpretation of the microscopic effect of the HFIHS has not been achieved. In this study, a new developed Mg/HAp nanocomposite with various HAp contents (0-10 wt %) were prepared using pure magnesium and HAp nanopowder as raw materials by using HFIHS. A grain size of less than 100 nm is the main objective. The mechanical properties of the fabricated nanocomposites are accessed.

2. Experimental procedures

The basic configuration of a HFIHS unit is shown in Figure 1. It consists of a uniaxial pressure device, a graphite die (outside diameter, 45 mm; inside diameter, 20 mm; height, 40 mm). A water-cooled reaction chamber that can be evacuated, an induced current, pressure-, position- and temperature-regulating systems are also presented. Temperatures can be measured by a pyrometer focused on the surface of the graphite die. The system is first evacuated to a vacuum of 5×10^{-3} Torr and a uniaxial pressure is applied. An induced current (frequency of about 50 kHz) is then activated and maintained until densification observed, indicating the occurrence of the sintering and the concomitant shrinkage of the sample. Sample shrinkage could be measured by a linear gauge measuring the vertical displacement. Magnesium powder of 98.5% purity obtained from Merck KGaA, Germany, was used as the matrix material. Hydroxyapatite (HA) nanopowder of 99% purity obtained from Sigma Aldrich Corporation, USA, was used as the reinforcement. The powder has been checked for purity by X-ray diffraction as shown in Fig 2. The measured crystal size for magnesium and hydroxyapatite were 51 and 41 nm respectively. The starting grain size has been established by inspection of grain mounts using a Field Emission Scanning Electron Microscope. Figure 3 shows FE-SEM micrographs of the used powders. Experiments have been conducted on the processing of nanostructured magnesium 1-10wt% hydroxyapatite nanocomposites and sintered by high-frequency induction heated sintering (HFIHS).

Pure magnesium and hydroxyapatite nanopowder were weighed carefully in an argon-filled glovebox. All powders were milled in FRITCSH-Germany, mechanical alloying machine using zirconia balls and a rotation speed of 200 rpm for 30 min with a ball-to-powder weight ratio of 5:1. The milling conditions were the same for all the powders. The blended powders were placed in a graphite die (outside diameter, 45 mm; inside diameter, 20 mm; height, 40 mm) and then introduced into the high-frequency induction heat sintering, shown schematically in Figure 1. The choice of induced current was based on preliminary experiments aimed at determining the highest heating rate (highest current) that results in sample temperatures below the melting point of the material used. The samples have been densified by heating to a sintering temperature of 550 °C, and then fast cooled to room temperature within a short time. A pressure of 60 MPa was applied in two steps. In the first step the pressure was 30 MPa and maintained constant during heating, increased to 60 MPa in the second step during holding time. A heating rate of 400 °C/min was applied. After sintering, the samples were ground and polished for subsequent indentation and microscopy studies. The density of the sintered samples was measured by the Archimedes principals. Theoretical density of the Mg-HAp nanocomposites were measured based on the following equation:

$$\rho_{ave} = \frac{100}{\frac{C_1}{\rho_1} + \frac{C_2}{\rho_2}} \quad (1)$$

$$\rho_1 = \rho\,(HAp) = 3.16\ g/cm^3 \quad (2)$$

$$\rho_2 = \rho\,(Mg) = 1.74\ g/cm^3 \quad (3)$$

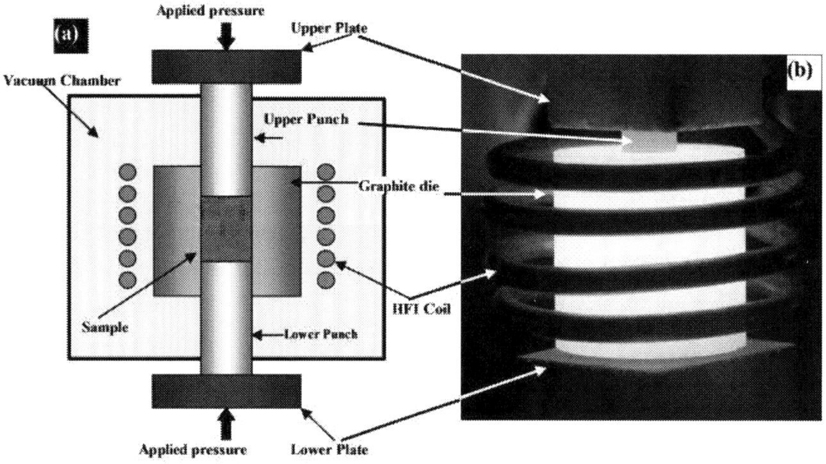

Figure 1. (a) Schematic diagram of high-frequency induction heated sintering apparatus, (b) Photo of the heated die.

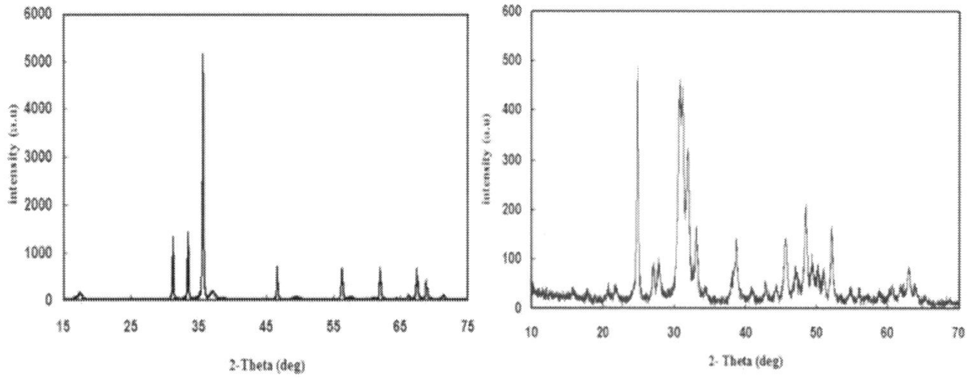

Figure 2. XRD pattern of the as received magnesium and hydroxyapatite powders.

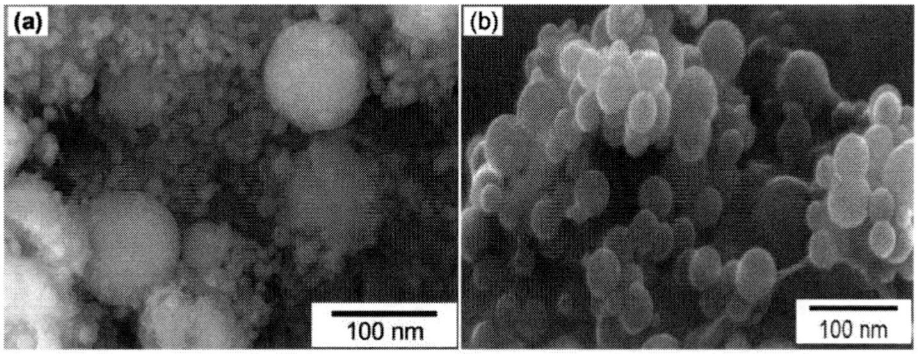

Figure 3. SEM micrographs of the hydroxyapatite and magnesium powders.

The microstructure observations of polished and coated fracture surfaces of the composite were investigated using a Field Emission Scanning Electron Microscope (FE-SEM) (JEOL FESEM 7600 F) coupled with EDS analysis. The surfaces were coated with platinum to avoid charging during the FE-SEM observation. Phase identification was performed on samples cut from the center of hot pressed bodies using X-ray diffractometry (XRD). X-ray diffraction analysis was carried out on the polished samples of Mg and Mg/HAp nanocomposites using an automated Shimadzu XRD-7000 diffractometer. The samples were exposed to Cu Ka radiation (k = 1.54056A°) at a scanning speed of 2 deg/min. The Bragg angle and the values of the interplanar spacing, d, obtained were subsequently matched with the standard values for Mg, and other related phases. The crystal size of the as received powder and sintered bodies has been also determined using XRD technique. The Full Width Half Maximum (FWHM) for each peak has been detected at specific 2θ degree and evaluated using known Scherer formula. Hardness test was carried out using a Vickers diamond indenter on an automated hardness tester (Buehler, Micomet5114, Microhardness tester, Akashi, Japan). Sintered specimens of pure nano-Mg and Mg-HAp with 1.0 to 10wt.% of HAp, were tested and evaluated for their hardness. During the hardness test, a load 0.3 Kgf with a loading time of 10 s was applied to ensure that no crack propagation path was developed. Specimens used in hardness testing had average dimensions of 20 mm in diameter and 5 mm in thickness. Three samples of each of the composition type were tested for their hardness at three different locations. The average of these readings were computed, reported and compared. This determination of hardness was at ambient temperature and followed the pattern ASTM C 1421-99. To determine mechanical strength under uniaxial compressive loading, the sintered Mg-HAp specimens were tested in a fully automated tensile tester from Instron (Model 3385H, Instron Co., USA), with a constant crosshead speed of 0.5 mm/min. Compressive strength of three samples was determined and the average for each composition were calculated and compared for analysis.

3. Results and Discussions

3. 1 Characteristics of the milled powders

Figure 4 shows morphologies of the mixture of Mg–1 and 5wt% HAp powder after milling for 30 min. The as received Mg and hydroxyapatite nanopowder shown in Figure 3 before milling shows multi-crystalline and agglomerations. Compared with the initial powder, the morphology of the milled powder is similar to a spherical particle and the specific surface area is somehow decreased. Milling results in a uniform distribution of Mg and HAp powders. The refinement of grain size is resulted from the collisions between powders and balls during milling process. Specific surface area is important in relation to several factors: particle size and particle size distribution, interior pores, particle shape, agglomeration, homogeneity of the elemental powder distribution and intrinsic characteristics of powder (frangible or ductile). HAp powder is frangible whereas Mg powders are ductile. The decrease of specific surface area can result in an increase of the apparent density and tapping density.

Figure 4. FE-SEM morphology of the Mg with (a) 1wt% (b) 5wt% HAp powder mixture after milling.

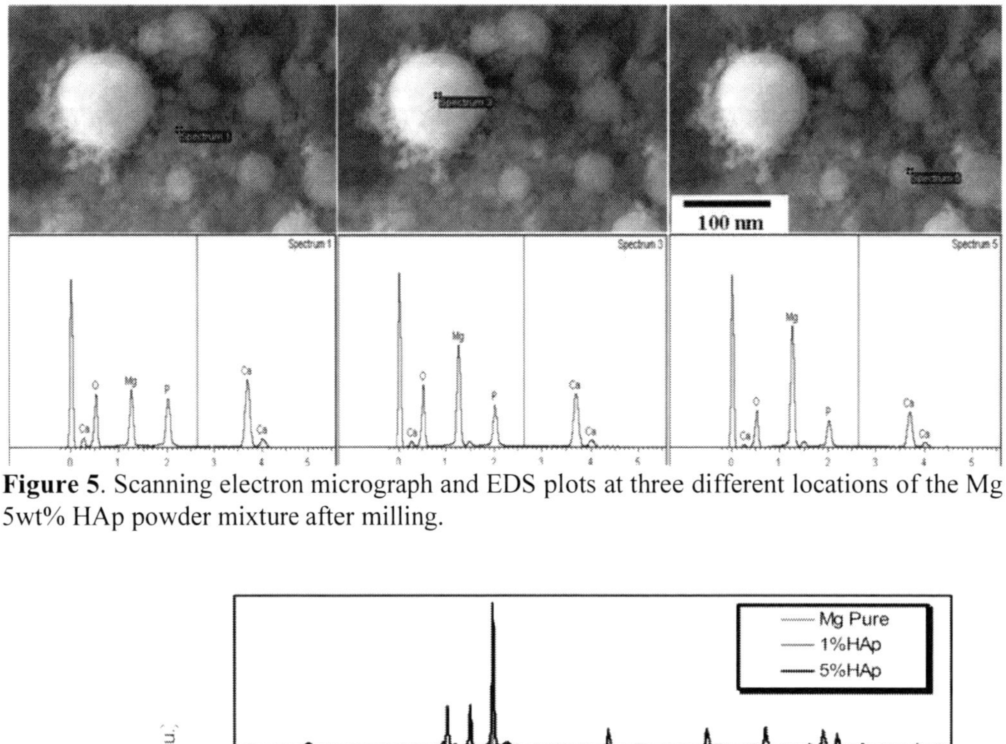

Figure 5. Scanning electron micrograph and EDS plots at three different locations of the Mg 5wt% HAp powder mixture after milling.

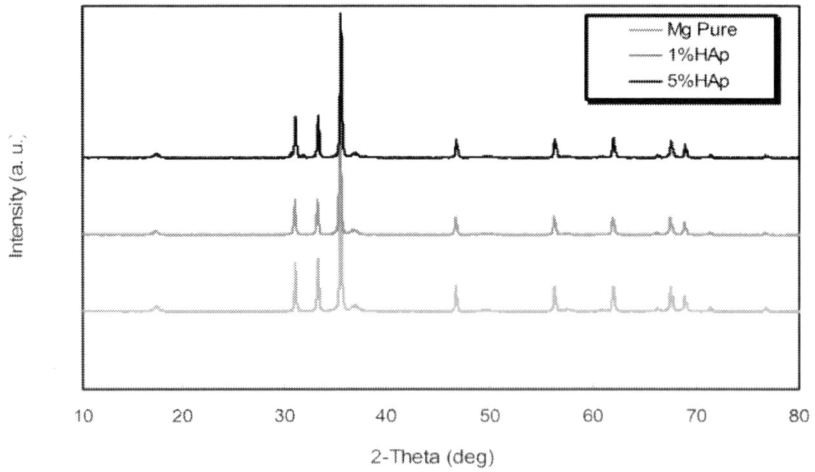

Figure 6. XRD pattern of the Mg pure and Mg with 1 and 5wt% HAp powder mixture after milling.

Figure 5 shows the EDS analyses of the milled powder. It can be observed that, each Mg particle whether it is a small or large is coated with HA particles. EDS analyses clearly showed the presence of

Ca, P and O of the HAp in the pure Mg structure. In addition, magnesium rich secondary phase was seen when observed at high magnification. Though the phase structure of HAp was not evident in our XRD analysis, we believe the presence of this phase was responsible for the improvement in physical and mechanical properties of Mg-HAp structures compared to pure Mg. Figure 6 shows the XRD pattern of the milled powder. It can be seen that the width of the diffraction peaks become wider than before milling suggesting decreasing in crystal size. As was measured by Scherrer formula, the crystal size of the samples containing 1 and 5wt% HAp were 34 and 31 nm, respectively. Compared with the XRD profiles of pure Mg-HAp, the Mg phase was the main constituent phase and even there are no peaks corresponding to HAp could be observed in the milled sample due to the lower weight fraction of the reinforcement material. There is no new crystalline phase formed during the milling indicating that no chemical reactions between Mg and HAp occurred.

3.2 Effect of HAp contents on the mechanical and microstructural properties

Mg/HAp nanocomposites have been consolidated very rapidly by HFIHS at constant heating rate around 400 °C/min, constant sintering pressure (60 MPa), temperature (550 °C) and time (3 minutes holding time) in order to develop a new biodegradable hard tissue substituent possessing the advantages of the both components. The mechanical properties of the Mg/HAp nanocomposite were discussed. The results of density and relative density measurements are shown in Table 2. The results revealed that near dense pure magnesium and composite materials can be obtained using the fabrication methodology adopted in this study.

Table 2. Results of density and relative density measurements.

Mg (C_2) (wt%)	*HAp (C_1) (wt%)*	*Theoretical density (g/cm^3)*	*Measured Density (g/cm^3)*	*Relative Density (%)*
100	0	1.740	1.727287	99.26939
99	1	1.748	1.742197	99.668
98	2	1.756	1.745886	99.42402
97	3	1.764	1.75187	99.31233
96	4	1.772	1.748808	98.69119
95	5	1.78	1.763222	99.05739

Figure 7 shows the density and microhardness of Mg-X wt% HAp sintered at sintering temperature of 550 °C, 60 MPa applied pressure and 400°C/min heating rate. The relative densities of the composites initially increased with increase the amount of HAp addition. We can see that, despite the short dwelling time when the current was applied, the relative density of the sintered samples reached as high as 99.7 % in the composite containing 1% HAp. The results of hardness measurements revealed an increase in microhardness values with increasing weight percentage of nano-size HAp reinforcement into the magnesium matrix to a certain level and then decreased. Among the nanocomposite formulations, magnesium reinforced with 1 to 3 wt% of HAp nanopowder showed the best overall improvement in both relative density and hardness. The hardness of Mg reinforced with 1 to 3 wt% HAp 70 HV as well as the relative density reached as high as 99.7%. The reason for the increase in Relative density and microhardness is probably due to the presence of HAp fragmented nanoparticles at the grain boundary regions. These nanosized inclusions can significantly enhance the mechanical properties of Mg. Remarkably, with increasing HAp content more than 3 wt% HAp, the relative density and microhardness decreased. Decreasing in relative density and microhardness of the nano composites with increasing HAp content can be attributed to two reasons. First, some degree of HAp nanoparticle clustering takes place. Second, the sintering temperature of pure HAp is almost around 1000 °C, with increasing HAp content in the Mg matrix, 550 °C the sintering temperature become not enough for densifications. Notably, increasing sintering temperature is not possible due to low melting temperature of Mg matrix. The hardness values for all of the Mg/HAp nanocomposite samples have narrow distribution ranges, indicating a homogeneous distribution of the HAp phase in the Mg matrix.

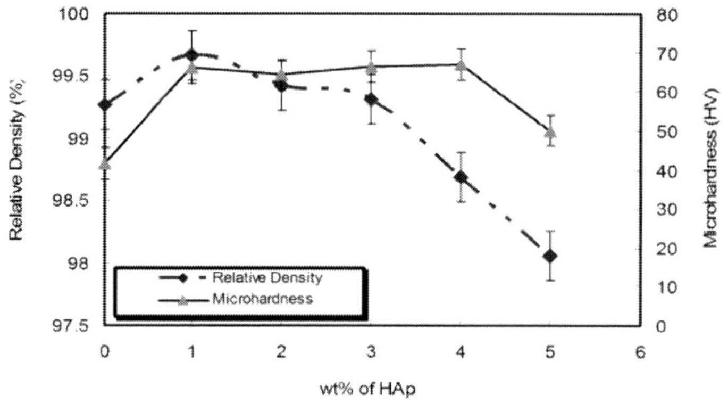

Figure 7. Effects of hydroxyapatite additions on the relative density and microhardness of Mg/HAp nanocomposites.

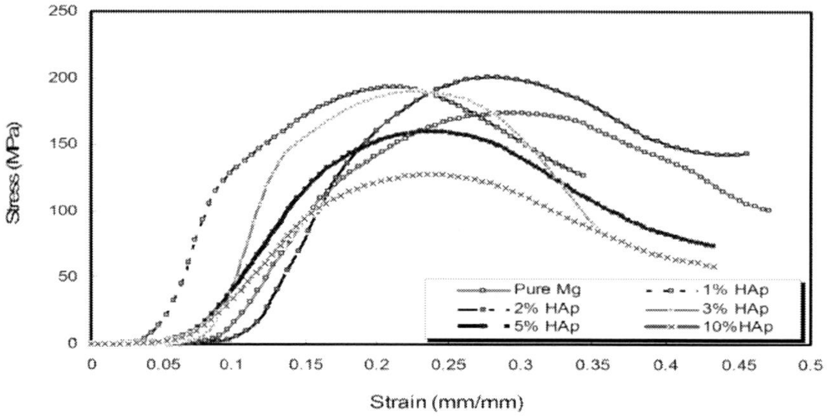

Figure 8. Typical stress–strain curve for the composites of Mg/HAp nanocomposites.

Pure and nanocrystalline Mg/HAp composites, sintered at 550 °C, were subjected to mechanical testing; to determine their failure strength under uniaxial compressive loading. Sintered specimens of compositions nano-HAp, 1.0 to 5 wt% HAp were evaluated for their compression strength in a fully automated, screw driven tensile tester as shown in Fig 8. It is clear that, in the sample of Mg contains 0 wt% HAp, the value of compressive strength was low about 173 MPa in structures of pure nano-Mg, which is higher compared to compressive strength of Mg structures of sintered by conventional sintering [22-26]. However, as the amount of HAp increased, the values of compressive strength increased due to the enhancing of the nanocomposite density. Figure 9 shows the maximum compressive strength of Mg/1-10 wt% HAp. The compressive strength of the Mg/2 wt% HAp is reached as high as 204 MPa. Thus, when adding higher than 2 wt% HAp to the Mg nanocomposites, the value was again decreased to about 127 MPa with 10 wt% of HAp. Mechanical properties of Mg have been widely investigated by many researchers. Varied results have been observed and reported. More or less, these properties depend on how the structures were processed. Pressure assisted sintering as well as isostatic pressing had shown superior results. In addition, sintering temperature, time, presence of additives and their amount has played a major role in mechanical properties of sintered Mg. Recently; the arena of nanotechnology has been explored to enhanced properties of HAp as well.

In a word we can conclude that, presence of HAp in the nano-Mg sintered by HFIHS further improved the mechanical strength. Addition of 1 to 3 wt% of HAp improved compression strength of Mg by 16%.

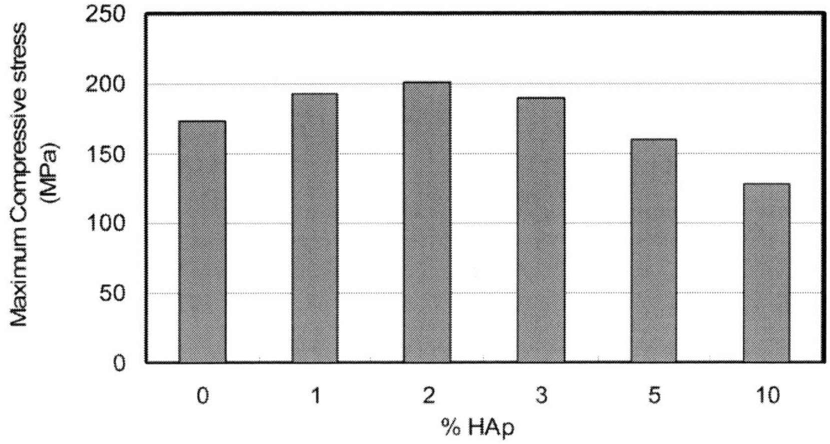

Figure 9. Maximum compressive strength of Mg with different HAp contents.

X-ray diffraction analysis was carried out on the polished samples of Mg and Mg/HAp nanocomposites using an automated Shimadzu XRD-7000 diffractometer. The samples were exposed to Cu Ka radiation (k = 1.54056A°) at a scanning speed of 2 deg/min. The Bragg angle and the values of the interplanar spacing, d, obtained were subsequently matched with the standard values for Mg, and other related phases. After filtering down the data, we can observe no any additional peaks in Mg composite reinforced with 1-5 wt% HAp that do not correspond with the pure Mg phase. Even HAp peaks were not detected in other composite formulations presumably because of its low weight fraction. The diffractograms of Pure Mg and Mg/HAp composite samples are shown in Figure 10.

To conform prohibiting of grain growth during sintering due to using ultra fast heating rate and a very short holding sintering time, the crystal size has been calculated by using Scherrer formula as mentioned before. The average crystal size from the strongest peak has been calculated. The results of crystal size measurements are plotted against wt% HAp contents as shown in Figure 11. It is clear that, addition of HAp decreases the crystal size of the nanocomposites. This is actually because the crystal size of the starting HAp is lower than that of Mg. However, these results prove that, sintering of Mg/HAp nanocomposites by HFIHS is very effective as long as the sintering parameters could be controlled and maintained. This indicates that, the sintering efficiency of this method is very high. This high efficiency could be explained from two aspects. First, the high pressure applied could accelerate the densification process. Second, the powder is heated in a very short time with a high-temperature exposure as a result of the heat being transferred to the product via electromagnetic waves. These results prove that HFIHS is a potential method for fabricating nanocomposites at much lower temperature and within an extremely short time [19].

Microstructural characterization studies were conducted to determine grain morphology, presence of porosity, pore morphology and distribution of reinforcement. Field Emission Scanning Electron Microscope has been used. The results of microstructural characterization conducted on HFIHS sintered samples at constant sintering temperature, pressure, time and heating rate are shown in Figure 12 through (a to f). It is clear that, Mg and Mg/1 to 3wt% HAp samples did not reveal presence of any microdefects. The outer surfaces were smooth and free of circumferential cracks. For Mg/3 to 5wt% HAp samples, the outer surfaces were comparatively rougher. The surfaces of extruded rods were smooth except for the last portion which showed the presence of shallow circumferential cracks. The results also revealed relatively uniform distribution of nano-sized HAp particulates with minimal presence of clusters (HAp particulates together but not fused with each other) in the case of samples containing low amount of HAp weight percentage (1 and 2).

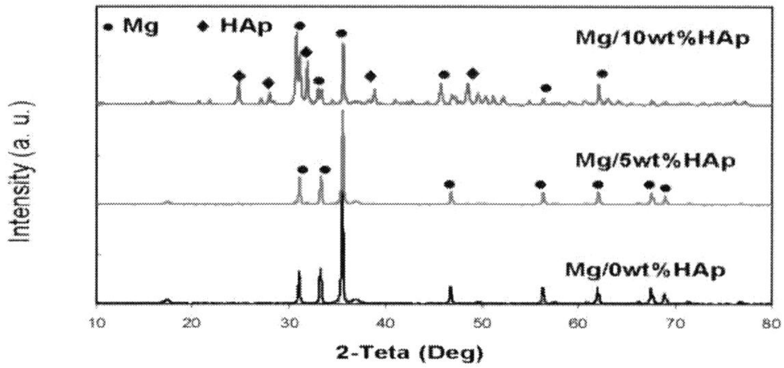

Figure 10. XRD pattern of the Mg pure and Mg with 5 and 10wt% HAp nanocomposites after sintering.

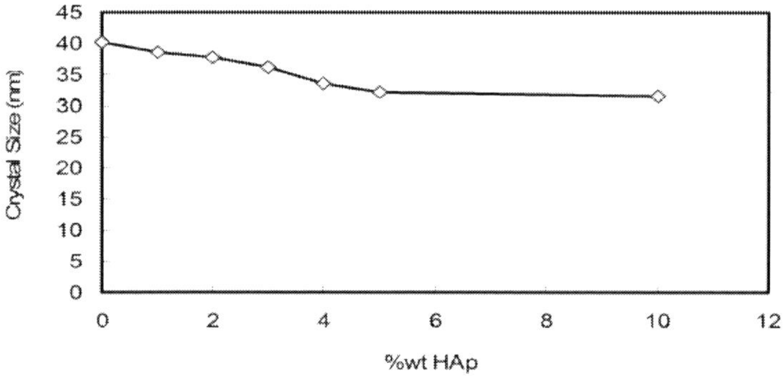

Figure 11. Crystal size measurements of the Mg/ X wt% HAp nanocomposites after sintering.

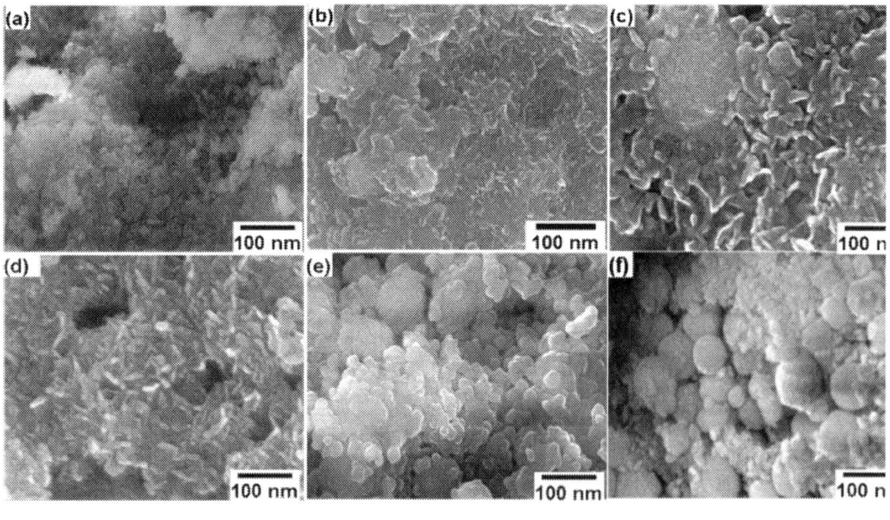

Figure 12. FE-SEM sowing the microstructure of (a) Mg-0.0%HA, (b) Mg-1.0%HA, (c) Mg-2.0%HA, (d) Mg-3.0%HA, (e) Mg-4.0%HA, and (f) Mg-5.0%HA sintered by HFIHS.

Figure 12 (a and b) and poorly distributed HAp particulates with presence of clusters and agglomerated region (HAp particulates fused with each other) in the case of samples containing high amount of HAp weight percentage (3 to 5) Figure 12 (d to f). It was observed from the FE-SEM micrographs (Figure 12) that particles were distributed everywhere inside the samples, either along the grain boundaries or inside the grains for samples sintered at both low and high HAp contents. In composite samples containing low content of HAp, HAp particles were well decorated along the particle/grain boundary.

4. Conclusions

Mg/HAp nanocomposites were successfully synthesized using High Frequency Induction Heat Sintering route. From the experimental results mentioned above, some conclusions were made as follows:

1. HFIHS was effective in the preparation of fine-crystals, nearly fully dense of Mg/HAp nanocomposites from the powder with a smaller crystal size.
2. The relative densities and microhardness of the composites initially increased with increase the amount of HAp addition. Despite the short dwelling time when the current was applied, the relative density and microhardness of the sintered samples reached as high as 99.7 % and 70 HV respectively, in the composite containing 1 to 3 wt% HAp. In the sample of Mg contains 0 wt% HAp, the value of compressive strength was low about 173 MPa in structures of pure nano-Mg. However, as the amount of HAp increased, the values of compressive strength increased due to the enhancing of the nanocomposite density. The compressive strength of the Mg/2 wt% HAp is reached as high as 204 MPa. Thus, when adding higher than 2 wt% HAp to the Mg nanocomposites, the value was again decreased to about 127 MPa with 10 wt% of HAp. Addition of 1 to 3 wt% of HAp improved compression strength of Mg by 16%.
3. Addition of HAp decreases the crystal size of the nanocomposites. A uniform distribution of HAp particles was observed along the boundary between matrix particles. The mechanical properties, i.e. hardness and compressive strength increased with increasing HAp content up to 2 wt%. However, when the HAp content was larger than 2 wt%, the compressive strength decreased due to the agglomeration of HAp particles. The agglomeration of HAp particles was found to lead to the degradation of the interfacial bonding strength between matrix and reinforcement. However, these results prove that, sintering of Mg/HAp nanocomposites by HFIHS is very effective as long as the sintering parameters could be controlled and maintained.

5. References

[1] Witte F, Kaese V, Haferkamp H, Switzer E, Meyer-Lindenberg A, Wirth CJ, et al. In vivo corrosion of four magnesium alloys and the associated bone response. Biomaterials 2005;26(17):3557–63.
[2] Witte F, Fischer J, Nellesen J, Crostack H-A, Kaese V, Pisch A, et al. In vitro and in vivo corrosion measurements of magnesium alloys. Biomaterials 2006; 27(7):1013–8.
[3] Xu L, Yu G, Zhang E, Pan F, Yang K. In vivo corrosion behavior of Mg–Mn–Zn alloy for bone implant application. J Biomed Mater Res 2007;83A(3):703–11.
[4] Wolf, B., Fleck, C., Eifler, D., 2004. Characterization of the fatigue behavior of the magnesium alloy AZ91D by means of mechanical hysteresis and temperature measurements. Int. J. Fatigue 26, 1357–1363.
[5] Fritze, C., Berek, H., Kainer, K.U., Mielke, S., Wielage, B., 1998. Fibre reinforced magnesium for automotive applications. In: Mordike, B.L., Kainer, K.U. (Eds.), Magnesium Alloys and Their Applications.Werkstoff-Information-sgesellschaft, Frankfurt, Germany, pp. 635–640.
[6] Avedesian, M., Baker, H. (Eds.), 1999. Magnesium and Magnesium Alloys. ASM International, Materials Park, OH, USA, pp. 12–25.

[7] Ferkel, H., Mordike, B.L., 2001.Magnesium strengthened by SiC nanoparticles. Mater. Sci. Eng. A 298, 193–199.
[8] Hassan, S.F., Gupta, M., 2005. Development of high performance magnesium nano-composites using nano-Al2O3 as reinforcement. Mater. Sci. Eng. A 392, 163–168.
[9] Maung, A.T., Lu, L., Lai, M.O., 2006. Mechanical properties of nanostructured Mg–5wt%Al–*x* wt% AlN composite synthesized from Mg chips. Compos. Struct. 75, 206–212.
[10] Tjong, S.C., Ma, Z.Y., 2000. Microstructural and mechanical characteristics of in-situ metal matrix composites. Mater. Sci. Eng. 29, 49–113.
[11] Halouani, R., Bernache-Assolant, D., Champion, E., and Abadou, A. (1994) Microstructure and related mechanical properties of hot pressed hydroxyapatite ceramics. J. Mater. Sci.: Mater. Med., 5, 563-568.
[12] Van Landuyt, P. Li, F., Keustermans, J. P., Streydio, J. M., Delannay, F., and Munting, E. (1995) The influence of high sintering temperatures on the mechanical properties of hydroxyapatite. J. Mater. Sci.: Mater. Med., 6, 8-13.
[13] Layrolle, P., Ito A., and Tateishi T. (1998) Sol-gel synthesis of amorphous calcium phosphate and sintering into microporous hydroxyapatite bioceramics. Journal of the American Ceramics Society, 81, 1421-28.
[14] Khalil Abdelrazek Khalil, Sug Won Kim and Hak Yong Kim, Consolidation and mechanical properties of nanostructured hydroxyapatite–(ZrO2 + 3 mol% Y2O3) bioceramics by high-frequency induction heat sintering, *Materials Science and Engineering A 456 (2007) 368–372.*
[15] Sug Won Kim and Khalil Abdelrazek Khalil, High frequency induction heating sintering of mechanically alloyed alumina-yttria stabilized zirconia nano-bioceramics, *Journal of American Ceramic Society, Vol. 89, No. 4, p. 1280–1285 (2006).*
[16] Khalil Abdelrazek Khalil and Sug Won Kim, Effect of Processing Parameters on the Mechanical and Microstructural Behavior of Ultra-Fine Al2O3- (ZrO2+8%Mol Y2O3) Bioceramic, Densified By High Frequency Induction Heat Sintering, *Int. Journal of Applied Ceramic Technology, 3 [4], 322–330 (2006)*
[17] Khalil Abdelrazek Khalil and Sug Won Kim, High-Frequency Induction Heating Sintering Of Hydroxyapatite-(Zro2+3%Mol Y2o3) Bioceramics, *Materials Science Forum Vols. 534-536 (2007) pp. 601-604, online at http://www.scientific.net*
[18] Khalil Abdelrazek Khalil and Sug Won Kim, Synthesis and Densification of Ultra-Fine Al2o3-(Zro2+3%Mol Y2o3) Bioceramics by High-Frequency Induction Heat Sintering*, Materials Science Forum Vols. 534-536 (2007) pp. 1033-1036.*
[19] Khalil Abdelrazek Khalil, Sug Won Kim, and Hak Yong Kim, Observation of Toughness Improvements of the Hydroxyapatite Bioceramics Densified by High-Frequency Induction Heat Sintering, *Int. Journal of Applied Ceramic Technology, 4 [1] 30–37 (2007)*
[20] Khalil Abdelrazek Khalil, Sug Won Kim, N. Dharmaraj, Kwan Woo Kim and Hak Yong Kim, Novel Mechanism to Improve Toughness of the Hydroxyapatite Bioceramics Using High-Frequency Induction Heat Sintering, *Journal of Materials Processing Technology 187–188 (2007) 417–420*
[21] Khalil Abdelrazek Khalil and Sug Won Kim, Mechanical Wet-Milling and Subsequent consolidation of Ultra-Fine Al_2O_3-(ZrO_2+3%Mol Y_2O_3) Bioceramics by using High-Frequency Induction Heat Sintering, *Trans. Nonferrous Met. Soc. China, 17(2007) 21-26.*
[22] Wen Xie, Yue Liu, et. al., Influence of sintering routes to the mechanical properties of magnesium alloy and its composites produced by PM technique, Journal of Alloys and Compounds 431 (2007) 162–166.
[23] W.L.E. Wong, M. Gupta, Development of Mg/Cu nanocomposites using microwave assisted rapid sintering, Composites Science and Technology 67 (2007) 1541–1552.
[24] Z.R. Yang, S.Q. Wang, et. al., A new-developed magnesium matrix composite by reactive sintering, Composites: Part A 39 (2008) 1427–1432.
[25] S.F. Hassan, Khin Sandar Tun, M. Gupta, Effect of sintering techniques on the microstructure and

tensile properties of nano-yttria particulates reinforced magnesium nanocomposites, Journal of Alloys and Compounds 509 (2011) 4341–4347.
[26] K.S. Tun, M. Gupta, Development of magnesium/(yttria + nickel) hybrid nanocomposites using hybrid microwave sintering: Microstructure and tensile properties, Journal of Alloys and Compounds 487 (2009) 76–82.

IOP Conf. Series: Materials Science and Engineering **40** (2012) 012032 doi:10.1088/1757-899X/40/1/012032

Corrosion behavior of pristine and added MgB_2 in Phosphate Buffered Saline Solution

D Batalu[1], D Bojin[1], B Ghiban[1], G Aldica[2], P Badica[2]

[1]Polytechnic University of Bucharest, Materials Science and Engineering Faculty, Bucharest, 060042, Romania
[2]National Institute of Material Physics, Magurele, 077125, Romania

E-mail: dan_batalu@yahoo.com

Abstract. We have obtained by Spark Plasma Sintering (SPS), dense samples of MgB_2 added with Ho_2O_3. Starting composition was $(MgB_2)_{0.975}(HoO_{1.5})_{0.025}$ and we used addition powders with an average particle size below and above 100 nm. For Mg, pristine and added MgB_2 samples we measured potentiodynamic polarization curves in Phosphate Buffered Saline (PBS) solution media at room temperature. MgB_2 based composites show corrosion/ degradation effects. This behavior is in principle similar to Mg based alloys in the same media. Our work suggests that the different morphologies and phase compositions of the SPS-ed samples influence the interaction with corrosion medium; hence additions can play an important role in controlling the corrosion rate. Pristine MgB_2 show a significant improvement of the corrosion resistance, if compared with Mg. The best corrosion resistance is obtained for pristine MgB_2, followed by MgB_2 with nano-Ho_2O_3 and μ-Ho_2O_3 additions.

Keywords: MgB_2; Ho_2O_3; spark plasma sintering; corrosion

1. Introduction

MgB_2 is a material mainly studied because of its superconducting properties [1]. It is also attractive due to its lightness, availability and low price. Aswal [2] studied the degradation of MgB_2 in water and ambient atmosphere and found that the material is strongly hygroscopic. This observation leads to the idea of having a biodegradable potential or MgB_2 might be of interest for various applications where degradation is necessary. Among such applications are temporary medical implants and disposable medical tools, but studies on MgB_2 potential in this regard are not available.

In the last years the interest for Mg based materials used for biomedical applications increased [3-9] and authors emphasized in their works the corrosion behavior. This is because of some negative issues such as high corrosion rate. At the same time, the hydrogen release and pH values of solution containing Mg-based alloys are too high for bio applications [3, 9]. Some authors proposed various methods for the decrease of the corrosion rate through alloying [3, 5, 8, 9], roughness control [4] and coating [3, 6].

In our work we studied corrosion resistance of pristine and Ho_2O_3 added MgB_2.

2. Experimental

Three different samples, pristine MgB_2 (Alpha Aesar, 99.5 %) and Ho_2O_3 added MgB_2 (μ-Ho_2O_3 - Fluka, 99.9 % and nano-Ho_2O_3, Aldrich, 99.9 %) were produced by Spark Plasma Sintering (SPS). We noted "nano" (n) the samples with added powder of Ho_2O_3 with particle size lower than 100 nm, and "micro" (μ) those with particle size of added Ho_2O_3 higher than 100 nm (Fig. 1).

MgB_2 and the addition powders, $(MgB_2)_{0.975}(HoO_{1.5})_{0.025}$, with different morphological characteristics (Fig. 1) were manually mixed in an agate mortar. Powder mixtures were loaded into a 2 cm diameter graphite die and processed by SPS (FCT Systeme GmbH – HP D 5, Germany) at 1150 °C for 3 minutes. Heating rate was 160 °C/min. Temperature was monitored with a pyrometer, through an axial hole in the upper punch, ended at 2 mm above the sample. The uniaxial pressure applied on punches was 95 MPa. The initial vacuum in the furnace was 30 Pa. A pulsed current with a pattern of 12-on/2-off pulses was applied, with a 3 ms period. The total time of one sequence (cycle) was ~0.04 s. The operating voltage and the peak current were below 5 V and 1600 A, respectively.

Apparent densities, ρ_a^{SPS}, of the sintered samples were determined by Archimedes method using a KERN ALT 220-4M density balance. Theoretical density, ρ_t^{SPS}, of the sintered bulk was calculated according to ref. [10], considering that samples contain MgB_2 (2.63 g/cm^3), MgO (3.58 g/cm^3), MgB_4 (2.49 g/cm^3), Ho_2O_3 (8.405 g/cm^3) and HoB_4 (5.557 g/cm^3). The weight fraction of each phase was estimated by applying Rietveld refinement on the X-ray diffraction (XRD) data. We used MAUD software [11], version 2.31. The relative density, $R^{SPS}\,(\%)$, of the sintered samples, was calculated based on the ratio $\left(\rho_a^{SPS}/\rho_t^{SPS}\right)\cdot 100$.

The X-ray diffraction (XRD) was performed using Bruker-AXS D8 ADVANCE, ($Cu_{K_{\alpha 1}}$ radiation, λ = 1.5406 Å).

Scanning electron microscopy images (SEM) were taken with Zeiss EVO50.

Corrosion studies were carried out in PBS solution, at room temperature. Potentiodynamic polarization technique (Autolab potentiostat, model PGSTAT302N) was used to study the corrosion behavior of the samples. Potentiodynamic polarization tests were carried out at a scan rate of 0.5 mV/s. Electrodes were Pt as counter, Ag/AgCl as reference, and our samples as working electrode. NOVA 1.4 software was used for data acquisition and interpretation. Prior to corrosion tests, samples were polished up to 3000 grit, ultrasonically cleaned in isopropyl alcohol, and immersed for 1 h in PBS, in order to establish the free corrosion potential.

3. Results

The raw materials used for producing the pristine and added samples were MgB_2 and Ho_2O_3 micro and nano-powders (Fig. 1). Different shapes of the addition aggregates and particles can be observed and they proved to have a significant influence on the corrosion properties. Same quantity of micro and nano-Ho_2O_3 addition was used (9.54 wt. %).

The XRD patterns of the pristine and added samples after SPS show formation of MgB_4 and MgO phases. Also, XRD data indicate the reaction between MgB_2 and addition, with formation of HoB_4. For μ-Ho_2O_3 addition, some unreacted residual Ho_2O_3 was observed.

Samples show high relative density values: 93.8 % for pristine MgB_2, 93.4 % for μ-Ho_2O_3 added MgB_2 and 98.1 % for nano-Ho_2O_3 added MgB_2. The highest density is for the sample with nano-Ho_2O_3 addition.

SEM observations (Fig. 2) reveal two distinct regions for added samples. The first region consists of relatively clean MgB_2, composed of dense large blocks (dark regions in Fig. 2. b and c, backscattering images). The second one is a matrix which surrounds the "clean" MgB_2. "Dirty" matrix is rich in HoB_4 (light regions in backscattering images are due to Ho, Fig. 2. b and c).

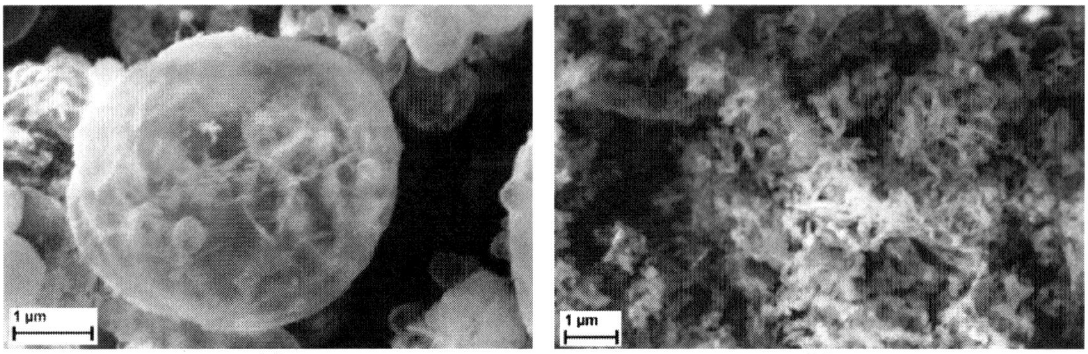

(a) (b)

Fig. 1. Raw rare earth oxide powders: (a) µ-Ho_2O_3 (particles size higher than 100 nm), (b) nano-Ho_2O_3 (particles size lower than 100 nm).

Contamination with oxygen during SPS also occurs. Around 7.3-11 wt. % MgO is present in the pristine sample, meaning that some residual oxygen is available during vacuum SPS processing and reacts with MgB_2.

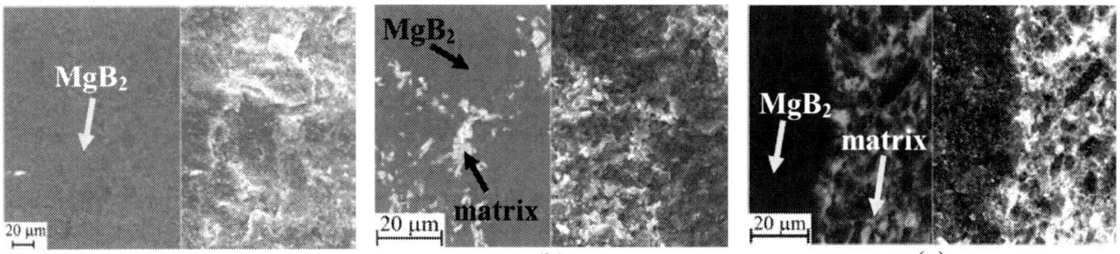

(a) (b) (c)

Fig. 2. SEM images for pristine MgB_2 (a), MgB_2 with µ-Ho_2O_3 addition (b) and nano-Ho_2O_3 addition (c) (left side is taken in backscattering mode).

The electrochemical polarization curves of samples (including Mg for comparison) are presented in Fig. 3. Corresponding electrochemical data are presented in Table 1. The corrosion potential (E_{corr}) values show that Ho_2O_3 additions shift the polarization curves towards less noble direction, decreasing the corrosion resistance of MgB_2. µ-Ho_2O_3 addition is shifting the corrosion potential towards the active direction, more than nano-Ho_2O_3 addition. The sample with µ-Ho_2O_3 has a low relative density and a mixt microstructure of clean and dirty areas, hence the corrosion resistance is supposed to be lower than for the pristine sample (with a homogenous microstructure) and lower than for nano-Ho_2O_3 added sample, which though it has a mixt microstructure, it is very dense. On the other hand, nano-Ho_2O_3 addition substantially increases the i_{corr} of MgB_2 sample. However, all MgB_2 samples have a better corrosion resistance than Mg. The corrosion potential of MgB_2 samples indicates a better corrosion resistance than for some reported Mg based alloys (e.g. AZ91, AZ91Ca or AZ61Ca [5]).

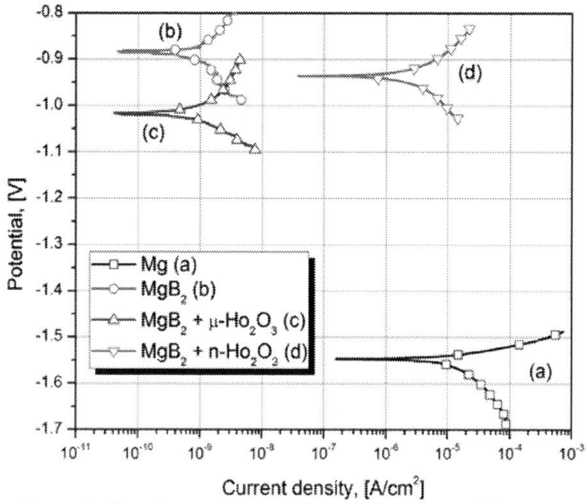

Fig. 3. Potentiodynamic polarization curves of samples tested in PBS at room temperature.

Table. 1. Electrochemical parameters of the alloys obtained from the polarization curves.

Sample	**Corrosion potential, E_{corr} (V)**	**Corrosion current density, i_{corr} (μA/cm^2)**
Mg	-1.544	7.26
MgB_2	-0.883	$2.89\cdot10^{-3}$
MgB_2 + μ-Ho_2O_3	-1.016	$7.20\cdot10^{-4}$
MgB_2 + nano-Ho_2O_3	-0.936	3.65

4. Discussions

The degradation behavior of MgB_2 is based on the reaction with water.

The reaction of MgB_2 with water is [2]:

$$MgB_2 + 2H_2O \rightarrow Mg(OH)_2 + 2B + H_2 \quad (1)$$

Further possible reactions in the ambient atmosphere are [2]:

$$4B + 3O_2 \rightarrow 2B_2O_3 \quad (2)$$

$$Mg(OH)_2 + CO_2 \rightarrow MgCO_3 + H_2O \quad (3)$$

Ref. [12] proposed the reaction of B_2O_3 with water:

$$1/2\,B_2O_3 + 1/2\,H_2O \rightarrow H_3BO_3 \quad (4)$$

Considering the Mg based secondary phases which are formed by SPS, two more reactions can be proposed:

$$MgO + H_2O \rightarrow Mg(OH)_2 \quad (5)$$

$$MgB_4 + 2H_2O \rightarrow Mg(OH)_2 + 4B + H_2 \quad (6)$$

However, literature indicates that solubility of MgO in water is very low (8.6 mg/ L at 30℃, [13]), while for MgB_4 supplementary confirmation is required.

Taking into account the presence of different chemical species in the PBS or the influence of the atmosphere during sample manipulation, as well as the influence of the impurity phases from the SPS-ed samples on chemical interactions, to understand the details of passivation and corrosion interactions observed in our work, a complex systematic study is required. This discussion does not take into consideration all the species present in PBS.

From the above reactions, we can calculate that H_2 release is 0.5 liter per 1 g of MgB_2, half than in the case of pure Mg. This is potentially a useful result for biomedical applications (e.g. biodegradable implants). We shall emphasize that added MgB_2 samples contain the same amount and type of addition; hence the main influence on corrosion behavior is given by the powder morphology, which, indeed, is very different (Fig. 1). The powder morphology is believed to strongly influence the rheology of mixed powders and the density of SPS-ed samples.

5. Conclusions

Pristine and Ho_2O_3 added MgB_2 samples were obtained by SPS technique. The additions of the same type, in same amount, but with different morphological characteristics have strong influences on corrosion properties. The characteristics of the raw powders additions (morphology, size and distribution) are useful tools for tailoring the corrosion properties of MgB_2. It was also found that there are significant differences between Mg and MgB_2 samples, with or without additions. MgB_2 sample has a higher corrosion resistance than Mg and other Mg based alloys, reported in literature. The corrosion behavior of MgB_2 samples can be further adjusted by changing the amount and type of additions or by porosity control.

Nano-Ho_2O_3 added MgB_2 has a higher corrosion potential than the μ-Ho_2O_3 added one, and a higher corrosion current density.

Acknowledgements

D Batalu recognizes financial support from the European Social Fund through POSDRU/89/1.5/S/54785 project: "Postdoctoral Program for Advanced Research in the field of nanomaterials"
G Aldica and P Badica acknowledge PCCE 9/2010, Romania.

References

[1] Nagamatsu J, Nakagawa N, Muranaka T, Zenitani Y, Akimitsu J 2001 *Nature* **410** 63
[2] Aswal D K, Muthe K P, Singh A, Sen S, Shah K, Gupta L C, Gupta S K, Sahni V C 2001 *Physica C* **363** 208
[3] Song G 2007 *Corros Sci* **49** 1696
[4] Denkena B, Lucas A 2007 *Annals of CIRP* **56** 113
[5] Kannan M B, Raman P K S 2008 *Biomaterials* **29** 2306
[6] Pereda M D, Alonso C, Burgos-Asperilla L, del Valle J A, Ruano O A, Perez P, Fernandez Lorenzo de Mele M A 2010 *Acta Biomater* **6** 1772
[7] Kannan M B 2010 *Mater Lett* **64** 739
[8] Niu J X, Chen Q R, Xu N X, Wei Z L 2008 *T Nonferr Metal Soc* **18** 1058
[9] Kubasek J, Vojtech D 2011 *Proc. Int. Conf. on Metallurgy and Materials* (Brno)
[10] Marks G W and Monson L A 1955 Ind Eng Chem **47** 1611
[11] Lutterotti L 2010 *Nucl Instrum Meth B* **268** 334
[12] Klepper C C, Williams J M, Truhan Jr. J J, Qu J, Riester L, Hazelton R C, Moschella J J, Blau P J, Anderson J P, Popoola O O, Keitz M D 2008 *Thin Solid Films* **516** 3070
[13] Mark A S 2006 The chemistry and technology of magnesia (New Jersey: John Wiley & Sons) pp 266

IOP Conf. Series: Materials Science and Engineering **40** (2012) 012033 doi:10.1088/1757-899X/40/1/012033

Conductive adhesives with self-organized silver particles

T Kowalik[1], M Amkreutz[1], C Harves[1], A Hartwig[1] and S J Aßhoff[2]

[1] Fraunhofer Institute for Manufacturing Technology and Advanced Materials; Wiener Strasse 12, 28359 Bremen, Germany.
[2] MESA+ Institute for Nanotechnology, University of Twente, Hallenweg 15, 7522 NB Enschede, Netherlands.

E-mail: thomas.kowalik@ifam.fraunhofer.de

Abstract. Common available conductive adhesives are filled with about 70-90 weight percent of silver particles. There is the technical need to lower the filler content for decreasing the viscosity and the pressure to finally decrease the costs for the highly expensive silver. Our approach to decrease the silver content with at the same time similar conductivity was to use nano-sized silver particles with different modified surfaces to obtain self-organized distribution of the silver particles in the liquid adhesive film followed by conventional curing techniques like temperature or UV curing. Through the self-organization, a bridging of the particles should be achieved which gives a similar conductivity as in highly filled commercial adhesives which follows percolation theory. The influence of the surface modification on the strength of the interaction between differently modified silver particles and the distribution in the liquid adhesive was simulated with Dissipative Particle Dynamics (DPD). Surface modification was verified with surface enhanced Raman spectroscopy (SERS) and X-ray photoelectron spectroscopy (XPS). The distribution of the silver particles in the cured adhesive film was investigated with scanning electron microscopy (SEM).

1. Introduction

Conductive adhesives are getting more and more in focus for all various kinds of microelectronics uses, in photovoltaic applications or for printable RFID antenna structures. In the last recent years, there were several developments performed to be able to enable the application of the mostly silver filled epoxy or acrylate based systems with different kinds of processing, as like inkjet-, offset-printing or sintering [1]. In all cases, there is the technical need to lower the filler content for decreasing the viscosity and the pressure to finally decrease the costs for the highly expensive silver filled adhesives. Common available adhesives are filled with about 70-90 weight percent of silver particles. The shapes of the particles were the main topic in investigations to decrease the silver amount and resulted e.g. in silver flakes with a highly irregular structure [2]. Other metals beside of silver as filler in conductive adhesives are copper, gold, aluminum and nickel, but, because of the formation of oxides on the particles surface, the long term stability of the conductivity is often a problem in commercial uses. There are some approaches to use carbon nanotubes (CNTs) in conductive adhesives which show through the orientation of the CNT anisotropic conductivity [3]. Isotropic conductivity with CNTs is very low because of the limited concentration of CNT in a resin. With around 5-10 % CNT, the filled resin is not processible anymore. A further example for an anisotropic conductivity through orientation is given in [4], where silver particles modified with self-assembled carbon acids were used. Similar behavior of self-organization has been observed in adhesive formulations with non-conductive fillers based on silicon dioxide and silicone particles [5]. The particles were modified in an acid and in a basic medium and an orientated structure has been observed. The conclusion was that this kind of structure was induced through electrostatic interactions between the acid and basic surface modification of the particles. The self-organization of metallic particles is not well described in the literature. Most of the work covers the behavior of gold particles in very low viscous and/or water based media [6-8]. Because of the special needs for an adhesive, it should be easy to process and to store over a defined period, permanent and strong interactions between the particles are not usable.

Our approach to decrease the silver content with at the same time similar conductivity was to use nano-sized silver particles with different modified surfaces to obtain self-organized distribution of the silver particles in the liquid adhesive film followed by conventional curing techniques like temperature or UV curing. Through the self-organization, a bridging of the particles should be achieved which gives a similar conductivity as in highly filled commercial adhesives. Investigated were three kinds of interactions: acid –basic interactions, ionic interactions and hydrogen bridges.

The influence of the surface modification on the strength of the interaction between differently modified silver particles and the distribution in the liquid adhesive was simulated with Dissipative Particle Dynamics (DPD). Surface modification was verified with surface enhanced Raman spectroscopy (SERS) and X-ray photoelectron spectroscopy (XPS). The distribution of the silver particles in the cured adhesive film was investigated with scanning electron microscopy (SEM).

2. Experimental

2.1. Experimental work

All used formulations were based on commercially available raw materials and were used without further purification. Two different silver particles were used, Metalor K1081P and Inframat® Advanced Materials™ 47MN-03. Both silver powders possess an unknown organic coating which was exchanged as follows: The amount of necessary bifunctional organic modifier was calculated over the theoretical BET surface of the particles. All organic modifiers were bought from Sigma-Aldrich. Silver particles and organic modifier were heated under reflux for one hour in Ethanol solution, followed by hot filtering and washing with Ethanol. The resulting silver particles were dried at 60 °C in an oven and the success of exchange was evaluated with SERS and XPS. SERS was measured with a Bruker IFS 66 with Raman FRA106s module, using 500 scans. XPS was measured on a Kratos Ultra. As basic adhesive resin an UV curable acrylate resin from the company Panacol, Vitralit Y-391125, was used. To adjust the viscosity a reactive thinner, Miramer M300, also delivered by Panacol, was taken. The modified silver particles were mixed into the resin using in a first step a Speed Mixer™ DAC 150 FVZ (3500 U/min, 1.5 minutes), followed by ultrasonic deagglomeration with a homogenisator from Bandolin Sonopuls. The electrical conductivity was measured after curing with a Hönle UV Cube 2000 in air on a printed circuit (PC) board. The application of the adhesives on the PC board was performed as follows: A wet 30 μm thin adhesive film was applied and stored 10 minutes before UV curing. After UV curing of the first film, a second 30 μm thin film was applied, stored for 10 minutes and cured. A final third 30 μm thin film was applied in the same way. After every curing, the electrical conductivity was measured with a four point measurement. The distribution of the silver particles in the cured film was investigated with SEM, using a Gemini 1530.

2.2. Dissipative Particle Dynamics (DPD) Simulation

To investigate the interaction of the surface modified silver particles with each other and with the surrounding liquid adhesive, Dissipative Particle Dynamics (DPD) simulation has been applied [9]. This is a mesoscale simulation method which has been chosen due to the size of the silver particles. The approach is based on a bead model, with beads being the smallest structural elements in the simulation. The beads are used to build up the systems components (silver particles, liquid adhesive) and the interaction between the different beads determines the arrangement of the various components of the simulated system. The aim has been to predict the structure formation and silver particle distribution in the liquid adhesive in dependence of the surface modification. Therefore, several periodic unit cells have been generated containing the corresponding number of the two different kinds of organic modified silver particles and adhesives components according to the given formulation (see Figure 1 and Figure 2). Both surface modified particles and adhesive's components were built up from beads. These unit cells served as structural input models for the simulation of the interactions and the structure formation. For all the interactions, the necessary A parameters for the DPD simulation have been derived from corresponding Hansen solubility parameters. These take into account the energy from dispersion bonds, from dipolar intermolecular forces, and from hydrogen bonds between molecules [10]. With this, insight into the structure formation was expected and thus self-organization and bridging of the silver particles could be predicted depending on the different surface modifications of the particles. This will support the formulation work and helps to find the optimal modifications.

3. Results and Discussion

Table 1 gives a summary of the used organic modified silver particles and the different solubility's calculated based on the Hansen solubility parameters. The colours of the columns give the colour code for Figure 1 and Figure 2.

Table 1. Overview of the used modifications and corresponding solubilities calculated based on Hansen solubility parameters. X = not soluble, √ = soluble, --- = at the borderline between soluble to not soluble.

Particle A Modification	Particle B Modification	Solubility		
		A in B	A in resin	B in resin
NaOH	Citric acid	---	X	X
NaOH	Formic acid	√	X	X
Ethylene diamine	Citric acid	X	√	X
Ethylene diamine	Formic acid	X	√	X
NaOH	Mercapto-benzoic acid	X	X	√
Ethylene diamine	Mercapto-succinic acid	√	√	√
4-Mercapto-pyridine	Citric acid	X	X	X
Hexamethylene diamine	Mercapto-succinic acid	X	---	√
Ethylene diamine	Mercapto-benzothiazole	---	√	---

In dependency on the overall amount of modified silver particles (55 and 75 weight-%, divided 1:1 into the two different kinds of modified particles) the distribution of the different surface modified particles in the periodic box was simulated. Figure 1 gives the results for the mixtures of modified K1081P particles with 75 weight-% amount . All simulated systems show different distributions of the particles in the periodic resin box, most of them quite irregular. Only the mixture with hexamethylene diamine and mercapto-succinic acid shows a regular distribution of the acid-modified particles, whereas the basic diamine particles are distributed irregularly. This kind of behaviour has not been observed with an overall amount of 55 weight-% silver (Figure 2).

Figure 2 gives the results for the 55 weight-% overall amount. It is obvious that the distribution is quite strongly depending on the overall amount of silver particles. The reduction of the amount of silver particles results in a breakdown of the structures formed at 75 weight-%, especially in the case of the hexamethylene diamine – mercapto-succinic acid system.

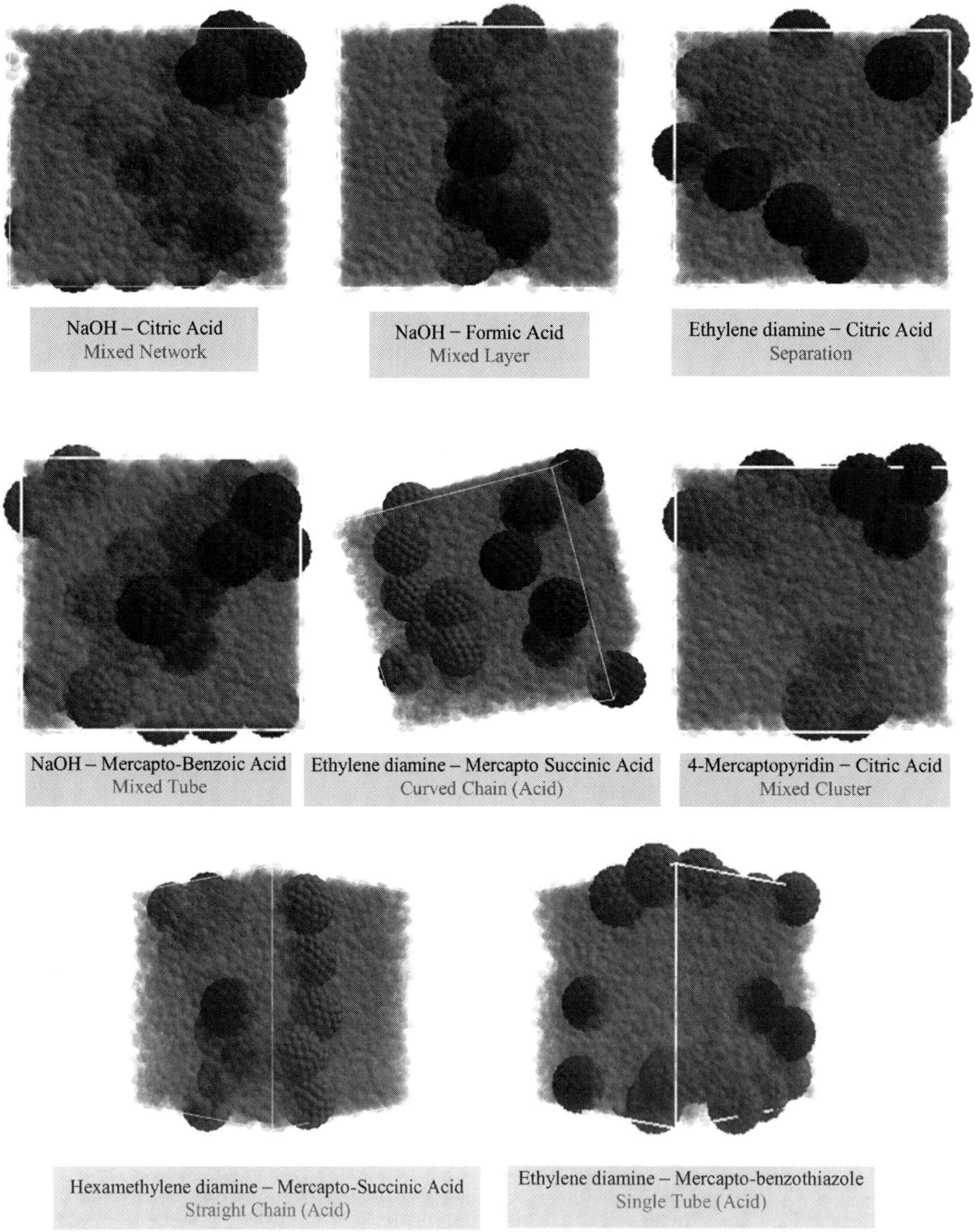

Figure 1. Distribution of the modified silver particles (75 weight-%) in thinned acrylate resin.

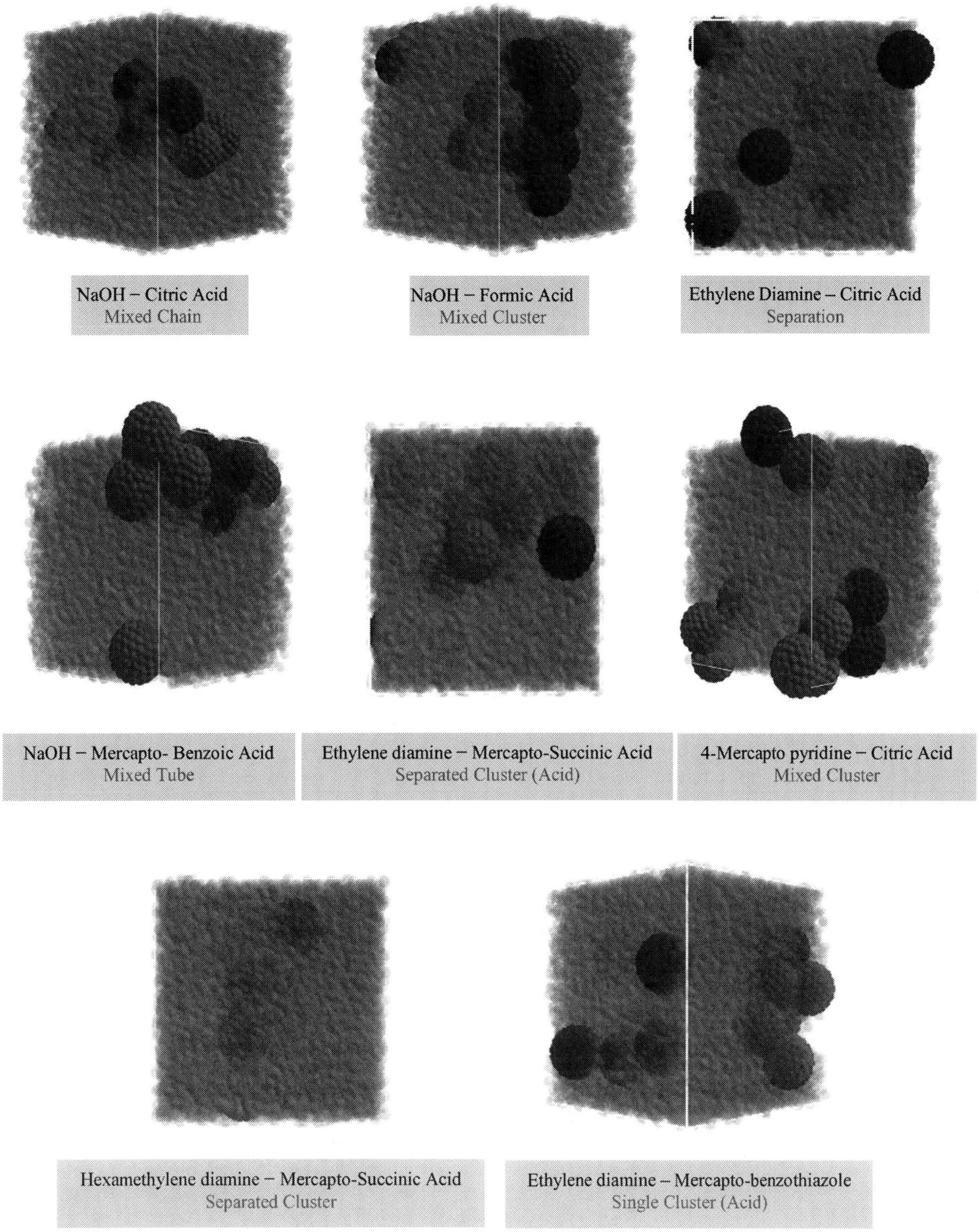

Figure 2. Distribution of the modified silver particles (55 weight-%) in thinned acrylate resin.

Surface modification was evaluated with SERS and XPS (not shown here). Figure 3 shows an example for the SERS evaluation. During modification, the original organic surface coating is removed and, in this case, the mercapto-benzoic acid has reacted with the silver surface through the mercapto group. This can be detected clearly through the loss of the peak at 2565 cm^{-1} which is assigned to the –SH-group.

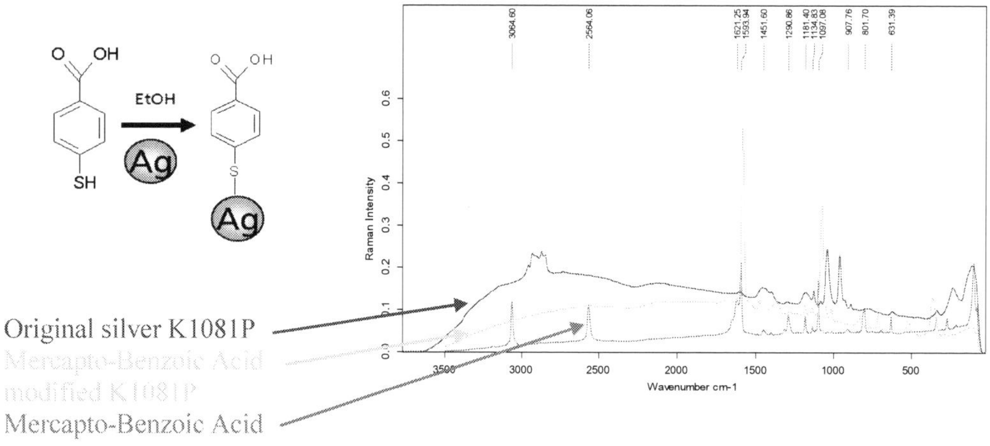

Figure 3. SERS evaluation of successful modification of the particle with mercapto-benzoic acid.

The validation of the simulated distribution of the modified silver particles has been performed through scanning electron microscopy. As an example, Figure 4 shows the non-modified K1081P silver, well distributed in the resin, the separated structure of the mixture of the ethylene diamine and citric acid modified K1081P formulation, and the theoretical polymer chain configuration of the ethylene diamine and mercapto-succinic acid K1081P formulation for comparison. It is to consider that in practise it was not possible to obtain real nano-sized silver particles. So, it is not to expect that the experimental validation shows exactly the same clear behaviour as predicted by simulation, but the SEM images seem to confirm the simulated behaviour as the mixed particles show particular orientation.

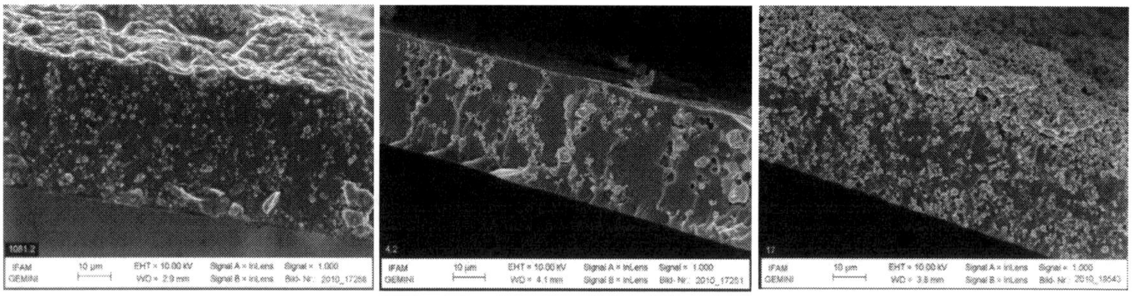

Well distributed original K1081P

Separated mixture of ethylene diamine and citric acid modified K1081P

Partly arranged mixture of ethylene diamine and mercapto succinic acid modified K1081P

Figure 4. Comparison of distributions of modified and non-modified silver particles in cured resin.

The results can also be confirmed by electrical conductivity measurements where the original K1081P shows the best conductivity, the partly arranged mixture a slightly lower value and the separated mixture a strong decrease in electrical conductivity.

4. Summary

Aim of the work was to reduce the necessary amount of silver particles in electrical conductive adhesives. The approach to reach the reduction was to use weak interactions (electrostatic, ionic or hydrogen bonding) between surface-modified nano-sized silver particles so that the particles undergo a self-structuring/orientation process in the liquid adhesive film. This is fixed through a later following curing process. Through the orientation, an isotropic conductive pathway should be obtained, so that with decreasing amount of silver particles a similar conductivity could be reached. The behaviour of different surface modified particles in the resin was simulated with Dissipative Particle Dynamics (DPD). It is shown that the distribution depends strongly on the kind of surface modification and the overall concentration of the particles. One simulated system showed a promising orientation behavior in the simulation, all other investigated systems did not. The experimental validation of the systems showed the successful replacement of the original organic coating of the silver particles. Unfortunately, it was not possible to disperse the silver particles in the resin to nano size, so that the distribution in the later cured film is in principle as simulated but not as proper as expected. Also the electrical conductivity measurements support the assumption that there is the tendency for a self-orientation. But it is not as structured to build up a conductive pathway to allow decreasing the amount of silver particles.

5. Acknowledgements

We like to thank our project partners Panacol and microTEC for their support and the VDI and the Federal Ministry for Education and Research BMBF for the financial support in the frame of the NanoChance program.

References

[1] Kowalik T, Worch S, Hartwig A, 2007, Macromolecular Symposia, **254**, 1, 300
[2] Lyons A M, 1991, Pol. Eng. Sci., **31,** 445
[3] European Patent EP1809716
[4] Li Y, Moon K S, Wong C P, 2005, J. Electr. Mat., **34,** 1573
[5] Trautmann J, 2006, Dissertation Universität Clausthal Zellerfeld, Papierflieger Verlag, ISBN 3-89720-900-4
[6] World patent WO 1996007487
[7] Dutta J, Sugunan A, 2004, ICSE Proc., Kuala Lumpur, Malaysia
[8] Brust C, Kiely J, 2002, Colloids and Surfaces A: Physicochem. Eng. Aspects **202** 175
[9] Hoogerbrugge P J, Koelman J M V A, 1992, Europhys. Lett., **19,** 155
[10] Hansen C M, Hansen Solubility Parameters – A User's Handbook, Second Edition, CRC Press, Taylor & Francis Group, 2007

IOP Conf. Series: Materials Science and Engineering **40** (2012) 012034 doi:10.1088/1757-899X/40/1/012034

Photoluminescence properties of gadolinium oxide nanophosphor

B Rudraswamy and N Dhananjaya

Department of Physics, J.B. Campus, Bangalore University, Bangalore, INDIA

E-mail: brudraswamy@gmail.com.

Abstract. In this paper, we report low temperature solution combustion synthesis and characterization of rare earth oxide (Gd_2O_3) nanocrystalline phosphor. During the synthesis, glycine was used as a fuel. The final product was examined by well characterized techniques such as powder X-ray diffraction pattern (PXRD), UV-Visible absorption spectroscopy, photoluminescence (PL) measurements etc. The average crystallite size of the Gd_2O_3 nanophosphor was estimated using Scherrer's formula and found to be 36 nm. The UV-Visible absorption peaks observed at ~243 and ~305 nm. Further the band gap of 900 °C for 3 h calcined Gd_2O_3 nanophosphor were estimated by Wood and Tauc relation and it was found to be ~5.4 eV. A broad PL emission band at ~380 nm was observed at 230 nm excitation. According to Wang's proposal, this emission band may be attributed to recombination of a delocalized electron close to the conduction band with a single charged state of surface oxygen vacancy.

1. Introduction

Trivalent state rare earth oxides (RE_2O_3) are most stable compounds and used in high performance luminescence devices, magnet, catalysts and other functional materials because of their optical, chemical and electronic characteristics resulting from the 4f electronic shell [1]. Different synthetic routes have been reported in literatures for the preparation of oxide nanoparticles such as sol-gel, hydrothermal, combustion, co-precipitation etc [2, 3]. The energy consuming solid-state reaction method could not produce desirable RE_2O_3. A low temperature solution combustion method is useful due to the advantage of high purity, easy to prepare, low cost, less processing time, good homogeneity etc [3].

In the present study, we have prepared Gd_2O_3 nanophosphor by a low temperature solution combustion method using glycine as a fuel. The prepared sample was characterized for their structural and luminescent properties.

2. Experimental

The starting materials used in the experiment were of analytical grade, without further purification. gadolinium nitrite [$Gd(NO_3)_3$] used as oxidizer and glycine [$C_2H_5NO_2$] were used as fuels for low temperature solution combustion synthesis. An aqueous solution containing stoichiometric amounts of $Gd(NO_3)_3$ and glycine were taken in a crystalline Petri dish of ~300 mL capacity. The excess water is allowed to evaporate by heating over a hot plate until a wet powder is left out. Then the crystalline Petri dish is introduced into a pre heated muffle furnace maintained at ~300 °C. The reaction mixture undergoes thermal dehydration and ignites at one spot with liberation of gaseous products such as oxides of nitrogen and carbon. The flame temperature was found to be ~1000 °C, and it was measured by an optical pyrometer placed inside the crystalline petri dish without touching the solution. The phase purity has been recorded on a PANalytical X'Pert Pro powder X-ray diffractometer using CuK$_\alpha$ (λ=1.541Å) radiation with a nickel filter. The infrared spectroscopy of the as prepared was examined using a Perkin-Elmer spectrometer (spectrum 1000) with KBr pellet. The UV-Visible absorption of the samples was recorded on SL 159 ELICO UV–VIS Spectrophotometer. The photoluminescence (PL) excitation and emission spectrum were recorded with a Shimadzu Spectroflourimeter equipped with Xe lamp as an excitation source at room temperature.

3. Result and discussions

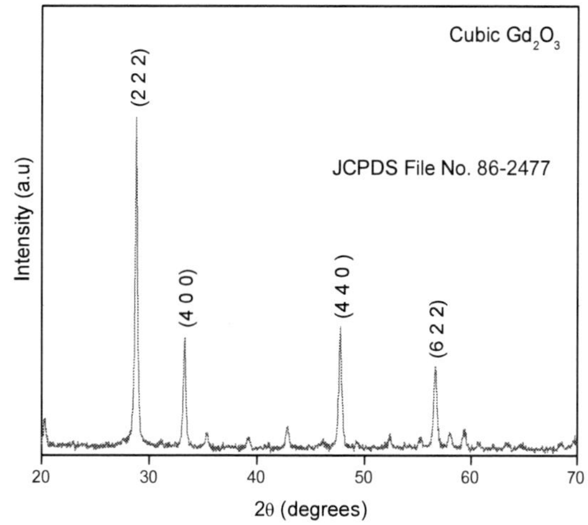

Figure 1. PXRD patterns of 900 °C calcined Gd_2O_3 nanophosphors.

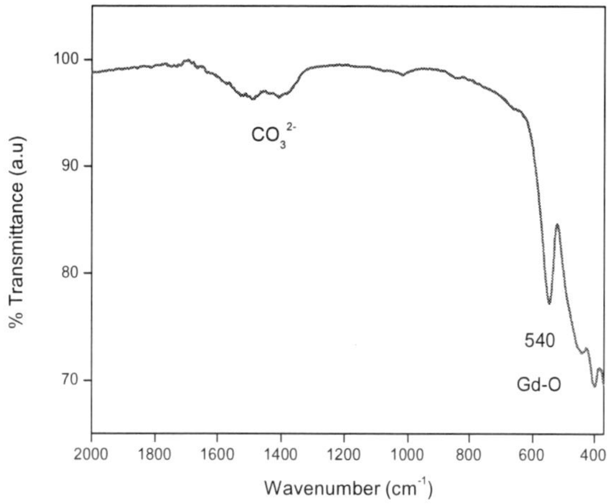

Figure 2. FT-IR spectrum of 900 °C calcined Gd_2O_3 nanophosphors.

The powder X-ray diffraction (PXRD) was used to investigate the phase structures of the resulting powder. According to the PXRD pattern of the 900 °C for 3 h calcined Gd_2O_3 nanophosphors (Figure 1) shows cubic [Ia3 (206)] phase (JCPDS File No. 86-2477). The diffraction patterns are well matching with the literature [2]. No crystalline impurity apart from rare earth oxides can be determined from the PXRD analysis. The average crystallite size (t) of Gd_2O_3 nanophosphor was estimated from the full width at half maximum (FWHM) of the diffraction peak of the powders, using

Scherrer's formula [4]. $t = 0.9\lambda/(\text{FWHM}) \cos\theta$, where λ is the wavelength of X-ray used, θ the Bragg angle. It is observed that the average crystallite size of the calcined Gd_2O_3 nanophosphors is ~36 nm. The lattice constant of the calcined product was calculated according to the equation $1/d^2 = (h^2+k^2+l^2)/a^2$, where d is the interplanar distance, h, k, l are the miller indices and a is the lattice constant. On the basis of the (2 2 2) crystal planes (d= 0.3106 nm) the lattice constant is a = 1.0759 nm, which is well compatible with the literature value of a= 1.0809(9) nm (JCPDS No. 86-2477). Figure 2 shows the FT-IR spectrum of 900 °C for 3 h calcined Gd_2O_3 nanophosphor. The strong absorption peak near 540 cm^{-1} is associated with the vibration of the Gd-O bond [5]. Further, the absorption bands in the range 1300-1600 cm^{-1} responsible for CO_3^{2-} anion groups.

UV-Visible absorption spectrum of 900 °C for 3 h Gd_2O_3 nanophosphors is shown Figure 3. Two absorption peaks at 243 nm (sharp) and 304 nm (broad) were recorded in Gd_2O_3 nanophosphors. The broad absorption peak at ~304 nm is attributed to $^8S_{7/2} \rightarrow {}^6P_{7/2}$ transitions and to the sharp absorption peak at 243 nm is attributed to $^8S_{7/2} \rightarrow {}^6I_{J/2}$. The optical band gap energy (E_g) was estimated using Wood and Tauc relation [6]. The band gap was found to be 5.4 eV and is well matched with the literature [7].

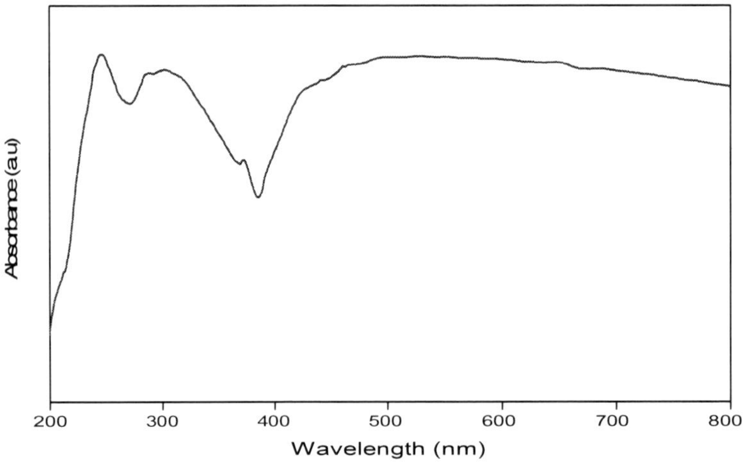

Figure 3. UV-Visible absorption spectrum of 900 °C calcined Gd_2O_3 nanophosphors.

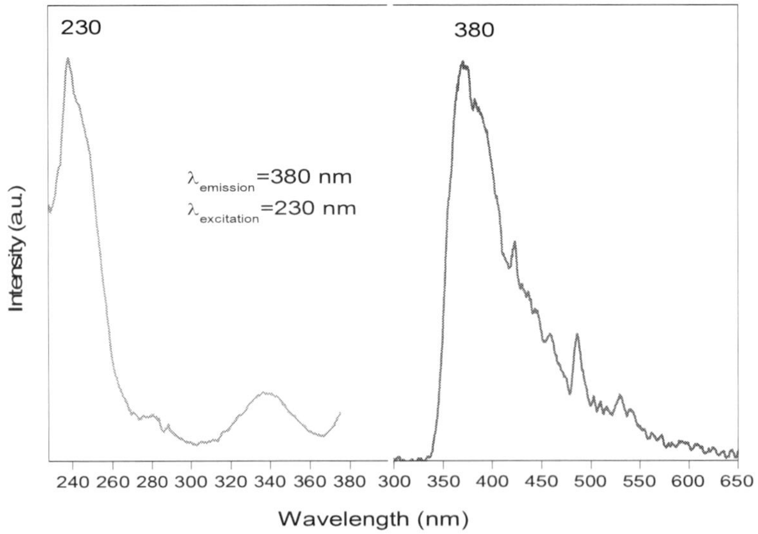

Figure 4. PL excitation and emission spectrum of 900 °C calcined Gd_2O_3 nanophosphors.

Figure 4 shows the excitation and emission spectra of 900 °C calcined Gd_2O_3 nanophosphors. The peak near 230 nm originates from the excitation of Gd_2O_3 host lattice (HL). The photoluminescence experiments at room temperature shows, a deep UV emission band at ~380 nm at excitation wavelength of 230 nm. This emission band may attribute to recombination of a delocalized electron close to the conduction band with a single charged state of surface oxygen vacancy, according to Wang's proposal [8]. For different fuels different density of oxygen vacancies might arouse differences of PL intensities.

4. Conclusions

In this paper, we studied the structural and luminescent properties of 900 °C for 3 h calcined Gd_2O_3 nanophosphors via low temperature solution combustion method. The crystalline Gd_2O_3 belongs to the cubic system. The crystallite size estimated from Scherer's method and was found to be 36 nm. UV-Visible absorption spectrum shows sharp peaks at 243 nm ($^8S_{7/2} \rightarrow {}^6I_{J/2}$) and broad 304 nm ($^8S_{7/2} \rightarrow {}^6P_{7/2}$). The band gap was found to be 5.4 eV. PL shows deep UV-band emission peaks at ~380 nm and it is promising for possible display applications.

References

[1] Yang J, Li C, Quam Z, Zhang C, Yang P, Li Y, Yu C, Lin J, J. Phys. Chem. C 2008, **112**, 12777.
[2] Xiao H, Li P, Jia F, Zhang L, J. Phys. Chem. C 2009, **113**, 21034.
[3] Li Y, Hong G, J. Lumin 2007, **124**, 297.
[4] Klug P, Alexander L E, X-ray Diffraction Procedure, Wiley, New York, 1954.
[5] You H, Nogami M, J. Phys. Chem. B 2005, **109,** 13980.
[6] Tauc J, in: F. Abeles (Ed.), Optical Properties of Solids, North-Holland, Amsterdam, 1970.
[7] Adachi G and Imanaka N, Chem. Rev. 1998, **98,** 1479.
[8] Hu C, Liu H, Dong W, Zhang Y, Bao G, Lao C, Wang Z L, Adv. Mater. 2007, **19,** 470.

IOP Conf. Series: Materials Science and Engineering **40** (2012) 012035 doi:10.1088/1757-899X/40/1/012035

Purification of nanoparticles by hollow fiber diafiltration

J Veeken

JM Separations BV

Abstract. Hollow Fiber Diafiltration (Hollow Fiber Tangential Flow Filtration) is an efficient and rapid alternative to traditional methods of nanoparticle purification such as ultracentrifugation, stirred cell filtration, dialysis or chromatography. Hollow Fiber Diafiltration can be used to purify a wide range of nanoparticles including liposomes, colloids, magnetic particles and nanotubes. Hollow Fiber Diafiltration is a membrane based method where pore size determines the retention or transmission of solution components. It is a flow process where the sample is gently circulated through a tubular membrane. With controlled replacement of the permeate or (dialysate), pure nanoparticles can be attained. Hollow Fiber Diafiltration can be directly scaled up from R&D volumes to production. By adding more membrane fibers and maintaining the operating parameters, large volumes can be processed in the same time with the same pressure, and flow dynamics as bench-scale volumes. Keywords: hollow fiber, Diafiltration, filtration, purification, tangential flow filtration.

1. Materials and Methods

A 100 kD polysulfone hollow fiber module (Spectrum Labs MiniKros® Sampler Pl us P/N M4AB-2 60-01 P) was selected based upon the unreacted by-product molecular weight (<50 kD) and the polymeric nanoparticle (250kD). The tubular geometry of the hollow fiber is beneficial to particle applications due to the phenomenon known as tubular pinch effect where particles migrate to the center of the hollow fiber where flow velocity is the highest.1 1200 mL of an 8% polymeric nanoparticle solution was used in a bench scale diafiltration. Discontinuous diafiltration was selected for the applicatio n due to the high viscosity, 500 cps. Initial dilution of the product would provide the best performance for permeate flow rate and process pressures and achieve better mixing of the diafiltration buffer with the polymeric solution. A bench scale Tangential Flow Filtration development system with data acquisition software and pressure indicator (Spectrum Labs KrosFlo® Research II P/N SYR2-S21-01N) was used for the process.

2. Results

1200 mL of the solution was poured into a 4 L TFF process reservoir (Spectrum Labs P/N ACTO- 4 P P-0 1 N) and diluted to 2400 mL with DI water. The feed pump was set to a feed flow rate of 950 mL/min to begin the process. This feed flow rate was selected to provide an even laminar flow process with a shear rate of 4000s-1.Figure 1 illustrates the relationship between flow/fiber and shear rate. Retentate backpressure was applied using a gauge down tubing diameter approach to reach a transmembrane pressure (TMP) of 20 psig with a feed-toretentate pressure drop of 7 psig.

$$P\,feed + P\,retentate$$

$$TMP = \frac{\_\_\_\_\_\_\_\_\_\_}{2} - P\ \textit{permeate}$$

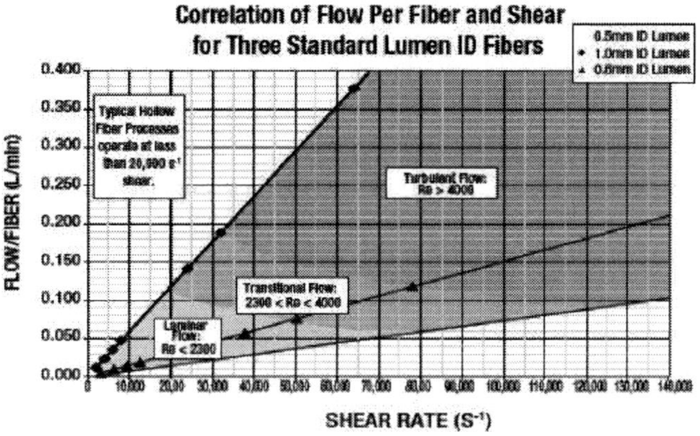

Figure 1

Permeate collection began in a graduated cylinder to monitor process volumes and permeate flux. When the collected permeate volume was equal to the initial volume (1200 mL), samples of the process were taken for analysis and the product was diluted twofold again to 4%. This dilution volume is referred as one Diafiltration volume. This procedure was repeated nine times. The data acquisition program collected pressure and flow rate data results, critical to the direct scaling to larger volumes, throughout the process and provided a process chart in real time. (Figures 2 and 3).

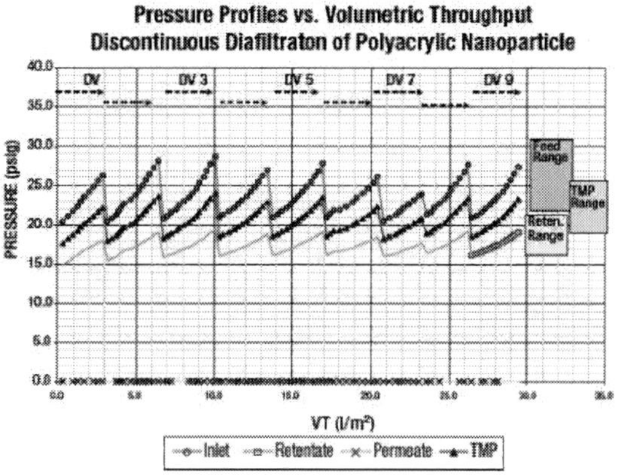

Figure 2

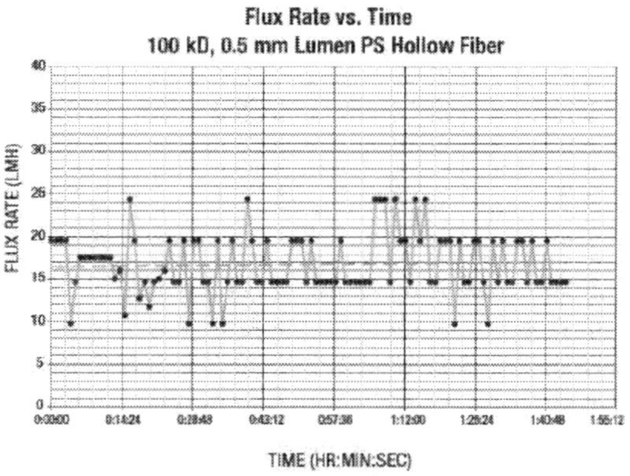

Figure 3

3. Discussion

As expected, the process pressures cycled with each dilution, climbing during the concentration from 4% to 8% and then falling back to their initial values with each dilution. This increase in pressure is caused by the increase in viscosity and polymer concentration at the membrane surface during the concentration and is typical for Tangential Flow Filtration processes. The permeate flow rate per unit of membrane area, also referred to as the process flux, ranged from 15-20 L/ M2/ hr following the concentration of the polymeric nanoparticle. As indicated by the sample analysis, the removal efficiency of the by-product (rejection coefficient) was nearly 50% higher due to the shift in concentration factor based on the viscosity of the process material therefore impacting the permeability of the solute removed as expressed below:

$$Y = 1-(VO / VR)\ 1+n(R-1)$$

Y= % Permeate Solute Removed

VO / VR =Volume Change Ratio

n=number of stages.

R=Rejection Coefficient

(2)

The removal efficiency of diafiltration is dependent upon how easily the small molecular weight by product passes through the membrane. The concentration of the retained material at the membrane surface during the TFF (Tangential Flow Filtration Process) plays a significant role in the transmission of the smaller molecules. Transmission is calcula ted by measuring the concentration of the byproduct in the permeate stream and dividing it by the concentration of the byproduct in the retentate stream.

$$\%T \quad \frac{\text{conc } PERM}{\text{conc } FEED} * 100\ \%$$

(3)

Analysis by HPLC and functional assays showed that the transmission of the by-product in this experiment started at 59% and decreased with each diafiltration volume to 43%. Many diafiltration processes with less viscous starting solutions exhibit transmission of 90 to 100% and require only 4-5 diafiltration volumes to reduce small MW molecules to minimum levels. The results of several R&D scale tangential flow filtration diafiltration experiments prompted the rapid scale-up of the process to a 12 L pilot scale batch volume. To determine the membrane area required for the scale-up, assuming product concentrations were to be identical at both scales, a linear factor of process volume/membrane area was used.

Therefore, for the 12 L pilot scale volume, 100 kD polysulfone hollow fiber module with 0.57 M2 of membrane (Spectrum Labs MiniKros® P/N K5AB-050-01P) was used. With a count of approximately 1800 fibers, and 20 cm fiber length a shear of 4000 sec-1 was again utilized at a feed flow rate of 5.2 LPM.

Table 1

Area	Shear (S^{-1})	ΔP @ 4%	ΔP @ 8%	Flux (LMH)
615 cm^2 (bench)	4000	4	8	16
0.57 M2 (pilot)	4000	6	11	13

Permeate flux, process time and final product yield of the 12 L process reproduced the bench scale measurements. The only anticipated difference was that slightly higher feed-toretentate ΔP was measured at the pilot scale due to the longer fiber length (20 cm vs 12 cm at the R&D scale). The following table presents the scalability results.

References

[1] A.G. Rinzler et.al " Large-scale Purification of single-wall carbon nanotubes, process, product, and characterization" Appl. Phy. A 67,29-37 (1988)
[2] Scott Sweeney et. al. "Rapid Purification and size separation of gold nanoparticles via diafiltration" J. Am.Chem Soc. 128, 3190-3197 (2006).
[3] Mark C. Porter, Handbook of Industrial Membrane Technology, Noyes Publications. pp 186-187. (1990).

Effect of silica on the thermal properties of LDPE-WF-silica composites

Ajay Kumar Mishra[1]
Department of Chemical Technology, University of Johannesburg, Doornfontein Campus, Private Bag 17011, Johannesburg, 2028, SOUTH AFRICA

E-mail: amishra@uj.ac.za

Abstract. The influence of nano-silica, synthesized and mixed with LDPE through a sol-gel process, on the thermal and mechanical properties of LDPE and LDPE/WF composites was investigated. SEM analyses of the samples showed the silica particles in the matrix to be nano-sized, and FTIR results indicated strong interaction between nano-silica and the LDPE matrix. Both wood fibre and nano-silica improved the thermal stability of the LDPE matrix, while the presence of nano-silica reduced the crystallinity of LDPE. The tensile strength of the samples decreased with increasing wood fibre content, while the tensile modulus substantially increased.

1. Introduction

Organic-inorganic hybrid nanocomposites have enormous potential applications [1]. The hybrid materials are prepared from organic polymers and silicon alkoxides (mainly tetraethoxysilane, Si(OC2H5)4) by a sol-gel process. This has been widely studied, because the reaction speed is easily controllable. The sol-gel processed hybrid materials have interpenetrating polymer networks in the glassy network formed by polymerization [2-5].

Natural fibre is used as reinforcement because of its advantages over the other established materials. Plant fibres are renewable, biodegradable, low-cost, low density raw materials with high stiffness and strength compared to other conventional products [6-7]. Natural fibres contain various organic materials (mainly celluloses as well as hemicelluloses and lignins) and their thermal treatment leads to a variety of physical and chemical changes. Thermal degradation of these fibres leads to poor properties and especially to deterioration of mechanical properties [8-9].

The matrix-filler interface plays a critical role in ensuring that the properties of each component contribute optimally to the bulk properties of the final product. The strength of fibre reinforced composites is highly dependent on the degree of interfacial adhesion, whereas the modulus is controlled by the fibre orientation. It is reported that the modulus of cellulose-filled composites normally increases significantly and their tensile strength remains unaffected, while elongation and impact strength are significantly reduced when compared to the unfilled polymer [10]. Plant fibres are not easily dispersed in non-polar polymers, because of their extensive intermolecular hydrogen bonding. A silane coupling agent may alter the hydrophilic properties of the interface.
To overcome the drawbacks and to prepare material with enhanced properties, many polyethylene nanocomposites with several inorganic nanofillers have been prepared [11-16]. The polymer

[1] To whom any correspondence should be addressed.

nanocomposites significantly improved the stiffness and tensile strength over the unmodified polymers. One of the disadvantages associated with the use of nanofillers is their high cost. However, this negative effect of high cost is counterbalanced by the small amount of nanofillers needed compared to traditional fillers.

The present study deals with the preparation of LDPE-wood fibre-silica composites. The nanosilica was prepared by a sol-gel technique. The composites were characterized and their thermal and mechanical properties determined.

2.Experimental

2.1 Materials

Low-density polyethylene (density 0.924 g cm-3) was supplied by Sasol Polymers, South Africa. Wood fibre was supplied by a local vendor. Tetraethoxy ortho silicate (TEOS) with 98 % purity and commercial grade dichloromethane (DCM) were supplied by Sigma Aldrich, South Africa.

2.2 Pre-treatment of wood fibre

The impurities of the wood fibre were removed by initial washing with distilled water, followed by soaking in petroleum ether and drying in an oven at 100 ± 5 OC for 6 hours.

2.3 Preparation of nano-silica

The nano-silica was prepared by a sol-gel technique using tetraethoxy ortho silicate (TEOS) as precursor [1], as shown in the reaction scheme (Scheme 1). TEOS, water and dichloromethane were mixed in a 1:4:4 molar ratios and stirred at room temperature (25°C).

Hydrolysis:

$$Si(OR)_4 \xrightarrow[OH^-/H_2O]{H^+/H_2O \text{ or }} Si(OH)_4 + 4\ ROH;\ \text{R is alkyl group}$$

Polycondensation:

$$Si(OH)_4 + Si(OH)_4 \longrightarrow (OH)_3Si–O–Si(OH)_3 + H_2O$$

Scheme-1 Reaction scheme for the silica formation from tetraethoxyorthosilane

The mixture was kept at room temperature to remove the solvent followed by drying in an oven at 100 ± 5 °C overnight.

2.4 Preparation of composites

Molten LDPE was mechanically mixed with a silica suspension, after which the solvent was evaporated in a vacuum oven at 60°C. The wood fibre was mixed with the LDPE/silica nanocomposites using Thermo Scientific Haake Rheomex OS equipped with roller type rotors. The compounding was performed in a mixing chamber of 64 cm3 at a screw speed of 30 rpm and at a temperature of 160°C for a period of 15 min. After mixing, the samples were immediately chopped into small chips which were then moulded into strips using the extruder.

IOP Conf. Series: Materials Science and Engineering **40** (2012) 012036 doi:10.1088/1757-899X/40/1/012036

2.5 Scanning electron microscopy (SEM)
The morphology were examined by a JOEL Scanning Electron Microscopy (SEM) instrument with a field-emission gun operated under the following conditions: 15 kV acceleration voltage, working distance (WD) 20 and spot size 21. Specimens were deposited on double-sided carbon conductive scotch tape and examined on the surface after double coating with gold for charge accumulation.

2.6 Thermogravimetric analysis (TGA)
Thermogravimetric analysis (TGA) was performed in a Perkin Elmer TGA 4000 Analyser equipped with Pyris software. The sample mass was between 8 mg and 10 mg and the temperature ranged between 30°C and 600°C with a heating rate of 15°C/min. The analyses were performed under a nitrogen atmosphere at a flow rate of 20 mℓ/min.

2.7 Differential scanning calorimetry (DSC)
Calorimetric analyses were performed on a DSC822e Deferential Scanning Calorimeter (DSC). The experiments were carried out at temperatures ranging from 25°C to 200°C at a heating rate of 10°C/min where nitrogen was used as a purge gas at a flow rate of 20 mℓ/min. The mass used for analysis was 5.0 mg to 6.0 mg.

3. Results and Discussion

3.1 Scanning Electron Microscopy Analysis
SEM picture of the polymer (LDPE) and wood fibre shows a single phase whereas polymer and wood fibre clearly show the presence of fibres in the polymer matrix (Figure 1a). Various percentage of silica shows (Figure 1b) uniform dispersion of nano-silica in the polymer-WF system. The particle sizes of the silica particles were found to be in the range of 10-50 nm, however some agglomerates were also observed.

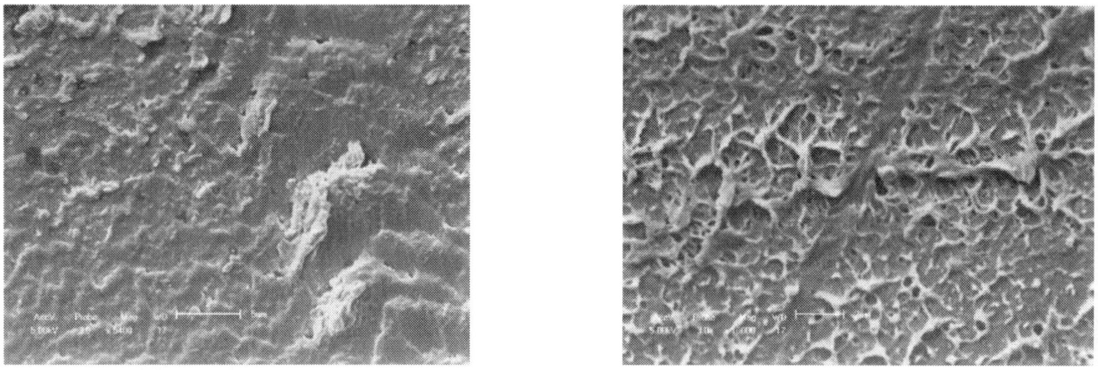

Figure 1a **Figure 1b**

Figure 1: SEM images of LDPE-WF-SiO_2 composites

3.2 Fourier Transform Infrared Spectroscopy Analysis
The FTIR spectra of the polymer-WF-silica can be seen in Figure 2. The FTIR spectrum of pure LDPE shows characteristic peaks at 728 (rocking-CH_2), 1308 (wagging-CH_2), 2835(-CH stretch), 2920 (-CH_2 stretch) and at 893 cm^{-1} (-CH=CH_2) [17-18]. The polymer-silica spectrum shows all the characteristic peaks of LDPE at shifted values. The additional broad peak at 1070 cm^{-1} is attributed to Si-O-Si and Si-O-C bonds [1]. In addition, the broad peak at 3367 cm-1 corresponds to bonded –OH of polysilanol. A small peak at 1056 cm^{-1} indicates the polysiloxane deposition onto the polymer. The spectrum of the LDPE-WF-silica composite shows all the characteristic peaks corresponding to LDPE

and polysilanol. The peak shift from 1056 cm^{-1} to 1090 cm^{-1} refers to good interactions of polysilanol with the lignocellulosic fibre and the deposition of the polysilanol onto the lignocellulosic fibre instead of the polymer.

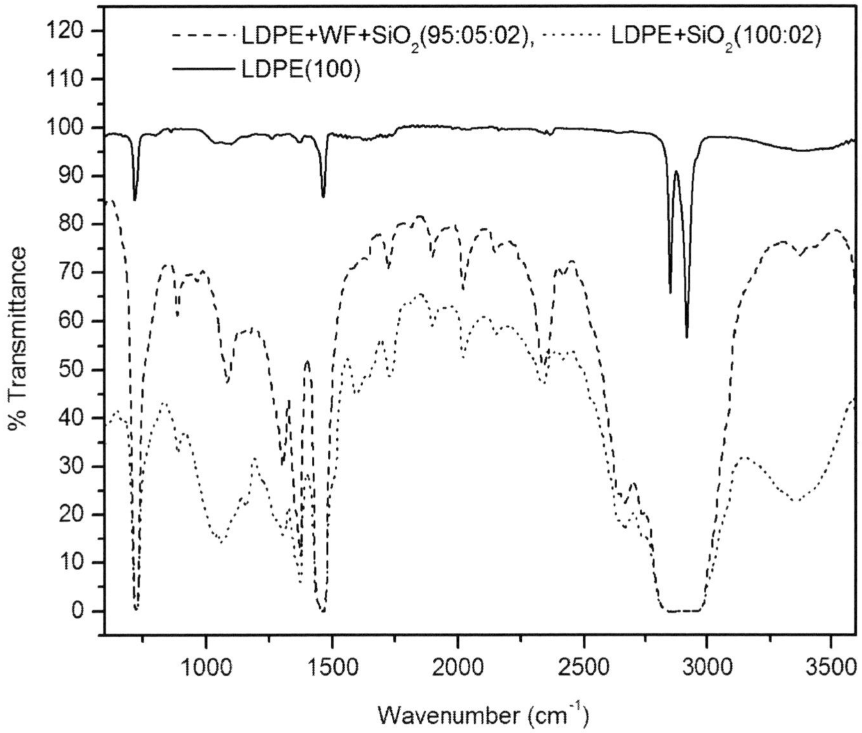

Figure 2: FTIR spectra of LDPE-WF-SiO_2 composites

3.3 Thermal Studies

3.3.1 Thermogravimetric analysis (TGA)
The TGA curves of the different LDPE-WF-silica composites are shown in Figure 3. Various amount of silica substantially influences the decomposition behaviour of the composite. The presence of wood in the polymer matrix improves the thermal stability of the composites which was even further improved by incorporation of nanosilica. The presence of nano-silica increases the onset temperature of degradation. This is probably due to the fact that the silica, which is fairly well dispersed in the matrix and which has a large surface area, is thermally stable in nature and hence there is slow diffusion of heat towards wood fibre and polymer matrix [18].

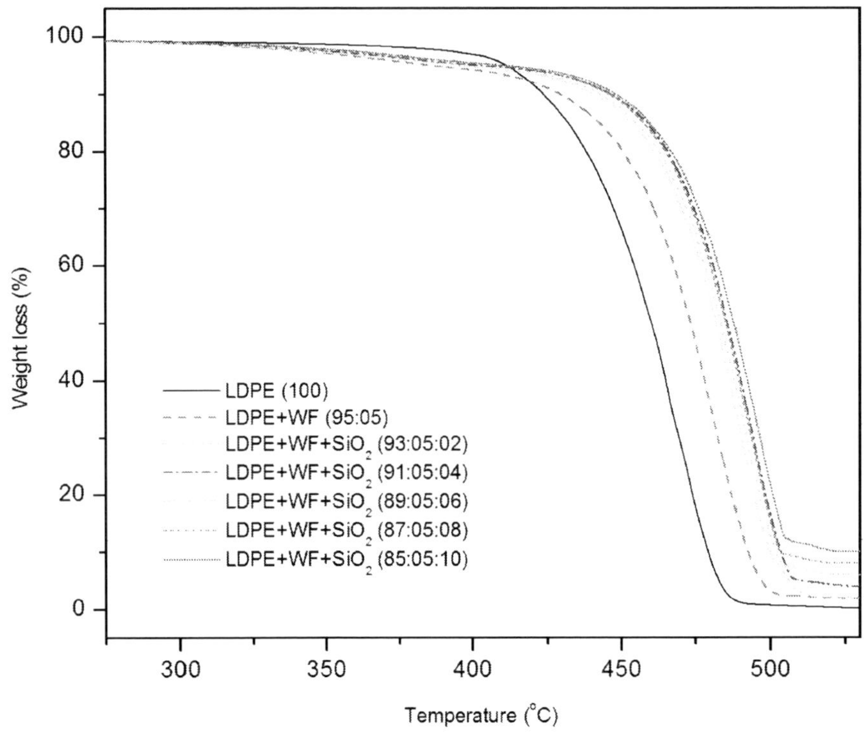

Figure 3: TGA curves of LDPE-WF-SiO_2 composites

3.3.2 Differential Scanning Calorimetry Analysis (DSC)

The typical DSC curves in Figure 4 shows the melting curves of LDPE-WF-silica composites. The onset and the peak temperatures of polymer and polymer-wood fibre composites are found to be more or less similar while incorporation of nanosilica influences on the crystallite size and distribution. Silica nano-particles are known to nucleate polymer crystallization, increasing its crystallinity, decreasing crystallite thickness and influencing the orientation of the lamellae in the crystallites [19]. Thus the variations of nano-silica influence the polymer crystallinity. It is therefore can be attributed that the presence of WF improves LDPE crystallization, while the presence of variations of nano-silica enhances more crystallization behaviour may be due to the interaction between LDPE-wood fibre-silica, which immobilizes the LDPE chains in the vicinity of nano-silica particles.

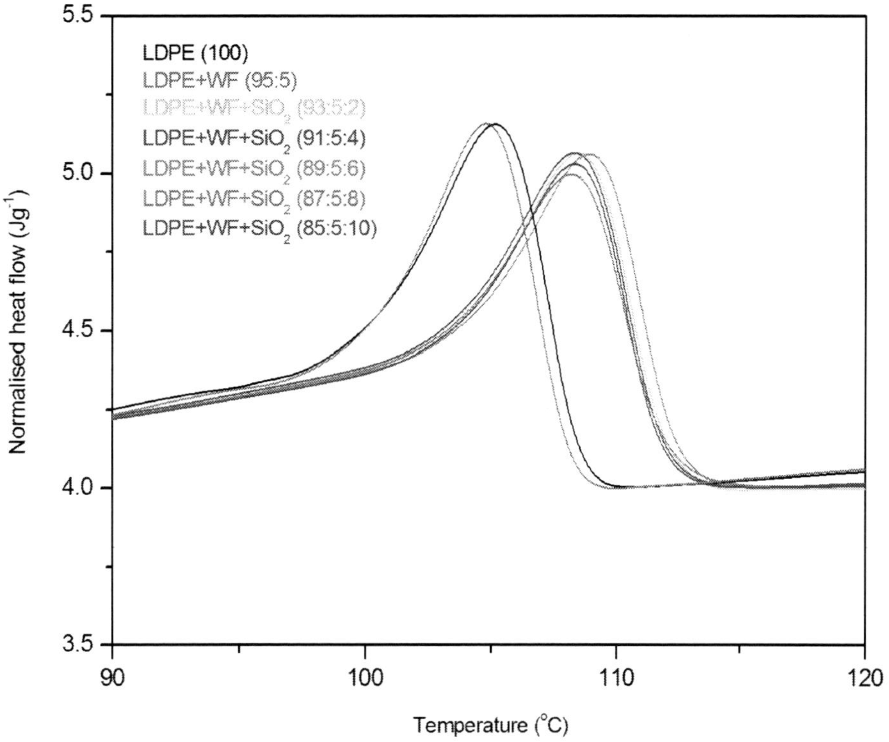

Figure 4: DSC curves of LDPE-WF-SiO_2 composites

3.4 Mechanical Properties

It is often expected that the silica nanoparticles will stiffen and toughen the polymer matrix [20-21]. The tensile strength and young modulus of the LDPE-WF composites were slightly higher than the LDPE-WF-silica composites (Table 1). It may be attributed that LDPE-WF composite show higher stiffness and good interaction between hydrophilic wood fibre and hydrophobic polymer. However, variation of nanosilica in the composite further reduces the tensile strength and young modulus probably due to the higher crystallinity behaviour of the LDPE-WF-silica composite. Thus it can be concluded that the presence of nano-silica in the polymer-wood fibre matrix substantially influence the polymer chain mobility.

Table 1: Tensile properties of LDPE-WF-SiO_2

Sample Name	Tensile strength	E-Modulus (MPa)
LDPE-Pure	7.9	99.8
LDPE+WF (95:05)	9.1	146.0
LDPE+WF+SiO_2 (93:05:02)	8.6	131.2
LDPE+WF+SiO_2 (91:05:04)	8.6	128.5
LDPE+WF+SiO_2 (89:05:06)	8.6	124.9
LDPE+WF+SiO_2 (87:05:08)	8.6	127.7
LDPE+WF+SiO_2 (85:05:10)	8.7	129.5

4. Conclusions

Nano-silica was synthesized via sol-gel method which was incorporated in the polymer-wood fibre composite. The spectral, thermal and mechanical properties of LDPE-WF-silica composites were investigated. The silica particles in the polymer matrix were found to be nano-sized. Due to the interaction and crystallinity behaviour of nano-silica with polymer-wood fibre the thermal stability of the resulting composites were improved. The tensile strength and young modulus of the polymer-wood fibre-silica composites decreased. Nano-silica incorporation in the composite decreases both tensile strength and tensile modulus which may be probably due to the result of the crystallinity behaviour.

5. References

[1] Raman V, Bhatia G, Mishra AK, Bhardwaj S and Sood 2006 KN *Mater. Lett.* **60** 3906.
[2] Ravaine D, Seminel A, Charbouillot Y and Vincens M 1986 *J. Non-Cryst. Solids* **82** 210.
[3] Nakanishi K and Soga N 1991 *J. Am. Ceram. Soc.* **74** 2518.
[4] Saegusa T and Chujo Y 1990 *J. Macromol. Sci. Chem.* A **27** 1603.
[5] Motomastu M, Takahashi T, Nie H, Mizutani W and Tokumoto H 1997 *Polymer* **1** 177.
[6] Georgopoulos ST, Tarantili PA, Avgerinos E, Andreopoulos AG and Koukios EG 2005 *Polym. Degrad. Stabil.* **90** 303.
[7] Nabi Saheb D and Jog JP 1999 Adv. *Polym. Technol.* **18** 351.
[8] Bledzki AK and Gassam J 1999 *Prog. Polym. Sci.* **24** 221.
[9] Bengtsson M and Oksman K 2006 *Composites* A **37** 752.
[10] Maldas D and Kokta BV 1993 *Trends Polym. Sci.* **1** 174.
[11] Kontou E and Niaounakis M 2006 *Polymer* **47** 1267.
[12] Bikiaris DN, Vassiliou A, Pavlidou E and Karayannidis GP 2005 *Eur. Polym. J.* **41** 1965.
[13] Wu CS 2005 *J. Polym. Sci.* A **43** 1690.
[14] Liao HT and Wu CS 2004 *J. Polym. Sci.* B **42** 4272.
[15] Zhang F, Li S, Karaki T and Adachi M 2005 *Jpn. J. Appl. Phys.* 1 **44** 658.
[16] Osman MA, Rupp JEP and Suter UW 2005 *Polymer* **46** 1653.
[17] Bartczak Z, Argon AS, Cohen RE and Kowalewski T 1999 *Polymer* **40** 2367.
[18] Mishra AK and Luyt AS 2008 *Polym. Degrad. Stabil.* **93** 1.
[19] Salemane MG and Luyt AS 2006 *J. Appl. Polym. Sci.* **100** 4173.
[20] Brooks NWJ, Duckett RA and Ward IM 1998 *J. Polym. Sci.* B **36** 2177.
[21] Lucas JC, Failla MD, Smith FL, Mandelkern L and Peacock A 1995 *J. Polym. Eng. Sci.* **35** 1117.

IOP Conf. Series: Materials Science and Engineering **40** (2012) 012037 doi:10.1088/1757-899X/40/1/012037

Silver nanocluster/silica composite coatings obtained by sputtering for antibacterial applications

M Ferraris[1], C Balagna[1], S Perero[1], M Miola[1], S Ferraris[1], F Baino[1], A Battiato[2], C Manfredotti[2], E Vittone[2] and E Vernè[1]

[1]Institute of Materials Physics and Engineering, Department of Applied Science and Technology, Politecnico di Torino, 10129, Torino, Italy

[2]Experimental Physics Department /Centre of Excellence "Nanostructured Interfaces and surfaces", University of Torino, and INFN sez.Torino Via P. Giuria 1, 10125 Torino, Italy

E-mail: cristina.balagna@polito.it

Abstract. Silver nanocluster silica composite coatings were deposited by radio frequency co-sputtering technique on several substrates. This versatile method allows tailoring of silver content and antibacterial behaviour of coatings deposited on glasses, ceramics, metals and polymers for several applications. Coating morphology and composition as well as nanocluster size were analyzed by means of UV-Visible absorption, X-ray diffraction (XRD), Field Emission Scanning Electron Microscopy (FESEM), electron dispersive spectroscopy (EDS), X-ray Photoelectron Spectroscopy (XPS) and Atomic Force Microscopy (AFM). The antibacterial effect was verified through the inhibition halo test against standard bacterial strain, *Staphylococcus aureus,* before and after sterilization process. Tape test demonstrated a good adhesion of the coatings to the substrates.

1. Introduction

Materials with antibacterial properties are widely requested in several fields where the risk of microbial contamination is considered a relevant issue, such as biomedical implants, agricultural/food industry, facilities in crowded places (bus, telephones), personnel protective systems and also space structures [1-3].

Silver is the most known and documented antimicrobial agent and its powerful action can be expressed in several forms as metallic, nanoparticles and ions [4, 5]. Silver antibacterial properties can be conferred to glasses, ceramics, metals and polymers by means of several techniques as ions-exchange, sol-gel method and sputtering [6–10]. Sputtering is one of the most versatile coating methods, suitable for most of substrates, because it does not reach high temperatures which could compromise the mechanical and thermal properties of materials to be coated (e.g. polymers) [10].

Preparation and characterization of materials with antibacterial properties are topics well covered by the authors' research group [11–17].

In this paper, antibacterial silver nanocluster silica composite coating were deposited on several substrates, using radio frequency (RF) co-sputtering technique [15–17] and setting several process parameters as a function of the substrate. Silica was chosen as a matrix for the composite coating because it provides good mechanical and thermal resistance, whereas silver nanoclusters confer antibacterial activity. In addition, silver nanoclusters embedded in a silica matrix allows the reduction of silver amount, if compared with pure silver coatings, with advantages in terms of toxicity [18, 19].

A characterization of the composition and morphology of the coating was also reported. The antimicrobial behavior was analyzed through the evaluation of the inhibition halo against standard bacterial strain, *Staphylococcus aureus* (ATCC29213) [20], before and after sterilization processes (gamma rays and EtO). Adhesion of the coatings was evaluated through standard tape test [21].

2. Materials and methods

RF co-sputtering (Microcoat™ MS450) was the technique used for the deposition of a silver nanocluster silica composite coating on several substrates. Silver (Sigma–Aldrich™ 99.99 %) and silica (Franco Corradi s.r.l. 99.9 %) were used as targets.

The morphology, composition and structure of coatings were evaluated by means Field Emission Scanning Electron Microscope (FESEM, SUPRATM 40, Zeiss™), UV-Visible absorption spectra (UV-Vis, Varian™ Cary 300 Bio), X-Ray Diffraction (XRD, X'Pert Philips™ diffractometer equipped with X'Pert High Score™ software analysis), Electron Dispersive Spectroscopy (EDS, SEM-FEI- Quanta™ Inspect 200 and EDAX PV 9900), X-ray Photoelectron Spectroscopy (XPS, VSW TA10 non-monochromatic Al Kα (1486.6 eV) X-ray source) and Atomic Force Microscopy (AFM, XE-100 with XEI 1.8.0 software).

The antibacterial potential of the silver nanocluster silica composite coatings on different substrates was determined against the standard bacterial strain, *Staphylococcus aureus*, by the inhibition zone evaluation in accordance to National Committee for Clinical Laboratory Standards (NCCLS) [20]. Several samples were tested before and after a sterilization treatments (gamma rays and EtO).

Tape test was performed on the coated samples according to the ASTM D3359 standard [21] in order to evaluate the adhesion of the coatings to the several substrates.

3. Results and discussion

Silver nanocluster/silica composite coatings were deposited on several substrates as glasses (silica, soda-lime), metals (aluminium, steel) and polymers (polypropylene, polyurethane, aramidic fibers and a commercial multilayer polymer composed of ethylene vinyl alcohol (EVOH), polyamide and polyethylene as the outer layer) by means of RF co-sputtering technique. These substrates were chosen among materials available for different applications in food industry, biomedical implants, space structures, personal protective systems. The main process parameters, optimized after a series of preliminary depositions and fixed for all the substrates, are reported in Table 1.

Table1: Sputtering Process Parameters

Sputtering Process Parameters	Set Values
Pre-Deposition Pressure	70 μPa
Working Pressure	5.5dPa
Power on silica target (6 inch)	200 W RF
Power on silver target (1 inch)	1 W DC

On the contrary, the deposition time was changed between 8 and 80 minutes depending on the substrates and the final application. In particular, the increment of deposition time allows the increment of coating thickness (from 15 to 300 nm), total silver content in the coating and consequently antibacterial effect. Even if the temperature reached in sputtering chamber is not so high (about 80 °C), the maintenance of this temperature for long time can damage thermo-sensitive substrates, such as polymers. For these materials, short sputtering treatments (8-15 min) have been selected. In addition, the silver amount has to be strictly optimized for materials in direct contact with human organs and tissues in the biomedical field in order to obtain an antibacterial activity without inducing cytotoxic effects. The correlation between deposition time and coating thickness and main properties is schematized in figure 1.

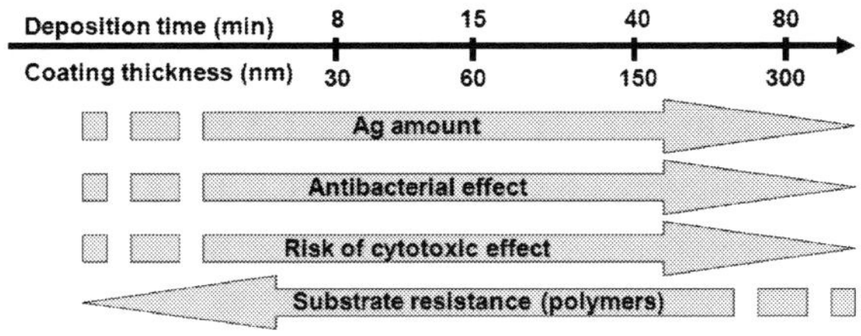

Fig. 1 : Schematic representation of the influence of deposition time on silver amount and its effects.

The coating morphology was observed by means FESEM (Fig. 2). The cross-section of a silica substrate coated with the antibacterial layer is reported in Fig. 2a. Silver nanoclusters are well visible as bright dots. The magnification in Fig. 2b shows the typical porous structure of the sputtered silica with embedded silver nanoclusters. EDS analyses showed the presence of silver, silicon and oxygen confirming coating presence on all tested samples.

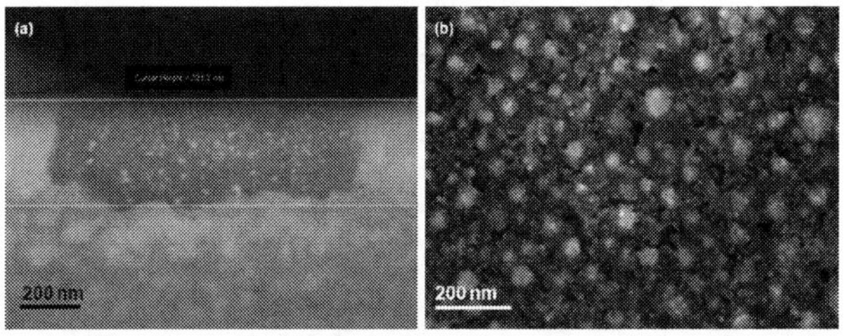

Fig. 2 : FESEM images relative to coating morphology: (a) cross-section of a coated silica substrate and (b) a micrograph of silver nanoclusters embedded into a silica matrix.

The UV–VIS absorption spectra relative to coatings with different thicknesses for comparison are reported in Fig. 3. In this case, the multilayer polymer was used as control. A large absorption with the maximum peak at about 396 – 398 nm was detected in all coated samples even if it is barely visible between 335 and 500 nm in the thinner coating (30 nm). The absorption could be ascribable to the localized surface plasmon resonance (L–SPR) typical of silver metal nanoparticles with dimensions of about 20 nm [22]. The coating thickness increment increases the maximum peak intensity, but it does not affect silver nanoclusters dimensions which depend on the position of the maximum peak (almost equal for all the coated samples).

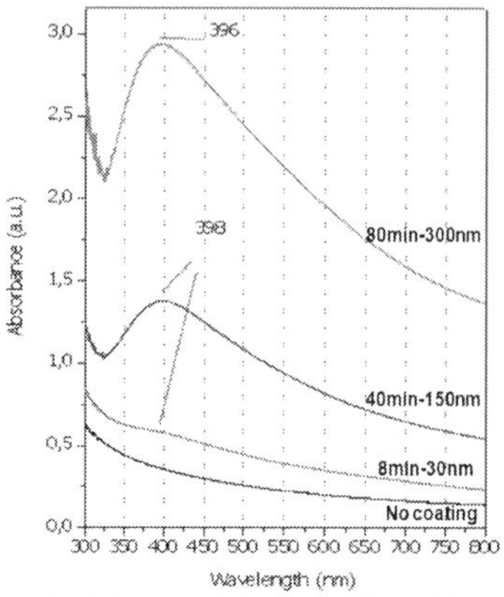

Fig.3: UV-Vis analysis on coatings with different thickness (30, 150 and 300 nm) compared with uncoated multilayer polymer used as substrate.

XPS analysis of the silver nanocluster/silica composite coatings, with two different thicknesses (300 nm vs 60 nm) and deposited on soda-lime and silica glass substrates, respectively, is reported in Fig. 4. In both the spectra, the principal peaks of silver (doublet Ag 3d), oxygen (O 1s), silicon (Si 2p) and carbon (C 1s) are revealed. The carbon is mainly due to surface contamination. Other secondary peaks (Si 2s and Auger peaks of Ag and O) are indicated in the graphs. The ratio Ag/Si remained equal to 0.1 for both the samples, thereby indicating that the thickness and the substrate material do not influence the coating composition.

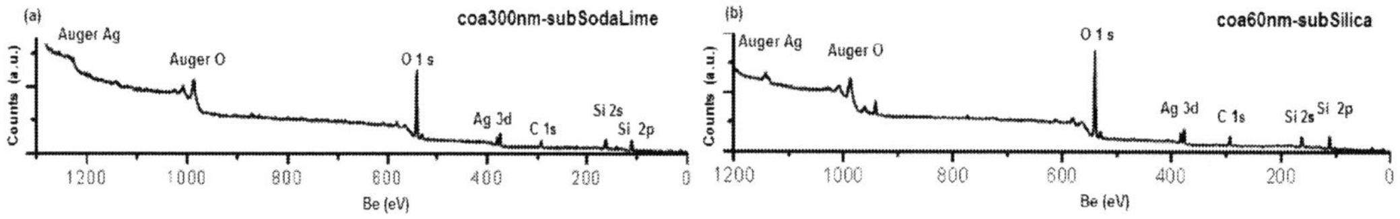

Fig. 4 : XPS analysis relative to (a) 300nm-thick coating on soda-lime glass substrate and (b) 60nm-thick coating on silica glass substrate.

AFM image of a deposited sputtered layer on a silicon wafer substrate, having a presumed thickness of 300 nm, is shown in Fig. 5. The root mean square roughness (RMS) value of about 6 nm and the roughness average (R_a) value of about 5 nm confirmed the nano-structured nature of the analysed layer; the overall uniformity and flatness of the image established that the sputtering technique allows a homogeneous growth of the coating and the uniformity of its surface morphology.

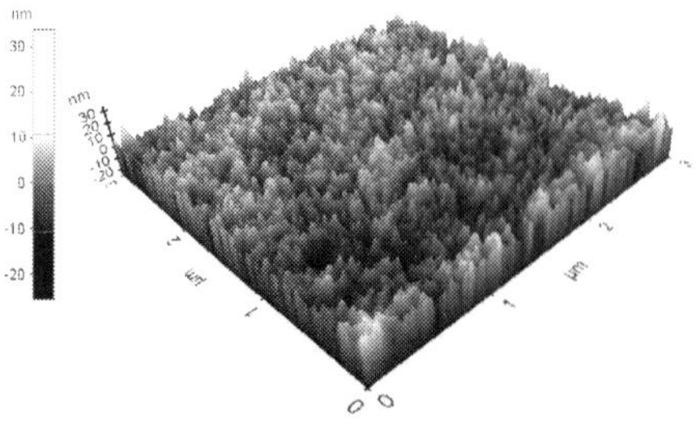

Fig. 5: AFM topography signal 3D image of an as deposited sputtered layer on a silicon wafer substrate

XRD analysis in Fig. 6 shows the presence of a wide peak around 20° not completely reported here: its nature can be connected with the amorphous silica present in the matrix and in the glass substrate; two low intensity peaks due to metallic silver are visible at 38 and 45° .

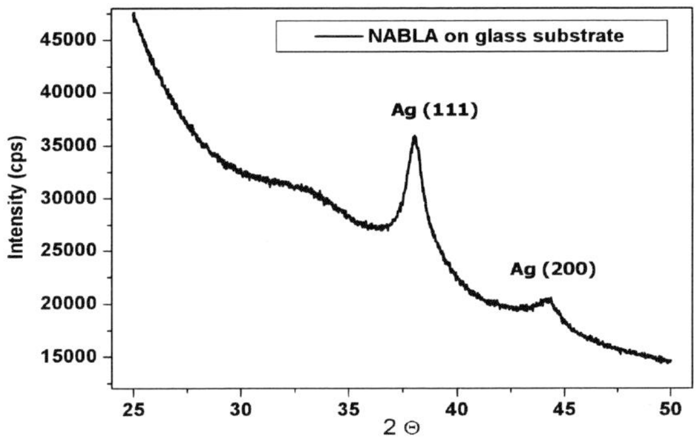

Fig. 6: XRD analysis of a 300nm-thick coating on glass substrate (constant grazing incidence angle: 2°)

The antibacterial activity against *S. aureus* of the silver nanocluster/silica composite coatings deposited on different substrates was verified by the observation of inhibition halo development. Fig. 7 reports some examples of the test performed with several substrates. A well visible halo of about 2-3 mm formed around all tested samples demonstrating that the antimicrobial behaviour can be conferred to all the material classes by means of the prepared sputtered silver nanocluster/silica composite coating. The same antibacterial effect was observed also for samples after sterilization by EtO or gamma rays.

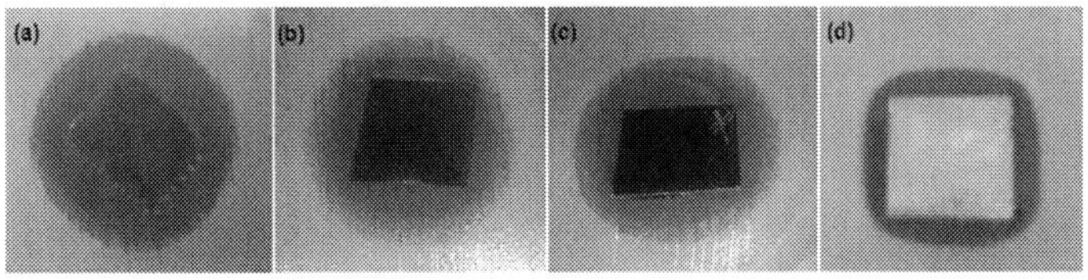

Fig. 7: Inhibition halo test with S. aureus strain for (a) fabric, (b) multilayer polymer, (c) silica and (d) steel

After the tape test (ASTM D3359 – 97 standard [21]), the silver nanocluster silica composite coating resulted well adherent and no sign of coating damage on the sample or coating detachment from the substrate on the tape was visible on the tape (not reported here).

4. Conclusions

The co-sputtering technique allowed the successful deposition of a coating composed of silver nanoclusters embedded in a silica matrix on glasses, ceramics, metals and polymers. The process parameters were set as a function of substrate material and final application. In particular, coating thickness and silver amount can be modulated by variation of the deposition time. Silver nanoclusters were observed well embedded in the silica matrix and XRD signal confirmed their presence. The coating thickness seems not to influence the nanoclusters dimensions and the Ag/Si ratio. The coating surface morphology shows a typical nano-structure. The coating antibacterial activity against *S. aureus* was verified with the formation of a well visible inhibition halo and it is not affected by the substrate and the sterilization procedure. The coating resulted well adherent to all the analysed substrate.

Acknowledgment

This work was partially funded by "NABLA-Nanostructured Anti-Bacterial LAyers"-Regione Piemonte, Italy and "NASLA- Nanostructured Anti-SepticaL coAtings"- EU Seventh Framework Programme.

This research was supported by Thales Alenia Space IT, Reply, Dipromed S.r.l., Alce Calidad, Aero Sekur SpA, Easreth.

References

[1] Tunc K *et al* 2006 *J. Infect.* **53** pp 140-143
[2] Shia X *et al* 2009 *Trends Food Sci. Technol.* **20** pp 407-413
[3] Gu J 2007 *Int. Biodeterior. Biodegrad.* **59** pp 170-179
[4] Rai M *et al* 2009 *Biotechnol. Adv.* **27** pp 76–83
[5] Guzman M *et al* 2012 *Nanomed. Nanotechnol. Biol. Med.* **8** pp 37–45
[6] Kawashita M *et al* 2000 *Biomaterials* **21** pp 393-398
[7] Di Nunzio S *et al* 2004 *J. Eur. Ceram. Soc.* **24** pp 2935-2942
[8] Di Nunzio S *et al* 2006 WO2006/058906
[9] Sant S B *et al* 2007 *Acta Biomater.* **3** pp 341-350
[10] Dowling D P *et al* 2011 *Thin Solid Films* **398–399** pp 602–606
[11] Vernè E *et al* 2008 *Key Eng. Mat.* **361-363** pp1195-1198
[12] Miola M *et al* 2009 *J. Mater. Sci. Mater. Med.* **20** pp 741-749
[13] Vernè E *et al* 2008 *Adv Appl. Ceram.* **107** pp 245-253
[14] Vitale Brovarone C *et al* 2008 *Chem. Eng.* J. **137** pp 129-136
[15] Ferraris M *et al* 2008 TO2008A000098.
[16] Ferraris M *et al* 2010 *Mater. Chem. Phys.* **120** pp 123-126
[17] Ferraris M *et al* 2010 *Adv. Eng. Mater.* **12** pp B276-B282
[18] Beer C *et al* 2012 *Toxicol. Lett.* **208** pp 286-292
[19] Wijnhoven S *et al* 2009 *Nanotoxicology* **3** pp 109 -138
[20] NCCLS M2-A9 Performance Standards for Antimicrobial Disk Susceptibility Tests, Approved Standard- Ninth Edition 2006
[21] ASTM D3359 – 97, Standard Test Methods for Measuring Adhesion by Tape Test
[22] Sánchez-Ramírez J F 2008 *J. Nanomater.* **2008** pp 1-9

IOP Conf. Series: Materials Science and Engineering **40** (2012) 012038 doi:10.1088/1757-899X/40/1/012038

Potential of silicon nanowires structures as nanoscale piezoresistors in mechanical sensors

M Messina and J Njuguna

Cranfield University, Cranfield Campus, Bedfordshire, UK

E-mail:j.njuguna@cranfield.ac.uk

Abstract. This paper presents the design of a single square millimeter 3-axial accelerometer for bio-mechanics measurements that exploit the potential of silicon nanowires structures as nanoscale piezoresistors. The main requirements of this application are miniaturization and high measurement accuracy. Nanowires as nanoscale piezoresistive devices have been chosen as sensing element, due to their high sensitivity and miniaturization achievable. By exploiting the electro-mechanical features of nanowires as nanoscale piezoresistors, the nominal sensor sensitivity is overall boosted by more than 30 times. This approach allows significant higher accuracy and resolution with smaller sensing element in comparison with conventional devices without the need of signal amplification.

1. Introduction

Silicon nanowires have a very large piezoresistance effect [1-5], capable of enhancing the mechanical sensors performance, which is now actively being explored to improve silicon transistors [6], [7]. Silicon nanowires are also attractive for applications in the filed-emission devices, photonics, chemical sensors and spintronics [8]. The piezoresistive effect is used for transducing, for example, acceleration in an electrical output. After an inertial force is applied to the sensor the strain on the piezoresistive material (silicon) changes its electrical resistance proportionally, therefore the correspondent voltage change is a measure of the acceleration by less than a constant of proportionality.

In the last decade experimental studies on the piezoresistance effect of SiNWs agreed that SiNWs under uniaxial stress offer an enhanced piezoresistance effect with respect to the bulk counterparts [1-5], [9], [10], [11]. The origin and behavior of this phenomenon called in the literature "Giant Piezoresistance", is currently not clearly understood and research is at infancy stage. This effect would be of enormous impact on the performance of mechanical sensor. To date, relatively few reports on the development of silicon nanowire-based sensors are available [12], [13]. However, p-type single crystalline SiNWs have been studied for sensor applications [4], [9], [10], [14], [16].

Toriyama et al. [9] studied silicon nanowire piezoresistors fabricated by separation of implanted oxygen (SIMOX), thermal diffusion, electron beam (EB) direct writing, and reactive ion etching (RIE). In their study longitudinal and transverse piezoresistive coefficients, $\pi_{l<110>}$ and $\pi_{t<110>}$, were both dependant on the cross sectional area of the nanowires. The $\pi_{l<110>}$ of the nanowire piezoresistors increased (up to 60%) with a decrease in the cross sectional area, while $\pi_{t<110>}$ decreased with a increase in the aspect ratio of the cross section. The enhancement behavior of the $\pi_{l<110>}$ was explained qualitatively using 1-D hole transfer and hole conduction mass shift mechanisms. The reduction in the $\pi_{t<110>}$ with increase in the aspect ratio of the cross section is explained due to decreased stress transmission from the substrate to the nanowire. The maximum value obtained of $\pi_{l<110>}$ of 48 × 10^{-11} Pa^{-1} at a surface concentration of 5 × 10^{19} cm^{-3}, indicate sufficient sensitivity for sensing applications.

Initial experimental studies undertaken by He and Yang [1] reported a very high piezoresistive effect (increased up to 3,776% with respect to the higher dimension counterparts) of self-assembled single crystal silicon nanowires in the <111> crystallographic orientation. Reck et al. [2] later used a lift-off and an electron beam lithography (EBL) technique to fabricate silicon test chips and studied the piezoresistive properties of crystalline and polycrystalline nanowires as a function of stress and temperature. They found that the piezoresistive effect in the <110> direction increased significantly as the silicon nanowire diameter decreased (up to 633%), consistent with the results from He and Yang [1]. Finally, Passi et al. [5] recently obtained an increase of piezoresistance of up to 2,140% respect to the bulk-Si in the same direction, the <110>.

To date, available published literature [1-5], [15], agree that low doping and surface-to-volume ratio represent the main parameters that boost the piezoresistance effect of SiNWs. Some hypotheses have been speculated on the origin of such phenomenon. Recently the major culprit has been indicated to be surface-state induced effect for nanowires smaller than 70nm width, and enhanced strain modulation of carrier mobility for larger nanowires [3].

Roylance and Angell introduced the first fully integrated piezoresistive micromachined accelerometers in 1978 for biomedical applications [15], [16]. An excellent literature review of micromachined piezoresistive accelerometers was provided by Barlian et al. [13] and interested readers are referred to their paper. Today, accelerometers are heavily commercialized MEMS application. They are widely used in automotive and space (crash detection, stability control and navigation [25]), biomedical (activity monitoring [27-29], surgical instrument tracking [24]), consumer electronics (portable computing, cameras lens stabilization, cellular phones), robotics (control and stability [26]), structural health monitoring, and military applications.

This paper presents the design and simulation of a single square millimeter 3-axial accelerometer for bio-mechanics detection [17], [18], [20]. The main requirements of this application are miniaturization and high measurement accuracy to allow the accelerometer to be implanted. In order to fulfil these requirements nanowires as nanoscale piezoresistive devices have been chosen as sensing element, due to their high sensitivity and potential in miniaturization.

2. Accelerometer Design

Initial technical specifications have been specifically designed in order to address the particular application requirements (impact measurement) [19], see table 1.

Table 1. Sensor technical specifications.

Range	*±250 G*
Sensitivity	4 mV/G
Frequency response	0 to 1,000 Hz
Shock limit	±1,000 G
Resolution	<10 mG
Non-Linearity	<1% FSO
Cross-Sensitivity	<5%
Dimension	<2×2×1 mm^3

The device range measurement is for medium-G impacts and the frequency response goes down to DC since bio-mechanics events, such as head injuries may have long duration transient.

The model of the sensing chip is shown in figure 1. The modeled geometry is an accelerometer with a highly symmetric micromechanical structure surrounded by four beams (figure 1) already available in the literature [23]. These types of geometries allow the minimizing of the overall cross-sensitivity due to a self-cancelling feature. The proof mass in red is suspended by four surrounding beams that resemble springs, while the damper is air in the model. Each beam is clamped-clamped to the fix frame and the middle point is connected to the seismic mass. The overall geometric dimension of the chip is 1310 μm × 1310 μm × 400 μm. When acceleration is applied to the chip, the proof mass is displaced due to inertial force, resulting to beam deformation (figure 1) [23].

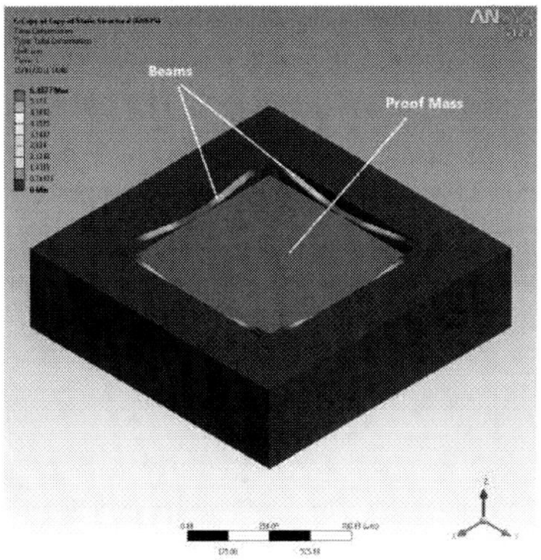

Figure 1. Model of the accelerometer under stress due to external forces. The red color corresponds to the maximum displacement of the proof mass. In blue is the surrounding frame which remains undeformed.

As it can be seen the beams change color gradually from blue to red due to deformation [23]. The beam deformation, due to the applied stress (σ) (figure 1), leads to a change in resistance (ΔR) of nanoscale piezoresistors proportional to the applied external inertial force (A). The fractional resistance change in the measurement circuit is proportional to the output voltage drop as well, therefore representing a measure of the acceleration.

2.1. Finite Element Structural Analysis

The material selected for the mechanical structure of the sensing element is anisotropic single crystal silicon (SCS), which has been chosen for its mechanical properties (good stress tensile strength and high gage factor) [22]. The modeled geometry has been analysed by about 350,000 meshing nodes and tetrahedron elements with the use of ANSYS 12.1 commercial software. First of all the inputs of the numerical model (geometry and material) have been validated against a readily available previous experimental work on a mechanical sensor demonstrator [23]. Table 2 presents the inputs parameters used in the developed FEM.

Table 2. Inputs geometric and material parameter used in the developed FEM.

Mechanical Structure material	Single Crystal Silicon Anisotropic
Density (kg/m^3)	2330
Young's Modulus [110] (GPa)	170
Poisson's Ratio [100]	0.28
Bulk Modulus (GPa)	128.8
Shear Modulus (GPa)	66.4
Longitudinal piezoresistive coefficient (Pa^{-1})	35×10^{-11} [23]
Length × width × thickness (mm)	2 × 2 × 0.45 [23]
Beams - Length × width × thickness (mm)	1 × 0.06 × 0.01 [23]

Because silicon is such an important economic material, these values have been investigated thoroughly, and the three independent stiffness coefficients of principal crystallographic orientations <100>, <010> and <001>, gives the following stiffness matrix [30], which has been used in the FEM developed:

$$[C] = \begin{bmatrix} 1.66 & 0.64 & 0.64 & 0 & 0 & 0 \\ 0.64 & 1.66 & 0.64 & 0 & 0 & 0 \\ 0.64 & 0.64 & 1.66 & 0 & 0 & 0 \\ 0 & 0 & 0 & 0.80 & 0 & 0 \\ 0 & 0 & 0 & 0 & 0.80 & 0 \\ 0 & 0 & 0 & 0 & 0 & 0.80 \end{bmatrix} \cdot 10^{11} Pa \tag{1}$$

The value of piezoresistive coefficient used for the sensitivity calculation is 35×10^{-11} Pa^{-1}, which is the value adopted by Dao et al. [23]. The simulation results of this study are on the same order of magnitude of the value of sensitivity obtained in the demonstrator device by Dao et al. [23]. Sensitivity under X-axis acceleration is 0.2 mV/g vs. 0.48 mV/g [23]; similar value is obtained under Y-axis acceleration. Under Z-axis acceleration the sensitivity is 0.314 mV/g vs. 0.56 mV/g [23]. Figure 2 compares the two studies.

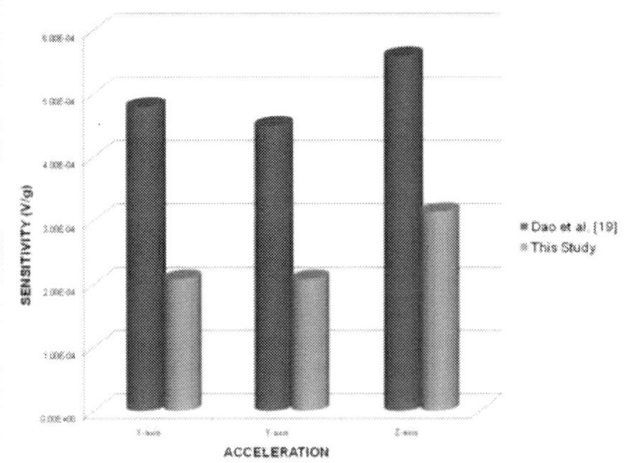

Figure 2. Validation of the model inputs by comparing to the demonstrator performance of Dao et al. [23].

After validation of our device against a real demonstrator we are able to further investigate on the design optimization of our model.

The optimization method uses an objective function which is to maximize the stress, using two constraints, the die size and the natural frequency. The results of the optimization process are as follows: 1st mode shape (ω_{oz}) of 5,277.7 Hz, the maximum equivalent stress (σ_{eq}) at 250G acceleration in the X or Y-axis are 44.837 MPa and 66.041 MPa in the microscale and nanoscale piezoresistors respectively. In the Z-axis σ_{eq} at 250G acceleration is 66.581 MPa and 99.243 MPa in the microscale and nanoscale piezoresistors respectively. By taking advantage of structure symmetry only the equivalent stress under X-axis acceleration as been analysed. The results for the Y-axis acceleration are equivalent.

Nanoscale piezoresistors, due to stress concentration region (i.e. the width is 10 times smaller than conventional piezoresistors), show an equivalent stress that is 47% higher than the conventional counterparts under 250G. This increase represents an initial sensitivity enhancement for geometrical reasons. Clearly progressively reducing the dimension will upraise the stress concentration effect.

2.2. Nanoscale Piezoresistors Measurement Circuit

A total of 16 nanoscale piezoresistors were placed in strategic locations on the top surface of the mechanical structure (figure 3). Basically, in order to maximize the device electrical sensitivity, the piezoresistors were placed at the highest stress regions that were identified by FE stress distribution analysis.

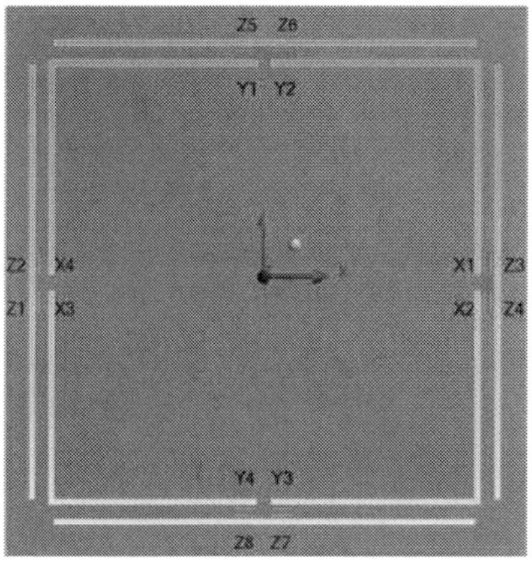

Figure 3. Piezoresistors location. 16 nanoscale piezoresistors, in orange, are placed in strategic location where is present the maximum stress in order to maximize the sensitivity and minimize the cross-sensitivity (top view).

The piezoresistors are interconnected in a measurement circuit by three full Wheatstone bridges, one for each axis-sensing, as in figure 4.

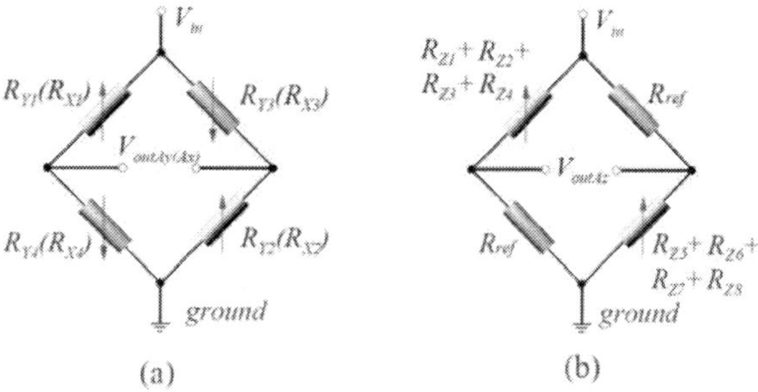

Figure 4. (a) A_x-, A_y-bridge and (b) A_z- Wheatstone Bridge measurement circuit

Hence, the change in resistance of piezoresistors is measured as output voltage drop by these bridges. The advantage of using a bridge is that as the four resistors are identical, the effect of the temperature coefficient can be cancelled out by the balanced configuration. Moreover, the highly symmetric geometry chosen for the structure allows a self-cancellation of part of the cross-axis acceleration since the piezoresistors are intentionally positioned symmetrically.

A detailed image of a nanoscale piezoresistor placed in one of the chip corner is presented in figure 5. Each piezoresistor is geometrically identical with a length of 3 micrometer and a width of 100 nanometers.

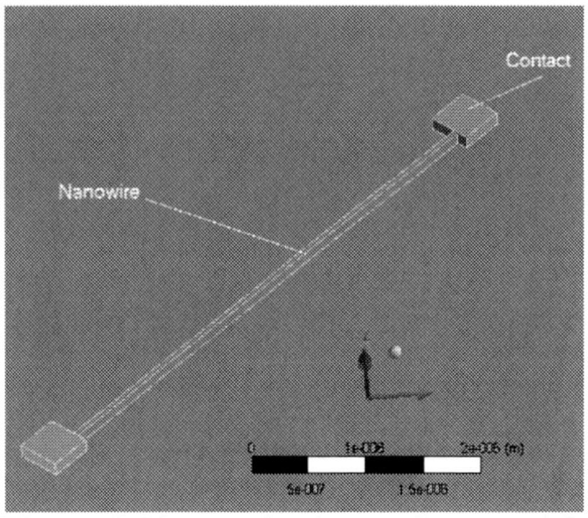

Figure 5. Nanoscale piezoresistor model.

3. Performance Calculation

3.1. Electrical sensitivity and cross-sensitivity

Table 3 summarizes the sensors performance calculation:

Table 3. Voltage noise, sensitivity, cross-sensitivity, resolution

	Microscale		Nanoscale	
	Ax, Ay	**Az**	**Ax, Ay**	**Az**
V_{jn} (μV)	1.01	1.42	3.2	4.52
$V_{1/f}$ (μV)	0.4	0.28	13.14	9.29
V_{Noise} (μV)	1.2	1.97	13.53	10.41
Sensitivity (mV/g)	0.677	0.898	20.9	28.9
Cross-sensitivity (%)	-	-	0.061	0.0058
	-	-	2.779	0.0139
Resolution (mg)	1.78	2.2	0.64	0.36

The calculated electrical sensitivity of nanoscale piezoresistors in the <110> direction at low boron concentration at 37°C, increases of approximately 3,000% in comparison to the conventional ones, see table 3.

Cross-sensitivity is below 5% for all axes due to the highly symmetric geometry selected. The cross-axis sensitivities under the X-axis acceleration $S_{(Ax\text{-}Ay)\%}$ and $S_{(Ax\text{-}Az)\%}$ are detected respectively in the output of the A_y, A_z-bridge for the nanoscale piezoresistors.
These values are consequence of the highly symmetric geometry selected, the symmetric localizations of the piezoresistors on the top surface of the device and the measurement circuit (figure 3 and 8). Similar results are obtained for the microscale piezoresistors.

3.2. Noise

Noise is any output voltage that occurs when there is no acceleration applied to the sensor. There are three typical noise sources existing in all piezoresistive sensors, including the *Johnson noise* (noise floor or white noise), *Hooge's noise* (or $1/f$ noise, also called pink noise), and the *thermo-mechanical noise* (brown noise or Brownian noise).

The calculated Johnson noises corresponding to nanoscale and microscale piezoresistors are shown on the table 3. Johnson noise does not depend on the frequency. The noise in the nanoscale piezoresistors is increased by around 200% due to one order of magnitude greater resistance than conventional microscale ones. The total $1/f$ noise voltage corresponding to each measurement bridge is calculated to be approximately 30 times greater in the nanoscale devices as clearly summarized on table 3. This is mainly due to the very low carrier concentration.

The thermo-mechanical noise corresponding to each component of acceleration at 37 °C with the calculated bandwidths at 5,539 Hz are in the order of few μg.

The total noise is calculated as the square root of the sum among the square power of each noise, as:

$$V_{Noise}^{Ax} = V_{Noise}^{Ay} = \sqrt{\left(V_{Jn}^{Ax}\right)^2 + \left(V_{1/f}^{Ax}\right)^2 + \left(V_{TM}^{x}\right)^2} \qquad (2)$$

Similar equation has been applied for the Z-axis.

Noise analysis indicates that the total voltage noise in the accelerometer is increased by 10 times in the X and Y-axis, and 4 times in the Z-axis, see table 3.

The main contributor to the noise signal is the Hooge noise due to very low number of carriers in the nanoscale piezoresistors.

3.3. Resolution

The resolution of an accelerometer determines the minimum acceleration that can be measured. Resolution is defined as the noise divided by the sensitivity, therefore the resolution of the accelerometer is defined by (3) and (4) [23]:

$$R_{A_{x,y}} = \frac{V_{Noise}^{A_{x,y}}}{S_{Az}} \qquad (3)$$

$$R_{A_z} = \frac{V_{Noise}^{A_z}}{S_{Az}} \qquad (4)$$

Table 3 summarizes the sensor resolution calculated according to (26) and (27).

The resolution of the accelerometer with nanoscale piezoresistors is improved respect to the conventional microscale one of around 60-80% in all axes, due to their much higher sensitivity, see table 3. However, the noise level of the nanoscale piezoresistors is worse than the conventional microscale ones because Johnson and Hooge noises are dependent on the resistance value and numbers of carriers respectively. Therefore, given the doping concentration, smaller size means higher resistance and lower number of carriers.

4. Discussion

Table 4. Sensor comparison

	Microscale		Nanoscale	
	Present work	**Dao et al. [23]**	**Present work**	**Dao et al. [11]**
Range	Medium-G	Low-G	Medium-G	Low-G
Sensitivity (μV/g)	677	450	20,900	400
Resolution (mg)	1.78	0.9	0.64	14
Cross-Sensitivity (%)	2.78	5.5	2.78	-
Dimension (mm^3)	1.3×1.3×0.4	2×2×0.45	1.3×1.3×0.4	0.5×0.5×0.35

The results obtained from the combined FE modeling, simulation of the mechanical structure and performance calculation of both MEMS and nanostructure are interesting for this field of study. Input values used as piezoresistive coefficient for calculating the sensitivity and cross-sensitivity are obtained from experiments of Passi [5] for the nanostructure and Smith [21] for the microstructure. This study raises the attention on silicon nanowires structures as devices being embedded into mechanical sensors as piezoresistors. The improved resolution of the designed accelerometer (less than 1mg on all axes) compared to the conventional ones with 60-80% increase (see table 3) suggests that nanowires have the credential to be the sensing element of the future NEMS. This level of accuracy and precision of measurement is comparable to the capacitive counterparts' sensors.

To date only few laboratory prototypes have been fabricated, as reported by Dao et al. [11] suitable for low-G measurements. However, their device [11] sensitivity for each axis is only about 400 μVG^{-1}, and the resolution of 14 mg which requires further signal conditioning, see table 4.

This study instead presents a sensor with higher sensitivity obtained by calculations (20.9 mVG^{-1}), mainly because a higher stress on the nanowires. This is thanks to the optimization process undertaken that maximizes the stress on the beams of the mechanical structure, under the constraints specified (see table 1). The resolution and the sensitivity in this work is improved considerably compared to the work of Dao et al. [11] (see table 4) due to the fact that the sensor of Dao et al. is significantly smaller.

The designed sensor results suitable for the bio-mechanical application, such as head injuries monitoring. Finally, the total noise of nanoscale piezoresistors results are much higher than the conventional ones, however, since in the nanoscale structures the sensitivity grows faster than the noise level, the overall resolution is significantly improved. It is worthwhile to point out that our work is based on previous published experimental works on nanowires structures used as piezoresistors. Experimental studies are currently underway and will be reported in the next future.

5. Conclusion and Future work

This paper presents the design enhancement of a motion sensor for biomechanical measurement within a space-constrained environment. Due to the exploitation of electro-mechanical features of nanowires as nanoscale piezoresistors the nominal sensor sensitivity is overall boosted by more than 3,000%. This technology avoids the signal amplification but allows a higher resolution with the advantage of a smaller sensing element. Therefore, in comparison with conventional devices, the measured accuracy is considerably improved. Therefore, this study highlights the potential of silicon nanowires structure as piezoresistors embedded in future mechanical sensors.

6. Acknoledgement

The authors are grateful for EPSRC funding support to M Messina under SEEDA EPSRC Case Award Voucher No.0900013338.

References

[1] He R and Yang P Giant piezoresistance effect in silicon nanowires 2006 *Nature Nanotechnol.* **1** issue 1 42–46

[2] Reck K, Richter J, Hansen O and Thomsen E V Piezoresistive effect in top-down fabricated silicon nanowires 2008 *IEEE 21st Int. Conf. on Micro Electro Mechanical Systems* pp 717–720

[3] Barwicz T, Klein L, Koester S J and Hamann H Silicon nanowire piezoresistance: Impact of surface crystallographic Orientation 2010 *Appl. Phys. Lett* vol 97 issue 2

[4] Dao D V, Nakamura K, Bui T T and Sugiyama S Micro/nano-mechanical sensors and actuators based on SOI-MEMS technology 2010 *Adv. in Nat. Sciences: Nanoscience and Nanotechnology*

[5] Passi V, Ravaux F, Dubois E, Raskin J-P Backgate bias and stress level impact on giant piezoresistance effect in thin silicon films and nanowires 2010 *IEEE MEMS 2010* pp 464 – 467

[6] Lee M L, Fitzgerald E A, Bulsara M T, Currie M T and Lochtefeld A Strained Si, SiGe and Ge channels for high-mobility metal-oxide–semiconductor field-effect transistors 2005 *J. Appl. Phys.* **97** issue 1 011101-1

[7] Haugerud B M, Bosworth L A and Belford R E Mechanically induced strain enhancement of metal-oxide–semiconductor field-effect transistors 2003 *J. Appl. Phys.* **94** issue 6 4102–4107

[8] Chen L J Silicon nanowires: the key building block for future electronic devices 2007 *J. of Mat. Chem.* **17** issue 44 4639–4643
[9] Toriyama T, Tanimoto Y and Sugiyama S Single crystal silicon nano-wire piezoresistors for mechanical sensors 2002 *J. Microelectromech. Syst.* **11** issue 5 605–611
[10] Toriyama T, Funai D and Sugiyama S Piezoresistance measurement on single crystal silicon nanowires 2003 *J. of Appl. Phys.* **93** issue 1 561–565
[11] Dao D V, Toriyama T and Sugiyama S. Noise and frequency analyses of a miniaturized 3-DOF accelerometer utilizing silicon nanowire piezoresistors 2004 *Proc. of IEEE, Sensors 2004* vol. 3 pp 1464–1467
[12] Shao M-W, Shan Y-Y, Wong N-B and Lee S-T Silicon Nanowire sensors for bioanalytical applications: Glucose and hydrogen peroxide detection 2005 *Adv. Funct. Mat.* **15** issue 9 pp 1478–1482
[13] Barlian A A, Park W-T, Mallon J R Jr, Rastegar A J and Pruitt B L Review: Semiconductor Piezoresistance for Microsystems 2009 *Proc. of the IEEE* vol 97 issue 3
[14] Okamura A, Dao D V, Toriyama T and Sugiyama S Fabrication of an ultra small accelerometer utilizing Si nanowire piezoresistors 2005 *Proc. of the 22nd Sensor Symposium* pp 203-206
[15] Roylance L M A miniature integrated circuit accelerometer for biomedical applications 1978 Ph.D. Electrical Engineering Department, Stanford University
[16] Roylance L M and Angell J A Batch-fabricated silicon accelerometer 1979 *IEEE Trans. Electron Devices* vol 26 issue 12 pp 1911–1917
[17] Stephen O E, Knox T and Cohn K A The development of a method to measure head acceleration and motion in high-impact crashes 2004 *Neurosurgery* vol 54 pp 672-677
[18] Nassiopoulos E and Njuguna J Finite element dynamic simulation of whole rallying car structure: towards better understanding of structural dynamics during side impacts 2011 *8th European LS-DYNA Conf.* p 31
[19] Wilson J S and Knovel 2005 *Sensor technology handbook* (Amsterdam; Boston: Elsevier)
[20] Unser M Motorsports accidentology 2009 MSc thesis School of Applied Sciences, Cranfield University, Cranfield, UK
[21] Smith C S Piezoresistance effect in germanium and silicon 1954 *Phys. Rev.* **94** issue 1 42-49
[22] Maluf N 2004 *An Introduction to Microelectromechanical Systems Engineering*, 2nd ed. (Boston: Arttech House)
[23] Dao D V, Okada S, Dau V T, Toriyama T and Sugiyama S 2004 Development of a 3-DOF Silicon Piezoresistive Micro Accelerometer *Proc. of the 2004 Int. Symp. on Micro-Nanomechatronics and Human Science and The Fourth Symp. Micro-Nanomechatronics for Information-Based Society*
[24] Ren H and Kazanzides P 2010 *IEEE/ASME Trans. On Mechatronics* vol **PP** issue 99
[25] Fang J, Zheng S, Han B 2011 Amb vibration control for structural resonance of double-gimbal control moment gyro with high-speed magnetically suspended rotor *IEEE/ASME Trans. On Mechatronics*, 10.1109/TMECH.2011.2161877
[26] Liljeback P, Pettersen K Y, Stavdahl O, Gravdahl J T 2011 Snake robot locomotion in environments with obstacles *IEEE/ASME Trans. On Mechatronics*, 10.1109/TMECH.2011.2159863
[27] Lawrence E, Navarro K F, Riudavets J, Messina M 2006 Macroscopic Sensor Networks: Application Issues in the Healthcare Industry *J. of Computational Methods in Sciences and Engineering* **6** issue 5, 6 supplement 2 309-319
[28] Messina M, Lawrence E, Navarro K F, Lubrin E 2007 Towards assistive healthcare: prototyping advances in wireless sensor network (WSN) system integration and application *IADIS International Conference - Wireless Applications and Computing (Lisbon)*
[29] Messina M, Lim Y Y, Lawrence E, Martin D, Kargl F 2008 Implementing and validating an environment and health monitoring system *5th Int. Conf. on Information Technology: New Generation ITNG (Las Vegas)*

[30] Kaajakari V 2009 Silicon as an anisotropic mechanical material *Practical MEMS: Analysis and design of microsystems, MEMS sensors, electronics, actuators, rf mems, optical mems, and microfluidic systems* (Small Gear Publishing) chapter 4 p 51

New active control nano-system to use in composites structure

M R Arche

Passive and active control of structures using nanotechnology, CENIM-CSIC, Spain

ABSTRACT. The present abstract, is a brief description about our project (NEDEA). We considered this project as very important, because it reunites in his development, several basic technologies: electronics, communications, software and new materials, all very interesting in the European industry. The project is developed in the CSIC (Spanish Researcher Center). We are involved. Across the project, in the development of nano-sensors, specialized in detecting defects, difficulties or problems in structures of composed materials. These materials are being used, and in the future more, in applications where a high degree of security is necessary. Some fields in the system usage are Aeronautical and military applications whit a necessary high security degree. The development proposed, is based in nano-sensors and active devices. They are installed into the material structure. The information from sensors is transmitted by optical fibers, to a radio transmitter, equally installed into the material. An external receptor picks up those data and transmits them to an external device. This external device presents/displays all the information across an interface GUI, in real time, to the supervisor. He can see than is happening in the material, in real time. Alarms can be programmed, by the supervisor. Is possible a tracking for the problem. All the devices and software are in develop in our laboratories. We think that this development will be used by the industry of materials, and that gradually, it will have other applications in the transport area (like new vehicles, wagons of train and metro, etc.).

1. New structures of materials.

Every day is more and more extended, the use of new materials not metallic, for manufacture of pieces in different areas, such as vehicles, airplanes or trains, allow that these materials can be monitored in real time, with a suitable system of sensors. Our project NEDEA, is concentrated in the investigation, design and manufacture, of this type of structures.

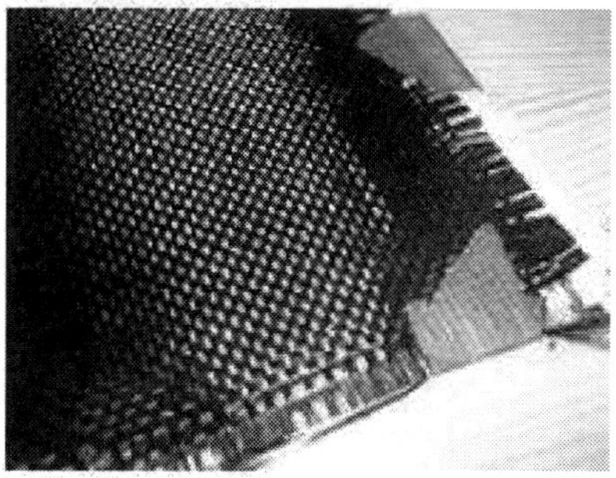

New materials structure

2. Nano-sensors.

The basic idea of this project is the development of "nano-sensors" that catch variations in the structure of the material. These sensors, are essentially of for type of sensors "extensometric", "seismic" "and "of vibration", and intend to catch the deformations and other aspects that can affect the structure, all in real time.

This is important in military airplanes, boats, trains and vehicles. The information that produces the sensors, in the form of data, is sent, to an optical fiber converter that send all to a radio-trasmisor. It could send this information to an external system of monitoring, by radio. The information goes codified and digitized, in 5 GHz band, and is transmitted. All the system, goes "contracted" in the own material.

Transmisor

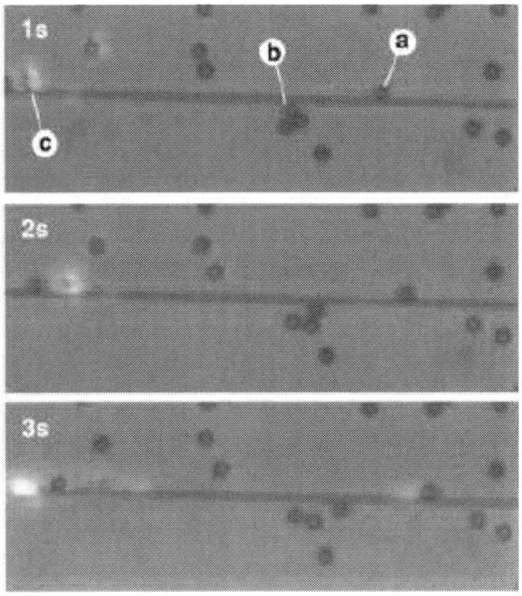

Nano-sensor network. a, b sensors. C. optical fiber

3. Control system.

Externally the control system goes has two parts:

- Receiver of the signals radio, that it is the subsystem in charge to pick up the signals of the contracted transmitter. Basically it is a receiver that also works in the band of 5 GHz, to take the data in real time. It cannot be to a distance superior to 25 ms of the trasmisor at present. For that reason the system of sensors must of being a network that it connects with a unique transmitter, located in the neighborhoods of the receiver. The receiver also includes hardware to be able to use the application software.
- Software specifically developed in java.

To be able to use the data. It is possible to do:

- Alarms.
- Parameters of measures.

- Graphs.
- Sounds
- The interface with the user is of type GUI (Graphics), on touch screen.

4. Consortium (NEDEA Project)

Número	Organización	País	Tipo de organización	Siglas
1. Coordinador	Agencia Estatal Consejo Superior de Investigaciones Científicas	España	National Research Center	CSIC
2	Baris Elektrik S.A.	Turquía	SME	BAR
3	Technological Institute of Optics, Colour and Imaging	España	Research Center	AIDO
4	FADA-CATEC Center for Advanced Aerospace Technologies	España	Research Center	CATEC
5	Riga Technical University	Letonia	University-Research Center	RTU
6	HLP DEVELOPEMENT S.L.	Francia	SME	HLP
7	TROYKA LTD	Turquía	SME	TRO
8	University of Chemical Technology and Metallurgy	Bulgaria	National Research Center-University	UCTM
9	Tampere University of Technology	Finlandia	University-Research Center	TUT
10	Centro Ricerche FIAT S.C.p.A. (CRF)	Italia	Research Center	CRF
11	SINTEF Group	Noruega	Research Center	SIN

Composite material pedestrian bridge for the Port of Bilbao

I Gorrochategui, C Manteca, A Yedra, R Miguel and F J del Valle

Corresponding author - Fundación Centro Tecnológico de Componentes (CTC), 39011, Santander, (Spain).

E-mail: igorrochategui@ctcomponentes.com

Abstract. Composite materials in comparison to traditional ones, steel and concrete, present advantages in civil works construction: lower weight, higher corrosion resistance (especially in the marine environment), and ease of installation. On the other hand, fabrication costs are generally higher. This is the reason why this technology is not widely used.

This work illustrates the process followed for the design, fabrication and installation of a composite material pedestrian bridge in the Port of Bilbao (Northern Spain). In order to reduce the price of the bridge, the use of low cost materials was considered, therefore polyester resin was selected as the polymeric matrix, and glass fibres as reinforcement.

Two material choices were studied. Currently in the market there is high availability of carbon nanoparticles: carbon nanotubes (CNT) and carbon nanofibres (CNF), so it was decided to add this kind of nanoparticles to the reference material with the objective of improving its mechanical properties. The main challenge was to transfer the CNT and CNF excellent properties to the polymeric matrix. This requires dispersing the nanoreinforcements as individual particles in the polymeric matrix to avoid agglomerates. For this reason, an advanced high shear forces dispersion technique (called "three roll mills") was studied and implemented. Also surface functionalization of the nanoreinforcements by chemical treatment was carried out.

Herein, a comparison is performed between both materials studied, the explanation of the employment of the reference material (without nanoreinforcement) as the one used in the fabrication of the pedestrian bridge is justified and, finally, the main characteristics of the final design of the structural element are described.

1. Introduction

Bridges are commonly fabricated in concrete or steel, but they also can be fabricated in composites. At present there are in the world more than 500 bridges totally made out of composites and approximately 200 bridges with a composite deck, mainly in USA and South Korea [1].

The advantages of composite materials against concrete and steel are lightness, corrosion avoidance and lower installation time. With regard to weight, a composite deck allows a reduction from 200 to 500 kg/m^2, making it possible to either increase the bridge load capacity, or build lighter and cheaper foundations. Corrosion resistance provides a longer bridge design life and reduces maintenance costs. Lower installation time causes lower work costs and limits traffic problems. A composite material bridge can be installed in hours or days, while a traditional one requires weeks or even months.

On the other hand, fabrication costs for composite material bridges are generally higher. This is the reason why this technology is not widely employed, being used in less than 0.1% of bridges. It is estimated that composite materials bridges of more than 10 m span are 70% more expensive than concrete ones. If maintenance and durability are considered, the difference is reduced, but they still are 37% more expensive. Consequently, this bridge type is only adequate in special environments where the composite materials are highly appreciated: high corrosivity, light foundations are needed, pedestrian bridges.

Also, it would be necessary that these materials were covered in codes and standardized for the development of the construction of composite materials bridges. The American Association of State Highway and Transportation Officials has already published "The Guide Specifications for Design of FRP Pedestrian Bridges" as a first step in this direction.

This paper addresses the material selection and characterization, the design, the fabrication and the installation of a composite pedestrian bridge for an aggressive marine environment in the Port of Bilbao.

2. Materials

Two low cost material choices were considered: a polyester resin polymeric matrix reinforced with glass fibres, with and without nanoparticles addition. These two options are presented in the next chapters.

2.1. Reference material

In the present work the composite material used is one of the most common for industrial applications, a polymer matrix reinforced with long fibres (FRP – Fibre Reinforced Polymer). FRP use to be employed as multilayer laminates, where each layer contains reinforcement fibres in the appropriate direction [2, 3, 4].

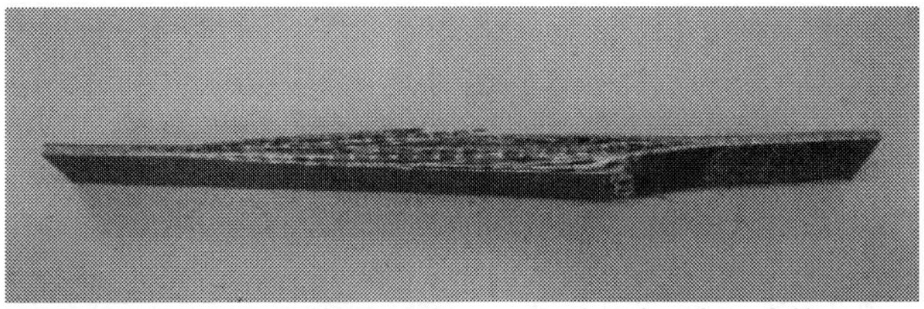

Figure 1 – Tensile specimen. Delamination failure.

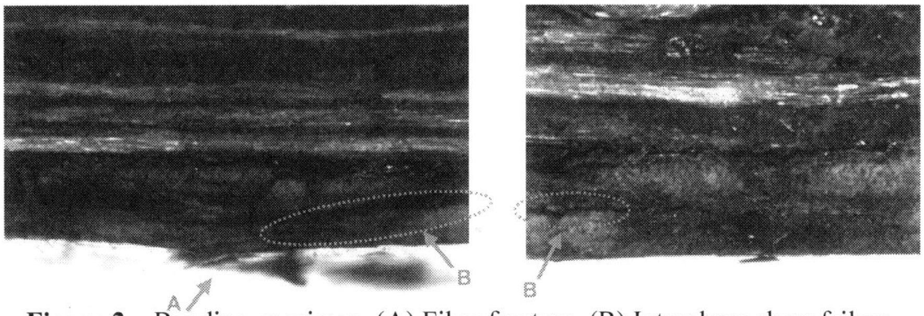

Figure 2 – Bending specimen. (A) Fibre fracture. (B) Interphase shear failure.

The polymer matrix selected for the composites synthesis was a non-saturated high reactivity low viscosity polyester resin.

The reinforcement material was glass fibre type E in different configurations. This is an inorganic fibre composed of 53-54% SiO_2, 14-15% Al_2O_3, 20-24% CaO+MgO, 6.5-9% B_2O_3 and low alkalis content.

Mechanical properties of this material were determined according to Spanish standards in the longitudinal and transversal orientations [5, 6].

Tensile characteristics were determined following UNE-EN ISO 527 [7] with 250x25x4 mm specimens. In all cases delamination failure was obtained (see Figure 1). Table 1 shows the obtained results.

For bending properties UNE-EN ISO 14125 [8] was the used standard with 800x5x4 mm specimens. Due to the material anisotropy two failure mechanisms were found: fibre tensile fracture and interphase shear failure (see Figure 2). Table 1 shows the obtained results.

Table 1 – Reference material mechanical properties.

	Tension		Bending	
	□$_u$ (Mpa)	**E** (GPa)	□$_u$ (Mpa)	**E** (GPa)
Longitudinal	410	27	363	10
Transversal	137	15	224	12

2.2 Nanocomposite

Thermoset multiscale composites have been considered as an alternate to the reference material. They consist of the use of carbon based nanostructures as nanoreinforcement, mainly multiwall carbon nanotubes (MWCNT's) but also carbon nanofibers (CNF's).

MWCNT's and CNF's have a huge potential as nanoreinforcements for polymeric matrices because they present unique and excellent mechanical, thermal and electrical properties with a competitive cost. Moreover, currently in the market there is enough availability of these nanomaterials to be used in industrial scale [9].

The main challenge was to transfer the MWCNT's (and CNF's) excellent properties to the polymeric matrix. This requires dispersing the nanoreinforcements as individual particles in the polymeric matrix to avoid agglomerates [10, 11]. For this reason, an advanced high shear forces dispersion technique called Three Roll Milling has been studied and implemented. Different number of dispersion cycles (3 and 5) and different nanoparticles content (from 0.1 to 1%) have been considered. Hand surface functionalization of the nanoreinforcements by chemical treatment has been carried out in order to improve the dispersion within the matrix and the interaction between the matrix and the nanoreinforcement.

A novel and detailed qualitative and quantitative characterization of the nanoparticles dispersion rate combining different techniques has been performed: viscosity measurements, Confocal Ramam spectroscopy and Atomic Force Microscopy. It has been observed a high MWCNT's/CNF's dispersion rate within the matrix using 3-roll-milling process. MWCNT's distribution map based on RAMAN intensity spectra (at G band) has been obtained. All samples present a high dispersion rate of the nanoreinforcements. Surface area where there are not nanoreinforcements is below 3%, where there are nanoreinforcements is more than 90% and where there are nanoreinforcements aggregates is below 14%. Furthermore, correlation between viscosity increase, nanoreinforcements content and dispersion rate has been observed.

Table 2 – Nanocomposites mechanical properties: improvement percentage (%).

	Tension		Bending	
	□	**E**	□	**E**
0.1%CNT	9	11	12	80
0.1%CNT-f	13	4	13	78
0.1%CNFa	12	0	23	84
0.1%CNFb	3	0	26	70

Finally the samples were characterized from the mechanical point of view [5, 6, 12]. This material presents better behavior than the original one. Depending on the nanoreinforcements type and content, and parameters used during the dispersion process, the mechanical improvements can be up to 13% in tensile strength and 15% in elastic modulus for longitudinal specimens, up to 12% in tensile strength and 13% in elastic modulus for transversal specimens, up to 26% in bending strength and 84% in elastic modulus for longitudinal specimens and up to 26% in bending strength and 13% in elastic modulus for transversal specimens.

Table 2 summarizes the results for longitudinal specimens with 0.1% content of CNT's, functionalized CNT's and two CNF's conditions using different dispersion techniques: (a) Three Rolls Mill and (b) Ball Mill.

2.3. Material selection

Although the mechanical properties of the nanocomposite were better than the ones of the reference material, the fabrication company and the Port of Bilbao Authority decided to build the pedestrian bridge out of the material without reinforcement. The main reason that justified their decision was the short period of time available to put the bridge in place that did not allow covering the following uncertainties:

- Selection of the optimum choice between the different nanocomposites studied.
- Detailed knowledge of the mechanical properties: more testing/statistics needed.
- Long term behaviour.
- Transfer of the material elaboration process from the laboratory to the factory.

Once the material was selected, standard beams fabricated by pultrusion were joined to construct the pedestrian bridge.

3. Pedestrian bridge design

Figure 3 provides a general view of the studied structure.

Figure 3 – General view of the pedestrian bridge.

The structure is 18 m long and 1 m wide with an effective span of 15.78 m. Four couples of U240 horizontal beams define the main frame. The U240 couples that support the platform are joined by U160 transverse beams (12 elements at the top flange of the U240 couples and 9 at the bottom flange of the U240 couples), keeping the width of 1 m. Additional U100 lateral beams join the upper couples of U240 with the lower couples of U240 along the whole structure, providing a high inertia to the assembly. Four lateral triangles increase the torsional stiffness of the assembly.

Eight supports are provided, six of them allow the structure to slide 60 mm in the longitudinal direction, as a provision for thermal expansion displacements.

The design procedure is similar to metallic structures, but in this case welding is not an option for joint design, thus all joints are bolted (preloaded bolted joints).

All the structural profiles were manufactured by means of pultrusion process, obtaining elements with cross sections similar to those used in the industry (IPE, HEB, UPN,...).

According to the load combinations defined by the Spanish Technical Code (wind, variable loads,...) [13-16] the efforts in each structural element are obtained (F_x, F_y, F_z, M_x, M_y, M_z). These efforts are approximately independent of the material properties of the elements, since they respond to the mechanical equilibrium of the structure (actually some differences can be obtained due to second order effects if deformation is considered in the mechanical equilibrium analysis).

Due to the low elastic modulus of the composite material, special attention must be paid to the displacements of the structure, in order to satisfy the criteria defined by the customer or design codes. The use of sections with higher inertia is expected than in the case of metallic profiles, in order to reduce the displacements of the structure. Furthermore, due to the low density of the composite material, the obtained design is lighter than the metallic alternative, although higher cross section profiles are used.

From the point of view of dynamic loads (wind, people,...), it is necessary to perform a linear modal analysis in order to obtain the natural frequencies of the structure. This is needed to avoid any possible dynamic resonance with the external actions. Natural frequencies are affected by the material properties. As it was indicated above, for the same cross section of a structural element, and due to the lower elastic modulus of the composite, a lower element stiffness is obtained. Consequently, the natural frequencies of the whole structure are lower than those for the same design with metallic elements. This effect is partially balanced by the lower weight of the structural elements.

Once the efforts in the structural elements are obtained, the analysis of the stress components is carried out. Typical failure criteria for steel elements (Von Mises, Tresca) are not considered, since the failure mechanisms of the composite material are quite different.

This was one of the challenges of the project. Due to the orthotropic behaviour of the composite material, a detailed analysis of the failure mechanisms was carried out.

From the analysis of the failure mechanisms, it was established that the main mechanism was the generation and propagation of flaws in the longitudinal plane, at the layer of pure resin located between layers of fibre plus resin. In the case of shear or bending loads, shear stresses are generated at the cross section and, according to Cauchy theorem, these shear stresses are also developed in the longitudinal plane. The maximum longitudinal shear stresses are obtained at the centre of the cross section (according to the static moment). When the value of this stress reaches the limit of shear stress in the resin, the brittle failure condition is reached. Depending on the mechanical properties of the resin, this condition can be reached after the fracture of the fibres in the flange of the element (tensioned region) or due to collapse of the resin in the flange in the region of compression. Thus, the detailed shear stress in the cross section of the structural elements is obtained, and compared with the allowed shear stress in the resin.

***Table 3** -Results summary*

	Calculated	Allowable
Tension	79.15 MPa	133.3 MPa
Compression	87.02 MPa	100 MPa
Shear	6.79 MPa	13.3 MPa
Deflection: Customer criterion	0.42/100	1/100
Deflection: CTE criterion	0.86/300	1 300
Flange buckling	2.00E06	2.05E06
Web buckling	13.8 MPa	832.11 Pa

The use of bolted joints provides an easier way to transport the structure to the final emplacement because some parts can be joined in situ. The lower weight of the structure allows an easier manipulation as well as installation (lower requirements for transportation and hoisting elements). In this particular case, the installation was done directly from a boat with a maximum hoist load crane of 2 tons. From the seismic point of view this type of material is interesting since the efforts transmitted to the civil works are lower, due to the lower weight of the assembly.

According to the obtained results from the mechanical analysis, it can be concluded that the structural integrity and stability of the assembly is assured. Table 3 summarizes the obtained results.

4. Conclusions

The main conclusion is that a composite material pedestrian bridge, which satisfies the requirements of the customer, has been designed and installed in the Port of Bilbao (Spain). Additional conclusions have been obtained for the development of this project:

- Low cost nanocomposites, able to be used for civil works applications, have been formulated. The nanoparticles have been dispersed in the matrix by an innovative Three Rolls Milling process. A very good dispersion has been confirmed.
- Different carbon nanostructures, CNT's and CNF's, have been employed in polymeric matrices. It has been found that the nanoreinforcements in all cases improve the mechanical properties of the composite, this improvement being dependant on the type and content of nanoparticles, even for contents as low as 0.1%.
- The designed pedestrian bridge satisfies the Spanish Design Codes.

References

[1] April 2010 *Composites Values* #**07**
[2] Mayugo J A 2003 *Estudio constitutivo de materiales compuestos laminados sometidos a cargas cíclicas* PhD Thesis Universidad Politécnica de Cataluña
[3] Chanda M and Roy S K 2007 *Industrial polymers, specialty polymers and their applications* CRC Press
[4] Rosato D V 1991 *Designing with plastics and composites. A handbook* Van Nostrand Reinhold
[5] Manteca C 2010 *Síntesis y caracterización de materiales compuestos de matriz polimérica con nanotubos de carbono multipared (MWCNT's) reforzada con fibra de vidrio* Tesis de Máster en Ciencia de Materiales Universidad de Cantabria
[6] Yedra A 2010 *Nanoreinforcements high dispersion within thermoset polymer matrix* Launching the NANOfutures European Initiative Gijón (Spain)
[7] *Plásticos. Determinación de las propiedades en tracción* Norma española UNE-EN ISO 527
[8] *Compuestos plásticos reforzados con fibras. Determinación de las propiedades a flexión* Norma española UNE-EN ISO 14125
[9] Company NANOCYL web: www.nanocyl.com // Company Grupo Antolín web: www.grupoantolin.com.
[10] Friedrich K Fakirov S and Zhang Z 2005 *Polymer composites: From Nano to Macro Scale* Springer
[11] Fiedler B et al. 2006 *Composites Science and Technology* **66** 3115-3125
[12] Yedra A Septiembre 2011 *Desarrollo de materiales nanocompuestos poliméricos de bajo coste con prestaciones mecánicas mejoradas* XXXIII Reunión Bienal de la Real Sociedad Española de Física Santander (Spain)
[13] CTE DB SE 1999 Código Técnico de la Edificación Documento Básico Seguridad Estructural
[14] CTE DB SE AE 1999 Código Técnico de la Edificación Documento Básico Seguridad Estructural Acciones en la Edificación
[15] CTE DB SE A 1999 Código Técnico de la Edificación Documento Básico Seguridad Estructural Acero
[16] CTE DB SU 1999 Código Técnico de la Edificación Documento Básico Seguridad de Utilización

IOP Conf. Series: Materials Science and Engineering **40** (2012) 012041 doi:10.1088/1757-899X/40/1/012041

Using artificial neural networks to model extrusion processes for the manufacturing of polymeric micro-tubes

N Mekras and I Artemakis

ANTER Ltd., Feidippidou 22 Str., 11527, Athens, Greece

Email: mnmekras@anter-net1.com; anter@anter.gr

Abstract. In this paper a methodology and an application example are presented aiming to show how Artificial Neural Networks (ANNs) can be used to model manufacturing processes when mathematical models are missing or are not applicable e.g. due to the micro- & nano-scaling, due to non-conventional processes, etc. Besides the ANNs methodology, the results of a Software System developed will be presented, which was used to create ANNs models for micro & nano manufacturing processes. More specifically results of a specific application example will be presented, concerning the modeling of extrusion processes for polymeric micro-tubes. ANNs models are capable for modeling manufacturing processes as far as adequate experimental and/or historical data of processes' inputs & outputs are available for their training. The POLYTUBES ANNs models have been trained and tested with experimental data records of process' inputs and outputs concerning a micro-extrusion process of polymeric micro-tubes for several materials such as: COC, PC, PET, PETG, PP and PVDF. The main ANN model of the extrusion application example has 3 inputs and 9 outputs. The inputs are: tube's inner & outer diameters, and the material density. The model outputs are 9 process parameters, which correspond to the specific inputs e.g. process temperature, die inner & outer diameters, extrusion pressure, draw speed etc. The training of the ANN model was completed, when the errors for the model's outputs, which expressed the difference between the training target values and the ANNs outputs, were minimized to acceptable levels. After the training, the micro-extrusion ANN is capable to simulate the process and can be used to calculate model's outputs, which are the process parameters for any new set of inputs. By this way a satisfactory functional approximation of the whole process is achieved. This research work has been supported by the EU FP7 NMP project POLYTUBES.

1. Introduction

In this paper a proposed methodology will be presented concerning the modeling of processes for the manufacturing of polymeric micro-tubes using Artificial Neural Networks (ANNs). Also the results of an ANNs processes modeling software application will be presented, which has been developed by the authors of this paper for the implementation of the proposed methodology. An existing micro-extrusion process for manufacturing polymeric micro-tubes will be used as case example. The physical process data were used by the authors of this paper at ANTER Ltd to create, train and test the micro-extrusion Neural Network. Artificial Neural Networks (ANNs) are considered a research area of Artificial Intelligence (AI) with significant applications in several domains, including the manufacturing processes domain [7-11]. ANNs are useful on several applications like for example on applications of systems modeling, materials modeling [1, 5, 12, 15], function approximation, forecasting & prediction [2, 4, 14], classification problems, pattern recognition, etc. Generally, ANNs can be used in cases that mathematical models are completely missing or are not adequate and accurate enough to represent a real world model of a physical process or system. In most cases ANNs are capable to learn and approximate real world models when adequate historical and/or experimental data sets of model's inputs & outputs are available for the training of the ANN.

In the following Figureure 1 the ANNs structure that was used for creating the POLYTUBES micro-manufacturing processes models is given. This structure represents the MLP (Multilayer Perceptron) feed-forward Neural Network model with 1 input layer, 2 hidden layers and 1 output layer, which uses the back-propagation training algorithm [3, 6]. In the cases of the POLYTUBES processes, the number of inputs and the number of neurons of the output layer correspond to the inputs and outputs of the physical process as these are defined by the relevant manufacturing theory and the needs of the process engineers. After the ANNs training is completed the ANNs model can be used to simulate the processes behavior and provide output results for any set of inputs provided by the user.

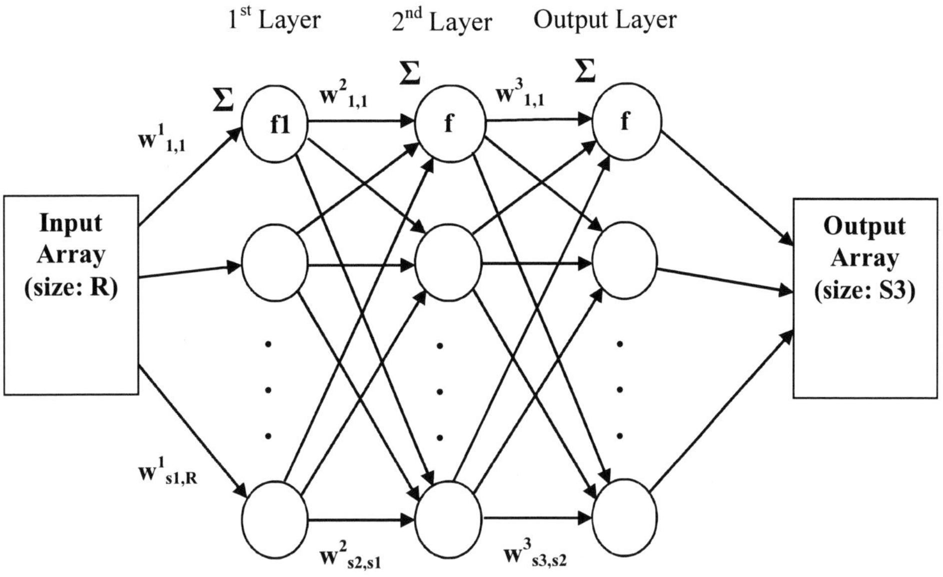

Figure 1. MLP Feed forward ANN ($R–S^1–S^2–S^3$) with 2 hidden layers.

2. Application case: Extrusion of polymer micro-tubes

The first step for modeling the micro-extrusion process concerned the selection of the Neural Network architecture and the selection of the number of the hidden neurons. The numbers of neurons of the 1st and 2nd hidden layers for the micro-extrusion process were estimated during the ANNs training and were adjusted through a test & trial procedure in an effort to minimize the training error for all process' outputs [13]. For the specific micro-extrusion ANN a 3-20-20-9 architecture was applied, with 20 neurons in the 1st hidden layer and 20 neurons in the 2nd hidden layer respectively. The Hyperbolic Tangent Sigmoid [3] was chosen as transfer function for both the 1st and the 2nd hidden layers and the linear function was chosen for the output layer. More specifically:

$$f1(x) = \frac{e^x - e^{-x}}{e^x + e^{-x}} \;,\; f2(x) = \frac{e^x - e^{-x}}{e^x + e^{-x}} \;,\; f3(x) = x \tag{1}$$

The extrusion process ANN model that was created had 3 inputs and 9 outputs. The 3 model inputs are: the micro-tube inner & outer diameters (mm), and also the material type, which is expressed by the material density (Kg/cm^3). The 9 model outputs are: the process die temperature (C), the melt density (Kg/m^3), the melt flow (rpm), the extrusion die inner diameter (mm), the extrusion die outer diameter (mm), the extrusion die pressure (bar), the die Nitrogen inside pressure (bar), the distance to the cooling media (water) (cm) and the draw speed (m/min). 30000 iterations were applied for the training of the ANN, with a learning rate of 0.005, and the training was stopped when the errors for all model's outputs, which was expressed by the absolute value of the difference between the training target values and the ANNs calculations results, were minimized to satisfactory levels for all the process outputs. The Mean Absolute Errors (MAE) achieved for the 9 process output parameters are:

Table 1. MAE achieved for the 9 process outputs of the 3-20-20-9 ANN model.

Output No	Process Parameters (Outputs)	Units	MAE	MAE (%)
1	Process die temperature	°C	1.12651	0.50182
2	Melt density	Kg/m^3	6.37156	0.63903
3	Melt flow	rpm	1.43085	5.93685
4	Extrusion die inner diameter	mm	0.00571	0.69026
5	Extrusion die outer diameter	mm	0.00596	0.35029
6	Extrusion die pressure	bar	0.78528	5.66754
7	Die Nitrogen inside pressure	bar	0.00325	0.31442
8	Distance to the cooling media	cm	0.03320	3.13089
9	The draw speed	m/min	1.90809	9.15724

In Figureure 2, we show 9 comparison charts, which were created after the training of the ANN that correspond to all the 9 process parameters (models' outputs). Each chart includes two curve lines. The first line represents the training target values (red dashed line) and the second line represents the Neural Network calculation results (blue line) for each of the corresponding outputs. For testing the behavior of the trained model and for creating the 9 comparison charts, the 100 records of the training inputs data set were used as inputs to the trained ANN, which provided the 9 corresponding calculation outputs (process parameters) for each triad of input values. The convergence of the micro-extrusion ANN training was satisfactory, considering the fact that our model was a complicated model that had only 3 inputs concerning tube features (outer and inner diameter and polymer material density) and is providing 9 outputs, which are the 9 process parameters. After the training and the testing of the micro-extrusion ANN, the model was capable to provide as outputs the required 9 process parameters for any set of inputs, which correspond to the 3 product features (outer & inner diameter, and material density).

Additionally and besides the above 3-20-20-9 model, the proposed methodology and the software tool was used and tested also for the inverse process model. In the inverse model, we used as inputs the above 9 process parameters and as outputs the above 3 product features. For this model a 9-15-15-3 feed-forward Neural Network was created and trained using the back-propagation algorithm. After training and testing this ANN, the Mean absolute Errors (MAE) achieved for all the 3 outputs are:

Table 2. MAE achieved for the 3 process outputs of the 9 Inputs - 3 Outputs ANN model.

Output No	Product features	Units	MAE	MAE (%)
1	Outer Diameter	mm	0.02371	2.16339
2	Inner Diameter	mm	0.02428	3.35821
3	Material Density	Kg/m^3	1.22700	0.10206

In Figureure 3, we show also the 3 comparison charts, which were created after the training of the second ANN model and which correspond to the above 3 product features (models' outputs). For testing the behavior of the 9-15-15-3 trained ANN model and for creating the 3 comparison charts, the training data set, consisting of the 100 records, was used with the 9 process parameters being used as inputs.

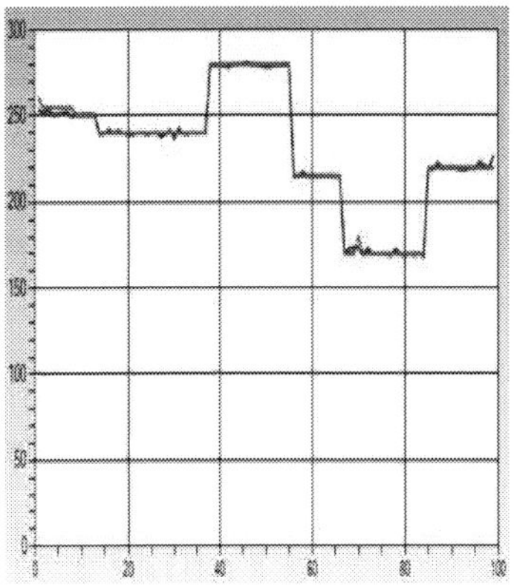

Figure 2.1. Output1- Process die temperature (°C).

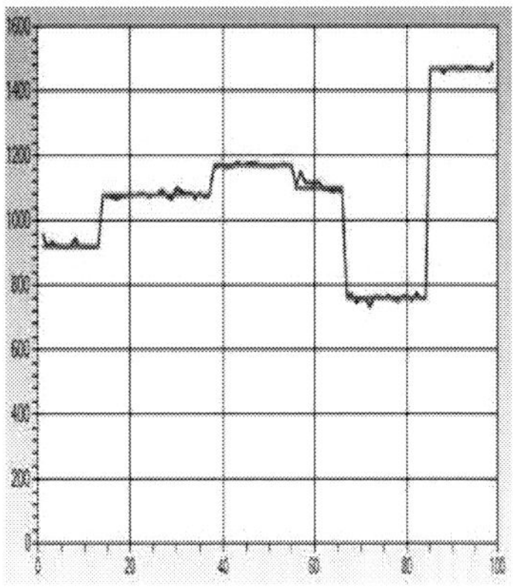

Figure 2.2. Output2 - Melt density (Kg/m^3).

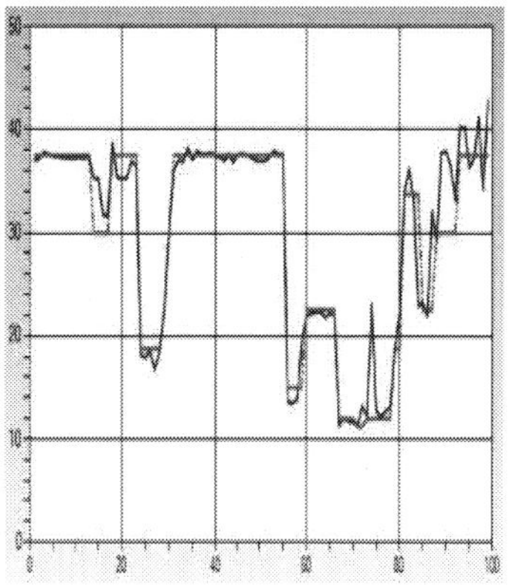

Figure 2.3. Output3 - Melt flow (rpm).

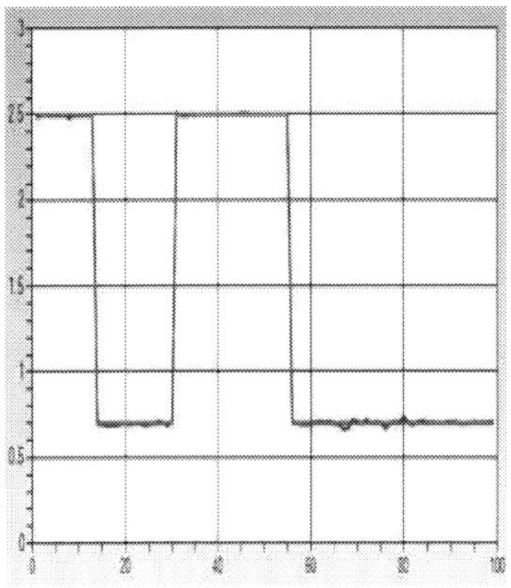

Figure 2.4. Output4 - Extrusion die inner diameter (mm).

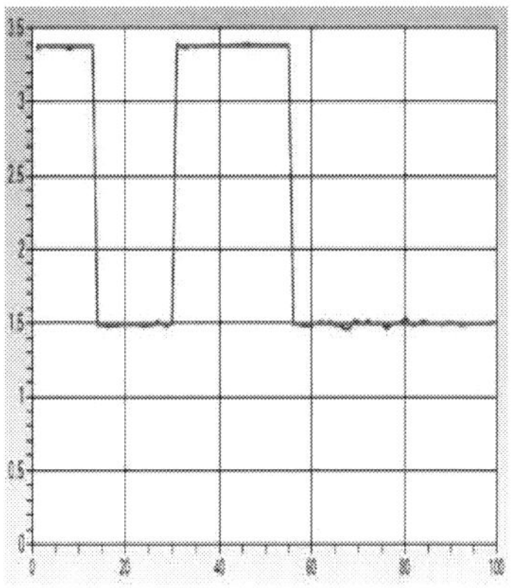

Figure 2.5. Output5 - Extrusion die outer diameter (mm).

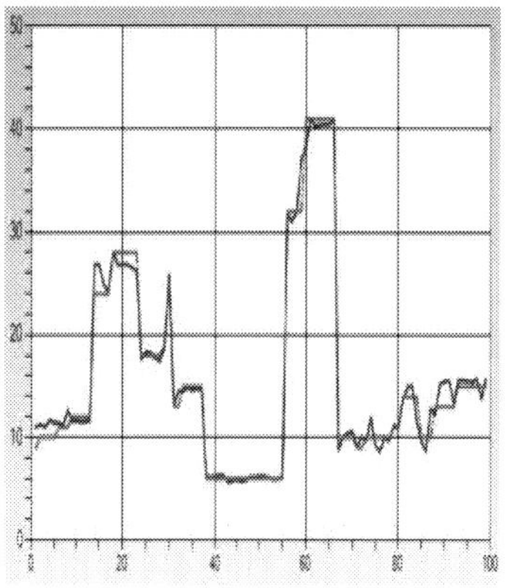

Figure 2.6. Output6 - Extrusion die pressure (bar).

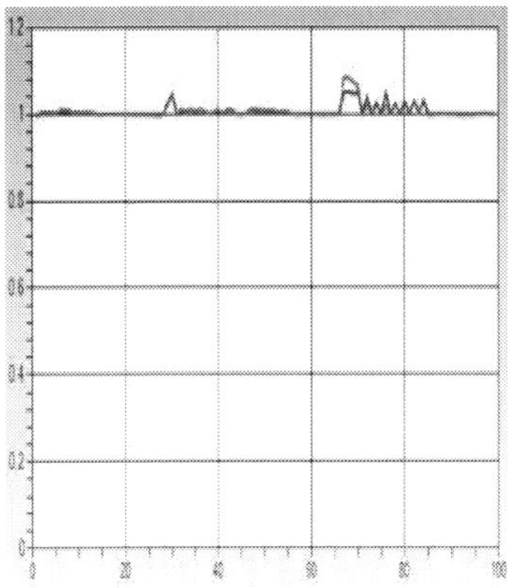

Figure 2.7. Output7 - Die Nitrogen inside pressure (bar).

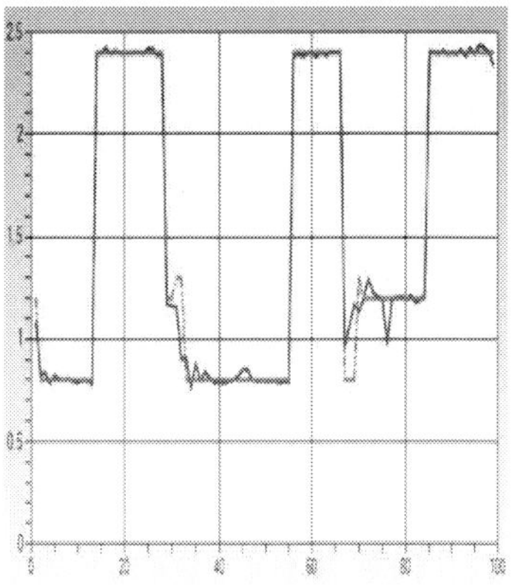

Figure 2.8. Output8 - Distance to cooling media (water) (cm).

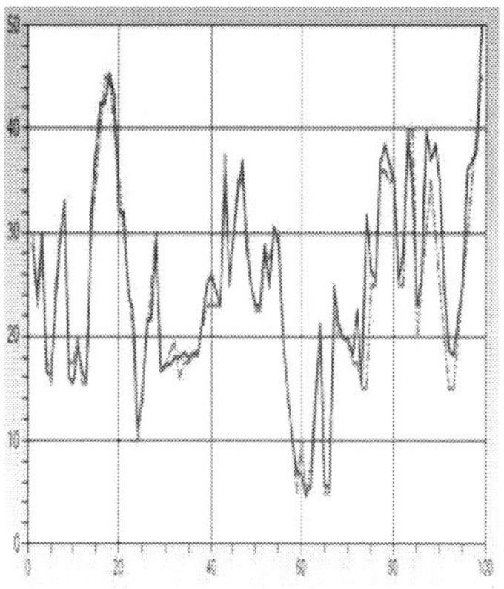

Figure 2.9. Output9 - Draw speed (m/min).

ANNs Results — Training Output Target Values

Figure 2. Process outputs comparison charts between ANN's results and Target Outputs of the 3-20-20-9 micro-extrusion ANN model.

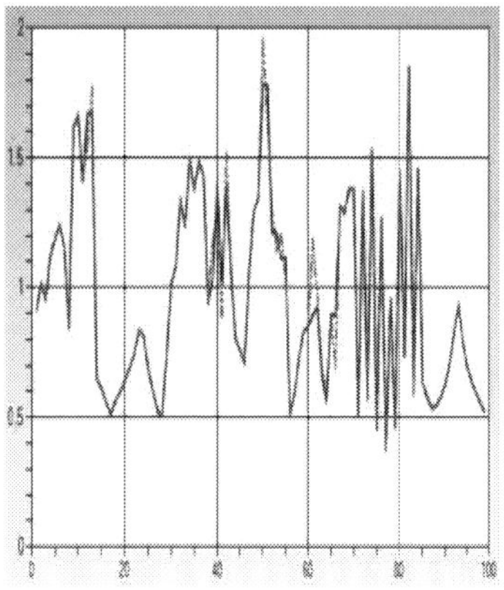

Figure 3.1. Output1- Outer Diameter (mm)

MAE1 = 2.16339 %.

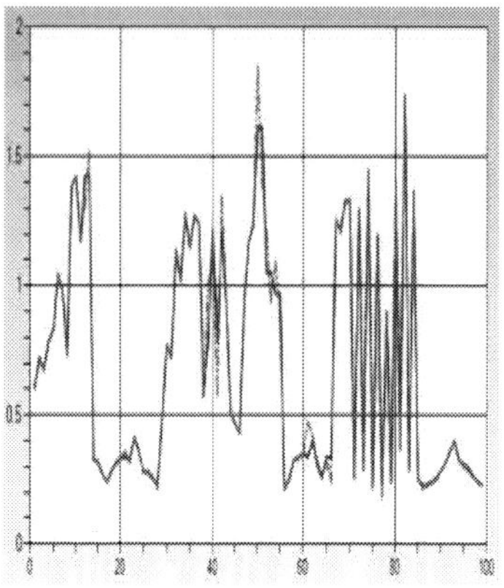

Figure 3.2. Output2- Inner Diameter (mm)

MAE2 = 3.35821 %.

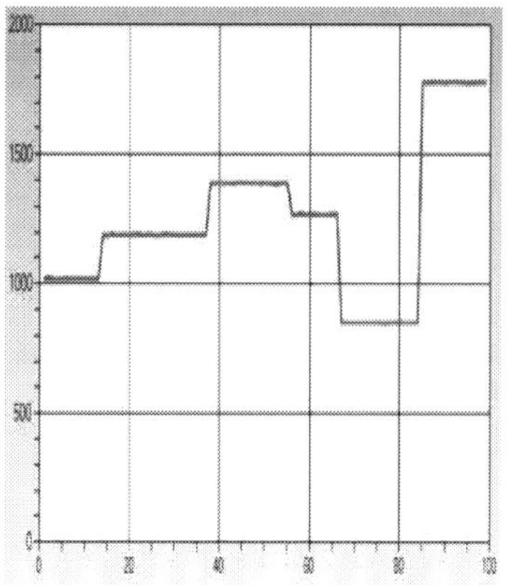

Figure 3.3. Output3- Material Density (Kg/m^3)

MAE3 = 0.10206 %.

——— ANNs Results — · — · — Training Output Target Values

Figure 3. Process outputs comparison charts between ANN's results and Target Outputs of the inverse 9-15-15-3 micro-extrusion ANN model.

3. Conclusions

The proposed Neural Network methodology was implemented as a Web application using the .NET programming environment of Visual Studio 2010. More specifically the programming languages C#, ASP.NET were used and also the MS-SQL Server as Database Management system. The whole research work was supported by the EU FP7 NMP project POLYTUBES (2009-2012) and the implementation case example, which was presented, concerned a functional approximation of the POLYTUBES extrusion process for the production of polymeric micro-tubes based on COC, PC, PET, PETG, PP and PVDF materials. Especially for the extrusion process, the main ANN model created included 3 inputs and 9 outputs, in an effort to support process engineers to calculate the required 9 process parameters when the 3 product features (tube's outer/inner diameter and material density) are provided as inputs. Even though the 3 Inputs / 9 Outputs ANN model is considered a complicated model and the achieved Mean Absolute training Error (MAE) varies among the 9 model's outputs, it is considered useful for process engineers since it gives the possibility to estimate with a satisfactory approximation of less than one 1% most of the process parameters, which are required for specific micro-tube's dimensions and material type. Additionally, the inverse 9 inputs / 3 outputs ANN model was created giving the possibility to the process engineers to cross check the output calculation results of both models.

4. Acknowledgments

The authors would like to thank the Swedish Research Center SWEREA IVF AB and more specifically Dipl. Eng. Daniel Wendels and Dr. Erik Perzon of SWEREA, who provided the experimental data concerning the micro-extrusion physical process of the polymer micro-tubes. Without their contribution, the training and the testing of the micro-extrusion ANNs models would not be feasible. Also the authors would like to acknowledge the support of the European Commission, which funded this research work through the EU FP7 NMP project POLYTUBES (http://www.polytubes.net).

References

[1] Chronakis I S, Mekras N D, Wiesauer K, Breuer E, Stifter D, Fuentes G F and Qin Yi 2009 MASMICRO micro-/nano-materials processing, analysis, inspection and materials knowledge

management *Int. J. of Advanced Manufacturing Technology* **47** 963-971
[2] Duce C et al 2006 Prediction of Polymer Properties from their Structure by Recursive Neural Networks J. Macromolecular Rapid Communications 27 711-715
[3] Hagan M T, Demuth H B and Beale M 1995 Neural Network Design (Boston: PWS Publishing Co) p 11.1
[4] Jiang Z et al 2007 Prediction on wear properties of polymer composites with artificial neural networks J. Composite Science and Technology 67 168-176
[5] Kadi H E 2006 Modeling the mechanical behavior of fiber-reinforced polymeric composite materials using artificial neural networks – A review J. Composite Structures 73 1-23
[6] Kasabov N 1996 Foundations of Neural Networks, Fuzzy Systems, and Knowledge Engineering (Massachusetts: MIT Press) p 267
[7] Liao T W and Chen L J 1998 Manufacturing Process Modeling and Optimization Based on Multi-Layer Perceptron Network J. Manuf. Sci. Eng. 120 109-119
[8] Mok S L et al 2000 An intelligent hybrid system for initial process parameter setting of injection moulding International Journal of Production Research 38 4565-4576
[9] Noor R A M, Ahmad Z, Don M M and Uzir M H 2010 Modelling and control of different types of polymerization processes using neural networks technique:A review Canadian Journal of Chemical Engineering 88 1065–1084
[10] Nascimento C, Giudici R and Guardani R 2000 Neural network based approach for optimization of industrial chemical processes J. Computers & Chemical Engineering 24 2303-2314
[11] Qi L H, Li H J, Hou J J and Cui P L 2000 Research on the Neural Networks Used for Shaping Tubes by the Liquid Extrusion Process J. of Materials Engineering and Performance 9 28-32.
[12] Sha W and Edwards K L 2007 The use of artificial Neural Networks in materials science based research J. Materials & Design 28 1747-1752
[13] Swingler K 1996 Applying Neural Networks (London: Academic Press Ltd) p 10
[14] Tzafestas S and Mekras N 1999 Industrial Forecasting using Knowledge Based Techniques and Artificial Neural Networks in Advances in Manufacturing: Decision, Control and Information Technology ed S. Tzafestas (Berlin: Springer-Verlang) p 171
[15] Zhang Z and Friedrich K 2003 Artificial neural networks applied to polymer composites: a review J. Composites Science and Technology 63 2029-2044

Mechanical performance of foam glass with nanoscale structure elements

O V Kazmina[1], B S Semukhin[2] and J Njuguna[3]

[1] National Research Tomsk Polytechnic University, 634028 Tomsk, 30 Lenin Ave.
[2] Institute of Strength Physics and Materials of the Siberian Division of the Russian Academy of Sciences, 634021 Tomsk, 2/4 Academichesky Ave.
[3] Cranfield University, Shrivenham, Swindon, SN6 8LA, United Kingdom

Abstract. The paper presents experimental data on studying the strengthening of glass foam material. Glass foam material is foamed glass of increased strength achieved due to nanoscale structural elements. Glass foam is obtained through thermal treatment of noncrystalline matrix containing crystalline material in the silica state. It is shown that strengthening of glass foam material is achieved by structuring the noncrystalline partition. New technique of strengthening is proposed due to the formation of nanoscale globules of noncrystalline matrix. Self-organization of spherical nanoscale globules. New technique for glass foam strengthening. For the first time the phenomenon of strengthening interpore partition of glass foam was established. The principle of the process consists in the formation of noncrystalline matrix with a specific globular structure.

1. Introduction

The phenomenon of materials strengthening has been studied and examined for a long time. Such classical studies are known as those of A.A. Bochvar, M.L. Bernstein, R. Zimmermann [1, 2]. They describe the techniques, schemes and classifications of strengthening for various materials. All strengthening techniques can be divided into two groups. The first group of materials (mostly metal) with a crystalline structure is well described by the classical dislocational model of strengthening. The second group of materials (non-metal) with a noncrystalline structure can be described by the non-dislocational model of strengthening. For example, Ref. [3] proposes the strengthening technique due to the clusters in the noncrystalline structure. Refs. [4, 5, 6] explain strengthening at inelastic deformation particularly by non-dislocational mechanisms. In particular, it is grain boundary sliding and movement of vacancies at boundaries. The work by Kozlov E.V. [7] directly indicates the observed critical grain size after which the amorphized material is strengthened in a dislocation-free mode due to the boundaries of nanostructural formations.

It is believed that the main cause of strengthening at dispersion hardening is elastic stress created by the separation of the second phase [4-6]. In case of noncrystalline bodies (glass) the Griffiths model is more common. It describes strength properties of glass with the use of energy fracture model. According to the model, the glass fracture is caused by the strain concentrators at the glass surface. In glass-ceramic materials where noncrystalline phase may reach up to 90 and 95%, the stress in the interface of crystalline phases and noncrystalline matrix cannot lead to the strengthening of the material as a whole. For the glass foam materials examined in this paper neither the first nor second model is applicable due to the following reasons: no separation of the second crystalline phase in the first crystalline phase; no continuous surface of the noncrystalline matrix, and presence of multiscale random pores.

The paper examines the process of glassceramic foam material strengthening. This material is manufactured of semiprocessed product – the quenched cullet synthesized at relatively low temperatures (< 950 °C). It is already at this stage when the quenched cullet contains glass phase in a noncrystalline state. It ensures the pyroplastic state of the material at the stage of repeated thermal treatment (sponging) of the quenched cullet. The quenched cullet is a noncrystalline matrix containing residual silica particles with the size up to 200 nm.

2. Results and discussion

The foam materials were produced according to the two-stage technology developed by the authors [8, 9]. Such crystal and noncrystalline silica-containing rocks as silica sand, diatomite, flint, silica clay, and perlite were sampled as a feedstock. The samples of foam material were obtained with the content of residual crystalline material from 5 to 20 % by volume. It was established that the strength of experimental samples increased from 1.8 to 3 MPa as the crystalline phase particle size decreased from 1,000 to 300 nm.

During the formation of submicron crystalline phase in the glass the change in its energy characteristics is observed [10]. The maximum strengthening effect should be expected if the microstructure of crystalline phase is significantly different from the equilibrium for this phase. The paper by academician N.P. Lyakishev shows that material with specific properties should consist of crystallites or mixture of nanoscale crystals and noncrystalline phase [11]. Therefore, the foam material synthesized from the quenched cullet is essentially "simulated" by these principles.

For the samples with fixed amount of residual crystalline material (5 % on average) and different size of particles the experimental dependence was established (Figure1). Assuming that the minimum critical size of crystalline phase particles is equal to 10 nm [7], the maximum theoretical strength of the foam material was calculated, which amounted to 5 MPa. In case of the glass foam the noncrystalline matrix of which contains no crystalline material, this value does not exceed 1.5 MPa [12].

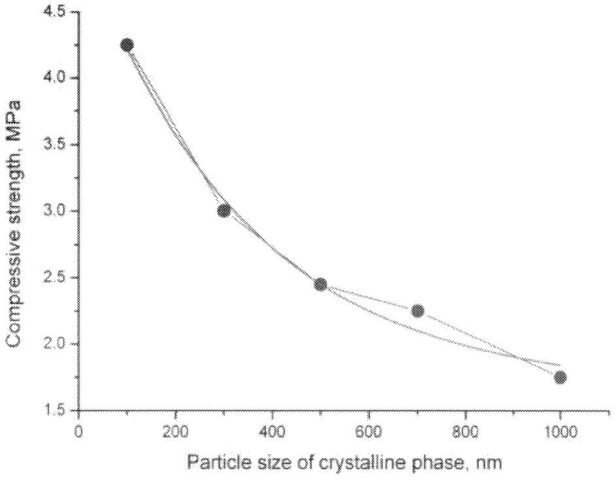

Figure 1. The strength of the foam material as a function of the crystalline phase particle size.

Since it proves to be impossible to strengthen the noncrystalline matrix due to the subsequent reduction of the particle size in the residual crystalline material, there is another option to alter the material strength. The strength of the partition can be changed due to the modification in the matrix internal structure. The strength values at compression determined for all experimentally obtained samples of foam material showed that the increased strength (4.3 MPa) is typical for those ones having the noncrystalline matrix of the interpore partition with the observed nanoscale spherical elements (spheroids). In electron micrographs of high resolution there are spheroids observed in the noncrystalline phase with the size of 60 to 160 nm. Such spheroids are not found in the interpore partitions of conventional glass foam and glassceramic foam material samples with the size of residual silica particles over 200 nm (Figure2). Figure2 distinctively shows a complex hierarchical arrangement consisting of both separate spheroids and their groups. X-ray dispersion analysis of the silicon content in the partition noncrystalline matrix shows its non-uniform distribution. The maximum concentration is observed at the partition boundary, whereas the minimum one – in the middle (Figure 3). The formation of spheroids leads to the redistribution of silicon in the partition. It implies that nanospheroids are main silicon-containing structural elements that are accumulated mostly by the partition boundaries.

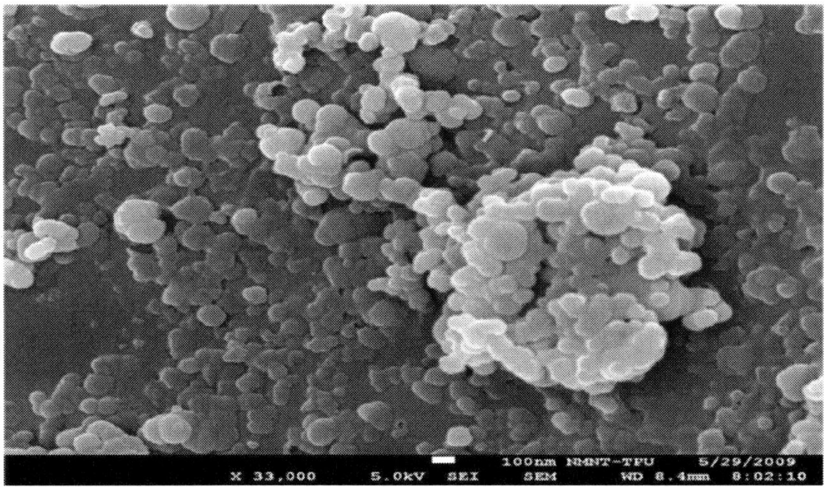

Figure 2. Electron micrograph of the spheroids in the foam material interpore partition.

Thus, the modification of the strengthened glass foam structure is determined not by the structure and presence of residual crystalline material, as it was expected, but by the change in the noncrystalline matrix structure. On the one hand, it is inconsistent with the scheme described in the scientific literature, and on the other hand, such behavior of glass-ceramic material agrees with the structure where the minimization of energy of the whole material is performed due to the energy minimization of basic (noncrystalline) component only. In the course of sponging the material by itself modifies the structure of the noncrystalline matrix. Spheroids are formed, which are referred to as "globules" in minerals-related literature. For example, Ref. [13] gives experimental evidence on the existence of globules in silicon oxide and proposes the mechanisms of their formation. The author believes that the spherical globule has different packing density of SiO_4 tetrahedrons in the centre and on the surface. The globule internal area has more incoherent packing of SiO_4 tetrahedrons, as compared to the nucleus. The general view of a globule is a three-dimensional particle with the structure similar to that of silica or crystobalite, on the surface of which there are one-dimensional Si,O-chains – dimmers, trimers, and monomers of silicic acid or sodium silicate. This paper assumes that globules have an intermediate position between crystalline and noncrystalline state. Other studies [14, 15] also give numerous experimental evidences on the formation of silicon oxide globules. According to the data of X-ray diffraction and thermography, the globules internal composition can be presented as a disordered set of nanocrystals and noncrystalline areas.

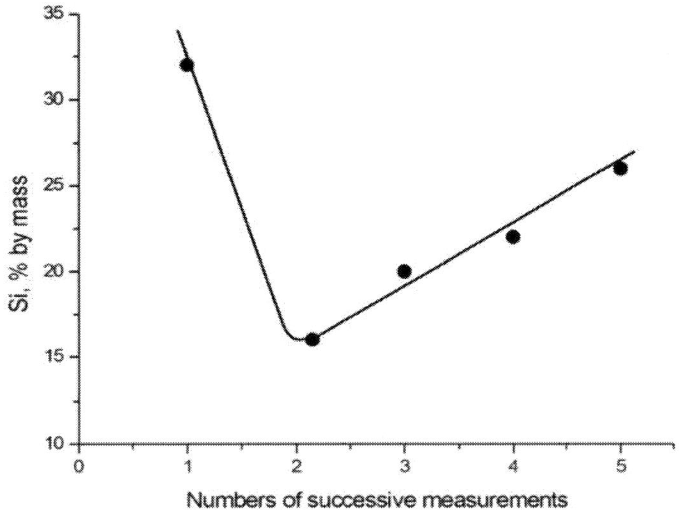

Figure 3. Distribution of silicon content in the interpore partition.

Based upon the IR spectroscopy data presented in Ref. [13] by the example of silica gels and noncrystalline synthetic alumosilicate, it was shown that the formation of globules in the materials spectra is accompanied by the appearance of the band 1200 cm^{-1}. This is indicative of the fact that fragments of SiO4 tetrahedrons appeared on the surface of crystals with the angles between them being equal 180o. These fragments can consist of two or three linked tetrahedrons that can be referred to as chains.

In our case, based upon the IR spectroscopy data of the glassceramic foam materials samples the appearance of a new band 1,249.6 cm^{-1} was established, which was not present in the IR spectra of glass foam. It is evident that these data with regard to micrographs can be indicative of the globular structure of the interpore partition noncrystalline component. Thus, we were first to discover the globules of noncrystalline matrix in the glass foam material. We consider that the sharp (abrupt) increase in the glass foam material strength is assigned specifically to them. Upon that, the microstructure of the high-strength foam material partition consists of the noncrystalline matrix with the present particles of residual silica (not more than 5% with the size of about 200 nm) and globules (100 nm and less in size).

The strength of nanoncrystalline materials is generally assigned to several mechanisms. The mechanisms of shift and local jumps are presented as basic ones. In the first case part of noncrystalline cluster (globule) is displaced at the deformation along the other one and dissipates the energy [16]. In terms of the second mechanism it is proposed to consider possible the abrupt displacement of individual atoms within a cluster (globule). More recent studies, in addition to these mechanisms, propose considering the formation of nanostructures in the deformation bands as a strengthening factor [17-21]. In our experiments the strengthening of foam material is related to two processes: 1) decreasing of the particle size in residual crystalline material; 2) formation of nanoscale spheroids – globules in the noncrystalline matrix of glass foam material. In the course of the first process – the change of the crystal size (Figure1) the strength of 3 MPa is reached, and at the introduction of the second process a theoretical strength of 5 MPa can be attained. The major energy dissipation at the deformation is performed by the areas with densely packed globules, with the volume content of which can amount to 95% in the material.

For the first time a new type of strengthening is observed in the noncrystalline material, that is the self-organization of nanospherical globules in the body of noncrystalline matrix. The strengthening model of noncrystalline materials with nanocrystal components proposed in Ref. [11] has been proved by the research conducted in our paper.

3. Acknowledgement
The work was performed with the financial support European grant of 7 Framework Program FP7-NMP-2008-SMALL-2 (Proposal No: CP-FP 228536-2 NEPHH).

References
[1] Mechanical properties of metals: study guide/ M.L. Bernstein, V.A. Zaimovsky. – M., 1979. – p. 495.
[2] Zimmerman R., Gunter K. Metallurgy and material science. Reference book, translated from German. M.1 Metallurgy, 1982. 480 p. (P.19-24, 90-108).
[3] Golovin Yu.I., Tyurin A.I. Non-dislocational plasticity and its role in mass-transfer and formation of imprint at the dynamic indentation / FTT, 2000. – V.42. Issue 10. p. 1818-1820.
[4] Andriyevsky R.A., Glezer A.M. Strength of nanostructures // Achievements of physical sciences. – 2009. – V.179, No.4. – P.337-358.
[5] Panin V.E., Sergeyev V.P., Panin A.V., Pochivalov Yu.I. Nanistructuring of surface layers and application of nanosctructural coatings – effective technique for strengthening constructional and instrumental matrials // FMM. – 2007. – V.104, No.6. – P.650-660.
[6] Dub S.N., Novikov N.V. Nano-hardness testing of solid bodies // Superhard materials. – 2004. – No.6. – P.16-33.
[7] Kozlov E.V., Koneva N.A., Popova N.A. Grained structure, geometrically necessary dislocations and particles of secondary phases in polycrystals of macro- and meso-levels // Physical meso-mechanics. - 2009, Vol.12, No.4 - p.93-106.
[8] Kazmina O.V., Vereshchagin V.I., Semukhin B.S., Abiyaka A.N. Low-temperature synthesis of quenched cullet from mixtures based on siliceous components for foamed material production // Glass and Ceramics. 2009. No. 10. P. 5 – 8
[9] Kazmina O.V., Vereshchagin V.I., Abiyaka A.N. Extension of the raw material base for foamglass ceramic material production// Construction materials. – 2009. – No. 7. – P. 54 – 56.
[10] Sarkisov P.D. Directional glass crystallization –s a basis for the production of multifunctional glass-ceramic materials. M., D.I. Mendeleyev RChTU, 218 p., 1997.
[11] Nanocrystalline structures as a new area of developing construction materials. Bulletin of Russian Academy of Sciences. V.73, No.5, p. 422 (2003). N.P. Lyakishev.
[12] Kazmina O.V., Vereshchagin V.I., Semukhin B.S. Structure and strength of foam glass ceramic materials from low-temperature quenched cullet // Physics and chemistry of glass. 2011. V. 37, No.4, P. 29 – 36.
[13] Chukin G.D. Chemistry of surface and structure of dispersed silicon oxide. M.: Paladin Press, LLC "Printa", 2008. – 172 p.
[14] Golubev Ye.A. Supermolecular structures of natural X-ray amorphous substances. Yekaterinburg: UrO RAS Press. 2006. 155 p.
[15] Golubev Ye.A. Globular structure of higher anthraxolites according to the data of scanning probe microscopy// Papers of Academy of Sciences. 2009. – V.425, No.4. – P.519–521.
[16] Golubev Ye. A. Scanning Probe Microscopy in Researches of Micro- and Nanostructure in Noncristalline Geomaterials // Microscopy & Microanalysis, 2003.Vol. 9 (S. 3). P. 304–305.
[17] V.I. Berezkin. Fullerenes as embryos of soot particles // FTT.-2000.-V.42, issue 43.-p.567-572
[18] Pampillo CA. J Mater Sci 1975;10:1194, Argon AS. Inelastic deformation and fracture of glassy solids. In: Mughrabi H, editor. Material science and technology. Weinheim:VCH; 1993. p. 63 Argon AS, Kuo HY. Mater Sci Eng 1979;39:101.
[19] Jiang WH, Atzmon M. Acta Mater 2003;51:4095 Chen H, He Y, Shiflet GJ, Poon SJ. Nature 1994; 367:541.
[20] Jiang WH, Atzmon M. Scripta Mater 2006; 54:333.
[21] Jiang WH, Pinkerton FE, Atzmon M. J Appl. Phys 2003;93:9287. Csontos AA, Shiflet GJ. Nanostruct. Mater. 1997; 9:281

Mechanical and corrosion resistance properties of TiO_2 nanoparticles reinforced Ni coating by electrodeposition

W Shao[1], D Nabb[2], N Renevier[3], I Sherrington[3] and J K Luo[1]

[1]University of Bolton, UK
[2]MSC Copperflow Ltd,UK,
[3]University of Central Lancashire,UK

Emails: W.Shao@bolton.ac.uk, J.Luo@bolton.ac.uk

Abstract. Coatings have been widely used in engineering and decoration to protect components and products and enhance their life span. Nickel (Ni) is one of the most important hard coatings. Improvement in its tribological and mechanical properties would greatly enhance its use in industry. Nanocomposite coatings of metals with various reinforced nanoparticles have been developed in last few decades. Titania (TiO_2) exhibit excellent mechanical properties. It is believed that TiO_2 incorporation in Ni matrix will improve the properties of Ni coatings significantly. The main purpose of the current work is to investigate the mechanical and anti-corrosion properties of the electroplated nickel nanocomposite with a small percentage of TiO_2. The surface morphology of nanocomposite coating was characterized by scanning electron microscopy (SEM). The hardness of the nanocoating was carried out using micromaterials nanoplatform. The sliding wear rate of the coating at room temperature in dry condition was assessed by a reciprocating ball-on-disk computer-controlled oscillating tribotester. The results showed the nanocomposite coatings have a smoother and more compact surface than the pure Ni layer and have higher hardness and lower wear rate than the pure Ni coating. The anti-corrosion property of nanocomposite coating was carried out in 3.5% NaCl and high concentrated 35% NaCl solution, respectively. The results also showed that the nanocomposite coating improves the corrosion resistance significantly. This present work reveals that incorporation of TiO_2 in nickel nanocomposite coating can achieve improved corrosion resistance and mechanical properties of both hardness and wear resistance performances, and the improvement becomes stronger as the content of TiO_2 is increased.

1. Introduction

Incorporation of ceramic, polymer and metal nanoparticles within metal matrix produces composite coatings which have many attractive properties such as wear resistance, corrosion resistance, particularly useful for engineering [1]. Majority of the work has been concentrated on nickel (Ni) coating. The reinforced particles could be SiC, Al_2O_3, titania (TiO_2), carbon nanotubes (CNTs) and polymers such as Polytetrafluoroethylene (PTFE) and Polyethylene (PE) [2-5]. Among these reinforcements, TiO_2 is one of the most important used in the engineering materials. The titanium-based material offers high strength, good corrosion and oxidation resistance [5], it is believed that incorporation of nanoparticles can greatly improve the mechanical, tribological properties and corrosion resistance, hence can be used in industry either in single layer or multilayer structure. In this paper, we will report the mechanical properties and corrosion resistance for the TiO_2 nanoparticles reinforced Ni coatings.

2. Experiments

2.1 Preparation of coatings

Mild steel plates with the dimension of 20mm×20mm×1mm were used as the substrates. Before electroplating, the steel substrates were cleaned by alkaline cleaning solution and followed by an acid etch process. De-ionized water was used to rinse after each step, and finally dried by dry N_2 gas. The electrolyte used in this study is a standard Watts bath, containing up to 12 g/l TiO_2 nanoparticles (Sigma, ~50nm) with suitable surfactant. Electrodeposition took place in the bath of 50 °C and at a current density of 3 A/dm^2 for 1hr.

2.2 Characterization of the coatings

A Philips XL30 ESEM (SEM) operating at an accelerating voltage of 15 KV was used to observe the surface morphologies of the coated samples.

Hardness tests were carried out using a three-sided flat pyramidal diamond flat tip with an angle of 104.3° at room temperature. The indenter was lowered at a rate of 50 nm/s to indent the matrix up to a depth of 500 nm and then to retrieve from the coating surface.

The wear tests were evaluated on a reciprocating ball-on-disk computer-controlled oscillating tribotester. A tungsten carbide (WC) ball with diameter 6 mm was used as the counter body. All the wear tests were performed using a load of 12 N at 12 mm/s sliding speed in dry condition at room temperature. The total sliding distance is 10000 mm.

PG581 potentiostat system was used for electrochemical measurements with three standard electrodes. The measurements were carried out in a 250 ml cell in 3.5 % NaCl solution at room temperature. All the potentials were recorded with respect to saturated calomel electrode (SCE). Prior to Tafel polarization, the specimens were allowed to immerse in the electrolyte while the open circuit potential (OCP) was measured until it reached a stable value. Potentiodynamic polarization curves were recorded from a starting potential of 500 mV below the OCP and were scanned towards the positive direction at a scan rate of 1 mVs^{-1}, until the potential reached 500 mV. Weight loss method was used to assess the corrosion resistance of coatings. The specimens were immersion into 35 % NaCl solution at room temperature in order to accelerate the corrosion rates. The weight differences were recorded after 10 days and 20 days immersion.

3. Results and discussion

3.1 Characterization of Ni nanocomposite coatings

Figure 1 shows the SEM morphologies of Ni coating and Ni-TiO_2 nanocoating containing 12 g/l TiO_2 nanoparticles. The microstructures between pure Ni coating and Ni-TiO_2 nanocoating are different significantly. The Ni coating has a rather regular structure with large grain size in the range of 2~4 μm (Figure 1a). The Ni-TiO_2 nanocoating shows smaller gain sizes of 0.1~1 μm (Figure 1b), which indicates that the co-deposited TiO_2 nanoparticles are much uniformly distributed in the Ni matrix. It shows that Ni-TiO_2 nanocoating has fairly uniform, continuous and compact morphology.

3.2 Hardness of Ni nanocomposite coatings

Figure 2 shows the microhardness of Ni coating and Ni-TiO_2 nanocoatings with the different contents of TiO_2 nanoparticles. It can be seen that the hardness of Ni-TiO_2 nanocoating increases with increasing the content of TiO_2 nanoparticles, and that with 12 g/l TiO_2 nanoparticles has the maximum hardness of 387 HV, clearly demonstrated its significant effect. This is because the co-deposited TiO_2 nanoparticles in the Ni matrix can restrain the growth of the Ni grains resulting in the very fine grain structure and dispersive strengthening effects which leads to the microhardness of Ni-TiO_2 nanocoatings increase with increasing the TiO_2 nanoparticles content [6].

Figure 1. (a) SEM of Ni coating, (b) SEM of Ni-TiO_2 nanocoating.

3.3 Anti-wear properties of the coatings

Figure 3 shows that the effect of TiO_2 particles on the wear rate of Ni matrix nanocoatings. There is a steadily decreasing trend with the increase of the content of the TiO_2 nanoparticles in the nanocoatings. The wear rate of the Ni-TiO_2 nanocoating with 12 g/l TiO_2 is 9.097×10^{-9} mm^3/Nm and that of Ni coating is 14.875×10^{-9} mm^3/Nm, improved about 48 %. The improvement is believed to be twofold: firstly an increase in surface fraction of TiO_2 nanoparticles reduces the direct contact between the Ni matrix and WC ball. When the TiO_2 co-deposited with the matrix metal, the fine smooth surface morphology of the nanocoating will increase the load-carrying area and reduce the stress between the friction couples which leads to the improvement of wear resistance of the nanocoating [2]. Secondly the TiO_2 nanoparticle has a much high wear resistance, hence reduces the overall wear rate of the nanocoating. Also the nanocoating is much denser with smaller grain sizes as it can be seen from Figure 1, this may also contribute to the improvement of the wear resistance. The SEM images of worn coating surfaces are shown in Figure 4. Clearly, the width and depth of the worn track of the pure Ni coating (Figure 4a) is much larger than that of the TiO_2 reinforced Ni nanocoating as shown in Figure 4b.

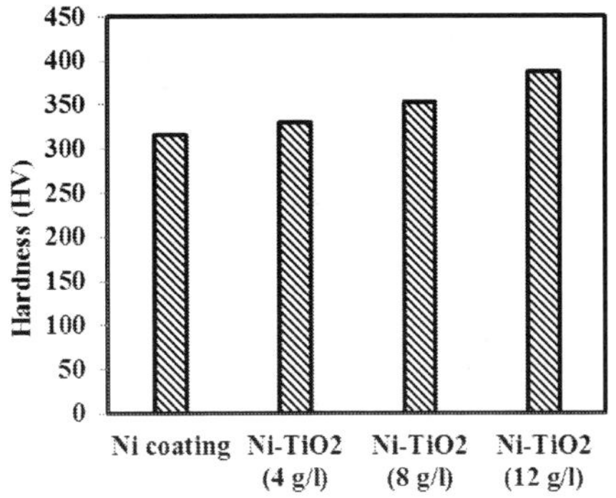

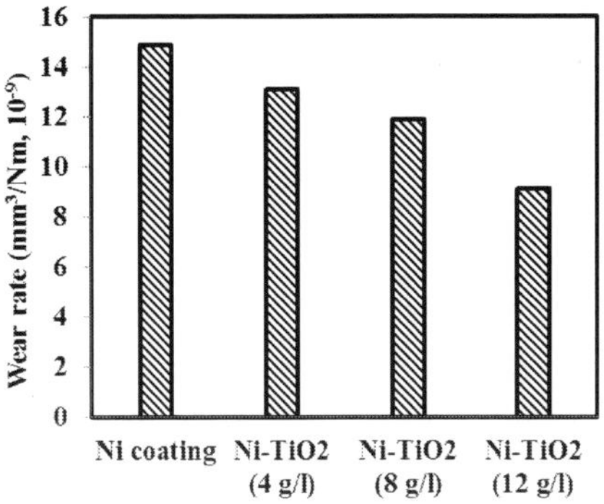

Figure 2. Hardness for Ni based nanocoatings. **Figure 3.** Wear rates for Ni based nanocoatings

3.4 Corrosion resistance of the coatings

Figure 5 shows the Tafel curves of Ni coatings (line 1), Ni-$TiO_2$with 12 g/l TiO_2 (line 2) nanocoatings. The electrochemical parameters such as corrosion potential (E_{corr}), corrosion current (I_{corr}) and the calculated corrosion rates (R_{corr}) are listed in Table 1. The corrosion rates of the 12 g/l TiO_2 nanoparticles reinforced Ni coating has improved 25 % compared with the pure Ni coating. With the increase of TiO_2 content in the nanocoating, the corrosion current decreases and the corrosion potential shifts to a more positive potential resulting in a decreased corrosion rate. Nanocoating has a much denser and more compact surface and has much less pores and cracks than pure Ni coating, which are attributed to contribute to the improvement of anti-corrosion property. Figure 6 shows the weight loss results for different Ni based coatings immersed in 35 % NaCl solution for 10 days and 20 days, respectively. It is evident that TiO_2 nanoparticles reinforced Ni has a good tendency for corrosion resistance compared with pure Ni coatings.

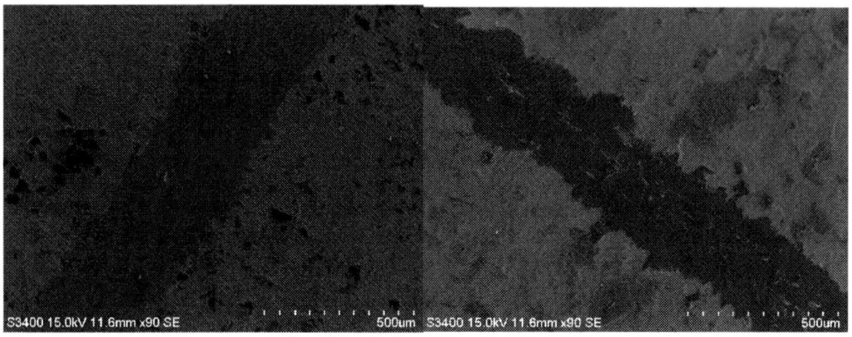

(a) (b)

Figure 4. SEM of worn surfaces (a) pure Ni coating, (b) Ni-TiO_2 nanocoating.

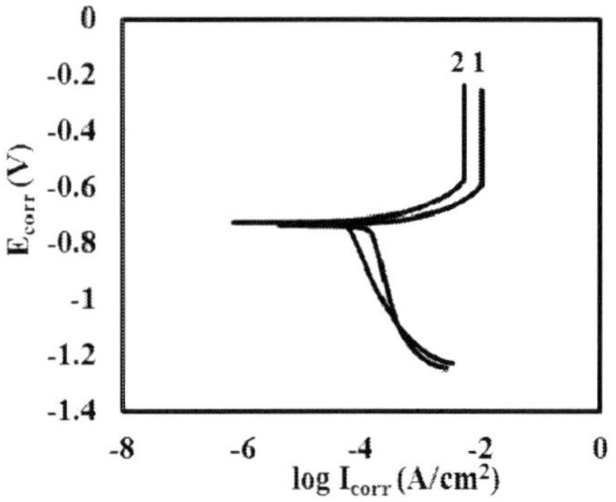

Figure 5. Tafel curves of Ni and Ni-TiO_2 nanocoatings.

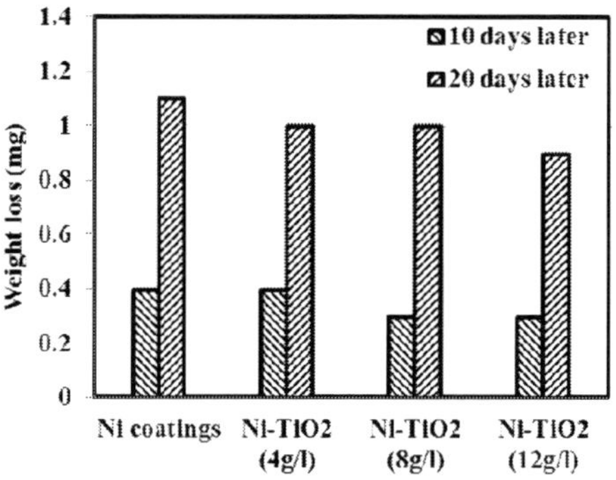

Figure 6. Weight loss results after 10 days and 20 days in 35 % NaCl solution.

Table 1. Electrochemical parameters and calculated corrosion rates for tested coatings.

Samples	E_{corr} **[mV]**	I_{corr}**[μA]**	R_{corr} **[μmpy]**
Ni coating	-755	333	15.3
Ni-TiO_2 (4 g/l)	-742	314	14.4
Ni-TiO_2 (8 g/l)	-733	296	13.6
Ni-TiO_2 (12 g/l)	-727	258	11.5

4. Conclusions

TiO_2 nanoparticles reinforced Ni nanocomposite coatings revealed a much smoother surface microstructure with excellent mechanical properties. The Ni-TiO_2 nanocomposite coatings exhibited higher hardness and higher wear resistance in comparison with pure Ni coating. Meanwhile, the corrosion resistance has been enhanced significantly.

5. Acknowledgement

The authors would like to acknowledge the financial support by Knowledge Transfer Partnership and TSB (KTP007867) and The Knowledge Centre for Materials Chemistry under Grant No. of X00680PR.

References

[1] C.S. Lin, C.Y. Lee, C.F. Chang and C.H. Chang: Surf. Coat.Technol., vol. 200 (2006), p.3690
[2] H.T.Ha, C.T.Anh, N.T.Ha and D.T.Cao: 2009 J. Phys.: Conf. Ser., vol.187 (2009), 012083
[3] C.T.J.Low, R.G.A. Wills and F.C.Walsh: Surf. Coat.Technol., vol. (2006), p.371
[4] R.H. Cayton: NSTI-Nanotech, vol. 3 (2004), p.312
[5] A.A.Aal: Mater. Sci. Eng., A, vol.474 (2008), p181
[6] L. Shi, C. Sun,P. Gao, F. Zhou and W. Liu: Appl. Surf. Sci., vol. 252 (2006), p. 3591

IOP Conf. Series: Materials Science and Engineering **40** (2012) 012044 doi:10.1088/1757-899X/40/1/012044

Replication of micro-feature using variety of polymer and commonly used mould at elevated temperature and pressure

S Nebo[1], Z Ali[2] and S Scott[2]

[1]Dept. of Mechanical Engineering, School of Science and Engineering, Teesside University Middlesbrough, Tees Valley, TS1 3BA, UK
[2]Graduate Research School, Teesside University Middlesbrough, Tees Valley, TS1 3BA, UK

Email: ebynebo@yahoo.com, z.ali@tees.ac.uk, s.m.scott@tees.ac.uk

Abstract. Due to the functional dependency of micro-systems technologies on stringent dimensions, masses and tolerances, strict quality requirement is paramount for mass producing micro-parts for use in micro-devices. The consistency of components produced could vary as a result of significant effects of process parameters or process noise, thereby necessitating the urge for experimental studies to determine the best process combination that will produce a variant free component that will meet the dimension, mass and other parameters requirement of its application use. In this paper, the influence of some process parameters on replication of micro-feature is explored. Parts with micro-feature designed are moulded using polyethylene and polypropylene. Design of experiment (DoE) approach was applied to correlate the fabricated parts quality to the processing parameters. Two level, half factorial 8-run (2^{4-1}) experimental plan was used to investigate the effect of four process parameters of melt temperature, injection pressure, mould temperature and holding time on the replicated quality of the moulded parts, with part mass as output parameter to reflect variability. Of all the investigated process parameters, result of experiment indicated melt temperature and holding time as most statistical significant processing parameter for parts produced using polypropylene.. While there are no statistically significant effects of process parameters for those parts produced using polyethylene, rather, variability is observed to have occurred as a result of process noise. Furthermore, the experiments show that variability in the response parameter during micro-injection moulding differs for different polymer. Therefore, appropriate polymer type should as well be made to match with suitable processing parameters so as to obtain quality replication of micro-feature when mass producing.

1. Introduction

As the application use of micro components becomes handful, and their utilization cutting across various fields such as Mechanical (Automotive sensors, printer heads, micro pumps, micro heat exchanger, etc.), Information technology (inkjet printer, reading caps for hard disk, etc.), High technological consumer products (mobile phones, hand held high definition display, etc.), Medical, Biotechnology and Pharmacy (bio sensors, gene chips, fuel cells, hearing aids, drug delivery systems, etc.) [6, 9], conformity to quality requirements becomes inevitable and is necessary for mass producing of micro parts for use in micro devices. Thus, the quest for an industrial scale micro technique that will be suitable for mass production of micro parts at low cost, reduced cycle time, with precision serving its intended purpose, and also economically especially for medical applications where micro components should be dispose for safety reasons to avoid cross contaminations, and for conditions with human limited access to micro devices due to anthropometrical reasons or technical resources to access, reset or recalibrate devices for multiple usage [1, 7].

However, among the different polymer manufacturing technology available, micro injection moulding have been of great interest to both researchers and manufacturers due to its potential to provide the required leveraged technology for micro structured component, as low cost disposable micro part becomes feasible [8]. Though the most suitable technique for micro component fabrication [2, 4], some factors impedes the scaling ability of this technique such as the moulding machine, tooling material, processing parameters, component handling and inspection, and these factors need to be addressed in order to meet the miniaturization need and standards of those fields in dear need of such technology.

In this study, the influence of some process parameters (melt temperature, mould temperature, injection pressure, and holding time) which constitute a far reaching effect on micro component replication has been explored using polyethylene and polypropylene at elevated temperature and pressure, and applying Design of Experiment (DoE) approach to correlate the fabricated part quality to the processing parameters. Rectangular high aspect ratio samples were fabricated on a Mark 2L Manumoud Machine with the temperature of the injection metering zone of the plasticizer set to 240°C.

Various research [1, 4, 5] showed that replication of micro-feature is done at elevated temperature, thus, explaining the important of incorporating variotherm system in micro injection moulding technology which is to help in achieving complete filling of mould cavity before the solidifying layer begins to block the cavity. The fabrication herein is based on the glass transition temperature of the polymer used, by introducing the molten polymer into the mould cavity at a temperature below the glass transition of the material (240°C). Replication imperfection is a critical issue since the functional reliability of micro component depends on stringent quality requirement. Therefore, the intelligent manipulation of process parameters could avert the danger of compromising quality. The authors have established successfully appropriate processing parameter mix and conditions to achieving replication of micro-feature for certain polymer using Mark 2L Manumould Machine, hence the critical process parameters [1, 4, 10].

The high aspect ratio sample was measured for its weight using a sensitive scale and compared against a benchmarked standard. Qualitative approach was adopted to determine the effect of the different factors using factorial design and to also determine the extent of influence of these factors on response (mass) [3]. The mould temperature and injection pressure has no statistical significant effect on the process parameters for parts produced using polypropylene, while for parts produced using polyethylene, variability is statistically observed to have occurred as a result of process noise.

The influence of some process parameter has been explored so as to ascertain the best conditions to transfer the micron precision of micro structures. Thus, the Design of Experiment approach applied to correlate terms.

2. Experimental setup

2.1. Geometry of part

Figure 1 presents the schematic diagram of the sample part design using NX-IDEAS software, with their critical dimensions. The moulded test part is a high aspect ratio (volume to width) rectangular shaped component with dimensions as follows; 76000μm length, 25400μm width and 800μm thick. It is designed for experimental purpose to serve as test specimen for the study of micro-feature replication of micro-injection moulding.

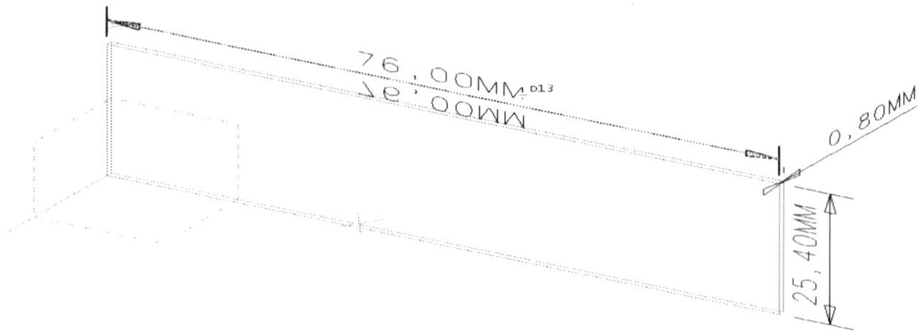

Figure 1. Schematic diagram of the specimen part designed using NX-IDEAS 6Note: The dimensions of the micro-features are shown in millimeter

2.2. Mould manufacture

The mould used to produce the test specimen for the experiment is aluminum mould with a micro structured brass insert attached in order to form a complete set. It is manufactured using the Teesside university micro-machining equipment. The engraving of the micro-feature on the mould surface was done on half of the mould with runner and gate system incorporated to it, while the other half has a plane surface as well as runner and gate system. The surface of the mould lapped flat to a ± 1μm over the 76000μm length polished to a tolerance of 0.2μm finish. Figure 2 showed the detail description of the complete mould design.

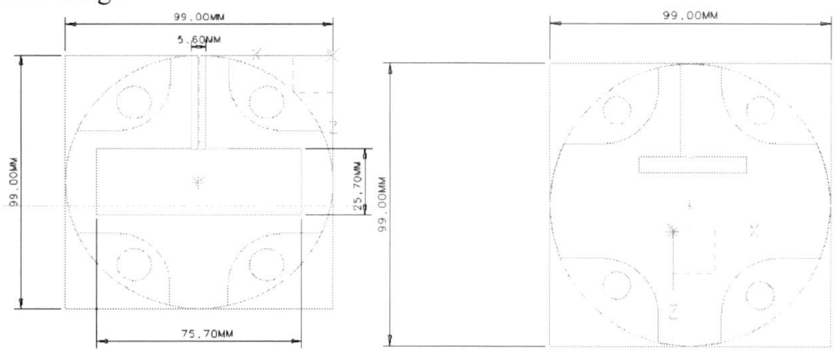

a) Cavity side of the mould b) Plain side of the mould

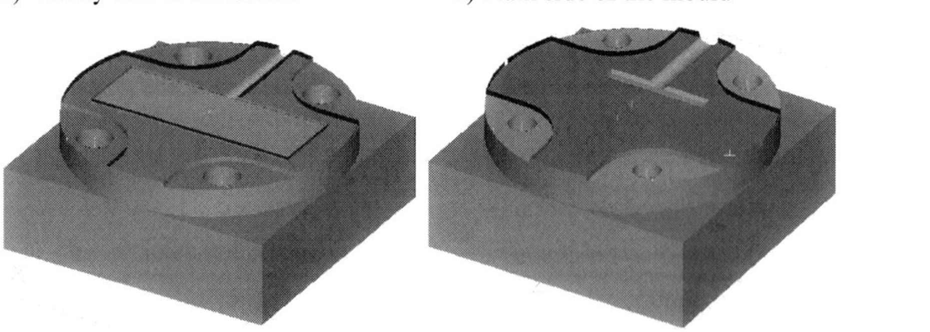

c) Solid model of the cavity side of the mould d) Solid model of the plain side of the mould

Figure 2. Design description of the single cavity mould for experimental specimen fabrication.

Mark 2L Manumould micro-injection moulding machine was used for specimen fabrication while a sensitive weighing machine (Mettler AJ100, AE200 type) was used to measure and determine variability in part mass.

3. Experimental design

A high aspect ratio feature slit component was selected for this study to investigate the influence of some processing parameters on replication quality using the same mould in order to eliminate variability that could occur as a result of difference in shape.

Two types of polymers were used for this experimental study namely; polypropylene (Moplene HP500N homo-polymer grade) and polyethylene (high density polyethylene, HDPE grade). The influence of four process parameters (input factors) of melt temperature(T_p), Injection pressure (P_h), mould temperature (T_m) and holding time (T_h) on the chosen polymers were studied with mass of the part serving as the quality parameter (response factor). In order to determine the effect of these processing parameters, alongside their interaction with each other on the replication potentials of the process to establish an optimize parameters for the process for the given mould and resin so as to achieve parts of the highest standard, the authors adopted the Design of Experiment (DoE) approach under the objectives of familiarization, screening and optimization [3], which aided in the drawing of relevance to the proposition of this study. The screening two level half factorial 8-run (2^{4-1}) experimental plan with three centre points was used for the materials and provided substantial information regarding the single factors and their interaction with each other

The experiments were carried out in a randomized order with the standard operation conditions (the centre point) for the four factors defined using process parameter mix composition as recommended by the authors [3]. The processing parameters and Design of Experiment (DoE) matrix used for the two polymers during the experiment is presented in table 1.

Table 1. Design of Experiment (DoE) and processing parameters matrix for the two polymers.

Polypropylene (PP)									
	Design Matrix				Experimental Matrix				
Process order	T_P (°C)	P_h (bar)	t_h (sec)	T_M (°C)	T_P (°C)	P_h (bar)	t_h (sec)	T_M (°C)	Average Mass (g)
1	-	-	-	-	220	32	5.0	37.0	1.5527
2	+	-	-	+	240	32	5.0	44.0	2.3957
3	-	+	-	+	220	42	5.0	44.0	2.2405
4	+	+	-	-	240	42	5.0	37.0	2.3572
5	-	-	+	+	220	32	8.0	44.0	1.5531
6	+	-	+	-	240	32	8.0	37.0	1.8368
7	-	+	+	-	220	42	8.0	37.0	1.7382
8	+	+	+	+	240	42	8.0	44.0	1.8371
9	0	0	0	0	230	37	6.5	40.5	2.0367
10	0	0	0	0	230	37	6.5	40.5	2.1716
11	0	0	0	0	230	37	6.5	40.5	2.0133
Polyethylene (PE)									
	Design Matrix				Experimental Matrix				
Process order	T_P (°C)	P_h (bar)	t_h (sec)	T_M (°C)	T_P (°C)	P_h (bar)	t_h (sec)	T_M (°C)	Average Mass (g)
1	-	-	-	-	220	32	5.0	37.0	1.7684
2	+	-	-	+	240	32	5.0	44.0	2.2239
3	-	+	-	+	220	42	5.0	44.0	1.9489
4	+	+	-	-	240	42	5.0	37.0	1.9559
5	-	-	+	+	220	32	8.0	44.0	1.7178
6	+	-	+	-	240	32	8.0	37.0	2.0725
7	-	+	+	-	220	42	8.0	37.0	1.7582
8	+	+	+	+	240	42	8.0	44.0	1.8130
9	0	0	0	0	230	37	6.5	40.5	1.8930
10	0	0	0	0	230	37	6.5	40.5	1.9552
11	0	0	0	0	230	37	6.5	40.5	2.2365

After achieving stabilization of process for each run during fabrication, ten samples of the produced parts are randomly selected while their average masses recorded. Upon completion of the experimental plan, the acquired data were processed in Minitab software and analyzed using statistics and regression model supported by the eqution;

$$y = \beta_O + \beta_1 X_1 + \beta_2 X_2 + \beta_3 X_3 + \beta_{12} X_1 X_2 + \beta_{13} X_1 X_3 + \beta_{23} X_2 X_3 + \varepsilon \qquad (1)$$

The coefficients of regression obtained are thereafter plotted in a bar chart, and comparison made between the value of response calculated from the regression model and the corresponding actual experimental value so as to evaluate the accuracy of the fit. Thereafter, the main effects plots and Pareto charts are plotted.

4. Results and discussion

4.1. Main effect of factors

Variability in micro-parts produced was analyzed using the main effects plot of Minitab for the average mass of the two polymers processing parameters. Figure 3 shows the main effects plot for polypropylene factors 3 (a) and that for polyethylene in 3 (b). From the figures, the significant effects of holding time and melt temperature for parts produced using polypropylene are glaring by the fact that the centre point shown with red block is within the confines of the two corners marked with black dot, and is evident even during the experiment. They are so pronounced that part produced at lower level of holding time exhibited signs of swell, not minding the levels of other factors, thus leading to increase in mass of the parts. This suggest the importance of holding time during injection moulding operation as it allows for appreciable setting time of the injected polymer under a given injection pressure prior to commencement of ejection and demoulding of part. Hence, the existence of conformance of practical finding to numerical calculated analysis of fractional factorial design.

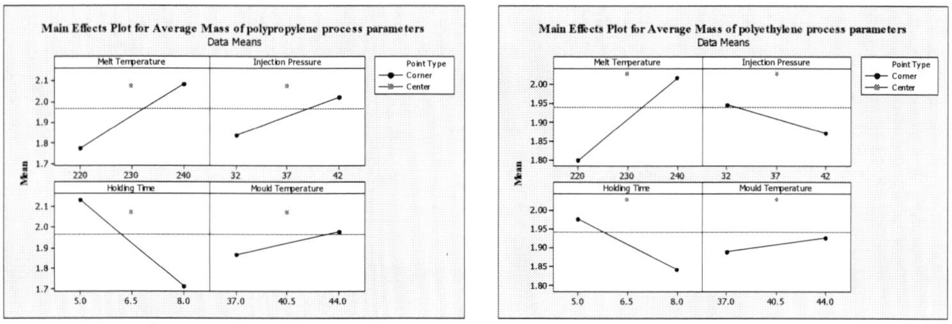

(a) Polypropylene factors plot (b) Polyethylene factors plot

Figure 3. Main effects plot for parts produced using polypropylene and polyethylene.

4.2. Pareto chart of standardised effects

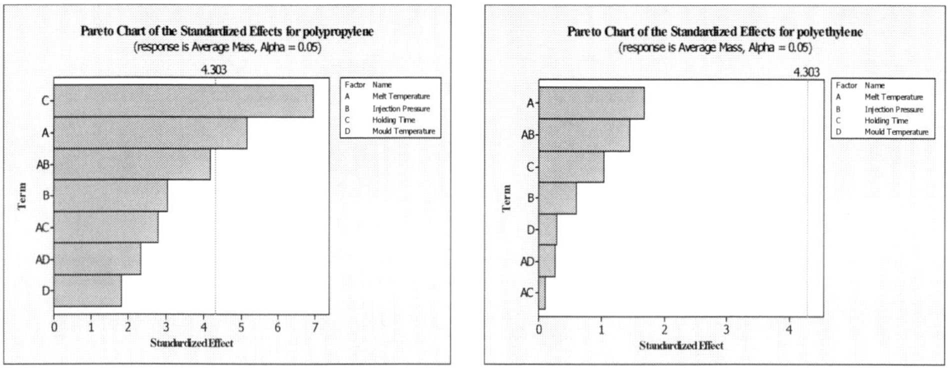

(a) Polypropylene effect plot (b) Polyethylene effect plot

Figure 4. Pareto chart of standardized effect plot for polypropylene and polyethylene moulded part.

The significant level for the analysis of the factorial design is determined by the alpha (α) value generated from Minitab t – distribution graph supported by equation (2)

$$t = 1 - (\text{alpha}/2) \text{ quartile of the distribution} \quad (2)$$

The Pareto chart of standardized effects plot which is a bar chart ordered in decreasing magnitude is shown for both polymers moulded parts in figure 4. The greater the bar magnitude the more the effect its interaction has on the response. Furthermore, the red vertical line in the chart indicates the magnitude of importance a factor is in determining variability, beyond which point, factors and interaction factors becomes statistically significant.

An alpha probability value of 0.05 was used in order to acquire a confidence limit of 95% with a 5% chance of finding effects that does not exist. For the experimental study, the authors observed from the Pareto chart that holding time and melt temperature has a significant effect for parts produced using polypropylene, as the length of the bar extends beyond a threshold value (4.303).

The authors further identified that melting temperature aided the flow of melt into the mould cavity. Hence its significant effect is in line with the findings of the authors [11, 12]. However, there are no observable effect of process parameter on part produced using Polyethylene. Figure 4 shows the factor interaction plot generated for the average mass of both materials using Minitab.

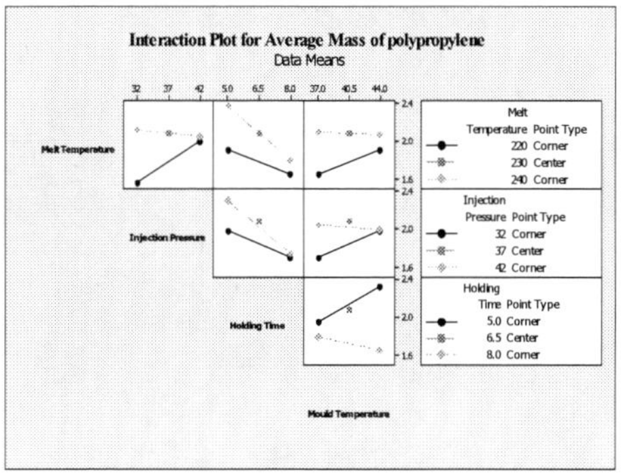

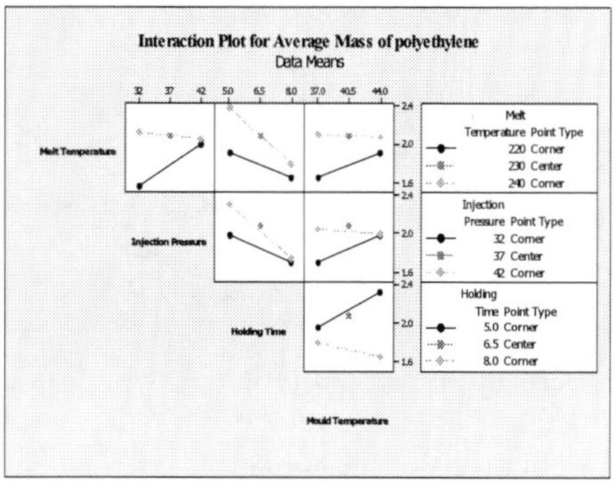

(a) Polypropylene interaction plot (b) Polyethylene interaction plot

Figure 5. Factor interaction plot for the moulded parts.

4.3. Defect in parts produced

Shown in table 2 is some performance defects that occurred during the process of micro-moulding of the test specimen as is evidence in figure 6, and the trouble shooting guide employed as a corrective measure to forestall the occurrence of such

Table 2 List of faults discovered in parts produced.

S/N	Fault	Corrective Measure
1	Flash	Increase clamp force, reduce holing pressure.
2	Shrinkage	Reduce mould temperature to allow for small cooling gap
3	Warping	Less pressure during demoulding of part, allow cooling of part within the mould before demoulding operation.
4	Sink Mark	Increase injection pressure, optimise holding pressure time.

The fabricated parts exhibited signs of defects, among which are shrinkage, warp, flash and sink marks. It is observed that in the fabricated part, shrinkage in dimension is anisotropic and varies in extent according to the crystallinity of the polymers used for the part. During the moulding trials of the fabricated parts, the heated polymers are allowed to cool rapidly at room temperature, thereby affecting the crystallinity of the polymers and accounting for the significant dimensional shrinkage. It is also worth noting that as the cooling rate of polymer becomes smaller, sufficient time is allowed for the polymer chain to recoup and arrange their selves to a higher degree of crystallinity, thereby increasing the percentage shrinkage.

Figure 6. Defect in parts produced.

Warpage was also observed in the fabricated part which could be as a result of the uneven cooling of the moulded part, resulting to residual stress and unbalanced shrinkage. In addition, during de-moulding process, the fabricated part is pulled with a force sufficient enough to separate it from the mould cavity. However, the part is elastically stretched thereby causing a ductile fracture at the weakest point in the part. The result of the fracture elongates the part and makes it visible on the side from which the part is turned off.

Sink marks were observed on parts, which occurrence is attributed to the injection speed, holding time and injection pressure of the operation process. This defect can be intelligently averted by a careful manipulation of the operation process parameters. Among other defects that occurred while fabricating the parts, flash is one of the prominent observable defects which surfaced as a result of insufficient clamp force, reduction in holding pressure and high melt or mould temperature

4.4. Optimum setting

From the experimental study performed to determine the best replication conditions for micro injection moulding of micro-devices using polyethylene and polypropylene as polymer based material, and utilising the Mark 2L Manumoud micro injection moulding machine, the settings for the mould and polymers which best produce the desired part quality was the settings made for process order 5 for the two polymers with the following process conditions
Melt temperature – 220°C
Injection pressure – 32bar
Holding time – 8sec
Mould temperature – 44°C

5. Conclusions

In conclusion, the Main effects plot, Pareto chart and interaction plot for average mass have been established for the experimental setup of this study. They are so graphed to aid the analysis of the factorial design and to create a link with experimental observation on the study of influence of some process parameters on replication using micro injection moulding technique. Based on the findings made, it can be inferred that

1. Holding time and melt temperature are the two vital factors when mass producing using Polypropylene, while for polyethylene, variability occurred as a result of process noise. Therefore, Polypropylene can be optimized with a predictive response of micro-replication.
2. Process parameters alone are not solely responsible for the replication of micro feature during micro injection moulding operation, rather, a whole combination of factors such as polymer type, sensitivity of the moulding machine, etc., are observed.

Therefore, the micro injection moulding capabilities and process technologies studied in this project has formed a technical background for further investigation of micro fabrication technology in industries, using the injection moulding machine to carve out a niche market for manufacturing possibilities.

6. Acknowledgments

The author wishes to express his acknowledgment to the Teesside university graduate research team for their unalloyed guidance, with special acknowledgment to the Lab technicians in the Stephenson building for their experimental contributions to the micro-injection studies.

References

[1] Attia U M, Marson S and Alcock J R 2009 Micro-injection of polymer micro fluidic devices *Micro fluid nanofluid* vol **8** pp 1-28

[2] Chun-sheng C, Shia-chung C, Won-hsion L, Rean-der C and Su-hsia L 2010 Micro injection moulding of a micro fluidic platform *J. of int. communication in heat and mass transfer* vol **37** pp 1290-1294

[3] Eriksson L, Johnson N, Kettaneh-wold N, Wikstrom C and Wold S 2008 Design of Experiment: Principles of Application 3rd ed Umetrics Academy Umea

[4] Giboz J, Coppennex T and Mele P 2007 Micro moulding of thermoplastic polymers: a *review J. of micromechanics and microengineering* vol **17** pp 96-109

[5] Griffiths C, Dimov S and Brousseau E 2008 Micro injection moulding: the influence of runner systems on flow behaviour and melt fill of multiple microscavities *Proc.of the inst. of mech. engrs. Part B engr. Manufacture* vol **222** no. 9 pp 1119-1130

[6] Hoyle R 2007 Creating micro and nano strructures in polymers by moulding *Medical device tech.* vol **18** no. 5 pp 18-20

[7] Jianhong Z, Mayes R, Ge C, Send C P, Hong X, Kay J Y and Ravi N 2002 Micro injection moulding process technical development *Simtech technical report (PT/02/004/APP)*

[8] Ran-Der Chien 2006 Micro moulding of biochip devices designed with microchannels *Sensors and actuator A-Physics* vol **128** no. 2 pp 238-347

[9] Theilade U A and Hansen H N 2007 Surface microstructure replication in injection moulding *Int. J. of advanced manufacturing technology* vol **33** pp 157-166

[10] Zhao J, Mayes R H, Chen G, Chan P S and Xiong Z J 2003 Polymer micro mould design and micro moulding process: Plastics, robber and composites, micromolecular engineering *The int. J. for processing properties and modelling of molecular and composite materials* vol **32** no.6 pp 240-247

[11] Jung J, Choi G and Kim D 2004 Experimental study on spray etching in micro fabrication of lead frame *KSME Int. J.* vol **18** no. 12 pp 2294-2302

[12] Sha B, Dimov S, Griffiths C and Packianather M 2007 Investigation of micro injection moulding: Factors affecting the replication quality *J.of materials processing technology* vol **187** no. 2-3 pp 284-296

Studies on incorporation of exfoliated bentonitic clays in polyurethane foams for increasing flame retardancy

J Quagliano* and Irma Gavilán García**

*Organic Synthesis Division, Applied Chemistry Department, Instituto de Investigaciones Científicas y Técnicas para la Defensa (CITEDEF), Ave Juan Bautista de Lasalle 4397, Buenos Aires, Argentina.

**Facultad de Quimica,Universidad Nacional Autonoma de Mexico

Corresponding author: E-mail:jquagliano@citedef.gob.ar

Abstract. In this contribution we report the results of studying the incorporation of exfoliated bentonitic clays into polyurethane foams. A suspension in water of a sodium bentonite from Argentine Patagonia was interchanged with cetyl trimethyl ammonium bromide (CTAB) for 4 h at 80°C, rendering an exfoliated clay, which is nanometric in only one dimension. This nanoclay, when dispersed in the polyurethane, resulted in the same fire retardancy rating (UL-94) than when polyurethane was treated with a commercial nanoclay. Scanning electron microscopy (SEM) at low augmentations of polyurethane samples treated with the synthethized nanoclay (2,5% w/w) showed no differences respect to untreated polyurethane, except for the irregularity of void edges.

1. Introduction

The use of nanoclays in polymers has been subject on numerous investigations, including their preparation, properties and uses of this new class of materials. Nanoclays are nanometric in only one dimension: they are presented in the form of sheets of few nanometeres thick to humdred thousands nanometers long [1]. In particular, the effect of nanoclays in polyurethane was investigated utilizing X-ray diffraction and SEM [2]. Exfoliated montmorrillonite was incorporated to polyurethane systems by in-situ polymerization, enhanced the thermal stability and flame retardancy of PU significantly [3]. Also, the nanoscale morphology of segmented polyurethane (SPU) nanocomposites containing various proportions of organomodified montmorillonite (MMT) was reported [4]. In this contribution we report the results of studying the incorporation of commercial and laboratory prepared nanoclays in a TDI-based polyurethane by incorporating them in one of the monomers at high shear mixing rate (*in situ* intercalative polymerization).

2. Materials and Methods

Sodium bentonitic clay (*Bentonita del Lago*, from Argentine Patagonia, gently provided by Argentine Geological Service) was interchanged with cetyl trimethyl ammonium bromide (CTAB) by heating bentonite suspension in distilled water at 80°C for 4 hours. This product was characterized by X-ray diffraction (SRD) utilizing a Phillips PW-3710 diffractometer, Cu-Kα. On the other side, the commercial nanoclay is exfoliated by a protected high-pressure extrusion process (Cloisite 20B, provided by Sothern Clay Products, US). Both nanoclays were incorporated into *in situ* blown samples of polyurethane, synthetized by reaction of a polyether polyol with hydroxyl number 56 and toluene diisocyanate (TDI). Cloisite 20B was incorporated first to the polyether polyol plus catalyst (1,25 and 2,5% w/w) by mixing with a 3 cm-semicircular blade impeller at 2000 rpm for 1 minute. Inmediately afterwards, TDI was incorported to this mixture with agitation at 1000 rpm until foaming starts. The rising foam was poured in a cardboard box (15 x 20 x 30 cm) and left overnight until it is completely cured at ambient temperature. Scanning electron microscopy was performed on these samples (x 600). Thermogravimetric studies were performed as well to evaluate thermal resistance of the modified polyurethanes respect to control ones.

3. Results

Local sodium bentonite was subjected to interchange with cetyl trimethyl ammonium bromide (CTAB) by heating a bentonite suspension at 80°C for 4 hours. To verify exfoliation, or achievement of clay platelets separation, X-ray studies were performed. The following Figure shows that interchanged bentonite experimented a shift to lower angles at low angles in its diffraction signals respect to crude bentonite, indicating that exfoliation was attained [1,5].

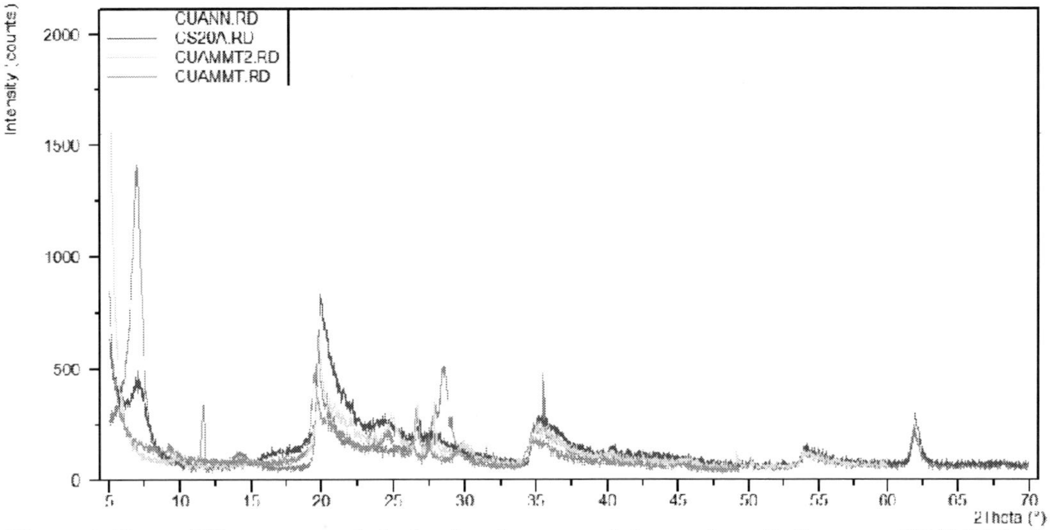

Figure 1. X-ray diffractogram of obtained and commercial nanoclays. References: CUANN (red): sodium bentonite (crude). CS20A (blue): Cloisite 20A (Southern Clay Products). CUAMMT (pink): interchanged bentonite with CTAB. CUAMMT2 (green): interchanged bentonite with CTAB, grinded in a marble mortar.

Interchanged bentonite with CTAB showed a diffraction signal at 2θ=5° respect to crude bentonite (2θ=7,4°). This is a result of the increased separation of clay platelets (increased interlayer spacing) which displaces the peak to low angles, according to Bragg´s Law: $\lambda=2dsin\theta$, where λ corrresponds to the wavelength of the X-ray radiation used in the diffraction experiment, d the spacing between diffractional lattice planes and θ is the mesured diffraction angle or glancing angle.

When subjected to the test UL-94 samples with the higher load of nanoclay displayed a better performance resulting in a V-1 rate (being V-0 the rating for maximal achievable performance, according to this test). However, this is a relative low load of fire retardant additive compared to standard polymer systems. Table 1 shows the results obtained:

Table 1. Results of UL-94 index for two nanoclay-treated polyurethane samples. 1. Exfoliated by both ammonium quaternary interchange and zirconia bead milling for 48 hours.

Sample	UL-94 index
PU + *Cloisite* 20A 1,25%	V-2
PU + *Cloisite* 20A 2,5%	V-1
PU + exfoliated bentonic clay[1]	V-1
PU control	V-2

Respect to thermogravimetric studies, nanoclay additioned polyurethanes displayed the same behaviour of control samples, loosing 50% of initial sample weight at around 350°C: irregular void edges (600x, pictures not shown). These behaviour could not be assigned to any particular characteristic of the process for treating foam samples.

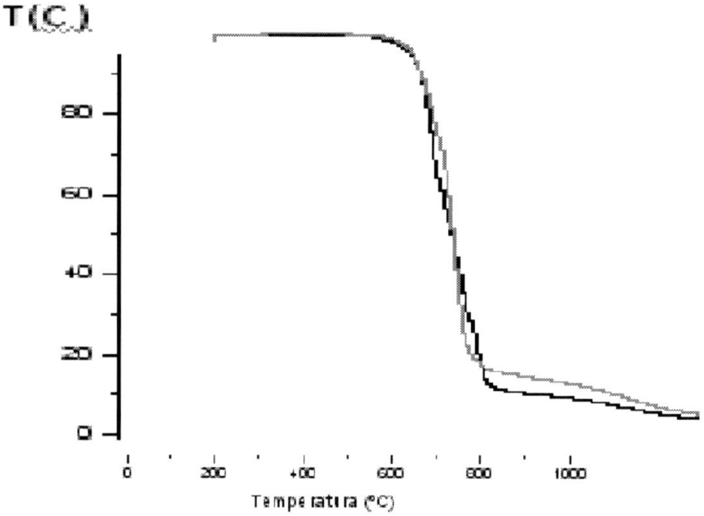

Figure 2. Thermogravimetric studies of control polyurethane sample (black line) and interchanged bentonite nanoclay-treated polyurethane (red line).

4. Conclusions

The addition of commercial nanoclay resulted in an improvement of fire retardancy UL-94 rating respect to untreated polyurethane samples. The local bentonitic clay exfoliated by the interchange procedure and incorporated at 2,5% w/w to the polyurethane gave the same rating as commercial nanoclay. This result suggested that the interchange procedure for exfoliating bentonitic clay gave comparable results respect to the exfoliation by high-pressure extrusion, as regards to its utilization as an active additive for imparting fire retardancy to TDI-based polyurethane. Thermal decomposition of interchanged bentonite nanoclay-treated polyurethane started almost at the same temperature as with untreated polyurethane, indicating that fire retardancy mechanism was not related to an increased temperature resistance but to other mechanism.

5. Acknowledgments
To Eng. Emilia Rapp from the Dpt. of Materials Sciences, Citedef, for performing the X-ray studies.

References
[1] Dubois P. y Alexandre M. 2000. "Polymer-layered silicate nanocomposites: preparation, properties and uses of a new class of materials" *Materials Science and Engineering*, 28, 1-63.
[2] Stefani P., Esposito L., Manfredi L. and Vazquez A. 2005. "Mechanical properties of montmorillonite filled polyurethane foams" *Conference on Science and Technology of Composite Materials COMAT* 169-170, Buenos Aires. ISBN: 987-544-162-7.
[3] Huang G., Jao J., Li Y., Han L. and Wang X. 2010. "Functionalizing nano-montmorillonites by modified with intumescent flame retardant: Preparation and application in polyurethane"*Polymer Degradation and Stability* 95 245–253.
[4] Mondal M., Chattopadhyay P., Setua D., Chas N. and Chattopadhyay S. 2006. "Influence of surface-modified nanoclay on the self-organized nanostructure of segmented polyurethane composites" *Journal of Fire Sciences*, 24, 5, 345-364.
[5] Moreno M., Gonzalez G., Sotomayor Torres C. and Benavente E. 2006. "Modified bentonite by intercalation with neutral and cationic surfactants" CONAMET/SAM 2006, VI Binational Congress of Metallurgy and Biomaterials, Chile.

IOP Conf. Series: Materials Science and Engineering **40** (2012) 012046 doi:10.1088/1757-899X/40/1/012046

Structural and nitrogen gas adsorption-desorption studies of bragg grating waveguide fabricated on porous silicon nanostructure

A A S M Radzi[1,2,*], S F M Yusop [1,2], M Rusop[1,3] and S Abdullah[1,2]

[1]NANO-SciTech Centre, Institute of Science
[2]School of Physics and Material Studies, Faculty of Applied Sciences
[3]NANO-ElecTronic Centre, Faculty of Electrical Engineering
Universiti Teknologi MARA, 40450 Shah Alam, Selangor D.E, Malaysia

*Email: afifsafwan7788@gmail.com

Abstract. Fabrication and structural analysis of Bragg grating waveguide on porous silicon were presented in this work. The Bragg grating structure was directly etched on porous silicon. Multilayer structure of porous silicon with alternating values of thickness and refractive index were determined. Nitrogen gas adsorption-desorption is a non-destructive method to determine the fabricated Bragg grating's interaction with gases. It is known that porous silicon is a high potential in-situ gas sensor due to the strong tube capillary effect and enormous surface area for molecular binding and detection. Thus, specific surface area and porosity of fabricated Bragg grating are also in the scope of this study. Good agreement of pore size distribution and multilayer thickness/refractive index achieved from adsorption-desorption technique and ellipsometer measurement respectively when comparing with estimations from field emission scanning electron microscopy. This Bragg grating waveguide has a big potential as gas sensor in many areas such as aerospace and in industries particularly.

1. Introduction

Porous silicon (PS) has been intensely investigated due to its optical and electrical properties. Since the discovery of room temperature photoluminescence emission (PL) by Canham in 1990 [1], substantial amount of research has been conducted regarding the potential applications of PS. Thin film porous silicon has several advantages such as very high internal surface area [2]. The porosity of PS determines its refractive index. By tuning the porosity during etching, alternating refractive index is attainable and proving that PS has the ability of tuneable refractive index [3]. Gas molecules passing through the porous surface will experience adsorption process. During this process, molecules will relax and combine together inside pore and this condition can alter the optical constant for that layer of porosity [4]. PS has several applications in photonics and optics such as Bragg grating waveguide (BGW), Fabry-Perot interferometer and Optical microcavity [5, 6]. Surface study is very crucial for PS since PL and UV-Vis spectra investigations shows that surface morphology and crystallites structure would influence the optical signal. Quantum confinement theory of PL emission still highly arguable but from literature [7] proves that they are highly affiliated.

Measuring the specific surface area of PS will be challenging because it possesses complex surface orientation and requiring complicated method. Nitrogen gas surface adsorption and desorption from

Brunauer-Emmet-Teller (BET) would be the best measurement for calculation of surface area, pore size distribution and volume-pore distribution [8]. Capillary effect also occurred due to cylindrical pore structure of the skeleton silicon. This effect is very interesting because it realise the gas sensing property of fabricated BGW. During adsorption process, layers of adsorbate will form on the skeletal structure of pores and fill the pores completely.

When desorption process occur, the pressure inside pores will increase while temperature will be extremely low, which is the best condition for adsorbates to change phase to liquid. The surface tension between those adsorbates will be less prompting liquid bridge phenomena within pores. Formation of liquid will happen if the pore depth is significant which is not shallow [9].

2. Experimental Details

PS was prepared using the electrochemical etching or anodisation on crystalline silicon wafer. Electrolyte for the chemical etching reaction is a mixture of hydrofluoric acid (HF 48%) and absolute ethanol (C2H5OH 99%) solutions. Mixture's ratio is composed of HF:Ethanol of 1:1. The silicon substrate used was highly doped p-type silicon wafer <1 0 0> oriented, and 525 μm thickness as the substrate. Electrochemical cell used was Teflon cell with aluminium plate as the base. A tungsten rod served as cathode of the electrolytic cell while aluminium plate as anode. Usually bottom of sample was coated first with aluminium, but we just use aluminium contact as replacement for the base of anode. Modulation of etching time and etching current density during the anodisation would materialize the formation of multilayer porous structure. 2 types of samples prepared for this study. Type 1 is PS with 20 layer alternating porosity thin films and type 2 is 40 layers with the same characteristics. Both types were first etched on the polished surface with current density of 30 mA for a period of time and continued with second current density of 10 mA with the same period of time. The etching times for each layer for both types are 10 seconds, 20 seconds and 30 seconds. Nitrogen adsorption-desorption isotherm process was performed at temperature 77.4 K. The process started by degassing followed by adsorption until the pressure reaches atmospheric pressure. Next the pressure drop back to high vacuum gradually to make way for desorption process to occur. For structural view, samples were characterised by Field Emission Scanning Electron Microscopy (FESEM JEOL JSM 7600F). For ellipsometric measurement, Variable Angle Ellipsometer L 116S (Gaertner Scientific Corp.) with fix wavelength He-Ne laser source of 632.8 nm was used. PL intensity (Horiba Jobin Yvon 79 DU420A-OE-325) and Reflectance Spectra from UV-Vis (PerkinElmer Lambda 750). For nitrogen adsorption-desorption measurement, BET (Quantachrome Autosorb-6B) was used.

3. Results and discussion

3.1 Structural analysis

3.1.1 Multilayer thickness and refractive index from ellipsometer measurement. Multilayer PS constructed of high and low refractive index layers. To determine the refractive index of each layer, ellipsometer measurement is the best method available. This measurement was base on effective medium approximation (EMA) of the ellipsometer.

Table 1. Thickness and refractive index values for Type 1 samples.

Etching time, (seconds)	Thickness, t (nm)	Refractive index, n	
		n_H	n_L
10s	t_1=164, t_2=163	1.495	1.669
20s	t_1=256, t_2=266	1.358	1.584
30s	t_1=336, t_2=312	1.478	1.758

t_1 = thickness for the first alternating layer, t_2= thickness for the second alternating layer.
n_H = refractive index of high porosity layer, n_L= refractive index of low porosity layer.

Table 2. Thickness and refractive index values for Type 2 samples.

Etching time, (seconds)	Thickness, t (nm)	Refractive index, n	
		n_H	n_L
10s	t_1=163, t_2=166	1.412	1.538
20s	t_1=284, t_2=293	1.453	1.847
30s	t_1=337, t_2=372	1.366	1.528

For 10s etching time, both types of sample have identical thickness values. This pattern is followed by 20s and 30s where the difference is marginally low between 1-20 nm. Refractive index pattern is obvious that longer etching time such as 30s will produce lower refractive index due to higher porosity structure.

*3.1.2 Surface morphology and thickness measurement from FESEM.*Surface study is important for this investigation since for an optical gas sensor, interaction between gas molecules and optical signal happens first on the surface. The pore size measurement is base on IUPAC standard for pore size [10]. From figure 1, it is clear that shorter etching time will produce a more uniform and well distributed pores with pore size of up to micropore size. Average pore size of the samples is in mesopore size, between 10-30 nm pore diameters. This is acceptable because to obtain high micropore size distribution is difficult with etching technique since this method is not micro scale precise during anodisation.

Figure 1. Average pore size diameter and surface morphology of a) Type 1 10s, b) Type 1 20s, c) Type 2 20s.

From cross section images, we can directly relate the thickness of FESEM measurement and ellipsometer. The thickness values from ellipsometer are almost accurate with estimation from FESEM hence giving a good agreement for both of the measurement methods. Figure 2a shows the average thickness for sample type 2 with etching time of 10s. The thickness measured is 160 nm, which is almost accurate comparing with ellipsometer measurement. This accuracy also applies for figure 2b and 2c where the etching time was increased and layer thickness retrieved is 230 nm and 310nm respectively. The realisation of multilayer structure is coherent with the crystallinity of the porous structure. Higher porosity layer will possess larger void area and vice versa. Due to different crystal composition, high contrast layers of alternating porosity can be seen and thicknesses were verified.
From FESEM estimations, whole thickness for each type 1 sample is 3.3 μm for 10s, 5.2μm for 20s and 6.5μm for 30s. For type 2, sample 10s is 6.6 μm, 20s is 11.5μm and 14.1μm for 30s.

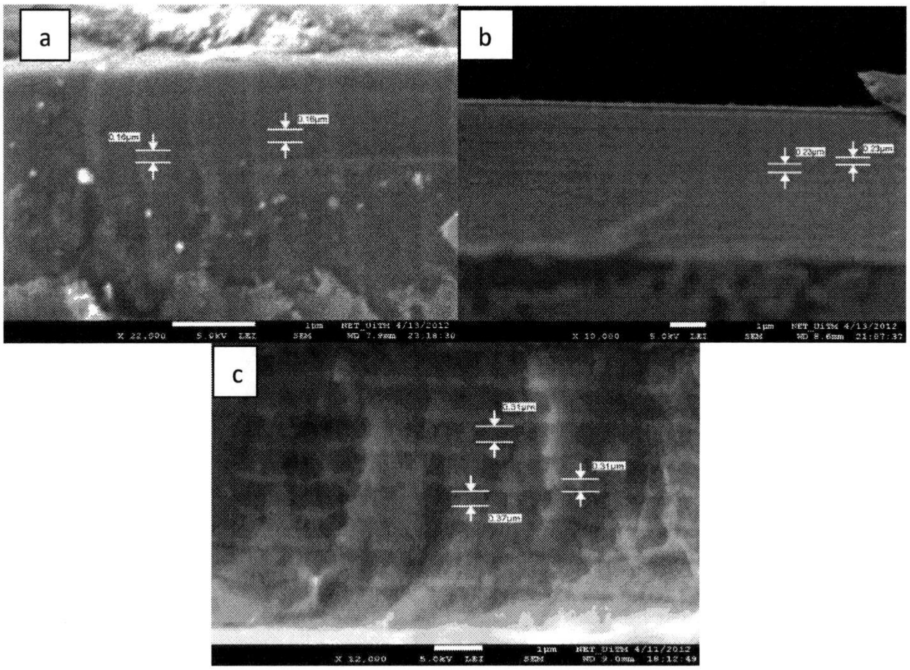

Figure 2. Cross-sectional view for a) Type 2 10s, b) Type 1 20s, c) Type 2 30s.

3.2 PL and Optical Properties.

Photo emission from PS surface has distinguished itself from bulk silicon in both optical and electrical properties. Bulk silicon has an indirect band gap, perturbing the photo emission to occur during excitation since it was converted into quantised lattice wave. PS on the other hand convert the excitation into photon energy and becomes a photoemitter. The efficiency and performance of photo emission from PS are widely studied mainly for its application such as in light emitting diode (LED). Structural morphology of PS and thin film layer contribute to the quality of emission. From Figure 3, the wavelength shift patterns differ very little from each type of samples. For type 1, blue shifted peaks are observed base on increasing etching time. This happen because porosity of longer etched PS will increase affecting the optical phonon response of the porous skeletal structure.

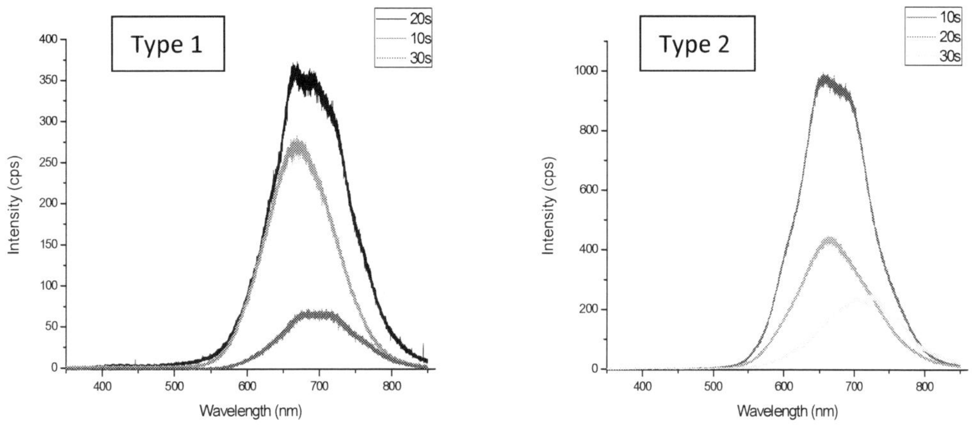

Figure 3. PL emission from sample Type 1 and Type 2.

Narrow skeletal structure will increase the quantum confinement effect on the surface thus better PL emission observed. For type 2, the intensity of PL emission decrease with respect to longer etching time. This happen because when light propagates through the multilayer structure, the alternating refractive index will confine it producing total internal reflection within the lattice. Since the light now confined, it is more prone to change its optical path length and moving deeper instead of reflecting back in-phase to the outer layers. The broad full width half maximum (FWHM) observed are referred to the grain size on surface of PS. Broader FWHM indicates that skeletal peaks are broader and wider.

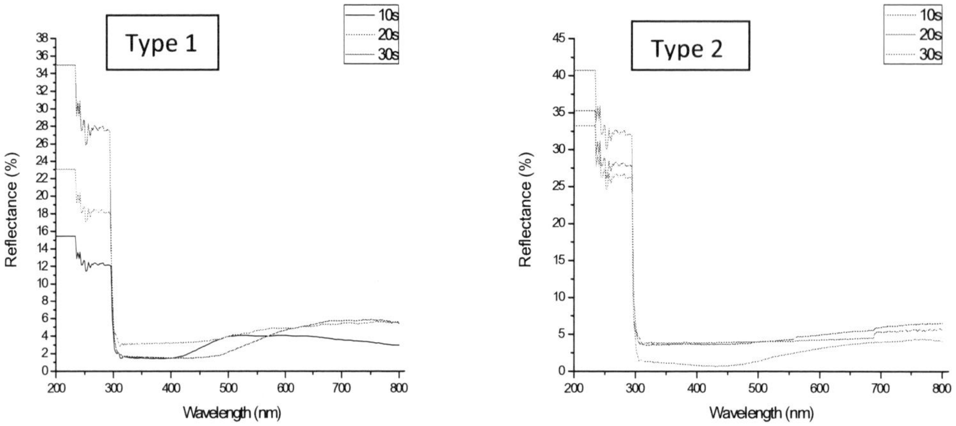

Figure 4. UV-Vis spectra from sample Type 1 and Type 2.

The reflectance spectrum from figure 4 shows that both types have the same low reflectance value in visible spectrum region. This is normal for a multilayer mirror structure because as explained earlier, light reflecting from the layers is confined within the structure itself, thus permitting very low amount of light to reflect and pass through the outer layers. From both types, 30s etched samples have the lowest reflection value and this is further strengthen the assumption that longer etching time will produce deeper thickness. This may cause the light itself trapped and changing its optical path towards deeper layers.

3.3 Gas interaction and surface area measurement.

N_2 adsorption-desorption from BET method was adopted for this measurement. From table 3, Langmuir surface area obviously increased with increasing multilayer thickness for each sample. With thicker thickness, the porosity will go further deeper and increasing the surface area for every pores. Total pore volume strongly related with surface area inversely [11].

Table 3. Surface area and pore volume results.

	Langmuir Surface Area (m^2/g)	Total Pore Volume ($x10^3$ cc/g)	BJH Cumulative Adsorption Surface Area ($x10^2$ m^2/g)
Type 1 30s	5.227	7.324	1.412
Type 2 10s	6.099	3.350	6.042
Type 2 20s	9.976	2.399	3.097

According to the isotherm plots in figure 5, it clearly indicates the existence of shallow pore features and low of micropores distribution. By presence of huge differences between adsorption and desorption in second region $0.3 < P/Po < 0.7$ enough to indicate that an existence of mesoporous with shallow pores depth and broad of pore diameter. DFT plot in figure 6 is supporting this evidence.

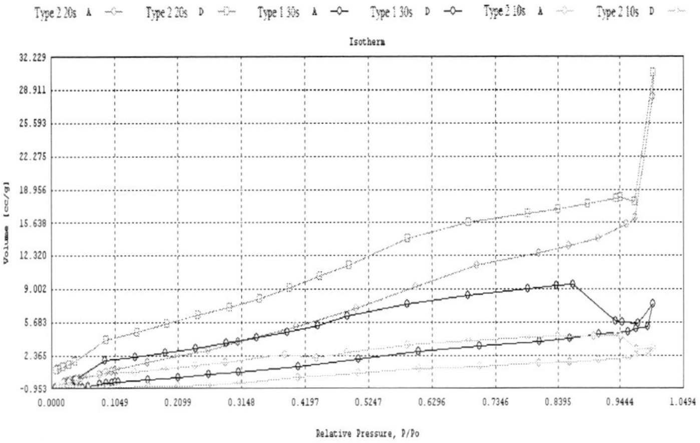

Figure 5. Isotherm adsorption-desorption for samples of Type 2 20s, Type1 30s, Type 2 10s.

Therefore, based on this matter, the Multipoint of BET method could not be figured out and single point is the best selection to determine the best pore of those isotherm peak points. The isotherm plot is classified as type IV hysteresis loop due to cylindrical shape pores and high mesopores distribution on the surface as seen in BJH plot in figure 7. From the isotherms we can determine the condensation effect within pores. But due to very high differences from adsorption-desorption, this effect is very small and difficult to occur. The reason why is because shallow pores will provide limited space for molecules to relax and bind together. When desorption pressure applied, molecules will easily be expelled from pores and the huge pressure difference will resulted in a negative pore volume values.

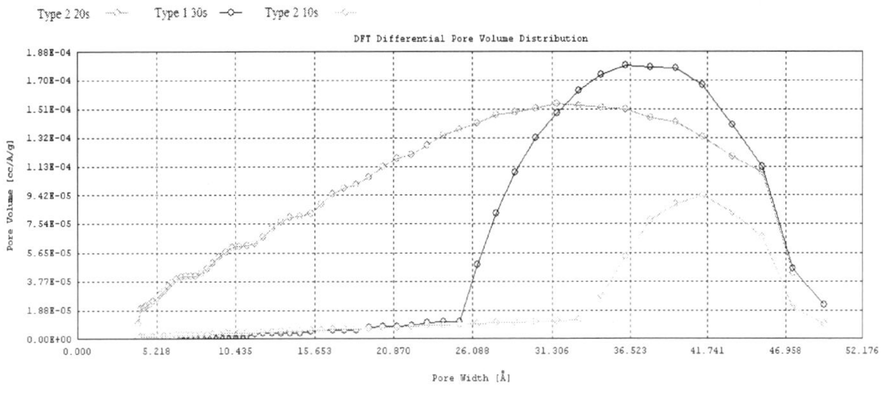

Figure 6. DFT pore size distribution for samples of Type 2 20s, Type1 30s, Type 2 10s.

DFT plot from figure 6 is showing the pore width distribution of pores in both micropore and mesopore sizes, while BJH plot in figure 7 displaying the distribution in mesopore size and above. In figure 6, the large broadening from sample type 2 20s indicates that high population of microsize pores on its surface compared to other samples.

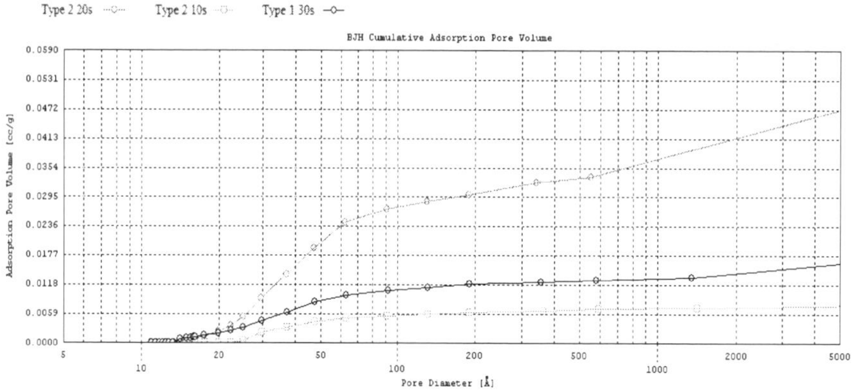

Figure 7. BJH pore diameter distribution for samples of Type 2 20s, Type1 30s, Type 2 10s.

The distribution is correlating with surface image of FESEM that showing a good distribution of mesopores and small numbers of micropores. Other samples showing discrete number of micropores and good distribution of mesopores in high pore volume region. In figure 7, the mesoporous structure is more and possesses higher pore volume depends on the etching time. This concludes that with higher etching time, the pore distribution will be better but will cause the micropore structures to decrease as a consequence.

4. Conclusion

Multilayer PS structure successfully fabricated. The thickness of each layer was determined and measurement from ellipsometer is in a good agreement with the estimation from FESEM. PL excitation shows that the surface structure would influence the PL shift and intensity blue shifted peaks are observed with increasing etching time. Multilayer structure will reflect low amount of light passing through it with light reflectance of around 2~10 %. The pore surface area and volume are inversely related as obtained from the BET measurement. Surface areas recorded are increase with increasing etching time as the PS thickness also increase. This work relates the surface study of PS with optical properties together with the gas sensing ability for further application as an optical gas sensor.

Acknowledgement

The author would like to thank Universiti Teknologi MARA (UiTM) Malaysia and Ministry of Higher Education Malaysia for the financial support under grant ERGS 600-RMI/ERGS/5/3(21/2011). Special thanks to my family and also to Nano-Optoelectronics Research and Technology Laboratory USM for the ellipsometer measurement.

References

[1] L T Canham, 1990 *Appl. Phys. Lett.* **57** 1046-4048
[2] L M Peter, D J Ripley, and R I Wielgosz, 1995 *Appl. Phys. Lett.* **66** 2355–2357
[3] H Munder, C Andrzejak, M G Berger, T Eickhoff, H Luth, W Theiss, U Rossow, W Richter, R Herino, and M Ligeon, 1992 *Appl. Surf. Sci.* **6** 56–58
[4] R B Bjorklund, S Zangooie, and H Arwin, 1996 *Appl. Phys. Lett.* **69**, 3001–3003
[5] G Vincent, 1994 *Appl. Phys. Lett.* **64**, 2367–2369
[6] V Torres-Costa, R J Martin-Palma, and J M Martinez-Duart, 2004 *Appl. Phys.* A **79**, 1919
[7] R S Dubey and D K Gautam, 2009 *Journal of Optoelectronic and Biomedical Materials* **1** 8-14
[8] H Elhouichet, M Oueslati, 2002 *Appl. Surf. Sci.* **191** 11-19

[9] Lei Z K, K Y-Lan, Q Yu, H Ming, C Hao, 2004 *Chinese Physical Society* **21** 1377-1381
[10] O Bisi, S Ossicini, L Pavesi, 2000 *Surface Science Reports* **38** 1-126
[11] J Sterte, 1986 *Clays and Clay Minerals,* **34** 658-64

IOP Conf. Series: Materials Science and Engineering **40** (2012) 012047 doi:10.1088/1757-899X/40/1/012047

Relationship between the Morphology and Physico-Mechanical Properties of Polyethylene/Clay Nanocomposites

R Rezanavaz and M K R Aghjeh

Institute of Polymeric Materials, Chemical Engineering Department, Sahand, University of Technology, Tabriz, Iran

Email: roya.rezanavaz@gmail.com

Abstract. Rheology, morphology and thermal behavior of HDPE/Clay nanocomposites were studied. The mechanical properties of these materials including tensile and creep behaviors were also taken into account. Different PE-g-MA samples with different MA contents and different rheological properties were laboratory synthesized and used as compatibilizer of PE and Clay. The results of X-ray diffraction in conjunction with the results of transmission electron microscopy (TEM) analysis indicated that, increasing in degree of grafted MA increases the penetration of PE chains onto the clay interlayer leading to intercalation and exfoliation. The intercalated and especially exfoliated nanocomposites exhibited higher viscosity and elasticity in particular at low frequency ranges, showing the formation of three dimensional networks with high interfacial interaction. The presence of such a network was evidenced by tanδ studies where the pseudo-solid like behavior was observed for exfoliated nanocomposites. From these results it was demonstrated that the linear viscoelastic properties of the nanocomposites have a reliable sensitivity to the extent of clay dispersion and they can be used as indirect method in the prediction of the morphology and therefore thermal and mechanical behavior of the nanocomposites. Incorporation of clay decreased the onset temperature of degradation due to the Hofmann elimination reaction, but increased remarkably the mid-point of the degradation temperature. Our laboratory synthesized intercalated nanocomposites displayed higher thermal stability than those of exfoliated samples. This was attributed to the barrier effect of clay layers to oxygen and volatile products, during the degradation of part of polymer chains which was intercalated in clay interlayer. Interestingly the results showed that the effect of Hofmann elimination reaction which decreases the onset temperature degradation of modified clay nanocomposites, can effectively be eliminated using a proper antioxidant in the mixing process of nanocomposites. The results showed that, incorporation of clay increases the young modulus and yield stress but decreases the elongation at break. In agreement to the results of thermal behavior, intercalated nanocomposites showed higher increase in properties than those of exfoliated. The PE nanocomposites showed lower creep strain and higher creep recovery than those of unfilled matrixes.

1. Introduction

Polymer/clay nanocomposites exhibit better mechanical, thermal, gas barrier and other physicochemical properties than neat polymer [1]. Major differences in behavior between conventional and nanostructured materials result from the fact that the latter have much larger surface (or interface)

area per unit volume. Since many important chemical and physical interactions are governed by surfaces, a nanostructured material can exhibit substantially different properties from a larger-dimension material of the same composition.

Successful nanocomposites have been made with high surface energy polymers such as polyamides, [2–4] in which considerable adhesion between the polymer and clay generated by the polarity and hydrogen-bonding capacity of these polymers favors the formation of an exfoliated structure. However, because of the poor adhesion between clay and polyolefins such as polyethylene, very little success has been achieved to date in preparing polyethylene-clay nanocomposite using conventional melt mixing. Researchers have attempted to enhance the interaction between clay and polyethylene by modifying clay with alkylammonium salts and utilizing a compatibilizer such as maleated oligomer in the mixing process. Still, most of the polyethylene-clay nanocomposites made by conventional melt-mixing exhibit only intercalated or semi exfoliated morphology. [5–12]

It has been generally established that the exfoliation of clay layers provides maximum interfacial interaction and is favorable for matrix reinforcement. However, for the polyethylene-clay system, the relationship between the properties, in particular mechanical properties, and morphology is rather complicated and inconsistent in open literature. It is known that the addition of clay can generally enhance yield and tensile strength of a polymer, but will generally decrease the ductility, [13-15] but zhang and sundararaj reported that all the mechanical properties including ductility increase with clay loading to LLDPE/PEMA. They found that the enhanced miscibility between LLDPE and PEMA, and more importantly, interfacial interaction between clay, PEMA, and LLDPE, are responsible for the distinct improvement in all the mechanical properties of the composite, and in particular for the marked improvement in ductility. [16]

In most studies on PE-PEMA-clay system, the effect of PEMA was found to be significant for the clay exfoliation at low clay loadings. The drastic discrepancy in mechanical property trends of different systems indicates that the relationship between the properties and morphology of polyethylene/clay nanocomposites depends on not only morphology but also on the particular polymer system. Therefore, it is imperative to look at different PE materials and compatibilizer systems to obtain more insight into the relationship between the polyethylene-clay structure and composite properties. [16]

In our previous study [17], rheology, morphology and thermal behavior of PE nanocomposites prepared using different PE-g-MAs, were studied. The results showed a strong relationship between the morphology and viscoelastic behavior and also thermal behavior. The objective of the present work was to study the effect of morphology on mechanical properties including tensile, creep and creep recovery.

2. Methods

Two different grades of polyethylene HDPE-5218 (MFI=20 g/10min; 2.16 Kg, 190°C) and HDPE-3840 (MFI=4.0 g/10min; 2.16 Kg, 190°C) both from Tabriz Petrochemical Company, Tabriz, Iran were used. They used as the base materials for synthesis of PE-g-MA compatiblizer and preparation of nanocomposites, respectively. Dicumyle peroxide (DCP) as a free radical initiator from AkzoNobel, Maleic Anhydride (MA) from Merck and Irganox 1010 from Ciba Geigy Co, as an antioxidant were used as received. The used montmorillonite was dimethyl dioctadecyl ammonium modified montmorillonite (CloisiteTM 15A) from Southern Clay Products Inc.

Synthesis of PE-g-MA compatiblizers- The grafting experiments were carried out in a 55-cc, roller-type rotor internal mixer (Brabender W50EHT).Using different reactants concentration and also different processing conditions, three compatiblizers varying in MFI values and degree of grafting were prepared. The specifications of compatiblizers are summarized in Table 1.

Preparation of nanocomposites- MMT was dried in a vacuum oven at 80°C for 24h in order to remove the absorbed moisture prior to mixing. Nanocomposites of PE and reference samples were prepared, using the same internal mixer at 170°C with a rotor speed of 60 rpm with compositions summarized in Table 2. The PBx codes represent the blends of PE/PE-g-MA (reference samples) and

PNCx codes correspondence to PE/PE-g-MA/Clay nanocomposites. For preparation of all nanocomposites, at first PE was filled into the chamber and after reaching the process to its steady state condition, PE-g-MA was added and after 3 min mixing, the clay was incorporated to the melt mixture and the process was continued up to 20 min. In all samples the clay content and the PE-g-MA/clay ratio was kept constant at 3 wt % and 5, respectively.

Table1. Specifications of synthesized compatiblizers.

Sample code	Degree of grafting (wt %)	MFI[a] (gr/10 min)	MFI[b] (gr/10 min)
1	1.129	0.73	4.59
2	1.026	1.19	7.2
3	0.532	4.59	16.69

a: 2.16 Kg, 190°C b: 5.0 Kg, 190°C

Table 2. Compositions of different nanocomposites and reference samples.

Samples	PE (wt %)	PE-g-MA type	PE-g-MA (wt %)	Clay (wt %)	Antioxidant (wt %)
PE	100	-	0	0	0
PE-A	100	-	0	0	0.1
PE-M	97	-	0	3	0.1
PB1	85	1	15	0	0.1
PNC1	82	1	15	3	-
PNC1-A	82	1	15	3	0.1
PB2	85	2	15	0	0.1
PNC2	82	2	15	3	-
PNC2-A	82	2	15	3	0.1
PB3	85	3	15	0	0.1
PNC3	82	3	15	3	-
PNC3-A	82	3	15	3	0.1

3. Results and discussion

Fig. 1 shows the variation of the mixing torque versus time obtained for neat HDPE-3840 processing in the internal mixer with and without antioxidant. The results show that, for sample without antioxidant, the mixing torque rises after about 6 min, which is a consequence of thermomechanical degradation of the PE chains, leading to chain branching and/or crosslinking and therefore increasing in mixing torque. On the other hand, it can be observed that, using antioxidant prevents this thermomechanical degradation. These results show that, the long time processing (20 min) at high temperatures has an influential effect on the material structure unless the effective stabilization is applied. Therefore in order to be able to have a real judge about the flow behavior of nanocomposites and reference samples, we decided to use antioxidant in all samples to eliminate the effect of thermomechanical degradation on rheological and mechanical properties of the nanocomposites.

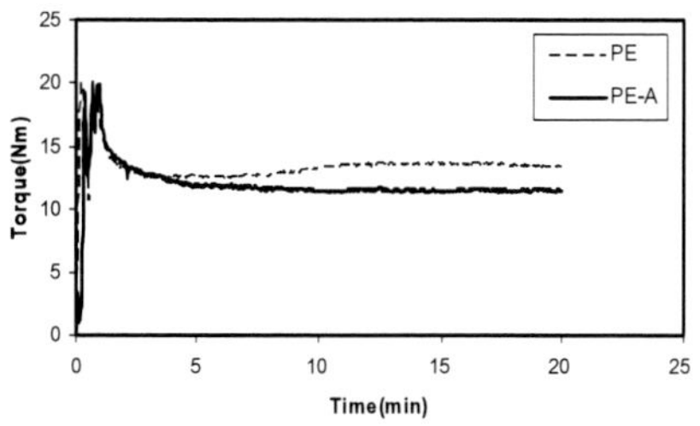

Fig. 1. Torque-time curves for PE processing with and without antioxidant.

Fig. 2 shows the complex viscosity and storage modulus of PNC-1 and PB-1 (reference sample) as function of angular frequency. Having considered to these results, one may notice that, incorporation of 3 wt% clay increases the viscosity and elasticity in particular at low frequencies. In this region, PNC-1 exhibit viscosity upturn behavior and its storage modulus tends to a plateau (non-terminal behavior) representative of fully dispersion of clay in PE matrix with a strong interfacial interaction. This is while; adding of 3 wt% of clay onto neat PE did not only increase its viscosity and elasticity, but also decreased them, due to the low interfacial interaction leading to melt slippage at the tactoid's surface. The same results were obtained for two other nanocomposites with difference in the extent of viscosity and elasticity increment. These differences may be due to the different state of clay dispersion in these three different nanocomposites. Higher the dispersion of the clay expected to result in higher increase in viscosity and elasticity.

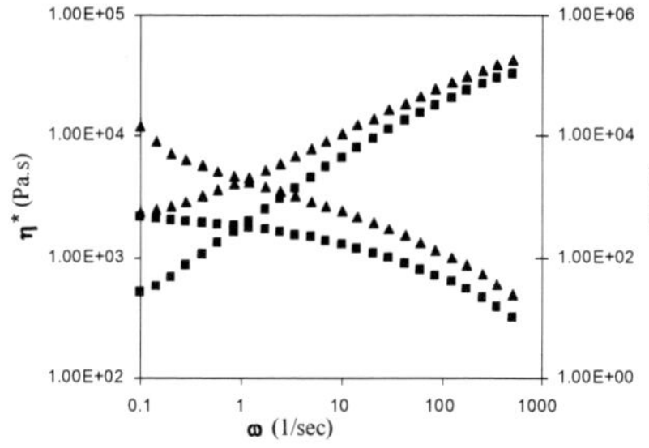

Fig. 2. Complex viscosity and storage modulus of PNC-1 (▲) and PB-1 (■) as function of angular frequency.

The XRD spectrum of MMT Cloisite 15A is presented in Fig. 3 along with the diffraction spectra of the 3 wt% clay nanocomposites prepared using different compatibilizers. The XRD trace of the Cloisite 15A clay give a layer spacing of 31° A. The XRD pattern of the PNC-1 did not show the characteristic basal reflection, suggesting an almost complete exfoliation of the clay layers in PE matrix. For PNC-2, there was a substantial reduction in the intensity of the diffraction peak, which is shifted to a lower angle than that of pure clay, suggesting a mixture of intercalated and exfoliated clay. The XRD pattern of PNC-3 showed a shift of the peak to a lower angle of about 2θ =2.48 and the basal spacing was increased to 3.56 nm. The increase in spacing is an indication of intercalation of polymer into the interlayer galleries of the clay. These results clearly show the strong effect of degree of maleation on the morphology of clay in the nanocomposites. In other words, these results indicate that, increasing in level of MA grafting facilitate the penetration of PE molecules into clay gallery, leading to its intercalation and exfoliation. Considering these results with the results of rheological studies, discussed above, clearly identifies the close relationship between the morphology and rheological properties of these nanocomposites. The results of XRD studies were evidenced using direct morphological studies by transmission electron microscopy (TEM). Figure 4 shows the TEM image of PNC-1 nanocomposit. The exfoliation of clay platelets can obviously be seen in this image. These results are in accordance with those of the similar samples with antioxidant.

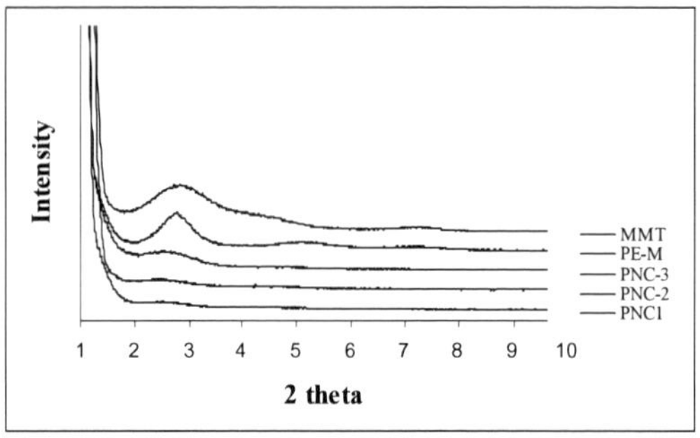

Fig. 3. X-ray spectrum of cloisite 15A and different nanocomposites.

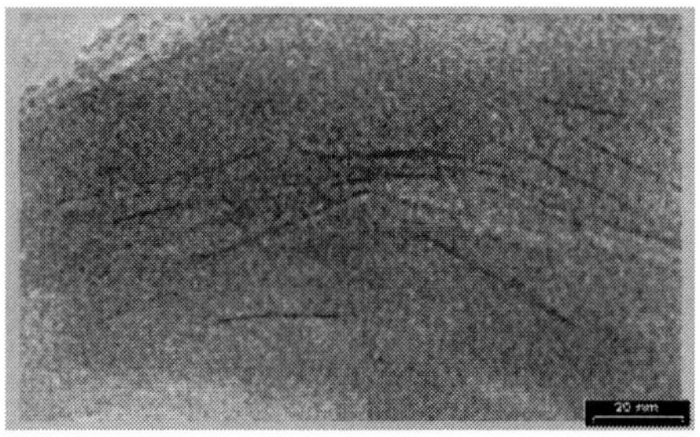

Fig. 4. TEM image of PNC-1.

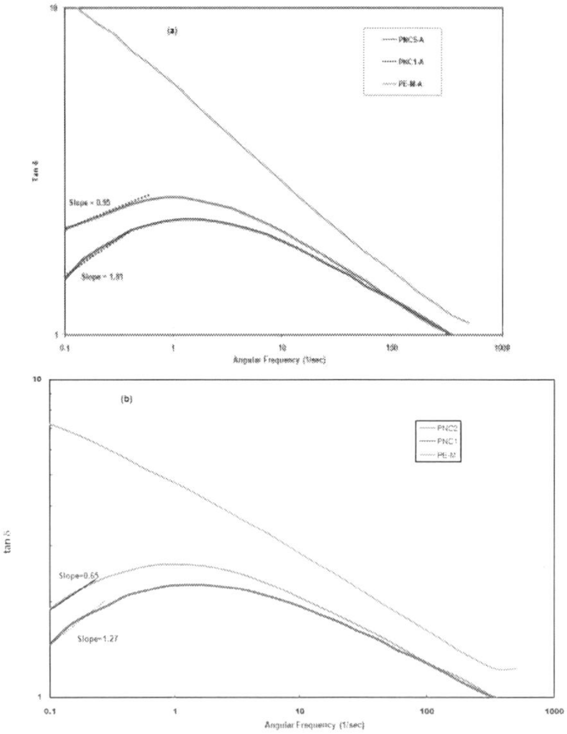

Fig. 5(a). Shows tan δ versus frequency for PE-M-A, PNC-1-A, PNC-2-A nanocomposites and **5(b)** Shows tan δ versus frequency for PE-M, PNC-1, and PNC-2.

Figure 5 (a) and (b) shows the results of tan δ as a function of angular frequency for PE-M and two different nanocomposites in presence and without the antioxidant, respectively. The results clearly show the different trend of variation of tan δ with frequency for PE-M (A) and two nanocomposites.

The tan δ decreases with increasing of angular frequency in the case of PE-M (A), whereas for two other samples it increases at low frequencies and decreases after passing through a maximum. The former usually represents a liquid like behavior and the latter corresponding to solid like behavior at low frequencies and liquid like behavior at higher frequencies. This is evidence to formation of three dimensional networks of dispersed clay platelets with a strong interaction to PE matrix. Comparing the tan δ of two nanocomposites reveals that, the nanocomposite with higher degree of exfoliation exhibit more solid like behavior than that with lower degree of exfoliation.

3.1 Thermal properties

TGA data for clay and its nanocomposites are collected in Table 3. The parameters listed are the onset temperature of the degradation, usually taken as the temperature at which 10% degradation occurs, $T_{0.1}$, the mid-point of the degradation, $T_{0.5}$, another indication of thermal stability and the non-volatile residue found at 600 °C, denoted as char. The nonvolatile fraction for modified clay is 62% at 600 °C, indicating the inorganic content. The 10% mass loss temperature of the clay is 264 °C but it does not have 50% mass loss temperature.

The onset temperature of PE is higher than that of the PE/clay nanocomposites, except PNC2. The mid-point of the degradation, T0.5, of all the PE nanocomposites is higher than that of virgin PE. The amount of char that is formed at 600 °C corresponds rather well to the inorganic clay loading in the system.

Clay layers have good barrier action, which can improve the thermal stability of polymer/clay nanocomposites. On the other hand, the alkylammonium cations in the organoclay could suffer decomposition following the Hofmann elimination reaction [18], and its product would catalyse the degradation of polymer matrixes [19]. Also, the clay can catalyse the degradation of polymer matrixes [20]. The latter two actions would reduce the thermal stability of polymer/clay nanocomposites. Thus, decrease of the onset temperature with clay loading is due to the Hofmann elimination reaction and the clay catalysed degradation. But at high temperatures the barrier effect is predominant, so the mid-point of degradation of nanocomposites is higher than that of virgin PE. This results is confirmed with TGA data for PE-A and PNC1-A. The presence of antioxidant inhibits Hoffman elimination reaction, thus the onset temperature of nanocomposite is higher than PE-A.

Comparing the TGA curves of different nanocomposites with different morphologies shows that the intercalated nanocomposite (PNC3) displays higher thermal stability than those of exfoliated (PNC1) and partially exfoliated (PNC2) samples after the mid-point of degradation. This can be attributed to the fact that the degradation of some part of polymer, located between clay platelets particularly for intercalated morphology, is retarded due to the barrier effect of clay layers to oxygen and volatile products.

Table 3. TGA data, in air, for PE and PE/clay nanocomposites.

Sample	$T_{0.1}$ (°C)	$T_{0.5}$ (°C)	Char at 600°C (%)
PE	408	427.2	0
Org-MMT	264	-	62
PE-M	371	425	1.91
PNC1	381	460	2.03
PNC2	408	465	2.13
PNC3	395	469	2.00
PE-A	396	425	0
PNC1-A	424	465	1.98

3.2 Tensile Properties

Figure 6 shows typical stress-strain curves for all of the nanocomposites prepared in this work in presence of antioxidant, together with the corresponding virgin polymers. The mechanical properties i.e., Young's modulus, tensile strength, strain at yield and elongation at break of the samples have been evaluated and the data are presented in Table 4. In order to clarify the effect of clay dispersion, the results of PBs and PNCs were compared. Apparently, addition of 3% clay into PE and PE/PE-g-MA mixture enhance the Young's modulus of them. As can be seen the Young's modulus for PNC-3-A is greater than that of PNC1, but the increase in the modulus of PNC1 relative to its of unfilled matrix (PB1) is sharper than that of PNC4. This fact is related to the better dispersion of clay in PNC1 which leads to more increscent in the value of elastic modulus.

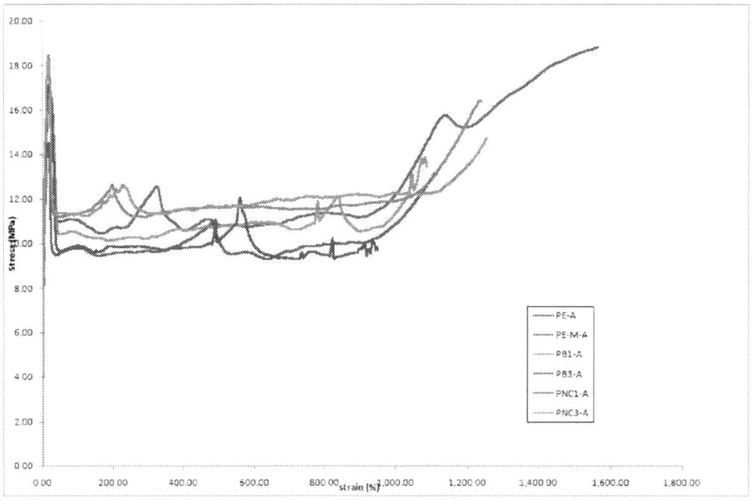

Fig. 6: stress-strain curves for PE-A, PE-M-A, PB1-A, PB3-A, PNC1-A and PNC3-A.

Table 4. Mechanical properties of nanocomposites and reference samples.

Samples	Young modulus (MPa)	Stress at yeild (MPa)	Yield strain (%)	Elongation at break (%)
PE-A	512	16.6	24.7	1790
PE-M-A	576	14.5	17.5	1200
PB1-A	545	16.3	24.1	1090
PB3-A	578	17.1	20.2	970
PNC1-A	651	18.5	18.6	1250
PNC3-A	663	18.3	17.2	1270

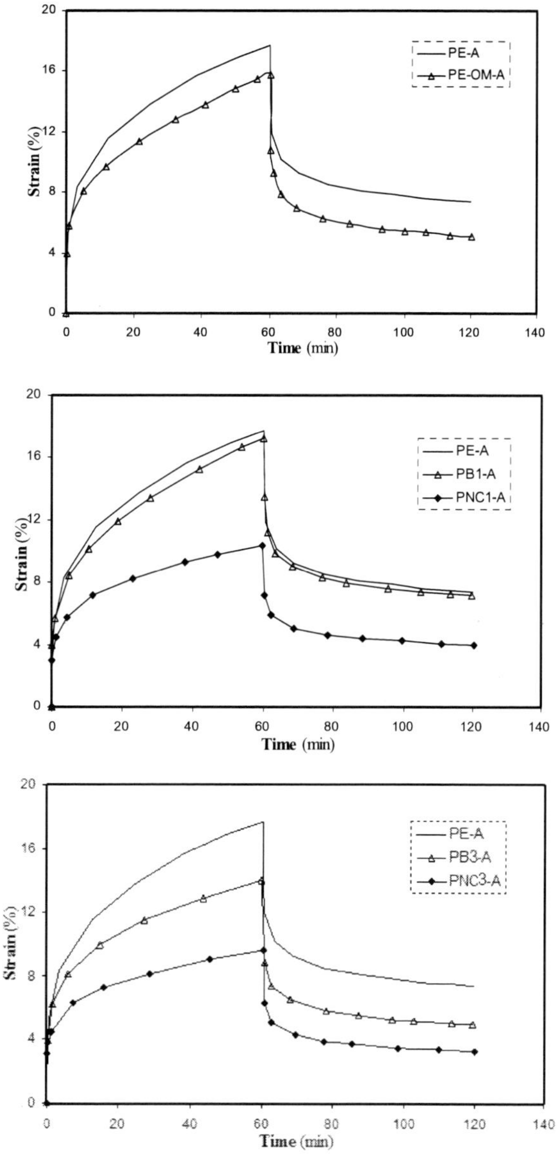

Fig. 7: Creep-recovery measurements of the nanocomposites and reference samples.

It can be seen from Table 4 that addition of clay into pure PE reduced the tensile strength but just as the modulus, the nanocomposites have higher tensile strength values than that of unfilled matrix, being slightly higher for the PNC1-A with exfoliated clay platelets .

It is rather interesting to see that elongation at break for the HDPE/clay composites was considerably reduced as compared to that of neat HDPE. Also elongation at break for PNC1-A reduced relative to that of unfilled matrix (PB1-A), but for PNC3-A with intercalated structure was increased.

As can be seen the PE/PE-g-MA blends (PB) have poor tensile properties than neat PE. We believe it results from immiscibility between PE and used PE-g-MAs. So the increased elongation at break for PNC3-A relative to its unfilled matrix can be related to its intercalated structure. In intercalated morphology of clay the simultaneous presence of PE and PE-g-MA chains in clay galleries leads to more interaction between them and therefore more miscibility between PE and PE-g-MA. But this phenomenon can not be seen for exfoliated structure. Therefore the intercalated structure leading to more miscibility between PE and PE-g-MA is thought to be responsible for the distinct improvement in elongation at break.

3.3 Creep behavior

Fig. 7 shows the creep behavior of PE nanocomposites at stress corresponding to 50% yield stress of pure PE. All the samples showed permanent irrecoverable strain. This behavior shows non-linear behavior of all samples. As can be seen Presence of PE-g-MA or clay decreases the strain but when they are added to PE together, the strain decreases significantly relative to the host PE. The decrease of %strain changes for each of compatiblizers. In order to clarify the effect of presence of PE-g-MA and clay dispersion, results of creep behavior of samples were normalized relative to that of pure PE and unfilled matrixes and are shown in table 5. This table also shows %creep recovery values of the samples.

Table 5. Normalized strain and (%) recovery of creep tests for nanocomposites and reference samples.

samples	Normalized strain	Normalized strain relative to unfilled matrix	Recovery (%)
PE-A	1	-	58.4
PE-M-A	0.89	0.89	67
PB1-A	0.97	-	58
PNC1-A	0.58	0.59	61
PB3-A	0.79	-	64
PNC3-A	0.54	0.68	66

As can be seen addition of clay decreased the normalized strain. This result showed that addition of clay cause to an increase in sample's strength. PNC1 nanocomposite showed more decrease in the value of normalized %strain relative to its unfilled matrix than PNC4 nanocomposite. This result is related to greater modulus and yield stress of PNC1-A compared to PNC3-A. Presence of solid-like networks containing dispersed platelets of clay which are interconnected with polymer chains, may be the main reason of more elastic behavior of PNC1-A nanocomposite than PNC3-A.
Table 5 shows that %recovery increased with addition of clay. Under creep, more the elasticity more is the likelihood of recovery. All nanocomposites have greater %recovery relative to pure PE and their unfilled matrixes. The %recovery of PNC1-A is more than PNC3-A.

4. Conclusion

Rheology, morphology and mechanical properties of HDPE/Clay nanocomposites were studied. The results of X-ray diffraction indicated that, increasing in degree of grafted MA increases the interlayer penetration of PE chains leading to intercalation and exfoliation. The intercalated and especially exfoliated nanocomposites exhibit higher viscosity and elasticity inparticullar at low frequency ranges, showing the formation of three dimensional networks with high interfacial interaction. The formation of such networks was evidenced using the results of tan δ, where the nanocomposites showed solid like behavior at low frequencies. Incorporation of clay decreased the onset temperature of degradation due to the Hofmann elimination reaction, but increased remarkably the mid-point of the degradation temperature. Intercalated nanocomposite displayed higher thermal stability than exfoliated samples, due to the barrier effect of clay layers to oxygen and volatile products, during the degradation of clay intercalated part of polymer. Interestingly the results showed that the effect of Hofmann elimination reaction which decreases the onset temperature degradation of modified clay nanocomposites, can effectively be eliminated using the antioxidant in the mixing process of nanocomposites. The tensile strength and yield stress of nanocomposites were higher than the corresponding matrixes. The PE nanocomposites showed lower strain and higher recovery than those of unfilled matrixes in the creep tests. The extent of increase in mechanical properties was higher for exfoliated than intercalated nanocomposite. It was demonstrated that, there is a close relationship between the morphology, rheology and mechanical properties of PE nanocomposites. It was also demonstrated that, the linear viscoelastic properties have reliable sensitivity to the state of dispersion of clay. Therefore they can be used as indirect methods for evaluation of the morphology and in prediction of the physic-mechanical properties.

References

[1] Arunvisut, S.; Phummanee, S.; Somwangthanaroj, A. *J. App. Polym. Sci.* **2007,** 106, 2210.
[2] Oya, Y. Kurokawa, H. Yasuda, J. Mater. Sci., 35 (2000), 1045-1050.
[3] G. Beyer, K. Eupen, Proc. Nanocomposites, Chicago, IL, 2001.
[4] J. Gilman, Appl. Clay Sci., 15 (1999), 31-49.
[5] Y. Kojima, A. Usuki, M. Kawasumi, A. Okada, T. Kurauchi, O. Kamigaito, J. Appl. Polym. Sci., 49 (1993), 1259-1264.
[6] P. Messermith, E. Giannelis, J. Polym. Sci., Polym. Chem., 33 (1995), 1047-1057.
[7] S. S. Ray, M. Okamoto, Prog. Polym. Sci., 28 (2003) 1539–1641.
[8] G. Brindley, G. Brown, London:Mineralogical Society, 1980, 227-232.
[9] J. Zhang, R. K.Gupta, C. A.Wilkie, Polymer, 47 (2006), 4537.
[10] J. Alexander, P. Dubois, Mater. Sci. and Engg., 28 (2000), 1-63.
[11] C. Zhao, H. Qin, F. Gong, M. Feng, S. Zhang, M. Yang, Polymer Degradation and Stability, 87 (2005), 183 .
[12] A. Bafna, G. Beaucage, F. Mirabella, S. Mehta, Polym., 44 (2003), 1103-1115.
[13] T. Fornes, P. Yoon, H. Keskkula, D. Paul, Polym., 42 (2001), 9929-9940.
[14] Y. Zhong, D. D. Kee, Polym. Eng. Sci. 2005, 45, 469.
[15] S. Hotta, D. R. Paul, Polymer 2004, 45, 7639.
[16] M. Zhang, U. Sundararaj, Macromolecular Matewrials and Engineering, 291 (2006), 697.
[17] R.Rezanavaz, M.K.Razavi Aghjeh, A.A.Babalou, POLYMER Composite JOURNAL, V. 31, Issue 6, P. 1028-1036, June 2010.
[18] R.A. Vaia, H. Ishii, and E.P. Giannelis, Chem. Mater., 5(1993), 1694.
[19] B.R. Christiani and M. Maxfield, U.S. Patent 5(1998),747, 560.
[20] B. Safadi, R. Andrews, and E.A. Grulke, J. Appl. Polym. Sci., 84(2002), 2660.